Advances in
Superconductivity

NATO Advanced Science Institutes Series

A series of edited volumes comprising multifaceted studies of contemporary scientific issues by some of the best scientific minds in the world, assembled in cooperation with NATO Scientific Affairs Division.

This series is published by an international board of publishers in conjunction with NATO Scientific Affairs Division

A	**Life Sciences**	Plenum Publishing Corporation
B	**Physics**	New York and London
C	**Mathematical and Physical Sciences**	D. Reidel Publishing Company Dordrecht, Boston, and London
D	**Behavioral and Social Sciences**	Martinus Nijhoff Publishers The Hague, Boston, and London
E	**Applied Sciences**	
F	**Computer and Systems Sciences**	Springer Verlag Heidelberg, Berlin, and New York
G	**Ecological Sciences**	

Advances in Superconductivity

Edited by

B. Deaver

and

John Ruvalds

University of Virginia
Charlottesville, Virginia

Plenum Press
New York and London
Published in cooperation with NATO Scientific Affairs Division

Proceedings of a NATO Advanced Study Institute on
Advances in Superconductivity,
held July 3–15, 1982,
at the Ettore Majorana Centre for Scientific Culture, Erice, Sicily, Italy

Library of Congress Cataloging in Publication Data

NATO Advanced Study Institute on Advances in Superconductivity (1982: Erice,
    Italy)
    Advances in superconductivity.

    (NATO advanced science institutes series. Series B, Physics; v. 100)
    Published in cooperation with NATO Scientific Affairs Division."
    "Proceedings of a NATO Advanced Study Institute on Advances in Supercon-
ductivity, held July 3–15, 1982, at the Ettore Majorana Centre for Scientific
Culture, Erice, Sicily, Italy"—T.p. verso.
    Includes bibliographical references and index.
    1. Superconductivity—Congresses. I. Deaver, Bascom Sine, 1930–   . II.
Ruvalds, J. III. North Atlantic Treaty Organization. Scientific Affairs Division. IV.
Title. V. Series.
QC612.S8N367 1982   537.6′23   83-11046

ISBN 978-1-4613-9956-8   ISBN 978-1-4613-9954-4 (eBook)
DOI 10.1007/978-1-4613-9954-4

©1983 Plenum Press, New York
Softcover reprint of the hardcover 1st edition 1983
A Division of Plenum Publishing Corporation
233 Spring Street, New York, N.Y. 10013

PREFACE

The Advanced Study Institute on "Advances in Superconductivity"
was held at the Ettore Majorana Centre for Scientific Culture
in Erice, Sicily, during July 3 to July 15, 1982. This Institute
was the third course of the International School of Low Temperature Physics, which was established at the Centre in 1977 with
the guidance and inspiration of T. Regge and A. Zichichi. The 1982
Course was centered on a topic which brought together fundamental
basic research and the most recent promising technological
applications. Accordingly, the participants represented a wide
spectrum of industrial and government laboratories, as well as
universities from various countries. The program of topics and
speakers was developed with the advice of the Organizing Committee,
composed of H. Fröhlich, T. Regge, B. Stritzker, and L. Testardi.

This Institute emphasized recent developments in the science
and technology of superconductivity. A historical perspective was
provided by H. Fröhlich, whose lectures recall the earliest
discoveries and theoretical attempts to understand superconductivity.
Ironically, his early suggestion of the electron-phonon coupling
as a key to superconductivity was met with initial widespread
skepticism. Later, the development of field theory methods for
solid state physics problems, and the evolution of the BCS theory
has led to a seemingly unanimous concensus regarding the electron-
phonon mechanism as the predominant source of superconductivity
in known materials.

Experimental studies of superconductivity exemplify the
strong interplay of science and technology in many ways. The
discovery in 1911 of zero resistance in Hg at temperatures below
4°K generated excitement in the potential applications of superconducting coils to be used as windings in magnets, but this hope
was set back for several decades by the realization that strong
magnetic fields destroy the superconducting state. As the availability of low temperature facilities grew, higher superconducting
temperatures were found, with T_c = 15°K of NbN discovered in 1941.
Naturally, the proximity of this T_c range to the reach of pumped
liquid hydrogen caused a flurry of interest. By 1970, approximately
1000 materials were known to become superconducting, with some

having T_c values near 20°K, and many exhibiting the capability of withstanding strong magnetic fields. A remarkable number of these technologically important materials were discovered by B. T. Matthias, using an intuitive approach which yielded empirical rules relating T_c to the electron/atom ratio in transition metal compounds. A quantitative theoretical basis for the Matthias rules remains a challenge.

At present, the technological advances are in the forefront of superconductivity research. Giant high field magnets are being developed for large scale applications such as fusion energy, high energy accelerators, and train levitation. At the other extreme, miniature superconducting circuits using the Josephson effect are being used as ultrasensitive magnetometers in biological systems and other weak signal detection applications. Finally, the superconducting microcircuitry now promises to be the key to the next generation of sophisticated high-speed computers.

The highlights of these technological applications, together with fundamental advances in the discovery and understanding of new superconducting materials forms the basis for this book.

Aside from the scientific programme of the Institute, it was memorable for its location and facilities. The Ettore Majorana Centre for Scientific Culture, with the gracious leadership of Director A. Zichichi, hosts schools and institutes in many disciplines each year, a large number of them returning periodically. The beauty of Erice, the warmth of its people, the comfort of the living arrangements and efficiency of the Centre's staff were important factors contributing to the success of the School. The administrative contributions and guidance of N. McLaughlin were especially appreciated by all participants.

We are grateful to several organizations for financial assistance. The Scientific Affairs Division of NATO awarded a grant that was crucial in bringing all of the lecturers and many of the seminar speakers and students from Europe and North America. The Italian government, through its Ministries of Education and of Scientific Research, the Italian Research Council, and the Sicilian Regional Government, gave support for several students' travel and local expenses, and provided for the Centre's facilities and staff. The U.S. National Science Foundation and the European Physical Society provided student fellowships.

Finally, it is a pleasure to express our appreciation to Jean Dozier at the University of Virginia, whose organization and preparation assistance played an essential role throughout the three years of planning and completion of the Institute and this volume.

B. Deaver and J. Ruvalds
Editors

CONTENTS

HISTORY OF THE THEORY OF SUPERCONDUCTIVITY

H. Fröhlich

Department of Physics
University of Liverpool
Liverpool U.K.

I. THE PROBLEM OF SUPERCONDUCTIVITY IN EARLY SOLID STATE PHYSICS

A good textbook usually presents developments in such a logical way that the student may wonder why they sometimes took so long and why so much fuss is often made about the achievements. Looking back, of course, every step in a path seems logical and unique. Looking forward, however, a hundred paths offer themselves and none of them may be correct as new concepts which only at a later stage will become obvious may be required for the development of a theory.

Superconductivity offers a splendid example for such delays. When Kammerlingh Onnes developed techniques to reach low temperatures, he discovered in 1911 that the electric resistivity of mercury completely disappeared at about $4^{O}K$ [1]. Other superconductive metals were discovered. From Ohm's law it was concluded that as the conductivity σ becomes infinite, the electric field E in a superconductor must vanish so that the current I remains finite,

$$J = \sigma E, \quad \sigma \to \infty \ , \quad E \to O \ , \tag{1}$$

From Maxwell's law of induction it then follows that the rate of change of the induction B must vanish,

$$\text{curl } E = \frac{1}{c} \dot{B}, \quad E = O, \quad \dot{B} = O. \tag{2}$$

This conclusion was generally accepted although it was found that a magnetic field above a critical one destroys superconductivity. Thus it was not until 1933 that the conclusion (2) was tested by Meissner and Ochsonfeld [2] and found to be wrong. They discovered, in fact, that a superconductor behaves as a perfect diamagnetic material. As a consequence, in constrast to (2), a magnetic field applied to a superconductor at a temperature T above its transition temperature T_c is expelled when the temperature is reduced below T_c. A superconductor, thus behaves differently from a perfect conductor ($\sigma \to \infty$).

Theoretical research in superconductivity has produced many basic mistakes, and it is characteristic that an author, publishing in a reputable journal in 1981 is unaware of the above mentioned result.

The discovery of the Meissner effect led to the two fluid model, normal and superconductive fluid, to the development of thermodynamics by Gorter and Casimir [3], and to the classification of the superconductive - normal transition as a phase transition of the second kind. Most remarkable was the extremely small energy difference between the normal and the superconductive state, obtained from measurements of the critical magnetic field H_c at temperatures below T_c. It turned out to be about a thousand times smaller than KT_c per atom.

Following these discoveries F. and H. London [4] presented their macroscopic equations of superconductivity: Divide the current I into a normal $j_n = \sigma E$, and a superconductive, j_s, which latter satisfied

$$c\Lambda \text{ curl } j_s = - B \tag{3}$$

where Λ is a material constant. Introduction of (3) into the Maxwell equations then yields the Meissner effect.

Furthermore introduction of the vector potential A by B = curl A yields

$$c\Lambda j_s = - A \tag{4}$$

if the so called London gauge for A is used. This involves div A = 0 and certain other conditions (cf [5], § 10) yielding a unique form for A.

We must now remember that in these years, 1935, mo-

dern solid state physics was well established. The free
electron model in metals was in fact highly successful
since the difficulties of the classical Drude model -
excessive paramagnetism and specific heat - had been
overcome by the introduction of Fermi statistics by Pauli
li and by Sommerfeld. A further difficulty, mean free
paths of the order of a lattice distance, was shown by
F. Bloch to be invalid in quantum mechanics. For he showed that an electron is not scattered by a perfect lattice but only by deviations from it through its thermal
vibrations i.e. through electron-phonon scattering. Fer-
mistatistics also made the low energy difference between the normal and the superconductive state more plausible as electrons in a range KT_c at the top of the Fermi distribution only would have to be affected, the rest
being contained through the Pauli principle.

Two problems thus arose; one to derive the London
equations through an appropriate model, the other to
find an interaction that could lead to a phase transition. Most efforts were directed towards "one".

Now in quantum mechanics, the velocity operator,v,
is obtained in terms of the momentum operator p as

$$v = \frac{1}{m} (p - \frac{e}{c} A) \tag{5}$$

For n electrons per unit volume, condition (4) can thus
be obtained by summing (5) over all electrons, provided

$$\sum_{j=1}^{n} <p_j> = 0, \text{ London gauge} \tag{6}$$

$$\text{and } \Lambda = \frac{e^2 n}{mc^2} \tag{7}$$

This also reflects on the corresponding energy terms

$$\sum \frac{m}{2} v_j^2 = \sum \frac{p_i^2}{2m} + E_p + E_d \tag{8}$$

where the diamagnetic energy term E_d is

$$E_d = \frac{e^2 n}{2mc^2} A^2 \tag{9}$$

while with (6) , the paramagnetic

$$E_p = - \frac{e}{mc} \sum p_j \ A = 0 \tag{10}$$

vanishes in the London gauge.

In normal materials in the ground state, the expectation values of the two terms E_p and E_d very nearly cancel though E_d then contributes in first whereas E_p contributes in second order only. Formally this takes the form

$$\langle E_p \rangle = \sum_n \frac{|M_{on}|^2}{U_o - U_n} \tag{11}$$

in terms of virtual transitions between the ground state, energy U_o, and appropriate excited states, with matrix elements M_{on}.

Clearly the near cancellation of two large terms can give rise to many mistakes, yielding incomplete cancellation and hence superconductivity. In the case of free electrons, this near cancellation can be proved with the help of sum rules. It is then found that the only remaining terms arise from the Landau quantisation of electronic orbits. Classically, as is well known, diamagnetism cannot arise because the relevant energy terms $\frac{1}{2} m v_j^2$ have the same form with or without magnetic field.

Inspection of (11) shows that in view of the continuous spectrum of free electrons, the largest contributions arise from near zero values of $U_o - U_n$. It was first suggested by Welker [6] that the magnitude of (11) could be strongly reduced by postulating the existence of an energy gap above the Fermi surface, so that E_p no longer compensates E_d, and the Meissner effect arises. This presupposes, however, that the matrix elements M_{on} are not changed which in general is incorrect. The concept of a gap, nevertheless, has proved of great importance. Thus it makes plausible the small energy difference per atom between the normal and the superconductive state as only the electrons near the Fermi surface are involved.

In the years up to 1950, many attempts to develop a theory of superconductivity failed, and the subject became an outstanding problem in physics. In view of this failure it was conjectured, at times, that new basic features which show themselves only in many body systems may be involved. The decisive step, however, turned out to be one of method, namely the introduction of field theory into solid state physics.

II. THE INTRODUCTION OF FIELD THEORY INTO SOLID STATE PHYSICS

This step was carried out by myself nearly single handedly, though the path towards it was anything but straight. Already in 1931 I had noticed [7] that the concept of field operators (connected with what was then known as second quantization) may be applied to the free electrons in metals. This was an isolated step, however, to be resurrected only twenty years later with the establishment of a Hamiltonian. The treatment of the lattice vibrations as a field whose sources are electrons proceded through the case of ionic crystals. Here I had noticed that in contrast to shat was then known about the interaction of electrons with lattice vibrations, quantitative calculations could be made with relative ease [8]. While applications of this possibility proceded, I got involved, following Yukawa, with Heitler and Kemmer in the development of a formal field theory for mesons and nuclear forces. After initial success this led to a number of disappointments, some of which arose from the various infinities that were met. Finally in 1948 I decided that electrons in ionic crystals were the place for a singularity free field theory, and this led to the theory of what is now known as the large polarons ([9], [10]). In this theory the polar vibrations are treated as a continuum: interaction with electrons then yields a natural cut-off so that no infinities in the self energy arise. An electron polarises its surrounding field, and, jointly with this polarisation behaves like a particle, the polaron, which has an effective mass different from the electronic mass.

This concept of an effective particle is, of course, very trivial nowadays, but it was far from being so at the time. In fact, earlier, it had been suggested that when an electron polarises its surroundings, it "digs its own hole" and thus gets trapped in the medium, contrary to the results of field theory (though lately, this possibility has been rescued as the "small polaron").

Since that time, the theory of the polaron has developed into a large subject in its own right. Already then, however, it was clear to me that application of field theoretical methods to metals may contain the key for dealing with superconductivity. For in general since the sources (particles) of a field through their interaction with it may absorb and emit quanta of this field, this same interaction yields a self energy of particles, and an interaction between them. Application of this idea with the use of second order perturbation theory to deal with the electron-phonon interaction led to the important result that the Fermi distribution becomes instable in a range $\hbar ws$ (w=largest wave number of phonons, s=velocity of sound) near its surface, provided the interaction parameter exceeds a critical value [11]. When this is the case then a rearrangement in this $\hbar ws$ range leads to a lowering of the total energy by about $(\hbar ws)^2/\zeta$ (ζ=Fermi energy) per electron. It was postulated then, that this new state is the superconductive state. Since s^2 is inversely proportional to the ionic mass M, it followed that the relevant energy difference is inversely proportional to M, the isotope effect.

The isotope effect in superconductors was in fact discovered during the same period. At that time I was at a 2-3 month visit to Purdue University, and submitted my paper on leaving on the 16.5.1950. I then spent a couple of days at Princeton and there, at my breakfast table, found the Physical Review with the two letters reporting the isotope effect [12], [13]. On checking I found my M-dependence confirmed and on the 19.5.1950 sent a letter [14] to claim confirmation of the basic idea, electron-phonon interaction.

The discovery of the isotope effect was not as straight forward as might appear now, for it was very generally believed that the ions in view of their heavy mass can have no influence on superconductivity. Also Kamerlingh Onnes in 1922 [15] found no isotope effect in radio lead.* As a consequence, a well known low temperature laboratory when offered isotopes refused to investigate them. E. Maxwell, one of the discoverers who at the time was a young research worker mentioned to me, that he had to do the work at night as his boss would not permit him to investigate isotopes.

At that time, I also paid a short visit to the Bell Labs. where John Bardeen, after hearing of the isotope effect, took up the problem [16] but avoided perturbation theory by using a variational method. He confirmed the result for the energy.

It should also be pointed out now that the self energy of a multiparticle system, if exactly calculated, contains all particle self- and interaction energies. In second order perturbation, however, only individual

---

*I am grateful to T. H. Geballe for this reference.

particle self energies are contained, though in view of
the Pauli Principle their total magnitude depends on
their distribution, thus providing an interaction in momentum space.

After the frustration of the previous twenty years,
the basic idea of the importance of the electron-phonon
interaction was at once widely accepted. It would be incorrect, however, to attribute this solely to the discovery of the isotope effect, for an important consideration arose from the fact that no extension of the basic
model of metals had to be made: simply another, forgot-
ten, aspect of Bloch's electron-phonon scattering had to
be considered.

Use of perturbation theory was suspect at an early
stage. Schafroth [17] on the suggestion of Pauli investigated the interaction with an electro-magnetic field.
He found that in no order of perturbation theory could
the Meissner effect arise. The conclusion was not, however, that the basic idea of the theory was incorrect,
but rather that perturbation theory was inapplicable.

An essential further step was initiated by Wenzel's
[18] objection that the required strong interaction
also leads to an instability in the lattice. It was recognized early [19] that this instability arises from
physical processes different from those that give rise
to the instability near the top of the Fermi distribution. By that time, modern field theory had introduced
the concept of renormalisation, and I decided that these
methods should now be used to reformulate the problem
[20] . This led to the renormised Hamiltonian, probably
the first time that both electrons and phonons are expressed in terms of field operators. The general form
of this Hamiltonian is now used very widely though of
course modified to account for special situations.

The renormalisation was carried out with the method
of canonical transformation. Apart from the usual terms
describing electrons and (renormalized) phonons and a
term for real transitions, the Hamiltonian now also contains explicitly an electron-electron interaction term
of the form

$$H_i = -\frac{1}{2} \sum_{w,k,q} \frac{D_w^2}{E_{q-w}-E_q+\hbar ws} a_k^+ a_{k-w} a_{q-w}^+ a_q \qquad (12)$$

Here a_k, a_k^+ are Fermi operators, E_k and $\hbar ws$ are elec-
tron- and phonon energies respectively. s is the renormalised velocity of sound, and D_w is a renormalised interaction parameter.

It will be noticed that for small differences of
electronic energy, this interaction is attractive. An
additional repulsive interaction was introduced by Bardeen and Pines [21] after Bohm and Pines had explicitly
introduced the treatment of Coulomb interaction.

The situation with regard to the development of a
theory of superconductivity was now drastically changed,
as will be seen from the discussion following an introduction to the subject at the meeting to celebrate the
centenary birthday of Lorentz and Kamerlingh Onnes [22].
From an obscure state when it was unclear which processes and interactions had to be considered the situation
was now clearly formulated in terms of a Hamiltonian.
The problem consisted in solving it without the use of
perturbation theory. Casimir at the discussion conjectured that a good guess was now required. This in fact
describes very nearly what took place after a further
three years.

Meanwhile, failing this, I was able to solve the
one dimensional problem [23]. The energy turned out to
have an essential singularity in the interaction parameter, a clear sign that perturbation theory to any order could not work. It will be known that during recent
years this one dimensional case has achieved prominence.

III. THE THEORY OF SUPERCONDUCTIVITY

Some more years passed without that the solution
for the three dimensional case was found. In 1955, Bardeen in his recollections [24] says, that he asked C.N.
Yang to recommend an expert in field theory who would
be willing to work in superconductivity. Thus Leon Cooper
was introduced to my Hamiltonian. He decided to investigate the simplest problem, that of two electrons
(on top ot the Fermi sphere) interacting through (12).
He found that they form a bound pair, and that the binding energy has an essential singularity in the interaction parameter [25].J.R. Schrieffer (personal communication), stimulated by his work on the theory of large
polarons, then suggested a wave function for the

multielectron problem in terms of electron pairs. Development of these ideas then led to the theory of superconductivity, the so called B.C.S. theory [26].

   To some extend, the requirement of electron pairing
had been anticipated by Schafroth [27] who, having earlier shown that a Boson gas becomes superconductive proposed that under the influence of interaction (12) electrons form resonant states of pairs. He did not, however,
carry this through quantitatively, as did Cooper.

   At this stage ends the history of the theory of superconductivity, but two questions should still be asked.
The first concerns the remark of the late B.T. Matthias
(discoverer of most known superconductive alloys) that
the theory has not predicted a single superconductor.
The theory does, of course, contain a condition for a
material to become superconductive. This condition arises from the value of certain parameters whose calculation is not a part of the theory. The critisism thus di-
rects itself against the prediction of the magnitude of
certain parameters, rather than against the theory of
superconductivity.

   Secondly, and more important, we may ask "what have
we learned now"? After 25 years of endeavour, one might
have expected to have learned more than a theory of superconductivity. The theory led, of course, to great activity (for early work cf. [28]), but one would have
hoped that some general features of the many body problem, beyond that of superconductivity, might be revealed. At least one should understand in a simple way the
similarity of substances so different as liquid helium
and the electron fluid of superconductors. After asking
this question in an article [29] C.N. Yang replied that
he had the answer [30]: It is the existence of "off diagonal long range order" - better denoted as the existence of macroscopic wave functions. With the use of this
concept it has been possible to show that the B.C.S. solutions are exact solutions of the relevant Hamiltonian
[31].

   Yang's concept, in fact, is of a very general nature and can probably be employed to describe coherent
properties quite beyond those met in simple physical systems. Stimulated by this I have suggested that active
biological systems should exhibit coherent features of
a different nature. Recent experiments confirm this -
but that is another story.

REFERENCES

[1] Kamerlingh Onnes,
       Leiden Comm. 122b, 124c, 1911
[2] W. Meissner and R. Ochsenfeld,
       Naturwissenschaften 21, 787, 1933
[3] C.J. Gorter and H.G.B. Casimir,
       Physica 1, 305, 1934
[4] F. and H. London,
       Proc. Roy. Soc. A, 149, 71, 1935
[5] F. London,
       Superfluids Vol. I, John Wiley, New York 1950
[6] H. Welker,
       Bayerische Akademie der Wissenschaften 1938,115
[7] H. Fröhlich,
       ZS. f. Phys. 71, 715, 1931
[8] H. Fröhlich,
       Proc. Roy. Soc. A 160, 230, 1937
[9] H. Fröhlich, H. Pelzer and S. Zienau,
       Phil. Mag. 41, 221, 1950
[10] H. Fröhlich,
       Phil. Mag. Suppl. 3, 325, 1954
[11] H. Fröhlich,
       Phys. Rev. 79, 845, 1950
[12] E. Maxwell,
       Phys. Rev. 78, 477, 1950
[13] C.A. Reynolds, B. Serin, W.H. Wright, and
     L.B. Nesbitt,
       Phys. Rev. 78, 487, 1950
[14] H. Fröhlich,
       Proc. Phys. Soc. A 63, 778, 1950
[15] Kamerlingh Onnes,
       Leiden Comm. 160, 1922
[16] J. Bardeen,
       Phys. Rev. 80, 567, 1950
[17] M.R. Schafroth,
       Helv. Phys. Acta 24, 645, 1951
[18] G. Wenzel,
       Phys. Rev. 83, 168, 1951
[19] Kun Huang,
       Proc. Phys. Soc. A 64, 867, 1951
[20] H. Fröhlich,
       Proc. Poy. Soc. A 215, 291, 1952
[21] J. Bardeen and D. Pines,
       Phys. Rev. 99, 1140, 1955
[22] H. Fröhlich,
       Physica 19, 755, 1953
[23] H. Fröhlich,
       Proc. Roy. Soc. A 223, 296, 1954

24   J. Bardeen
         in Impact of Basic Research on Technology,
         ed. Kursunoglu and Perlmutter, Plenum Press,
         p. 32
25   L.N. Cooper,
         Phys. Rev. 104, 1189, 1956
26   J. Bardeen, L.N. Cooper, J.R. Schrieffer,
         Phys. Rev. 106, 162, 1957; 108, 1175, 1957
27   M.K. Schafroth,
         Phys. Rev. 96, 1442, 1954
28   N.N. Bogoliubov, V.V. Tolmachev and D.V. Shirkov,
         A New Method in the Theory of Superconducti-
         vity, 1959, New York: Consultants Bureau
29   H. Fröhlich,
         Reports on Progress in Physics 24, 1, 1961
30   C.N. Yang,
         Rev. mod. Phys. 34, 694, 1962
31   H. Fröhlich,
         Riv. Nuov. Cim. 3, 490, 1973,
         Collective Phen. i, 173, 1974

FUNDAMENTAL LIMITS ON SQUID TECHNOLOGY

John Clarke

Department of Physics
University of California
            and
Materials and Molecular Research Division
Lawrence Berkeley Laboratory
Berkeley, California 94720

I. INTRODUCTION

In the last few years, the flux noise energy of the dc Superconducting Quantum Interference Device (SQUID) (Jaklevic et al.,
1964) has been lowered by about four orders of magnitude. The most
sensitive SQUID yet made is close to the limits set by fundamental
quantum processes. The purpose of this article is to review these
developments, and, in particular, to discuss at some length the
limiting quantum processes in both single junctions and SQUIDs.

The article begins with a very brief review of flux quantization and the Josephson effects (Josephson, 1962) which includes a
discussion of the resistively shunted junction (RSJ) in the presence
of thermal noise. Section III outlines the theory of quantum noise
in a single resistively shunted junction, and describes experiments
to test the theory. Section IV is concerned with the dc SQUID, and
starts out with a general overview of its properties and operation.
This is followed by discussions of thermal and quantum noise, including both theory and experiment. The next section describes the
use of a SQUID as an amplifier, and, in particular, the quantum
noise limit. The importance of magnetic coupling is outlined. The
following section deals with the subject of 1/f noise, and the restrictions that it places on the design of SQUIDs for low frequency
applications. Section IV concludes with a brief discussion of two
potential applications of quantum-limited SQUID amplifiers. Section
V contains a summary.

In this article, we shall focus rather narrowly on the intrinsic noise properties of SQUIDs, and not address at all the role of the flux-locked loop that imposes limitations on such important parameters as slewing rate, frequency response, and dynamic range. We shall also ignore the problem of long term drift in SQUIDs, which is a little-understood problem of considerable importance in certain applications such as gradiometers.

II. SUPERCONDUCTIVITY AND THE JOSEPHSON EFFECT

According to the microscopic theory of Bardeen, Cooper, and Schrieffer (1957), the zero resistance of superconductors arises from the existence of Cooper pairs of electrons. These pairs are in a macroscopic quantum state that can be described by a single wave function $\Psi(\vec{r},t) = |\Psi(\vec{r},t)|e^{i\phi(\vec{r},t)}$, where $\phi(\vec{r},t)$ is the phase. In a closed loop of superconductor, the requirement that $\Psi(\vec{r},t)$ be single-valued leads to the concept of flux quantization (London, 1951): The magnetic flux contained in the loop can take only the values

$$\Phi = n\Phi_o \quad (n = 0, 1, 2 \ldots), \tag{2.1}$$

where

$$\Phi_o = h/2e \simeq 2.07 \times 10^{-15} \text{ Wb} \tag{2.2}$$

is the flux quantum.

The Josephson effect (Josephson, 1962) involves the tunneling of Cooper pairs through an insulating barrier separating two superconductors. The supercurrent, I_s, flows through the barrier according to Josephson's current-phase relation

$$I_s = I_o\sin\delta, \tag{2.3}$$

where $\delta = \phi_1 - \phi_2$ is the difference between the phases ϕ_1 and ϕ_2 of the two superconductors. The critical current of the junction, that is, the maximum supercurrent it can sustain, is I_o. For currents less than I_o, the phase difference is time independent, and the voltage across the junction is zero. On the other hand, when the applied current exceeds I_o, there is a non-zero voltage, V, across the junction, and ϕ evolves with time according to Josephson's voltage-frequency relation

$$\dot{\delta} \equiv \omega = 2eV/\hbar = 2\pi V/\Phi_o. \tag{2.4}$$

($\dot{\delta} \equiv d\delta/dt$, where t is time.) Thus, the supercurrent oscillates at a frequency $\nu = \omega/2\pi$.

The current-voltage (I - V) characteristic of a Josephson tunnel junction is, in general, hysteretic, a property that is most
undesirable for SQUIDs. The hysteresis can be eliminated by shunting
the junction with an appropriate conductance (McCumber, 1968;
Stewart, 1968). Hence, it is useful to review the properties of the
resistively shunted junction (RSJ) model at this point.

The model, shown in Fig. 1, consists of a tunneling element with
critical current I_O in parallel with the self-capacitance of the
junction, C, and shunted with an external resistance, R. A current
noise source $I_N(t)$ is associated with R, and the applied current is
I. The equation of motion is

$$V/R + I_o \sin\delta + C\dot{V} = I + I_N(t).$$ (2.5)

Neglecting the noise term for the moment and setting $V = \hbar\dot{\delta}/2e$, we
obtain

$$\frac{\hbar C}{2e} \ddot{\delta} + \frac{\hbar}{2eR} \dot{\delta} = I - I_o \sin\delta = -\frac{2e}{\hbar} \frac{\partial U}{\partial \delta} ,$$ (2.6)

where

$$U = -\frac{\Phi_o}{2\pi} (I\delta + I_o \cos\delta).$$ (2.7)

The potential, U, is that of a "tilted washboard", and Eq. (2.6)
describes the motion of a ball moving on the washboard. The term
$\hbar C/2e$ represents the mass of the particle, while $\hbar/2eR$ represents
the damping. Figure 2(a) shows the particle in the free-running
mode $I > I_o$.

It is sometime convenient to rewrite Eq. (2.6) in the dimensionless form

$$\beta_c \ddot{\delta} + \dot{\delta} = i - \sin\delta,$$ (2.8)

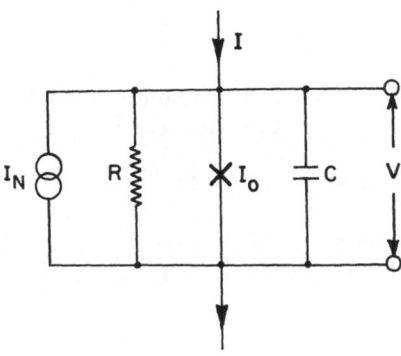

Fig. 1.   Resistively shunted junction model.

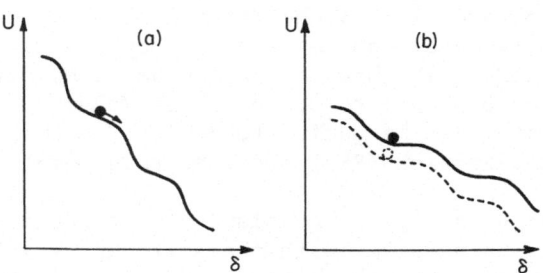

Fig. 2. The washboard model: (a) particle running down noise-free
         potential for the case $I > I_o$; (b) the case $I < I_o$ in the
         presence of noise — solid curve represents the mean poten-
         tial, dashed curve represents instantaneous potential due
         to fluctuation.

where $\beta_c \equiv 2\pi I_o R^2 C/\Phi_o$, $i = I/I_o$, and time is in units of $\Phi_o/2\pi I_o R$.
We can now consider two limiting cases. When $\beta_c \gg 1$ (heavy mass)
the I – V characteristic is hysteretic. As one increases the current from zero, the tilt increases until at $I = I_o$ (in the noise-
free case) the ball rolls out of its initial potential minimum and
starts to roll down the washboard. If one now reduces the current
to below I_o, the kinetic energy of the ball is sufficient to carry
it over successive potential maxima: Thus, the junction remains at
a non-zero voltage. One must reduce I to a value (depending on β_c)
below I_o to induce a transition to the zero voltage state. This
mechanism is the source of hysteresis. On the other hand, in the
overdamped limit $\beta_c \ll 1$, the term in $\ddot{\delta}$ in Eq. (2.8) is relatively
small. When one increases I to above I_o and then reduces it to just
below I_o, the damping is sufficient to bring the particle to rest,
so that the junction immediately returns to zero voltage. Thus,
there is no hysteresis. The cross-over from hysteretic to non-hysteretic behavior occurs at $\beta_c \sim 1$, the exact value depending on the
level of noise. For the case C = 0, Eq. (2.6) can be solved exactly
for the average voltage V:

$$V = R(I^2 - I_o^2)^{\frac{1}{2}}. \quad (\beta_c = 0) \tag{2.9}$$

The washboard model enables us to understand the influence of
noise on the junction. We rewrite Eq. (2.5) as

$$\frac{\hbar C}{2e}\ddot{\delta} + \frac{\hbar}{2eR}\dot{\delta} + I_o\sin\delta = I + I_N(t); \tag{2.10}$$

this is an example of a Langevin equation. In the thermal noise
limit, the Nyquist noise current has the spectral density

$$S_I(\nu) = 4k_B T/R. \tag{2.11}$$

It is evident that $I_N(t)$ causes the tilt in the washboard to fluctu-

ate with time. This fluctuation has two effects on the junction.
First, when I is less then I_0, from time to time fluctuations cause
the ball to roll out of one potential minimum into the next. For
the underdamped junction, this process induces a series of voltage
pulses randomly spaced in time. Thus, the time average of the voltage is non-zero even though $I < I_0$, and the I - V characteristic is
"noise rounded" (Ambegaokar and Halperin, 1969). The second effect
of the noise is to induce a voltage noise when the junction is biased
with a fixed current. In the limit $\beta_c \ll 1$, for $I > I_0$, and when the
frequency, ν_m, at which the noise is measured is much less than the
Josephson frequency, ν_J, the spectral density of the voltage noise
is (Likharev and Semenov, 1972; Vystavkin et al., 1974)

$$S_v(\nu_m) = \left[\frac{4k_BT}{R} + \frac{1}{2}\left(\frac{I_0}{I}\right)^2\frac{4k_BT}{R}\right]R_D^2. \qquad \left\{\begin{array}{c} \beta_c \ll 1 \\ I > I_0 \\ \nu_m \ll \nu_J \end{array}\right\} \qquad (2.12)$$

In Eq. (2.12), $R_D \equiv \partial V/\partial I$ is the dynamic resistance. The two terms
on the right hand side of Eq. (2.12) are illustrated in Fig. 3. The
first term, $(4k_BT/R)R_D^2$, represents Nyquist current noise generated
at the measurement frequency ν_m flowing through a dynamic resistance
R_D to produce a voltage noise. The second term, $\frac{1}{2}(I_0/I)^2(4k_BT/R)R_D^2$,
represents Nyquist noise at frequencies $\nu_J \pm \nu_m$ mixed down to the
measurement frequency by the Josephson oscillations. The factor
$\frac{1}{2}(I_0/I)^2$ is the mixing coefficient, which vanishes for sufficiently
large bias currents. The mixing coefficients for noise near harmonics of the Josephson frequency, $2\nu_J$, $3\nu_J$ ..., are negligible in the
limit $\nu_m/\nu_J \ll 1$.

   This concludes our introductory review of the Josephson effect.
We now turn to a discussion of quantum noise in a single junction.

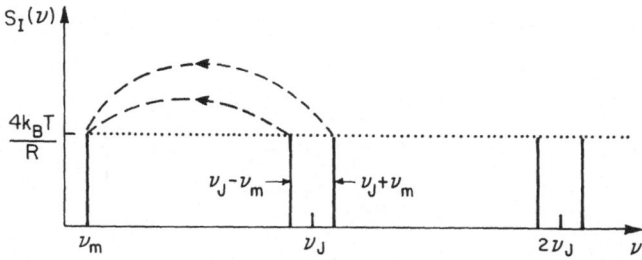

Fig. 3. Schematic representation of the terms in Eq. (2.12). The
         Nyquist noise generated in the resistor at ν_m contributes
         at ν_m; that generated at $\nu_J \pm \nu_m$ is mixed down to ν_m; the
         mixing coefficients for the noise generated near $2\nu_J$ and
         higher harmonics are negligible for $\nu_m/\nu_J \ll 1$.

III.   QUANTUM NOISE IN SINGLE JUNCTIONS

A.   Theory

Quantum effects become important when $h\nu_J \gtrsim k_B T$, because Eq. (2.11), the Nyquist formula, then breaks down. Koch et al. (1980) suggested that quantum effects could be taken into account by retaining Eq. (2.10), but using for the spectral density of $I_N(t)$ the Callen-Welton (1951) formula

$$S_I(\nu) = \frac{4h\nu}{R}\left[\frac{1}{e^{h\nu/k_B T} - 1} + \frac{1}{2}\right] \equiv \frac{2h\nu}{R}\coth\left(\frac{h\nu}{2k_B T}\right) . \tag{3.1}$$

Equation (3.1) has the limiting forms

$$S_I(\nu) \rightarrow \begin{cases} 4k_B T/R & (h\nu \ll k_B T) \tag{3.2} \\ 2h\nu/R. & (h\nu \gg k_B T) \tag{3.3} \end{cases}$$

In the case of Eq. (3.3), the noise arises from zero point fluctuations. When Eq. (3.1) describes the noise current in Eq. (2.10), the latter is an example of a quantum Langevin equation.

Following the procedure of Likharev and Semenov (1972), Koch et al. showed that in the limits $\beta_c \ll 1$, $I > I_o$, and $\nu_m \ll \nu_J$, the spectral density of the voltage noise for a current biased junction is

$$S_v(\nu_m) = \left[\frac{4k_B T}{R} + \frac{2eV}{R}\left(\frac{I_o}{I}\right)^2 \coth\left(\frac{eV}{k_B T}\right)\right]R_D^2 . \quad \begin{cases} \beta_c \ll 1 \\ I > I_o \\ \nu_m \ll \nu_J \end{cases} \tag{3.4}$$

Euqation (3.4) is illustrated in Fig. 4. It has been assumed that

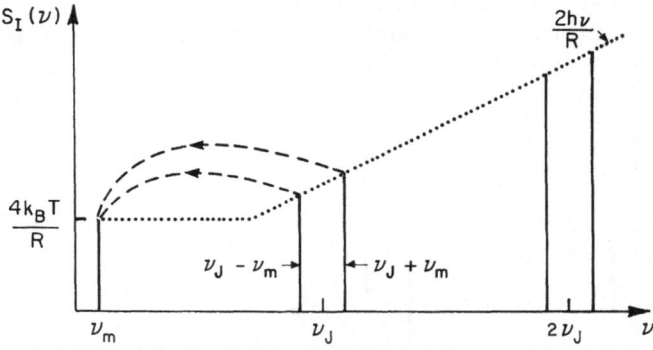

Fig. 4.   Schematic representation of the terms in Eq. (3.4). The Nyquist noise generated in the resistor at ν_m contributes at ν_m; the noise (including the contribution from the zero point term) generated at $\nu_J \pm \nu_m$ is mixed down to ν_m; mixing coefficients for the noise generated near $2\nu_J$ and higher harmonics are negligible for $\nu_m/\nu_J \ll 1$.

$h\nu_m \ll k_BT$ so that the first term on the right hand side is in the thermal limit. Even though $S_I(\nu)$ increases with frequency at high frequencies, the contribution from noise near $2\nu_J$, $3\nu_J$ ... is of order $(\nu_m/\nu_J)^2$, and hence negligible in the limit $\nu_m \ll \nu_J$. In the limit $eV = h\nu_J/2 \ll k_BT$, we can set $\coth(eV/k_BT) = k_BT/eV$, and Eq. (3.4) reverts to Eq. (2.12).

It is of interest to consider the conditions under which quantum corrections may be observable. Clearly, we require $eV \gtrsim k_BT$; however, this condition is not sufficient because if one simply increases eV/k_BT by increasing the bias current, the mixing coefficient will become very small, and the second term on the right hand side of Eq. (3.4) will be negligible compared with the first. To ensure that the second term is both in the quantum limit ($eV/k_BT \gg 1$) and larger than the first term, we require $eV \gg k_BT(I/I_o)^2$. With the aid of Eq. (2.9), it is easy to show that this condition is satisfied for some intermediate range of voltages provided

$$\kappa \equiv eI_oR/k_BT \gg 1. \qquad (3.5)$$

In the extreme quantum limit, Eq. (3.4) reduces to

$$S_V(\nu_m) = \frac{2eV}{R}\left(\frac{I_o}{I}\right)^2 R_D^2, \qquad (3.6)$$

and the noise arises solely from zero point fluctuations mixed down from near the Josephson frequency.

Figure 5 shows $S_V(\nu_m)/4k_BTR$ vs I/I_o. For $\kappa = 0.1$, the curve is almost identical with that predicted by the Likharev-Semenov result, Eq. (2.12). As κ is increased, the spectral density of the noise at intermediate currents progressively increases: This is the region in which quantum effects are observable. At high currents, all the curves tend asymptotically to unity, since the mixing coefficient tends to zero. At low currents, as $I \to I_o$ the curves also tend to the value predicted by Eq. (2.12). Notice, however, that the apparent divergence of the noise in this limit arises from the neglect of noise rounding in the theory, which implies that $R_D \to \infty$ as $I \to I_o$. In reality, noise rounding ensures that R_D remains finite so that as the current is lowered towards I_o, the noise goes through a maximum and then decreases.

We conclude this section with some comments on the validity of the quantum Langevin equation. First, we note that although we have put in quantum corrections to the noise in Eq. (2.10), the terms in ϕ remain firmly classical; thus, the treatment is "semi-classical". It is well known that such a treatment is correct for a simple harmonic oscillator (for example, Louisell, 1973), but one must use caution in assuming that it is valid for any non-linear equation.

Another prediction of the quantum Langevin equation is "quantum

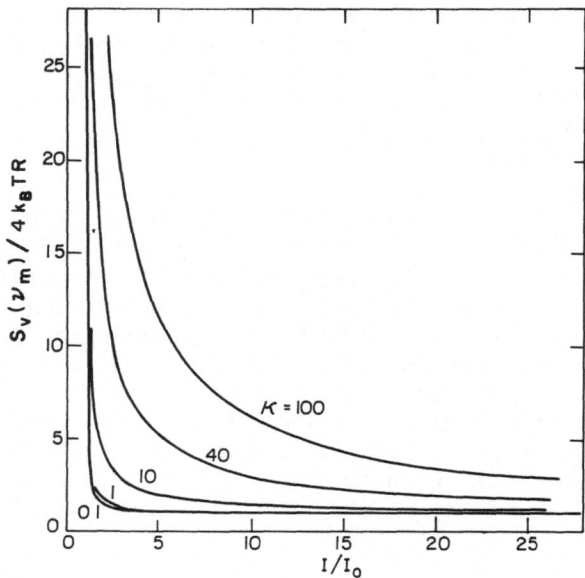

Fig. 5.   Calculated low frequency spectral density of the voltage
          noise across a single resistively shunted junction vs. bias
          current for 5 values of $\kappa \equiv eI_oR/k_BT$ and with $\beta_c \ll 1$ (Koch
          et al., 1980).

noise rounding". Thus, even at T = 0, for $I < I_o$ zero point fluctuations induce fluctuations in the tilt of the washboard, thereby en-
abling the particle to roll from one potential minimum to the next.
By analogy with the thermal case, one might call this process "quantum activation". However, there has recently been much interest in
an alternative process by which the junction may decay out of the
zero voltage state, namely "macroscopic quantum tunneling" (MQT)
(for example, Ivanchenko and Zil'berman, 1968; Leggett, 1978; Caldeira
and Leggett, 1981, 1982; Amgegaokar et al., 1982; Voss and Webb, 1981,
1981a, Jackel et al., 1981). In this approach, one allows for the
fact that the "particle" must actually be described by a quantum mechanical wave packet with a non-zero width, so that the particle can
<u>tunnel</u> between potential wells when $I < I_o$. Thus, as the bias cur-
rent is increased from zero, the junction will make a transition out
of the zero voltage state before I reaches the critical current. If
the junction is overdamped, this process will produce rounding of the
I - V characteristic at low voltages. Caldeira and Leggett (1981,
1982) have shown that the tunneling probability decreases exponentially from the WKB value as the damping of the junction is increased
from zero (i.e. as β_c is reduced from infinity). The increase in
damping causes the wave packet to narrow, thereby decreasing the tunneling probability. In the limit where the tunneling probability is
small, and where the damping is not too large, the quantum Langevin

equation predicts a substantially higher escape rate than MQT (Koch, 1981; Voss and Webb, 1981a). Furthermore, for relatively underdamped junctions, Voss and Webb (1981) found experimentally that the probability of decay out of the zero voltage state appears to be much lower than that predicted by Caldeira and Leggett. Thus, in this limit, the activation rate predicted by the quantum Langevin equation is in serious disagreement with the experimental data. This discrepancy is hardly surprising: In the limit in which quantum mechanical broadening of the wave packet is substantial, the picture of a point particle moving on a washboard cannot be correct. Nevertheless, the quantum Langevin approach may still be appropriate for overdamped junctions ($\beta_c < 1$) when I is not too much less than I_0, so that the probability of transitions from the zero voltage state is relatively high. It is clear that a more detailed investigation of the relationship between macroscopic quantum tunneling and quantum activation in the region $I \lesssim I_0$ is required. In this context, Schmid (1982) has shown that the quantum Langevin approach should be valid for $R \ll \hbar/4e^2 \approx 10^3 \Omega$. One would also like to have a clear-cut criterion for the limit in which one can neglect the quantum mechanical broadening of the wave packet.

For the remainder of this article we shall be concerned only with the limit $I > I_0$ in which the junction is free running, and MQT is irrelevant. Thus, our next task is to investigate the validity of Eq. (3.4) experimentally.

B. Experiment

Experiments to observe quantum corrections to the noise in resistively shunted junctions have been carried out by Koch et al. (1981a, 1982). To observe quantum noise effects, one requires junctions with $\kappa \gtrsim 1$. Writing $\kappa = (e/k_BT)(\beta_c\Phi_0 j_1/2\pi c)^{\frac{1}{2}}$, where j_1 is the critical current density and c is the capacitance per unit area, we see that junctions with high critical current densities are necessary to observe these effects in the liquid He^4 temperature range. At 4.2K, with $\beta_c = 0.2$, $j_1 = 10^4 A\ cm^{-2}$ and $c = 0.04pF\ \mu m^{-2}$ we find $\kappa \approx 1.1$. This is a convenient value of κ, since, as the temperature is lowered to near 1K, κ increases so that quantum effects become dominant.

Koch et al. fabricated $PbIn-In_2O_3-Pb$ tunnel junctions, resistively shunted with $CuA\ell$ films, on glass substrates using the photolithographic lift-off techniques described by workers at IBM (Greiner et al., 1980). The configuration is shown in Fig. 6. First, a 10 μm-wide Cu (0-3 wt.% $A\ell$) film 40 to 100 nm thick was deposited, followed by a 10 μm-wide, 250 nm-thick Pb (20 wt.% In) film at right angles to the $CuA\ell$ strip. After another resist patterning, a SiO oxide layer, 100 nm thick, was deposited and two windows were opened by lifting off the SiO to expose the PbIn and $CuA\ell$ films. After patterning the resist for the upper electrode, the exposed metal surfaces were cleaned by rf sputter-etching in Ar, the In_2O_3 oxide was

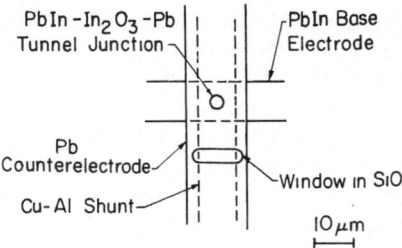

Fig. 6. Configuration of resistively shunted tunnel junction used
in quantum noise measurements (Koch et al., 1982).

grown thermally in a low pressure of oxygen, and the 400 nm-thick Pb
counter-electrode was deposited and lifted off. A final protective
layer of SiO was then evaporated. The diameter of the junction was
about 2.5 µm, and the critical current ranged from 0.1 to 2 mA (0.2
to 4×10^4 A cm^{-2}) at 4.2K, depending on the oxidation parameters.
The capacitance of the junction was estimated to be 0.5 pF. The resistive shunt was about 5 µm long and ranged in resistance from 0.05
to 0.7Ω, depending on the thickness and composition of the CuAℓ. The
Pb counterelectrode formed a ground plane for the shunt, reducing its
inductance, L_s, to about 0.2 pH. Leads were attached to the junctions with pressed In pellets. Junctions fabricated with these techniques omitting the resistive shunts displayed excellent tunneling
characteristics with little excess current at voltages below the sum
of the gaps.

The apparatus for measuring the noise is shown schematically
in Fig. 7. The junction was connected across two cooled LC-resonant
tank circuits, either of which could be coupled to a low-noise preamplifier. By connecting together the tank circuit leads at the top
of the cryostat a third, intermediate frequency could be obtained;

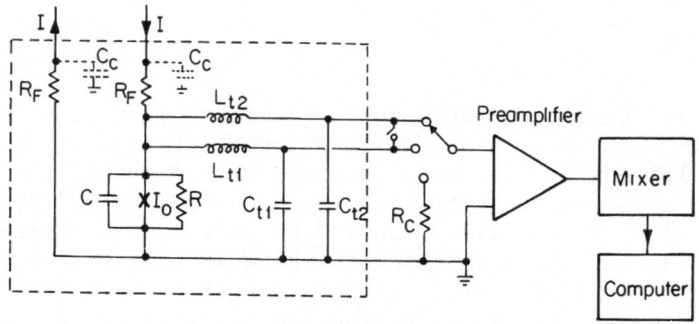

Fig. 7. Schematic of apparatus for measuring noise across a junction; the components enclosed in the dashed line are im-
mersed in liquid helium (Koch et al., 1982).

the three resonant frequencies were 183 kHz, 106 kHz, and 70 kHz. The output from the preamplifier was mixed down to frequencies below 500 Hz, and the spectral density was then measured with a computer. The junction was enclosed in a superconducting can, and the cryostat was surrounded by a mu-metal shield. The cryostat, bias supply and preamplifier were enclosed in a shielded room. The gain of the preamplifier-mixer-computer chain was calibrated frequently against the Nyquist noise in a resistor (R_c) at room temperature. The accuracy of the gain was estimated to be ± 2%.

The spectral density of the voltage noise produced by the junction across the tank circuit was $Q^2 S_V(\nu_m) = \omega_m^2 L_t^2 [S_V(\nu_m)/R_D^2]$, where L_t is the inductance. Thus the quantity

$$\frac{S_v(\nu_m)}{R_D^2} = \frac{4k_B T}{R} + \frac{2eV}{R}\left(\frac{I_0}{I}\right)^2 \coth\left(\frac{eV}{k_B T}\right) \begin{Bmatrix} \beta_c \ll 1 \\ I > I_o \\ \nu_m \ll \nu_J \end{Bmatrix} \qquad (3.7)$$

can be measured without knowing the Q of the tank circuit. In fact, there is a considerable advantage in comparing Eq. (3.7) rather than Eq. (3.4) with experiment. Real junctions may deviate from the simple RSJ model with $\beta_c = 0$ because the capacitance of junction, and, in some cases, the inductance of the shunt are non-negligible. Computer simulations, however indicate that the major effect of the deviations is to modify R_D, leaving the current noise terms on the right hand side of Eq. (3.7) relatively unchanged (Voss, 1981; Koch, 1981). Thus, for a real junction, the measured value of $S_V(\nu_m)$ may differ significantly from Eq. (3.4) while $S_V(\nu_m)/R_D^2$ is in good agreement with Eq. (3.7).

It was necessary to consider the following sources of error:

(i) The contributions of the measured voltage and current noises of the preamplifier were subtracted. The errors introduced in the spectral density of the junction noise ranged from about 5% at 4.2K to 15% at 1.6K.

(ii) Losses in the tank circuit (for example, due to the presence of stray resistance) were important only at the intermediate frequency. In this case, parts of the leads were at room temperature. The error after this noise contribution had been subtracted was 5%.

(iii) From the measurements at the three frequencies it was found that some junctions had a small 1/f noise component. The spectral density of this voltage noise was proportional to $(\partial V/\partial I_0)^2$, suggesting that the noise arose from fluctuations in the critical current (Clarke and Hawkins, 1976). The error introduced after the 1/f noise was subtracted was no more than 3%.

(iv) The noise measurements were made at voltages well below the

sum of the gaps of the superconductors. The small noise contribution from the quasiparticle current in this region was at most 1%
of the current noise from the shunt resistor, and was neglected.

(v) The power dissipation in the resistive shunt raised its
temperature above the bath temperature at higher bias voltages.
The heating effect was determined as a function of temperature by
reducing the critical current to zero with a magnetic field and measuring the Nyquist noise of the shunt for various bias currents.
For the junction described in this article, the heating was significant only at high voltages ($V \gg k_BT/e$) where the term $\coth(eV/k_BT)$
in Eq. (3.7) is nearly independent of temperature. Thus, it was
sufficient to correct only the term $4k_BT/R$ in Eq. (3.7) by subtracting $4k_B\Delta T/R$ from the measured noise, where ΔT is the temperature
rise. The error was no more than 3%.

(vi) Considerable care was taken to shield the experiment from
extraneous noise sources, such as 60 Hz pick-up, radio and television
stations, and interference from nearby computers and lasers. To
demonstrate that this shielding was sufficient, it was shown that the
noise measured in resistors in the range 1.5 to 4.2K was within 3% of
the predicted value. Furthermore, as we shall see, measurements on
junctions in the classical limit $eV \ll k_BT$ were in excellent agreement
with theory, and showed the correct temperature dependence.

We turn now to a discussion of the experimental results. We
shall confine our discussion to a single junction: Results from

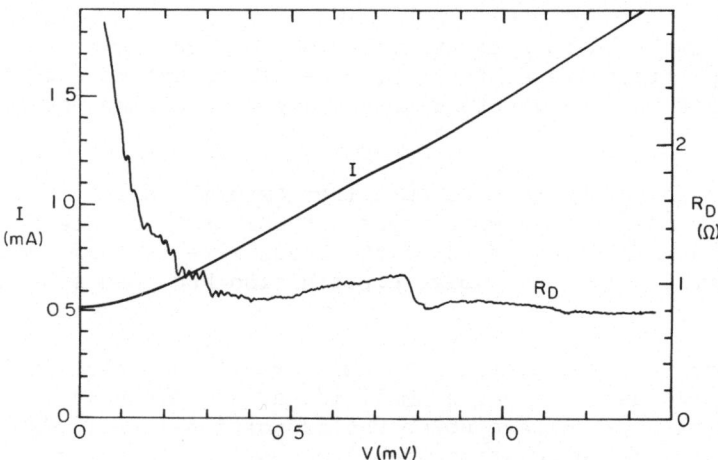

Fig. 8. I and R_D vs. V for resistively shunted junction at 4.2K with
$I_O = 0.51$mA, R = 0.70Ω (at 100μV — R was somewhat voltage-
dependent), C = 0.5pF, $\beta_c = 0.38$, and $\kappa = 0.99$ (Koch et al.,
1982).

other junctions can be found in the paper by Koch et al. (1982).
The relevant parameters are listed in the caption to Fig. 8; in addition, there was an inductance associated with the shunt resistor
of about 0.2 pH.

Figure 8 shows I and R_D vs. V at 4.2K for this junction. There
is a small dip in R_D at 800μV that is probably due to a self-reson-
ance of the junction capacitance and shunt inductance, and some fine
structure around 200μV of unknown origin. However, as mentioned
earlier, small deviations in R_D from the prediction of the RSJ model
are not expected to affect $S_v(\nu_m)/R_D^2$. In Fig. 9 we plot measured
values of $S_v(\nu_m)/R_D^2$ vs. voltage (open circles) after the preamplifier
noise has been subtracted. The solid circles are the noise after the
1/f noise subtraction and the heating correction have been made. At
low voltages the correction is entirely due to 1/f noise, while at
high voltages, the correction is largely due to heating. The solid
line through the solid circles is the prediction of Eq. (3.7) using
the measured values of R, I_o, I, V, and T. The upper dashed line is
the predicted noise in the absence of zero point fluctuations, that
is

$$\frac{S_v'(\nu_m)}{R_D^2} = \frac{4k_BT}{R} + \frac{4eV}{R}\left(\frac{I_o}{I}\right)^2 \frac{1}{\exp(2eV/k_BT) - 1} . \tag{3.8}$$

The triangles in Fig. 9 represent the measured mixed-down noise,
which was computed by subtracting $4k_BT/R$ from the solid circles.
The solid line through the triangles is the mixed-down noise predicted by Eq. (3.7), $(2eV/R)(I_o/I)^2\coth(eV/k_BT)$, while the lower
dashed line is the mixed-down noise predicted by Eq. (3.8) in the
absence of zero point fluctuations, $(4eV/R)(I_o/I)^2[\exp(2eV/k_BT) - 1]^{-1}$.

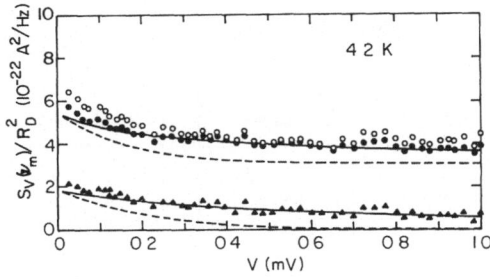

Fig. 9. $S_v(\nu_m)/R_D^2$ vs. V at 4.2K for junction shown in Fig. 8. The
         open circles show the total measured noise across the junc-
         tion; solid circles below show the noise remaining after
         correction for 1/f noise and heating. Upper solid and dash-
         ed lines are predictions of Eqs. (3.7) and (3.8). Solid
         triangles are measured mixed-down noise, lower solid and
         dashed lines are mixed-down noise predicted by Eqs. (3.7)
         and (3.8) (Koch et al., 1982).

It is evident that both the total measured noise across the junction
and the measured mixed-down noise are in excellent agreement with the
theory that includes a contribution from the mixed-down zero point
fluctuations, and are substantially higher than the predictions of a
theory that does not include this contribution.

Figure 10 shows the temperature dependence of the noise for
twelve bias voltages ranging from 50 µV to 550 µV. The temperature
$T = 2eV/k_B$ is indicated for the six lowest voltages; mixed-down noise
at temperatures well above $2eV/k_B$ is in the classical limit, while

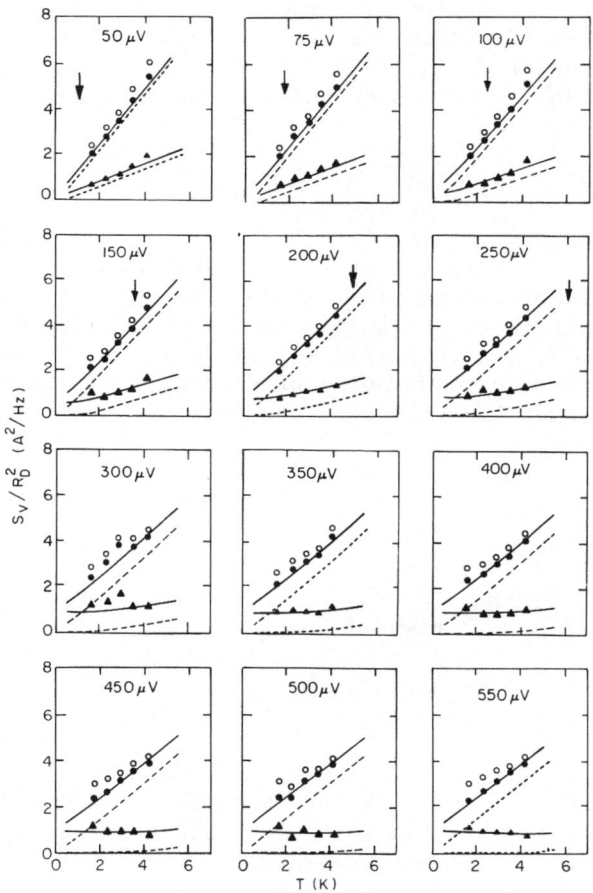

Fig. 10. $S_V(\nu_m)/R_D^2$ vs. T for junction at 12 bias voltages. Notation
          is as for Fig. 9. Arrows indicate $2eV = k_BT$ (Koch et al.,
          1982).

that at temperatures well below $2eV/k_B$ is in the quantum limit. The mixed-down noise at the six highest voltages is in the quantum limit at all temperatures measured. For all twelve voltages, the total junction noise is in good agreement with the predictions of Eq. (3.7), and substantially greater than the predictions of Eq. (3.8). (The discrepancy for 300 μV probably arises from the structure shown in Fig. 8.) The mixed-down noise at 350 μV and above is independent of temperature, and in excellent agreement with the value of Eq. (3.6), $S_V(0)/R_D^2 = (2eV/R)(I_0/I)^2$. As the voltage is lowered the mixed-down noise becomes increasingly temperature dependent, and remains in good agreement with the predictions of Eq. (3.7). At 50 μV, the mixed-down noise is in the classical limit for the whole temperature range, and proportional to T, as expected. This temperature dependence demonstrates that the contribution of any extraneous noise was negligible.

We can extract from our data the measured spectral density of the current noise $S_I(\nu)$ generated by the shunt resistance R at the Josephson frequency $\nu = 2eV/h$. We divide each value of the mixed-down noise by the mixing coefficient $(I_0/I)^2/2$, a procedure that converts the mixed-down noise in Eq. (3.7) into Eq. (3.1). The results are plotted in Fig. 11 for two temperatures. The solid lines are the corresponding predictions of Eq. (3.1) using measured values of ν =

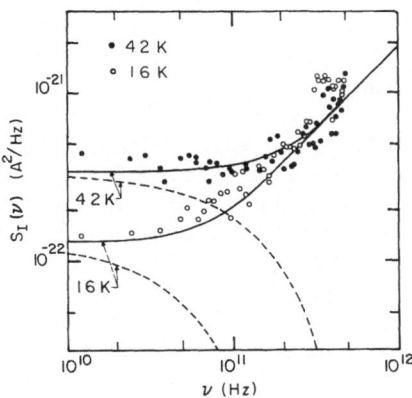

Fig. 11.   Measured spectral density of current noise in shunt resistor of junction at 4.2K (solid circles) and 1.6K (open circles). Solid lines are prediction of Eq. (3.1), while dashed lines are $(4h\nu/R)[\exp(h\nu/k_BT) - 1]^{-1}$ (Koch et al., 1982).

2eV/h, R, and T. The slight increase of the data above the theory at the highest voltages may reflect the presence of a resonance on the I-V characteristic. The agreement between the data and the predictions is rather good, bearing in mind that no fitting parameters are used. By contrast, the dashed lines represent the theoretical prediction in the absence of the zero point term, $(4h\nu/R)[\exp(h\nu/k_BT) - 1]^{-1}$, and fall far below the data at the higher frequencies. The existence of zero point fluctuations in the measured spectral density of the current noise is rather convincingly demonstrated.

In experiments on other junctions Koch et al. (1982) found that the noise in a junction at a fixed temperature exhibited the correct dependence on $\kappa \equiv eI_0R/k_BT$ when I_0 was changed by means of a magnetic field. Furthermore, they studied one junction in which the shunt inductance was anomalously large, and the dynamic resistance showed substantial structure at voltages for which the Josephson frequency was a subharmonic of the LC-resonant frequency. The measured noise, however, was in good agreement with a computer prediction in which these resonances were included. Because of the higher non-linearities of this junction, the mixing coefficients for noise at harmonics of the Josephson frequency ($2\nu_J$, $3\nu_J$ ...) are no longer negligible, and it is necessary to include these contributions in the computer simulation.

In concluding this section, it should be emphasized again that only measured parameters were used in comparing the data with theory in Figs. 9 to 11; there were no fitting parameters. Thus, the quantum Langevin equation accurately predicts the limiting voltage noise in a current-biased resistively shunted junction for $\beta_c < 1$ and $I > I_0$. The noise arises from zero point fluctuations mixed down from near the Josephson frequency. At the moment, there are no measurements of the noise in the noise rounded region $I < I_0$ —— quantum effects are quite negligible in this region for the junctions studied in the He^4 temperature range.

Finally, the fact that zero point fluctuations can be observed at frequencies as high as 500 GHz (Fig. 11) implies that a Josephson mixer using its self-oscillations as the local oscillator is an ideal quantum-limited device at these frequencies. Unfortunately, a mixer in this mode is of limited practical use because the frequency of the oscillator has a rather large linewidth induced by the voltage fluctuations. Likharev and Semenov (1972) have shown that the rms linewidth, $\Delta\nu_J$, is related to the spectral density of the low frequency voltage noise by $\Delta\nu_J = \pi S_V(\nu_m)/\Phi_0^2$. Using the quantum limited expression for $S_V(\nu_m)$, Eq. (3.6), we find $\Delta\nu_J = 2e\pi V(I_0/I)^2(R_D^2/R)/\Phi_0^2$. With the aid of Eq. (2.9), we can write this in the form $\Delta\nu_J = 2e\pi R^3 I_0^2/\Phi_0^2 V$. Taking $R = 1\Omega$, $I_0 = 1mA$, and $V = 1mV$, we find $\Delta\nu_J \approx 250$ MHz for a frequency of 500 GHz, a linewidth that is undesirably large for many applications.

IV. DC SQUIDS

A. Introduction

    The principal objective of this section on SQUIDs is to apply
the ideas of quantum noise developed in Sec. III to the dc SQUID, and
then to compare these results with experiment. Before becoming involved in the quantum limited case, however, it is convenient first
to outline the general principles of the dc SQUID, and to discuss the
thermal noise limit.

    The dc SQUID consists of two Josephson junctions connected in
parallel on a superconducting loop of inductance L as shown in Fig.
12(a). We assume that the junctions are identical and overdamped
($\beta_c < 1$) so that the I-V characteristic is non-hysteretic. When the
flux, Φ, threading the SQUID loop is changed, the critical current
and hence the I-V characteristic oscillate with period Φ_0, as indicated in Fig. 12(b). If the SQUID is biased with a current greater
than the maximum critical current, the voltage across the SQUID is
periodic in the flux, as shown in Fig. 12(c). Thus, the SQUID is
simply a flux-to-voltage transducer, with a transfer function $V_\Phi \equiv$
$(\partial V/\partial \Phi)_I$ that is a maximum near $(2n + 1)\Phi_0/4$. In most applications,
at least for low frequency measurements, the SQUID is operated in a
flux-locked loop. An oscillating flux is applied to the SQUID and
the resulting voltage across the SQUID is amplified by a cooled LC-
resonant circuit or transformer and then by a low noise amplifier.
The signal from the amplifier is lock-in detected, and the output
from the lock-in is fed back into a coil coupled to the SQUID. Thus,
feedback maintains a constant flux in the SQUID, and the dynamic
range can be as high as 10^3 Φ_0 or more. Details of flux modulation
schemes have appeared elsewhere (Clarke et al., 1976; Clarke, 1977,
1980).

    The simplest characterization of the sensitivity of a SQUID is
in terms of its equivalent flux noise, which has a spectral density

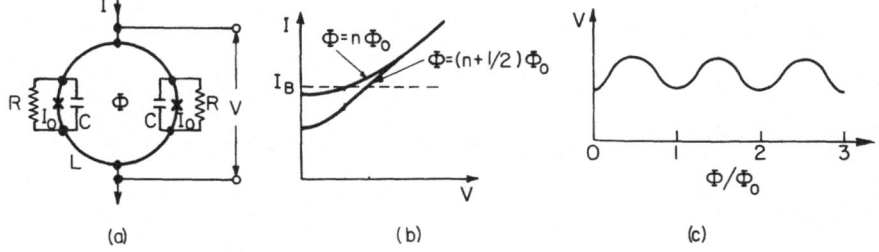

Fig. 12. (a) Configuration of dc SQUID; (b) current-voltage (I-V)
          characteristic with $\Phi = n\Phi_0$ and $(n + \frac{1}{2})\Phi_0$, where n is an
          integer; (c) V vs. Φ at constant bias current.

$$S_\Phi(f) = S_V(f)/V_\Phi^2. \tag{4.1}$$

In Eq. (4.1), $S_V(f)$ is the spectral density of the voltage noise across the SQUID. However, in order to compare SQUIDs with different inductances, it is more useful to define a flux noise energy per unit bandwidth

$$\frac{\varepsilon(f)}{1\text{Hz}} = \frac{S_\Phi(f)}{2L} . \tag{4.2}$$

As we shall see later, $\varepsilon/1\text{Hz}$ is not a complete specification of the SQUID noise sources; however, it is a convenient measure for comparing different devices.

The first dc SQUID for which detailed noise measurements were made and compared with theory was the cylindrical device shown in Fig. 13(a) (Clarke et al., 1976). The spectral density of the equivalent flux noise measured in a flux-locked loop is shown in Fig. 13 (b). The noise is nearly white at frequenices above 2×10^{-2} Hz; below this frequency the noise scales approximately as $1/f$. The roll-off at higher frequencies is due to a low-pass filter in the measurement circuit. In the white noise region, the equivalent flux noise is approximately $3.5 \times 10^{-5}\Phi_0$ $\text{Hz}^{-\frac{1}{2}}$; with an area of 7 mm^2, this corresponds to a magnetic field sensitivity of about 10^{-14} T$\text{Hz}^{-\frac{1}{2}}$. The corresponding noise energy, with an estimated SQUID inductance of 1 nH, is about 3×10^{-30} JHz^{-1}.

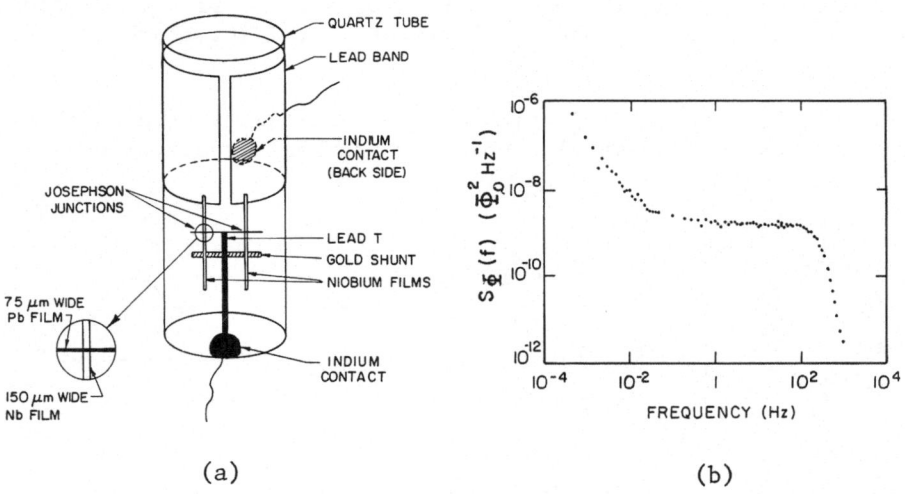

(a)   (b)

Fig. 13.   (a) Configuration of cylindrical dc SQUID; (b) spectral
           density of SQUID flux noise, $S_\Phi(f)$, for typical cylindri-
           cal SQUID (Clarke et al., 1976).

In the next section we compare these results with theoretical predictions, and see how subsequent devices have been fabricated with greatly improved noise energies.

B.   Thermal Noise in the dc SQUID

A model for noise calculations in the dc SQUID is shown in Fig. 14. The current in the SQUID loop is $J(t)$. There are two independent noise currents, $I_{N1}(t)$ and $I_{N2}(t)$, associated with the two shunt resistors. The phase differences across the two tunnel junctions, δ_1 and δ_2, obey the following equations (Tesche and Clarke, 1977; Koch et al., 1981)

$$V = \frac{\hbar}{2e}(\dot{\delta}_1 + \dot{\delta}_2), \tag{4.3}$$

$$J = \frac{\Phi_0}{2\pi L}\left(\delta_1 - \delta_2 - \frac{2\pi\Phi}{\Phi_0}\right), \tag{4.4}$$

$$\frac{\hbar C}{2e}\ddot{\delta}_1 + \frac{\hbar}{2eR}\dot{\delta}_1 = \frac{I}{2} - J - I_0\sin\delta_1 + I_{N1}, \tag{4.5}$$

and

$$\frac{\hbar C}{2e}\ddot{\delta}_2 + \frac{\hbar}{2eR}\dot{\delta}_2 = \frac{I}{2} + J - I_0\sin\delta_2 + I_{N2}. \tag{4.6}$$

Equation (4.3) relates the voltage to the average rate of change of phase, Eq. (4.4) relates the circulating current to $\delta_1 - \delta_2$ and to the applied flux Φ, while Eqs. (4.5) and (4.6) are two Langevin equations coupled via J. Tesche and Clarke (1977) solved the equations in the limit $C = 0$ with $\Gamma \equiv 2\pi k_B T/I_0\Phi_0 = 0.05$ and with a spectral density $4k_B T/R$ for I_{N1} and I_{N2}. They computed V_Φ and S_Φ vs. I for different values of applied flux and for various values of $\beta = 2LI_0/\Phi_0$. As in the case of the single junction, the noise was computed for frequencies much less than the Josephson frequency, and was found to have a white power spectrum in this region. To optimize the SQUID, they assumed the maximum value of β_c was unity, thereby deter-

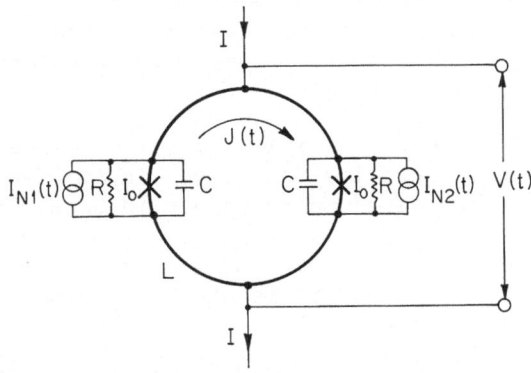

Fig. 14.   Model for noise calculations in the dc SQUID.

mining an upper limit on R for given values of C and I_0. It should be noted, however, that the introduction of a non-zero capacitance in Eqs. (4.5) and (4.6) undoubtedly changes the computed values of V_Φ and S_Φ somewhat, so that the results must be regarded as approximate.

The conclusions reached were as follows. The SQUID is optimized when $\beta = 1$ and (somewhat arbitrarily) $\beta_c = 1$, and has its best performance when Φ is near $(2n + 1)\Phi_0/4$. The optimized value of V_Φ is[*]

$$V_\Phi \approx R/L \tag{4.7}$$

while the spectral density of the optimized voltage noise is

$$S_V \approx 16k_B TR. \tag{4.8}$$

The corresponding optimized flux noise energy is

$$\varepsilon/1Hz \approx 9k_B TL/R. \tag{4.9}$$

It is convenient to eliminate R from Eq. (4.9) by setting $\beta_c = \beta = 1$ to obtain

$$\varepsilon/1Hz \approx 16k_B T(LC)^{\frac{1}{2}}. \quad \left\{ \begin{array}{l} \beta = 1 \\ \beta_c = 1 \\ \Gamma = 0.05 \end{array} \right\} \tag{4.10}$$

The numerical coefficients in Eqs. (4.9) and (4.10) should be treated with suspicion because of the neglect of the capacitance in the calculations. Nevertheless, the result gives a clear prescription for reducing the flux noise energy of SQUIDs: One should reduce T, L, and/or C.

The performance of the cylindrical SQUID can be compared with Eq. (4.10). Taking T = 4.2K, L $\approx$ 1nH, and C $\approx$ 1nF[†], we find $\varepsilon/1Hz \approx 10^{-30} JHz^{-1}$, a factor of 3 smaller than the observed value. We note, however, that we have neglected the effects of flux modulation, which increases the noise somewhat, by a factor that depends on the details of the modulation scheme. Thus, the agreement between theory and experiment is not unreasonable.

Since the prediction given in Eq. (4.10) appeared, many SQUIDs have been fabricated to test its validity. These have all been in a planar configuration; two examples appear in Secs. IV D and

---

[*]Bruines et al. (1982) have pointed out a numerical error of about 1.5 in the value of V_Φ obtained by Tesche and Clarke (1977). The corrected values have been used here.

[†]The estimation of C is difficult. We shall rather arbitrarily take 0.04 pF μm^{-2} for Pb-based junctions and 0.08 pF μm^{-2} for Nb-based junctions in this article.

IV F. The main emphasis has been to reduce the inductance, and, particularly, the capacitance of the SQUIDs by means of photo- lithography or electron beam lithography. The performance of a selection of devices is summarized in Fig. 15, where the measured

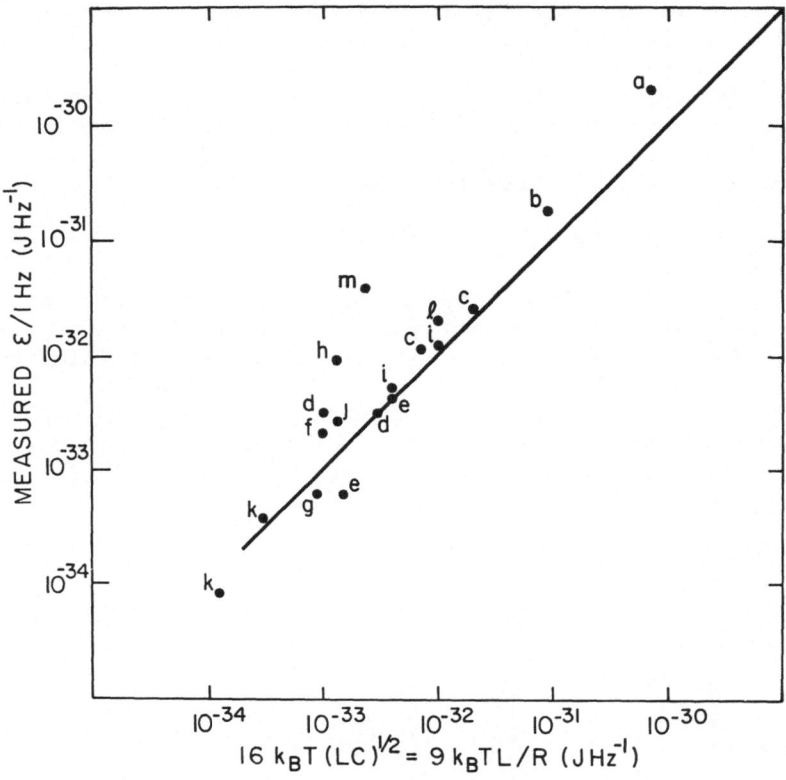

Fig. 15. Measured values of $\varepsilon/1\mathrm{Hz}$ vs. $16k_BT(LC)^{\frac{1}{2}} = 9k_BTL/R$ for a number of dc SQUIDs involving thin film tunnel junctions (ex-
cept f). The SQUIDS referred to are:
a Clarke et al., 1976 (4.2K)
b Koch and Clarke, 1979 (4.2K)
c Voss et al., 1980 (4.2K, 1.6K)
d Ketchen and Voss, 1979 (4.2K, 1.8K)
e Voss et al., 1981 (4.2K, 1.5K)
f Voss et al., 1981 (4.2K) - nanobridge
g Cromar and Carelli, 1981 (4.2K)
h Cromar and Carelli, 1981 (4.2K)
i Ketchen and Jaycox, 1982 (4.2K, 1.5K)
j Carelli and Foglietti, 1982 (4.2K)
k Van Harlingen et al., 1982 (4.2K, 1.4K) - 1/f noise subtracted
ℓ Martinis and Clarke, 1982 (4.2K)
m De Waal et al., 1981 (4.2K)

values of $\varepsilon/1Hz$ are plotted against $16k_BT(LC)^{\frac{1}{2}} = 9k_BTL/R$. No distinction has been made among devices that were operated in a flux-locked loop and those that were operated as small-signal amplifiers with no feedback. It should also be noted that in some cases the value of Γ was considerably less than 0.05, so that Eq. (4.10) cannot be expected to hold exactly. Nevertheless, although there is a general scatter of the results, it is evident that the measured noise energies follow the predicted values rather well.

It is also apparent that the performance of the most sensitive devices approaches $\varepsilon/1Hz \sim 10^{-34} JHz^{-1}$ or $\hbar$. One might expect quantum effects to be important in this range, and the following section is devoted to a discussion of quantum noise in SQUIDs.

C. Quantum Noise in the dc SQUID: Theory

Koch et al. (1981) have extended their calculations on quantum noise in single junctions to the case of the dc SQUID. They again solved Eqs. (4.3) to (4.6) on a computer, this time retaining the terms in C and using Eq. (3.1), $S_I(\nu) = (2h\nu/R)\coth(h\nu/2k_BT)$, for the spectral density of the noise currents. The computed values of V_Φ, S_V, and $\varepsilon/1Hz$ at frequencies much less than ν_J for a particular SQUID with low values of L and C are shown in Fig. 16(a), (b) and (c). The optimum value of $\varepsilon/1Hz$ is

$$\varepsilon/1Hz \approx \hbar, \tag{4.11}$$

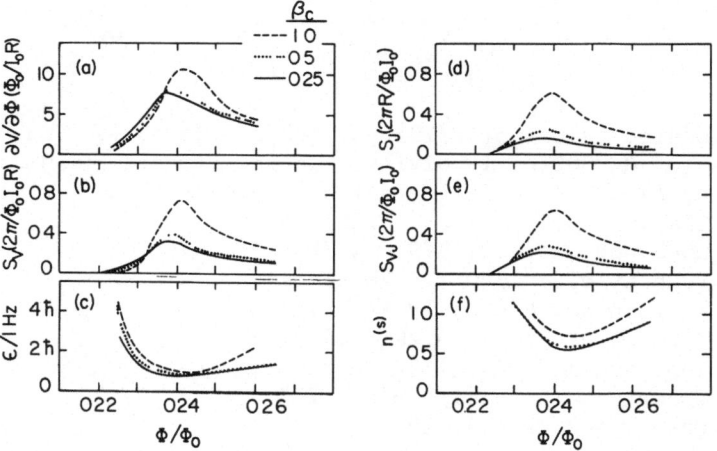

Fig. 16. Values of $\partial V/\partial\Phi$, S_V, $\varepsilon/1Hz$, S_J, S_{VJ}, and $n^{(s)}$ vs. Φ/Φ_0 computed at frequencies much less than the Josephson frequency for a SQUID at T = 0 with $\kappa\Gamma = 2e^2R/\hbar = 0.02$, $\beta = 1$, $I/I_0 = 1.63$, and for $\beta_c = 0.25$, 0.5, and 1 (Koch et al., 1981).

and is determined by zero point fluctuations. It cannot be emphasized too strongly, however, that there is no exact minimum value
of $\varepsilon/1Hz$ determined by quantum mechanics, and that the values computed for SQUIDs with different parameters are likely to be scattered
around $\hbar$ rather than exactly equal to $\hbar$.

Koch et al. (1981) extended their calculations to non-zero temperatures, and found that $\varepsilon/1Hz$ increased to about $3\hbar$ at 4.2K for
the parameters given in the caption of Fig. 16.

D. Quantum Noise in the dc SQUID: Experiment

To my knowledge, there have been three types of SQUIDs with
noise energies within a factor of six of $\hbar$ (Cromar and Carelli, 1981;
Voss et al., 1981; Van Harlingen et al., 1982). We briefly describe
the most sensitive, that of Van Harlingen et al. (1982). The configuration of their SQUID is shown in Fig. 17 (inset), and is very
similar to that of Cromar and Carelli (1981). Two Pb (20 wt.% In)
- In_2O_3 - Pb tunnel junctions with a nominal diameter of 2 μm and a
separation of 30 μm were defined by lifting off windows in a SiO insulating layer. Two additional windows determined the length of a
10 μm-wide CuAl shunt resistance for each junction. The capacitance
of each junction was estimated to be about 0.3pF, while the SQUID
loop had an inductance of about 2pH. Flux was applied to the SQUID
by means of a current along the counter electrode.

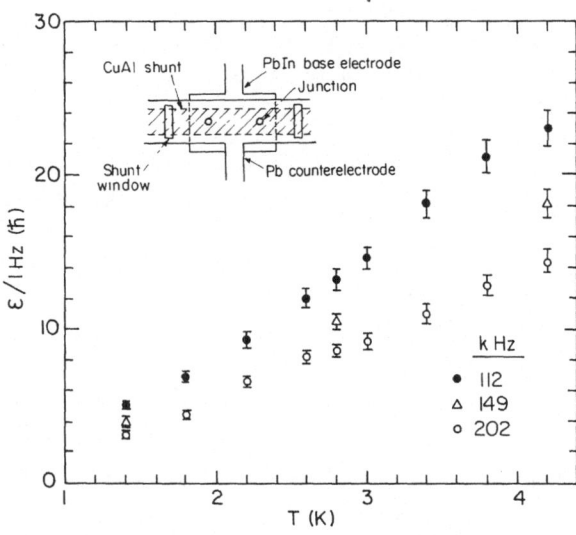

Fig. 17. Measured optimized values of $\varepsilon/1Hz$ vs. T at three frequencies for most sensitive SQUID of Van Harlingen et al.
(1982). Inset shows configuration of SQUID.

The spectral density of the voltage noise, $S_V(f)$, was measured using the apparatus shown in Fig. 7. Values of V_Φ were measured by applying a 1kHz-flux to the SQUID with an amplitude of typically 10^{-4} Φ_o, and measuring the voltage across the SQUID with a lock-in detector. At each temperature, the flux noise $S_\Phi(f) = S_V(f)/V_\Phi^2$ was measured with the flux near $(2n + 1)\Phi_o/4$ and with V_Φ maximized with respect to the bias current. The optimized values of $\varepsilon/1Hz$ vs. T for the most sensitive SQUID are shown in Fig. 17 for three measurement frequencies. At each frequency, the noise energy is roughly linear in the temperature. The best measured value of $\varepsilon/1Hz$ at 1.4K and 202kHz was $3.2\hbar \pm 0.2\hbar$, corresponding to an equivalent flux noise of $(1.7 \pm 0.1) \times 10^{-8}\Phi_o Hz^{-\frac{1}{2}}$. Clearly, there is a substantial frequency-dependent contribution to the noise. From measurements at the three frequencies, it was found that the spectral density of the total noise could be expressed as the sum of a white noise component and a 1/f noise component. The values of these components are listed in Table I. Unfortunately, the rather large relative magnitudes of the 1/f noise give rise to substantial error bars on the estimates of the white noise, and make it impossible to determine whether or not the noise tends to an asymptote at the lowest temperatures. The lowest extrapolated white noise energy (at 1.4K) was $(0.8 \pm 0.7)\hbar$. The best estimates of the white noise energy at 4.2K and 1.4K have been plotted in Fig. 15. Given the substantial errors, the measured values of the white noise energy are in reasonable agreement with the predictions (Koch et al., 1981): For a SQUID with $L \approx 2pH$, $C \approx 0.3pF$, $\beta_c \approx 1$, and $\beta \approx 1$, we expect $\varepsilon/1Hz$ to be about $3\hbar$ at 4.2K, decreasing to a little more than $\hbar$ at 1K.

From the measurements, one can conclude that it is possible to operate SQUIDs with white noise energies approaching $\hbar$, but the theory has clearly not been tested as rigorously as in the case of

Table I. Performance of Ultra Low-Noise SQUID[a] (at 4.2K, L = 1.9 pH, $C \approx 0.3$ pF, I_o = 0.55 mA, $R \approx 1.3\Omega$, β = 1.0, $\beta_c \approx 0.9$)

T	V_Φ	R_D	$\varepsilon/1Hz$ 202kHz	$\varepsilon/1Hz$ white	$\varepsilon/1Hz$ 1/f (100kHz)
(K)	(mV/Φ_o)	(Ω)	($\hbar$)	($\hbar$)	($\hbar$)
4.2	6.8	4.7	14.4	3.6 ± 3.0	21.6 ± 4.7
2.8	9.4	6.4	8.6	2.8 ± 1.8	11.6 ± 2.8
1.8	12.5	8.9	4.5	1.5 ± 1.0	6.0 ± 1.5
1.4	13.9	11.1	3.2	0.8 ± 0.7	4.8 ± 1.0

[a]Van Harlingen et al., 1982.

the single junctions reported in Sec. III. More accurate measurements with this type of SQUID would necessitate higher frequencies to avoid the errors introduced by subtracting the 1/f noise. Furthermore, the type of SQUID described here is of very low inductance so that it is difficult to couple it efficiently to an input coil of, say, 0.1 - 1 μH as is often required in typical applications. Other SQUIDs that have achieved noise energies below 20ℏ suffer from the same disadvantage. A discussion of techniques for coupling to planar SUQIDs appears in Sec. IV F.

E. SQUID Amplifiers

Within the reasonably near future, it is likely that SQUIDs will be fabricated in which the white flux noise is determined by zero point fluctuations. However, it does not automatically follow that such a SQUID can be used as an amplifier in which the measurement of a signal applied to the input is quantum noise limited. The purpose of this section is to discuss the design of such amplifiers, and to show that quantum limited performance should in fact, be possible if the input circuit is optimized appropriately.

When a SQUID is inductively coupled to an input circuit, the current noise circulating in the SQUID loop inevitably induces a voltage noise in this circuit. This additional noise source implies that $\varepsilon/1Hz$ is not a complete specification of the SQUID noise, as mentioned in Sec. IV A. The current noise has a spectral density, $S_J(f)$, that has been calcrlated by Tesche and Clarke (1979) in the thermal limit, and by Koc'ı et al. (1981) in the quantum limit. The current noise is partially correlated with the voltage noise across the SQUID, some of which arises from the flux noise generated by the current noise. The resulting cross spectral density, $S_{VJ}(f)$, must also be taken into account. The matter is further complicated because the calculated values of V_Φ, S_V, S_J, and S_{VJ} are for a bare SQUID, whereas the values for a SQUID coupled to an input circuit will be modified by the presence of that circuit. Unfortunately, corrections for this effect are by no means straightforward. At the signal frequency ($\ll \nu_J$) the model of an inductance coupled to the SQUID is an excellent approximation, but at the Josephson frequency (at least a few GHz) the model breaks down because of the presence of stray capacitance between the coil and the SQUID, and between the turns of the coil.[*] To my knowledge, there has not yet been a successful attempt to take these capacitances into acount in model calculations. For the present purpose, we shall assume that the coupling between the coil and the SQUID is weak, so that we can use parameters computed for the bare SQUID. The calculations follow those of Clarke et al. (1979) and Koch et al. (1981).[*]

---

[*] Hilbert and Clarke (1982) have found that for a planar SQUID tightly coupled to a thin film spiral coil (Ketchen and Jaycox, 1982) the frequency response extends to about 200MHz; see Sec. IV H.

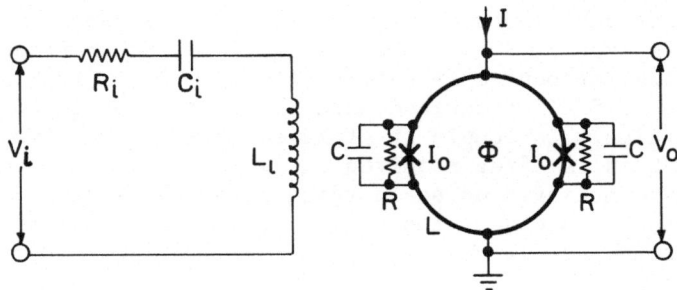

Fig. 18. Schematic of dc SQUID coupled to a tuned input circuit; V_i and V_o are the input and output voltages.

We consider only the case of the tuned amplifier shown in Fig. 18. Here, L_i is the inductance coupled to the SQUID, with mutual inductance M_i, C_i and R_i are the series capacitance and resistance, and V_i and V_o are the input and output voltages. The total impedance of the input circuit is

$$Z_i \approx R_i + j\omega L_i + \frac{1}{j\omega C_i} + \frac{\omega^2 M_i^2}{Z_s} , \qquad (4.12)$$

where Z_s is the SQUID impedance which consists of a resistive and inductive component in parallel (Koch, 1982). Since $M_i^2 = \alpha^2 L L_i$, and α^2 is small, we shall neglect the term $\omega^2 M_i^2/Z_s$ for low signal frequencies, ω. The current noise, $J_N(t)$, in the SQUID induces a voltage source $-M_i \dot{J}_N$ in the input circuit that generates a current $-M_i \dot{J}_N/Z_i$ and hence a flux $-M_i^2 \dot{J}_N/Z_i$ in the SQUID. Thus, the total voltage noise at the output of the SQUID is[†]

$$V_N'(t) = V_N(t) - M_i^2 \dot{J}_N V_\Phi/Z_i. \qquad (4.13)$$

Computing the spectral densities from Eq. (4.13) we find the spectral density of the noise at the SQUID ouput:

$$S_V'(f) = S_V(f) + \frac{M_i^4 \omega^2 S_J(f) V_\Phi^2}{|Z_i|^2} - \frac{2M_i^2 \omega S_{VJ}(f) V_\Phi(\omega L_i - 1/\omega C_i)}{|Z_i|^2} . \qquad (4.14)$$

---

*Koch (1982) and Tesche (1982) have each calculated the modifications to the SQUID parameters assuming that the simple model of a capacitance-free inductance coupled to a SQUID is correct. Koch demonstrates explicitly that his results reduce to those given here in the weak-coupling limit.

[†]In Clarke et al. (1979), the sign of the term $M_i^2 \dot{J}_N V_\Phi/Z_i$ was incorrectly given as positive: This error was pointed out by Koch (1982). Fortunately, this error does not affect the results for the tuned amplifier described here.

The corresponding mean square signal is given by

$$\langle V_o^2 \rangle = M_i^2 V_\Phi^2 \langle V_i^2 \rangle / |Z_i|^2, \tag{4.15}$$

and the signal-to-noise ratio in a bandwidth B is

$$\langle V_o^2 \rangle / S_V'(f) B. \tag{4.16}$$

In the thermal limit, we introduce a noise temperature, T_N, by setting $\langle V_o^2 \rangle / S_V'(f) B = 1$ with $\langle V_i^2 \rangle = 4k_B T_N R_i B$.

One can then optimize the noise temperature with respect to C_i and R_i, to find

$$T_N^{(opt)} = (\pi f / k_B V_\Phi)(S_V S_J - S_{VJ}^2)^{\frac{1}{2}}. \tag{4.17}$$

We note that $T_N^{(opt)}$ scales with frequency, and otherwise depends only on the SQUID parameters. Also, $T_N^{(opt)}$ is independent of α^2 (although the optimum values of R_i and C_i are not), implying that tight coupling between the input circuit and the SQUID is not essential for optimum noise performance.

A comment on the relationship between $T_N^{(opt)}$ and $\varepsilon/1Hz$ in the thermal limit is in order. In their simulation for an optimized SQUID with $\Gamma = 0.05$, Tesche and Clarke (1977, 1979) found $S_V \approx 16$ $k_B TR$, $S_J \approx 11 k_B T/R$, and $S_{VJ} \approx 12 k_B T$. Inserting these values into Eq. (4.17), one finds $T_N^{(opt)} \approx 18fT/V_\Phi$, or, with Eqs. (4.7) and (4.9), $T_N^{(opt)} \approx (2f\varepsilon/1Hz), k_B$. Thus, although it is true that $\varepsilon/1Hz$ does not completely specify the noise sources in the SQUID, nevertheless, to the extent that one has faith in the computer model, it is clear that as one improves $\varepsilon/1Hz$ one produces a corresponding improvement in $T_N^{(opt)}$.

We turn now to a consideration of the noise temperature in the quantum limit. The values of $S_V(f)$, $S_J(f)$, and $S_{VJ}(f)$ vs. flux for an optimized SQUID at $T = 0$ are shown in Fig. 16(b), (d), and (e) (Koch et al., 1981). These spectral densities are computed at frequencies much below the Josephson frequency, where the noise is white. Since the classical definition of T_N no longer holds, it is convenient to rewrite Eq. (4.17) in terms of $\langle V_i^2 \rangle$. We then define the quantity $n^{(s)}hf = \langle V_i^2 \rangle / 4R_i B$ as the mean photon power per unit bandwidth in the input circuit due to SQUID noise. The quantity $n^{(s)}$ is the mean number of photons induced in the input circuit by the SQUID.[*] We thus find

$$n^{(s)} = \pi (S_V S_J - S_{VJ}^2)^{\frac{1}{2}} / h V_\Phi. \tag{4.18}$$

---

[*]The quantity $n^{(s)}$ is identical to the quantity A ("added noise number") used subsequently by Caves (1982).

The variation of $n^{(s)}$ with flux according to Eq. (4.18) is shown in Fig. 16(f). The minimum value (Koch et al., 1981) is

$$n^{(s)} \approx \tfrac{1}{2}. \quad (T = 0) \tag{4.19}$$

The approximate equality arises from uncertainties in the computations. The minimum value for any linear phase preserving amplifier is exactly $\tfrac{1}{2}$. Inspection of Eq. (4.18) reveals that the establishment of a lower limit on $n^{(s)}$ does not impose a lower limit on S_v or, consequently, on $\varepsilon/1Hz$. Thus, the exact value of S_v (and $\varepsilon/1Hz$) when $n^{(s)} = \tfrac{1}{2}$ may depend on the parameters of the SQUID, so that there is no precise lower limit on $\varepsilon/1Hz$. Nevertheless, it is believed that the values of $\varepsilon/1Hz$ for a quantum limited SQUID are in the vicinity of $\hbar$.

Equation (4.19) can be rewritten as a noise temperature (Louisell et al., 1961) by adding the zero point energy in the resistance R_i, hf/2 , to the contribution from the SQUID, hf/2, to find a total energy hf. One then equates hf to the available noise energy available from R_i in the quantum limit, $hf/[\exp(hf/k_BT) - 1]$, to obtain the result

$$T_N = hf/k_B \ell n2. \tag{4.20}$$

Thus, Eqs. (4.3) to (4.6) using Eq. (3.1) for the noise driving term predict that a SQUID at T = 0 weakly coupled to a tuned input circuit also at T = 0 should behave as an ideal quantum-limited amplifier. As yet, no practical SQUID amplifier has come within orders of magnitude of this limit. At a temperature of 1K, $n^{(s)}$ will increase to about 2 for the SQUID discussed here (Koch et al., 1981); more important, the noise in the input circuit will be very much in the thermal limit except at unrealistically high frequencies. Two ways in which the quantum limit may perhaps be achieved are discussed in Sec. IV H.

F. Coupling to SQUIDs

Although the efficiency with which an input circuit is coupled to a SQUID does not affect the noise temperature for a tuned circuit, this is not the case for the untuned input circuit. Furthermore, even for the tuned case, very small values of α^2 may lead to unreasonable values of $R_i^{(opt)}$ (Clarke et al., 1979). Thus, from a practical standpoint, it is usually necessary to achieve values of α^2 that are not too small.

For the cylindrical SQUID [Fig. 13(a)], one could achieve values of α^2 of about 0.5 simply by winding a solenoid around the SQUID. The advent of the planar SQUID, however, a configuration that is virtually mandatory if one is to use lithography, has made good coupling a more subtle problem. This difficulty has been overcome by Jaycox and Ketchen (1981) who used a spiral coil deposited over the body of the SQUID, and by Cromar and Carelli (1981) and Carelli and Foglietti (1982) who used a SQUID with many large-inductance loops in parallel. We briefly describe the former device.

The essential idea (Jaycox and Ketchen, 1981; Ketchen and Jaycox, 1982) is shown in Fig. 19. The SQUID loop is square, and the width of the film, w, is somewhat greater than the width of the hole, d. In the limit $w \gg d$, the loop-inductance is given by L(loop) = 1.25 $\mu_o d$. The superconducting input coil is deposited as a spiral overlaying the SQUID loop, and separated from it by an insulating layer. Jaycox and Ketchen give the following expressions for the inductances and α^2:

$$L = L(loop) + L_j, \tag{4.21}$$

$$M_i = n(L - L_j), \tag{4.22}$$

$$L_i = n^2(L - L_j) + L_s, \tag{4.23}$$

and

$$\alpha^2 = (1 - L_J/L)[1 + L_s/n^2(L - L_j)]^{-1}, \tag{4.24}$$

where L_j is the parasitic inductance associated with the junctions, n is the number of turns in the input coil, and L_s is the stripline inductance of the coil. Among several versions of this SQUID, Ketchen and Jaycox (1982) te·ted one with measured values of d ≈ 50 μm, n = 50, L ≈ 90pH, L_i ≈ 190 nH, and α^2 = 0.86. These values were in excellent agreement with the predictions of Eqs. (4.21) to (4.24). Using a second SQUID to detect the voltage, they found a noise energy

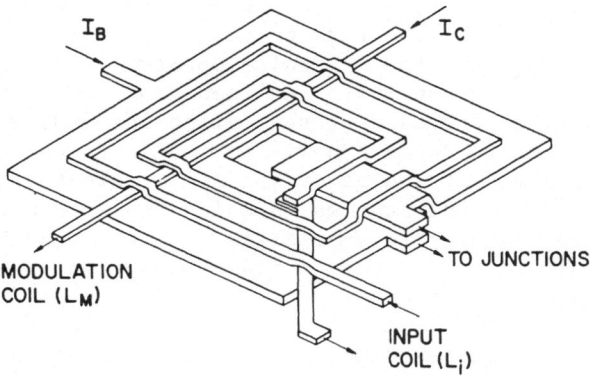

I_B I_C

MODULATION
COIL (L_M)

TO JUNCTIONS

INPUT
COIL (L_i)

Fig. 19. Planar SQUID with spiral input coil (only 2 turns are shown) and quarter-turn modulation coil (Ketchen and Jaycox, 1982).

referred to the input coil, $S_\Phi/2\alpha^2 L$, of about 120ℏ at 4.2K. At 1.5K, $S_\Phi/2\alpha^2 L$ decreased to about 50ℏ, the lowest noise energy referred to an input coil yet reported, although it was not obtained in a flux-locked configuration. Martinis and Clarke (1982) have fabricated a series of planar Nb-NbOx-Pb SQUIDs using an adaption of the configuration of Ketchen and Jaycox, and have achieved noise energies of about 200h at 4.2 K in a flux-locked loop.

G. Flicker (1/f) Noise in SQUIDs

No discussion of the noise limitations of SQUIDs can be complete without some reference to the question of 1/f noise. As is evident from Figs. 13(b) and 17, the spectral density of the flux noise at low frequencies scales as 1/f; all SQUIDs on which careful measurements have been made show similar behavior.

The 1/f noise in single resistively shunted Josephson junctions arises from fluctuations in the critical current which in turn are believed to arise from temperature fluctuations (Clarke and Hawkins, 1976). However, while the present model predicts the noise level in large area ($\sim 10^4 \mu m^2$) junctions quite well, the 1/f noise in the small area ($\sim 6 \mu m^2$) junctions studied by Koch et al. (1982) is substantially lower than that predicted by the model. This observation does not necessarily invalidate the temperature fluctuation model altogether, but certainly suggests that it is incomplete. Given the level of 1/f noise in single junctions, one then can estimate the noise that would be generated in a SQUID. In the simplest models (Clarke et al., 1976, Ketchen and Tsui, 1980), one simply assumes that the spectral density of the flux noise is $2L^2 S_{I_0}(f)$, where $S_{I_0}(f)$ is the spectral density of the critical current fluctuations in each junction, which contribute independently. (One also assumes that the SQUID is flux modulated, so that the 1/f voltage noise does not contribute.) Detailed computer models (Koch et al., 1982a; Tesche, 1982a) produce more exact numerical coefficients linking the critical current noise to the flux noise. The essential conclusion from these models is that the 1/f noise observed in tunnel junction SQUIDs is much greater than that predicted from the critical current fluctuation model; in the case of the cylindrical SQUID by two orders of magnitude. Thus, it appears unlikely that critical current fluctuations are the dominant source of 1/f noise. In fact, Koch et al. (1982a) used two different schemes in which the bias current and flux were modulated to demonstrate explicitly that the predominant source of 1/f noise could be regarded as a flux noise. At present, the origin of this noise is not known.

Koch et al. (1982a) examined the 1/f flux noise in 5 different types of tunnel junction SQUIDs made at Berkeley with inductances and capacitances that each ranged over 3 orders of magnitude, and areas that ranged over almost six orders of magnitude. They found that the spectral density of the 1/f flux noise at 4.2K was remarkably consistent, varying from $[(0.4$ to $3) \times 10^{-10}/f]\Phi_o^2 Hz^{-1}$, with an average of about $(10^{-10}/f)\Phi_o^2 Hz^{-1}$. The 1/f noise in the SQUID of Ketchen and Jaycox (1982) at 4.2K was about $(3 \times 10^{-10}/f)\Phi_o^2 Hz^{-1}$, while that of Carelli and Foglietti (1982) was lower than any of these values, about $(10^{-11}/f)\Phi_o^2 Hz^{-1}$.

It is important to note that the lack of dependence of the 1/f flux noise on the SQUID parameters gives a quite different criterion for optimizing the noise energy at low frequencies compared with that derived on the assumptions that the noise arises from critical current fluctuations and that the temperature fluctuation model is valid (Ketchen, 1981; Ketchen and Tsui, 1981). As an example, suppose that we have a SQUID that is quantum limited in the white noise region, and we require the 1/f "knee frequency" (at which the white and 1/f noise energies are equal) to be as low as possible given that $S_\Phi^{1/f}$ remains constant at $(10^{-10}/f)\Phi_o^2 Hz^{-1}$. Now $\epsilon^{1/f}/1Hz = S_\Phi^{1/f}/2L = [2 \times 10^{-31}/(L/1nH)f]JHz^{-1} \approx 2000\hbar/(L/1nH)f$. Equating this result to the white noise energy, $\hbar$, we find a knee frequency of $2kHz/(L/1nH)$. Thus, one should make L as large as possible, consistent with the requirement that the thermal or quantum flux fluctuations must not be so large as to destroy the periodicity of the SQUID. Thus, to obtain a knee frequency of 1kHz (for example, for a gravity wave antenna, see Sec. IV H), one must choose $L \gtrsim 2nH$. Unhappily, for the SQUID to be quantum noise limited in the white noise region at 1K, this choice of L requires a junction capacitance of no greater than $10^{-4}pF$ [from Eq. (4.10)]. This value is likely to present a significant challenge, since it implies a junction area of less than 1/400 μm^2. Of course, this discussion assumes that the 1/f noise arises from flux noise rather than critical current fluctuations even for such tiny junctions, and should be regarded with some suspicion until appropriate measurements can be made. Nevertheless, the argument illustrates the importance of establishing an appropriate model for 1/f noise, even if it is entirely empirical.

The problem of 1/f noise in SQUIDs remains an outstanding one. It now appears that "flux noise" is responsible for the predominant contribution, at least in a certain family of devices involving thin film tunnel junctions, but a good deal of light has yet to be shed on the origin of this noise.

H. Applications of Quantum-Limited SQUID Amplifiers

It should be realized that at the low signal frequencies at which SQUIDs are usually operated, the optimized noise temperature in the thermal limit, $T_N^{(opt)}$, is exceedingly small. For example, if we take R/L = 20 μV (corresponding to a frequency of $10^{10}Hz$), T = 4K, and f = 100Hz, we find $T_N^{(opt)} \approx 18fT/V_\Phi \approx 0.75$ μK. Thus, the thermal noise power in the input resistor overwhelms the SQUID noise by more than 6 orders of magnitude, implying that the SQUID noise is totally negligible even for very badly matched input circuits, and rendering

any discussion of the quantum limit redundant. In order to have any
chance of observing quantum limited performance one clearly requires
$k_B T \lesssim hf$. If T = 1K, we require $f \gtrsim 25$GHz, a frequency that is well
above the capability of any presently-designed input circuit. On the
other hand, if one were to cool the input circuit to 10 mK, one would
reduce the minimum required frequency to about 250 MHz. Hilbert and
Clarke (1982) have operated SQUIDs with a configuration similar to
that of Ketchen and Jaycox (1982) at signal frequencies of up to 200
mHz, and it seems likely that this frequency range can be extended ―
perhaps to about 1GHz. Although this type of SQUID is far from the
quantum limit of 4.2K, it would be relatively close at 10 mK if the
noise decreases as predicted by Eq. (4.10). Thus it may be possible
to achieve quantum limited amplifiers in a dilution refrigerator for
signal frequencies of a few hundred megahertz.

A second very exotic potential application of quantum limited
dc SQUIDs is as the transducer for gravity wave antennas. A number
of groups around the world are designing antennas consisting of massive cylindrical bars cooled to liquid He^4 temperatures. A gravity
wave excites a longitudinal oscillation in the bar, and one means of
detecting the motion is with a SQUID amplifier. For example, the detector at Stanford (Boughn et al., 1977), which has been operational
for some years, consists of a 4800kg Al bar cooled to 2K and with a
mechanical Q of order 10^6 at a resonant frequency of ν_a = 840Hz. A
transducer is attached to one end of the bar, and consists of a superconducting niobium diaphragm with two superconducting pancake coils
facing its two sides. The coils carry a persistant supercurrent and
are connected in parallel with the input coil of a SQUID. The diaphragm is a high-Q mechanical resonator the motion of which modulates
the inductance of the two pick-up coils. The current fed to the
SQUID is proportional to the displacement of the diaphragm from its
equilibrium position.

In order to achieve the quantum limit in the detector, at first
sight one might expect that the experiment should be cooled to temperatures below $840(h/k_B)K \approx 40$ nK! Fortunately, however, one can
take advantage of the very high Q of the system together with the
relatively short length of the gravity wave signal, $\tau_s \sim 1$ ms, to
achieve the quantum limit, at least in principle, in a detector that
is at a realizable temperature (Giffard, 1976). The decay time of
the antenna is very long, on the order of $Q/\nu_a \sim 10^3$s. Thus, one is
required to detect a short pulse, of length τ_s, on a background that
is changing very slowly with time. It can be shown that under these
circumstances, the effective antenna temperature is

$$T_{eff} \sim \frac{T\tau_s}{Q/\nu_a} , \qquad (4.25)$$

about $10^{-6}T$ for the values given above. Thus, if the bar is cooled
to 10 mK in a dilution refrigerator, $T_{eff} \sim 10$ nK, and the bar is
quantum limited. Needless to say, the achievement of this goal will

require a very dedicated effort, including the development of a suit-able quantum limited SQUID.

V. CONCLUDING SUMMARY

In this chapter, I have tried to show how one can introduce quantum noise effects into overdamped resistively shunted Josephson junctions. The central idea is to replace the classical current noise term in the resistor with the quantum term which has a spectral density $(2h\nu/R)\coth(h\nu/2k_BT)$. The equation of motion, however, remains the classical description of a particle moving on a tilted washboard [Eq. (2.10)]. This so-called quantum Langevin equation can be solved exactly for the case $\beta_c \ll 1$ and for $I > I_0$; at $T = 0$ it predicts that the voltage noise at low frequencies, $\nu_m \ll \nu_J$, has a spectral density $(2eV/R)(I_0/I)^2$, and arises from zero point fluctua-tions near the Josephson frequency, ν_J, mixed down to the measurement frequency, ν_m. Measurements on junctions designed to exhibit sub-stantial quantum corrections to the noise in the He^4 temperature range were in excellent agreement with the theory (Figs. 9 and 10). In particular, it was demonstrated that the noise driving term did indeed have a spectral density $(2h\nu/R)\coth(h\nu/2k_BT)$ (Fig. 11). These results indicate rather strongly that the quantum Langevin equation is a valid description of a resistively shunted junction in the free running mode $I > I_0$. The measurements also demonstrate that the Jo-sephson junction operated as a self-mixer is an ideal quantum-limited device at frequencies up to 500GHz. The formalism developed in Sec. III A could also be extended fairly readily to calculate the effects of quantum noise in a Josephson mixer with an external local oscilla-tor.

The quantum Langevin equation fails for underdamped junctions biased well below the critical current. In this regime, quantum me-chanical broadening of the wave packet describing the "particle" be-comes important, and the picture of a point particle moving on a tilted washboard is no longer valid. Instead, one should use the macroscopic quantum tunneling description. However, the quantum Langevin equation may still be appropriate for overdamped junctions biased just below the critical current, and may give an accurate des-cription of quantum noise rounding of the I - V characteristic. A clarification of the relationship between the quantum Langevin equa-tion and MQT in this regime is very much needed.

The theory of dc SQUIDs in the thermal limit (Sec. V B) predicts $\varepsilon/1Hz \approx 16k_BT(LC)^{\frac{1}{2}}$. A large variety of SQUIDs are in reasonable agreement with this formula, as shown in Fig. 15. The quantum Lange-vin approach has also been extended to the dc SQUID (Sec. IV C), but, unfortunately, only computer solutions are possible. For an opti-mized SQUID at $T = 0$, the theory predicts $\varepsilon/1Hz \approx \hbar$. It was empha-sized that although the noise energy per Hz is a convenient measure

with which to compare SQUIDs, it is not a complete description of the
noise in SQUIDs, and there is no exact fundamental lower limit on its
value. The best practical SQUID had a noise energy of about $3\hbar$ at
1.4K; much of this was $1/f$ noise, and the extrapolated value of the
white noise was $(0.8 \pm 0.7)\hbar$. It should be remarked that it is more
difficult to observe quantum noise in a SQUID than in a single junction using the same junction technologies. The SQUID imposes a further constraint, namely that it must be operated at a relatively low
voltage where V_Φ is a maximum. Thus, one must achieve a higher value
of $\kappa \equiv eI_0R/k_BT$ (for example, by operating at a lower temperature)
than would be necessary to observe quantum effects in a single junction.

Section IV E described the coupling of a SQUID to a tuned input
circuit to make an amplifier. In this situation, the current noise
in the SQUID, which is partially correlated with the voltage noise
across the SQUID and which induces a voltage noise into the input
circuit, must be taken into account. For weak coupling between the
SQUID and the input circuit and for a signal frequency, f, much less
than ν_J, the theory predicts that the optimized amplifier at $T = 0$
should have a noise temperature $T_N \approx hf/k_B\ell n2$. Alternatively, one
can say that the SQUID induces an average energy of $hf/2$ into the input circuit.

Although quantum limited amplifiers do not in principle require
very efficient coupling between the SQUID and the input circuit [Eq.
(4.17)], in practice it is undesirable for the coupling to become
too weak. At least two efficient coupling schemes for planar SQUIDs
have been described in the literature (Sec. IV F), and it would appear that this problem is largely solved.

The problem of $1/f$ noise in SQUIDs at low frequencies is peren-
nial (Sec. IV G). At least for SQUIDs based on tunnel junctions (as
opposed to artificial barriers, microbridges, or point contacts), it
now appears that the $1/f$ noise does not arise predominantly from critical current fluctuations, but from some unknown source of "flux
noise". The magnitude of the $1/f$ flux noise, $S_\Phi^{1/f}$, is remarkably constant (say, within one order of magnitude of $(10^{-10}/f)\Phi_0^2 Hz^{-1}$) for a
wide range of SQUID inductance, capacitance, and area. To the extent
that this conclusion holds true, it implies that to obtain the lowest
possible noise energy at low frequencies one must make the SQUID inductance as large as possible.

What are likely future developments? To incorporate a SQUID
into an amplifier capable of making a quantum limited measurement is
not going to be easy. In my opinion, it will be necessary first to
go to low temperatures, to make the ratio of signal frequency to temperature as large as possible. Next, one must either operate at high
frequencies, so that $hf/k_BT > 1$, or look for short pulses with high Q
circuits, as is the case for gravity wave detectors (Sec. IV G). In
either case the problems are formidable. Needless to say, the possi-

bility of observing gravity waves is an extremely exciting one, and likely to stimulate strenuous efforts to build quantum limited detectors in the coming years.

ACKNOWLEDGEMENTS

Much of the work described here was carried out at Berkeley over a period of some eight years, and I wish to express my grateful thanks to Wolf Goubau, Gil Hawkins, Claude Hilbert, Mark Ketchen, Roger Koch, John Martinis, Claudia Tesche, and Dale Van Harlingen for their dedicated efforts. I have also enjoyed collaborations with Robin Giffard, Bob Laibowitz, Colin Pegrum, Stan Raider, and Dick Voss in some aspects of the work. I thank P. Carelli, C. M. Caves, A. J. Leggett, A. Schmid, and C. D. Tesche for preprints of their work. Claude Hilbert and John Mamin kindly read the manuscript and made constructive suggestions for improvements.

This work was supported by the Director, Office of Energy Research, Office of Basic Energy Sciences, Materials Sciences Division of the U. S. Department of Energy under Contract Number DE-ACO3-76SF00098.

REFERENCES

Ambegaokar, V., and Halperin, B. I., 1969, Voltage due to thermal noise in the dc Josep .son effect, Phys. Rev. Lett., 22:1364.
Ambegaokar, V., Eckern, U., and Schön, G., 1982, Quantum dynamics of tunneling between superconductors, Phys. Rev. Lett., 48:1745.
Bardeen, J., Cooper, L. N., and Schrieffer, J. R., 1957, Theory of superconductivity, Phys. Rev., 108:1175.
Boughn, S. P., Fairbank, W. M., Giffard, R. P., Hollenhorst, J. N., McAshan, M. S., Paik, H. J., and Taber, R. C., 1977, Observation of mechanical Nyquist noise in a cryogenic gravitational-wave antenna, Phys. Rev. Lett., 38:454.
Bruines, J. J. P., de Waal, V. J., and Mooij, 1982, Comment on "DC SQUID: Noise and optimization" by Tesche and Clarke, J. Low Temp. Phys., 46:383.
Caldeira, A. O., and Leggett, A. J., 1981, Influence of dissipation on quantum tunneling in macroscopic systems, Phys. Rev. Lett., 46:211; 1982, Quantum tunneling in a dissipative system (unpublished).
Callen, H. B., and Welton, T. A., 1951, Irreversibility and generalized noise, Phys. Rev., B83:34.
Carelli, P., and Foglietti, V., 1982, Behavior of a multiloop dc superconducting quantum interference device (unpublished).
Caves, C. M., 1982, Quantum limits on noise in linear amplifiers, (unpublished).
Clarke, J., 1977, Superconducting quantum interference devices for

for low frequency measurements, in: "Superconductor Applications: SQUIDs and Machines," B. B. Schwartz and S. Foner, eds., Plenum, New York.

Clarke, J., 1980, Advances in SQUID magnetometers, IEEE Trans. Electron Devices, ED-27:1896.

Clarke, J., Goubau, W. M., and Ketchen, M. B., 1976, Tunnel junction dc SQUID, J. Low Temp. Phys., 25:99.

Clarke, J., and Hawkins, G., 1976, Flicker (1/f) noise in Josephson tunnel junctions, Phys. Rev., B14:2826.

Clarke, J., Tesche, C. D., and Giffard, R. P., 1979, Optimization of dc SQUID voltmeter and magnetometer circuits, J. Low Temp. Phys., 37:405.

Cromar, M. W., and Carelli, P., 1981, Low-noise tunnel junction dc SQUIDs, Appl. Phys. Lett., 38:723.

De Waal, V. J., van den Hamer, P., and Mooij, J. E., 1980, dc SQUIDs fabricated with niobium-niobium tunnel junctions, in: "SQUID '80, Superconducting Quantum Interference Devices and their Applications," H. D. Hahlbohm and H. Lübbig, eds., Walter de Gruyter, Berlin, p.391.

Giffard, R. P., 1976, Ultimate sensitivity of a resonant gravitational wave antenna using a linear motion detector, Phys. Rev., D14:2478.

Greiner, J. H., Kircher, C. J., Klepner, S. P., Lahiri, S. K., Warnecke, A. J., Basavaiah, S., Yen, E. T., Baker, J. M., Brosious, P. R., Huang, H.-C. W., Murakami, M., and Ames, I., 1980, Fabrication process for Josephson integrated circuits, IBM J. of Res. and Dev., 24:195.

Hilbert, C., and Clarke, J., 1982, (unpublished).

Ivanchenko, Yu. M., and Zil'berman, L. A., 1968, The Josephson effect in small tunnel contacts, Zh. Eksp. Teor. Fiz., 55:2395 [Sov. Phys. JETP, 28:1272].

Jackel, L. D., Gordon, J. P., Hu, E. L., Howard, R. E., Fetter, L. A., Tennant, D. M., Epworth, R. W., and Kurkijärvi, J., 1981, Decay of the zero voltage state in small area, high-current-density Josephson junctions, Phys. Rev. Lett., 47:697.

Jaklevic, R. C., Lambe, J., Silver, A. H., and Mercereau, J. E., 1964 Quantum interference effects in Josephson tunneling, Phys. Rev. Lett., 12:159.

Jaycox, J. M., and Ketchen, M. B., 1981, Planar coupling scheme for ultra low noise dc SQUIDs, IEEE Trans. Magn., MAG-17:400.

Josephson, B. D., 1962, Possible new effects in superconductive tunneling, Phys. Lett., 1:251.

Ketchen, M. B., 1981, DC SQUIDs 1980: The state of the art, IEEE Trans. Magn., MAG-17:387.

Ketchen, M. B., and Voss, R. F., 1979, An ultra-low noise tunnel junction dc SQUID, Appl. Phys. Lett., 35:812.

Ketchen, M. B., and Tsui, C. C., 1980, Low frequency noise in small-area tunnel junction dc SQUIDs, in: "SQUID '80, Superconducting Quantum Interference Devices and their Applications," H. D. Hahlbohm and H. Lübbig, eds., Walter de Gruyter, Berlin, p.227.

Ketchen, M. B., and Jaycox, J. M., 1982, Ultra-low-noise tunnel junction dc SQUID with a tightly coupled planar input coil, Appl. Phys. Lett., 40:736.

Koch, R. H., 1981, (unpublished).

Koch, R. H., 1982, Quantum noise in Josephson junctions and dc SQUIDs, Ph.D. thesis, University of California, Berkeley (unpublished).

Koch, R., and Clarke, J., 1979, Small area tunnel junction dc SQUID, Bull. Amer. Phys. Soc., 24:264.

Koch, R. H., Van Harlingen, D. J., and Clarke, J., 1980, Quantum noise theory for the resistively shunted Josephson junction, Phys. Rev. Lett., 45:2132.

Koch, R. H., Van Harlingen, D. J., and Clarke, J., 1981, Quantum noise theory for the dc SQUID, Appl. Phys. Lett., 38:380.

Koch, R. H., Van Harlingen, D. J., and Clarke, J., 1981a, Observation of zero-point fluctuations in a resistively shunted Josephson junction, Phys. Rev. Lett., 47:1216.

Koch, R. H., Van Harlingen, D. J., and Clarke, J., 1982, Measurements of quantum noise in resistively shunted Josephson junctions, Phys. Rev., B26:74.

Koch, R. H., Clarke, J., Goubau, W. M., Martinis, J. M., Pegrum, C. M., and Van Harlingen, D. J., 1982a, Flicker (1/f) noise in tunnel junction dc SQUIDs, submitted to J. Low Temp. Phys.

Leggett, A. J., 1978, Prospects in ultralow temperature physics, J. de Phys., C6:1264.

Likharev, K. K., and Semenov, V. K., 1972, Fluctuation spectrum in superconducting point junctions, Pis'ma Zh. Eksp. Teor. Fiz., 15:625 [JETP Lett., 15:442].

London, F., 1951, "Superfluids," vol. 1, Dover Publications, Inc., New York.

Louisell, W. H., 1973, "Quantum Statistical Properties of Radiation," John Wiley and Sons, New York.

Louisell, W. H., Yariv, A., and Siegman, A. E., 1961, Quantum fluctuations and noise in parametric processes, I., Phys. Rev., 124:1646.

Martinis, J. M., and Clarke, J., 1982 (unpublished).

McCumber, D. E., 1968, Effects of ac impedance on dc voltage-current characteristics of superconductor weak-link junctions, J. Appl. Phys., 39:3113.

Schmid, A., 1982, On a quasiclassical Langevin equation, (unpublished).

Stewart, W. C., 1968, Current-voltage characteristics of Josephson junctions, Appl. Phys. Lett., 12:277.

Tesche, C. D., 1982, Analysis of tightly coupled SQUID systems, Bull. Am. Phys. Soc., 27:266

Tesche, C. D., 1982a, Parameter fluctuations and low frequency noise in the dc SQUID, Appl. Phys. Lett., 41:99.

Tesche, C. D., and Clarke, J., 1977, dc SQUID: Noise and optimization, J. Low Temp. Phys., 27:301.

Tesche, C. D., and Clarke, J., 1979, dc SQUID: Current noise, J. Low

Temp. Phys., 37:397.

Van Harlingen, D. J., Koch, R. H., and Clarke, J., 1982, Superconducting quantum interference device with very low magnetic flux noise energy, Appl. Phys. Lett., 41:197.

Voss, R. F., 1981, Noise characteristics of an ideal shunted Josephson junction, J. Low Temp. Phys., 42:151.

Voss, R. F., and Webb, R. A., 1981, Macroscopic quantum tunneling in 1-μm Nb Josephson junctions, Phys. Rev. Lett., 47:265.

Voss, R. F., and Webb, R. A., 1981a, Pair shot noise and zero-point Johnson noise in Josephson junctions, Phys. Rev., B24:7447.

Voss, R. F., Laibowitz, R. B., Raider, S. I., and Clarke, J., 1980, All Nb low noise SQUID with 1μm tunnel junctions, J. Appl. Phys., 51:2306.

Voss, R. F., Laibowitz, R. B., Broers, A. N., Raider, S. I., Knoedler, C. M., and Viggiano, J. M., 1981, Ultra low noise Nb dc SQUIDs, IEEE Trans. Magn., MAG-17:395.

Vystavkin, A. N., Gubankov, V. N., Kuzmin, L. S., Likharev, K. K., Migulin, V. V., and Semenov, V. K., 1974, S-c-s junctions as nonlinear elements of microwave receiving devices, Phys. Rev. Appl., 9:79.

JOSEPHSON COMPUTER TECHNOLOGY

Hans H. Zappe

IBM Thomas J. Watson Research Center
P. O. Box 218
Yorktown Heights, New York   10598

> Les lumières de la géometrie, de la physique et de la mécanique m'en fournirent le dessein, et m'assurèrent que l'usage en serait infaillible si quelque ouvrier pouvait former l'instrument dont j'avais imaginé le modèle. Mais ce fut en ce point que je rencontrai des obstacles aussi grands que ceux que je voulais éviter, et auxquels je cherchais un remède.
>
> *Pascal, La Machıne d'Arithmétique*

I. INTRODUCTION

Spurred by an insatiable thirst for machines with ever growing computing power, increasing attention is being given to today's fastest systems, collectively known as supercomputers [1]. Here we shall concentrate on high performance machines of a different class, powerful future computers expected to exceed present day capabilities by over one order of magnitude. The discussion will center on one possible approach that promises to achieve the required technological breakthroughs. It is a superconductive technology based on the Josephson effect.

Computing has been an integral part of human life ever since the first farming settlements were firmly established some 10,000 years ago. Then, the bookkeeping of transactions was performed with the use of incised tokens of clay that seemed to have been standardized in a vast area extending from the Nile river to the Indus Valley [2]. Initial improvements were slow. Yet, one should not be surprised to find, upon detailed study, that the development of computing aids shows with time an exponential growth curve similar to one that measures the complexity of human life. Tokens gave way to calculi on counting boards, the abacus was invented and is still in extensive use in such advanced eastern countries as Japan. Pascal used gears and invented the concept of the display. Babbage increased the complexity of the mechanical unit and brought us nearer to the concept of the modern machine [3]. The punched card was invented to serve the role of a modern token both for data processing and as an inexpensive, disposable memory substrate. Finally, electronic computers, first using vacuum tubes, then transistors, have evolved explosively into the modern classes of systems. It is not the intent here to retrace a detailed history of computing, yet the past is important to extrapolate into the future some broad outlines of the lessons learned so far.

First we find that the most successful and widely used machines are digital, and that modern machines are implemented in the simplest digital form, namely the binary system which ensures the largest discrimination between states and safeguards the data best against degradation from noise. For this reason, we shall concentrate on binary digital circuits in this discussion. Analog Josephson device studies are being pursued. But these are not directly related to computer applications and will be mentioned only briefly in section VIII.

The second lesson from history is that there seems to be no foreseeable limit to the need for increased computing power. With the advent of the modern electronic business computer in the early 1950's, it was estimated that the U.S. market would be saturated with perhaps 50 such systems [4]. This was a serious miscalculation and all subsequent attempts to predict a limit to the world's computing needs have been equally wrong. In fact, if we take the evolution of IBM high end machines from the IBM 360/65 to the IBM 3081 as an indication, we find up to now a performance improvement trend of one order of magnitude per 13 years [5]. Thus, the answer to the question on whether ᵣ ıll larger and faster computers are needed must be affirmative.

Finally, one can observe that as computing power increased, the complexity of the systems increased as well. However, the number of operations that could be performed per unit of time was never drastically improved by simply increasing the number of components. Performance improvements were ultimately always related to an increase in raw speed of logic and memory elements and to a proportional decrease in the time required to communicate between interconnected elements. In other words, performance improvements were in the past intimately related not only to technological improvements but also to successive changes in basic technology [6]. The change from electromechanical accounting machines to vacuum systems resulted in a performance improvement of four orders of magnitude. Semiconductor technology has permitted us to increase that performance further by more than a factor of one thousand, as exemplified by today's supercomputers [7]. Future improvements will be made, but as we shall see, the limits of semiconductor technology are not as far away as they used to be. For this reason, it is legitimate to ask whether the step to the next generation of computers may not have to involve yet another technological change.

We shall use this question as the starting point of our discussion and examine the design limitations that face a modern hardware designer. It will become apparent that many of the difficult technological problems that exist today can be alleviated with Josephson technology.

The initial Josephson circuit experiments by J. Matisoo in 1966 [8,9] caused, in time, a flurry of activities in this new field. At IBM it lead to a relatively large research and development program under the guidance of W. Anacker [10,11] and to similar but smaller programs mainly at Bell, the National Bureau of Standards and the University of California at Berkeley. After a decade of activities, IBM has clearly remained in the lead. It demonstrated not only experimental logic and memory chip technologies, but has recently performed a system level experiment to establish that Josephson chips can be integrated into a dense high performance package. Present activities in the Josephson field are expanding in the United States, in Europe and in Japan. It is Japan's intention to mount a sizeable program under the auspices of MITI, its Ministry of International Trade and Industry.

One question often asked is why it takes so long to develop a Josephson computer. The answer is that it is not an easy technology. Materials, devices, device operation, device speed, are very different from established technologies. Operating margins require tight control of the device threshold, a problem reminiscent of FET's. A totally new packaging concept has to be developed to maintain the performance advantage on the system level and, being a superconductive technology, it must operate in liquid helium at 4K. However, the biggest stumbling block seems to be the necessity to build an entire Josephson system. So far, no way has been found to introduce Josephson gradually by mixing, for example, a few Josephson chips with the rest of a semiconductor system, as semiconductors had done by initially retaining certain functions in vacuum tube technology [6] and as GaAs is doing presently with silicon.

Having demonstrated that logic and memory chips can be operated and integrated into a high performance package at the laboratory level, it is now necessary to demonstrate that a reasonably large number of operational chips and package parts can be fabricated again and again. Thus, at least at IBM, the Josephson effort has entered a new phase under the leadership of J. Logue. Much work remains, and is being done. But whether the Josephson technology will be a technological and a commercial success can only be answered in the future.

> Natura non rompe sua legge.
> *Leonardo da Vinci,*
> *Collections de l'Institut de France*

II. DESIGN PROBLEMS IN HIGH PERFORMANCE SYSTEMS

A. *ELEMENTS OF A SYNCHRONOUS COMPUTER AND BASIC DEFINITIONS*

The basic elements and functions of a synchronous stored program computer are symbolically shown in Fig. II-1. The system is composed of two basic parts, logic and memory. Logic is composed of elementary circuits performing, for instance, the OR, AND and INVERT functions, all three of which are necessary and sufficient to make any logical decision. Since it is easy to show that a NOR circuit, composed of an OR followed by an INVERT, can produce the AND function, a machine can, in principle, be entirely constructed out of NORs. The same is true of NANDs. To simplify the machine design, other more specialized functions such as the Exclusive-OR, Shift Registers, etc. are almost always implemented. These circuits are interconnected into networks which accept a large number of combinations of binary inputs and generally yield a lesser number of combinations at the output.

The data stream is sequenced through such a system in an orderly fashion from register to register. Registers are composed of latches, fast memory units, that can receive and hold information for the purpose of resynchronization. This leads to the definition of the cycle time, one of the basic performance parameters of a machine. It is the time designated to propagate data from latch to latch. At the beginning of the cycle, the information is released from a register into a logic network, typically ten circuit levels deep, and it is captured into another register at the end of the cycle.

The central processing unit, or CPU, is composed of two intertwined logic networks. One, the data or execution unit, processes data obtained from memory and feeds results back into memory. The other, the instruction unit, controls which parts of the CPU are operational and in which sequence. One should note that only a relatively small portion of the logic is functional at any one time. The instruction unit also controls memory from which it receives addresses and instructions. In addition, it receives branch controls from the execution unit. Finally, superimposed over the entire system, is a timing network that determines, in conjunction with the instructions, at which instant the various elements have to be activated.

A more conventional representation of a computing system is shown in Figure II-2. The figure clearly indicates the overwhelming amount of required memory. It also shows the memory hierarchy that is necessary to keep the CPU going. The problem is that a CPU needs, in principle, a truly large memory that can be accessed in one or two cycles and there is no sufficiently economic single technology in which this can be achieved. Memories are either fast and small or large and slow. It has been found, however, that it is possible to organize a storage hierarchy

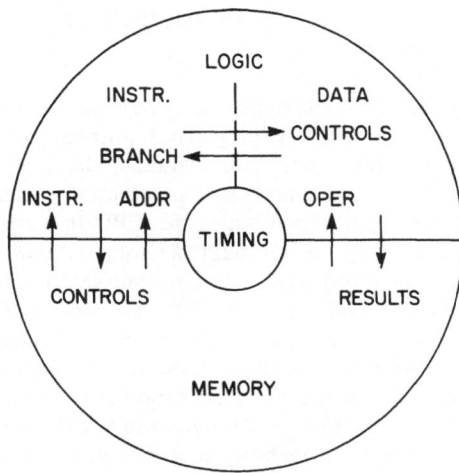

Fig. II-1:   Schematic diagram of a computer illustrating the interrelationship of its components.   Note that the predominance of memory is not properly represented.

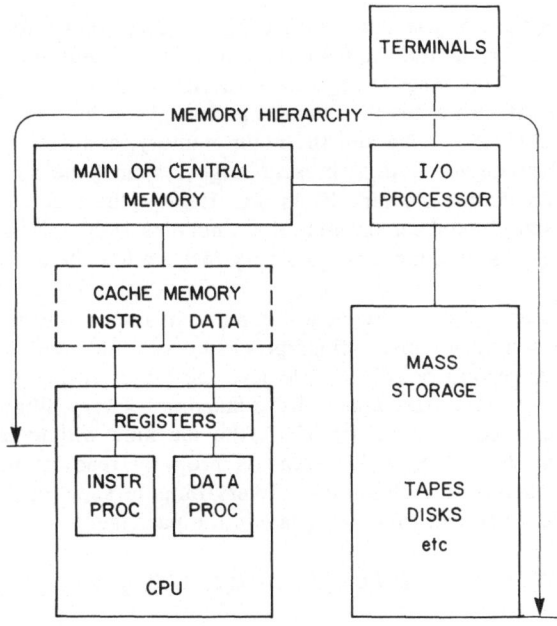

Fig. II-2:   Block diagram of a modern general purpose computer.   The memory hierarchy which cascades data into and out of the processor is shown. Specialized vector processors may dispense with the cache memory.

in such a way that, most of the time, the CPU has the impression of being connected to a very fast, yet large, memory.

The fastest memory units are registers. They form the bottom of the hierarchy. Then follows a cache memory in general purpose machines. Specialized machines may dispense with this unit. For instance, the Cray-1 does not have a cache. It was designed to be a vector processor in which it is known beforehand that long ordered data streams or vectors will enter the CPU in a specified sequence. In this case one can connect directly to the main memory without major penalty. But to reach peak performance such processors require that vector lengths be in excess of 10^3 or system performance degrades rapidly [7]. To process a variety of problem sets efficiently, general purpose computers use the cache which receives data from the main memory in the form of specified blocks. Its operation is based on a statistical property of stored data that may be formulated as follows: given a sufficiently large block of information (32 to 128 bytes) in which data requested from the CPU is contained, there is a high probability that subsequent data will be inside this block and a high probability that some of the data will be requested several times. If the information is not contained in this block, the cache misses and the slower main memory is accessed. The block in which the newly requested data is found is then transferred from main into the cache. When properly designed and when block sizes are large enough, a cache looks to the CPU as a large and fast memory 90% to 99% of the time, but the interplay between the two memory units clearly indicates that the size and speed of both have to be carefully tuned.

It also implies that, as the speed of a CPU increases, not only the speed but also the size of both the cache and the main have to be increased. The reason is that the rate of data transfer from mass storage to main via the I/O processor will remain essentially constant in the foreseeable future. A Josephson system will, therefore, have to include the CPU, the cache and the main memory, connecting in and out of liquid He at the I/O processor. As in present large systems, the latter would be a rather powerful semiconductor machine in its own right. However, one can initially consider a medium size Josephson system that includes only the cache. Such a system could in the interim be connected to a very fast semiconductor main memory.

With this conceptual structure of a computer in mind, one may ask what it takes to increase the number of instructions processed per unit of time, measured in million instructions per second or MIPS. Clearly, one factor is directly proportional to performance, the basic cycle frequency of the machine. Everything being equal, if the cycle time is decreased by a factor of N, the machine will have N times the performance, assuming the memory hierarchy is properly readjusted. This is the approach taken in Josephson technology. Everything is sped up, logic circuits, memory elements and the transit time of signals in the package.

B. SYSTEM DIMENSIONS AND POWER DISSIPATION

Should we find a technology with circuits that switch in zero time, the inclusion of such circuits into an existing system will improve its performance by not more than about a factor of 2. The reason is that systems performance depends on circuit speed and on time of flight delays through the package [12]. In today's machines, circuit and package delays are nearly equal. This balance must be approx-imately maintained, otherwise one has a problem. Either the circuits are too good and therefore too expensive for the package or the package is too good and therefore too expensive for the circuits. Both cases are wasteful. This means that, to improve system performance by reducing the machine cycle time, one has to improve both the circuits and the package. It also implies that from an economical point of view, any restriction or fundamental problem with one component roughly determines the value of the other. We shall consider in the following some fundamental problems in package design related to the constancy of the speed of light.

In a computer, signals are transmitted through lines that are embedded in an insulator with dielectric constant ε. Since magnetic materials play no role, the velocity $\bar{c}$ of the signals is related to the speed of light in vacuum by $\bar{c} = c\varepsilon^{-1/2}$. This allows us to define a geometric system scale factor $\Delta X = ct_c\varepsilon^{-1/2}$ which is the distance a signal propagates in a time equal to the cycle time t_c of the machine. In this definition we have purposely neglected circuit delays to make sure that we do not equate the linear machine size with ΔX. All the scale factor implies is that the system size will have to be roughly measured in units of ΔX. In this context one should note that some paths through a computer involve a large number of circuits, so-called silicon paths, others are almost only package delays. Also, the relation between ΔX and size will depend on the machine architecture. No doubt, one has to use this yardstick with caution.

Figure II-3 depicts the relation between the cycle time t_c and ΔX for three values of dielectric constant, approximately spanning the range that one might expect with common insulators. The cycle times of four machines are also indicated. We find that the size of the Cray-1 with a cycle time of 12.5 ns has to be measured in terms of a few meters since the effective dielectric constant has been engineered to be of the order of 2. If we now take this machine as a starting point and postulate a performance improvement of one order of magnitude, the resulting supermachine must be measured in terms of a few tens of centimeters, if the same dielectric constant can be realized.

The first question is how we might artificially increase ΔX. Parallel process-ing is one possibility. With some problem sets this approach works well, but not in general. There is always a certain amount of the total execution time in which one needs to process serially or one must take care of overhead to keep the processors in concert [7]. Assume a problem in which this time is 20% of the total execution time. Then, four parallel processors will give a performance improvement of 2.5, eight processors a factor of 3.3, and sixteen processors a factor of 4. This can easily be derived since only 80% of the execution time can be reduced in proportion to the number of processors. With an infinite number of processors we just reach a factor of 5, hardly an economical approach. Unfortunately, as the number of parallel processors increases, the sensitivity of the system to the overhead is increased. Thus, with 1024 parallel processors and as little as 1% overhead, the performance is not increased by 1024 but by a mere factor of 91. Performance improvements through parallel processing will certainly find applications by tuning a system to a problem set, but since our argument is rather basic, such processors do not represent a general solution.

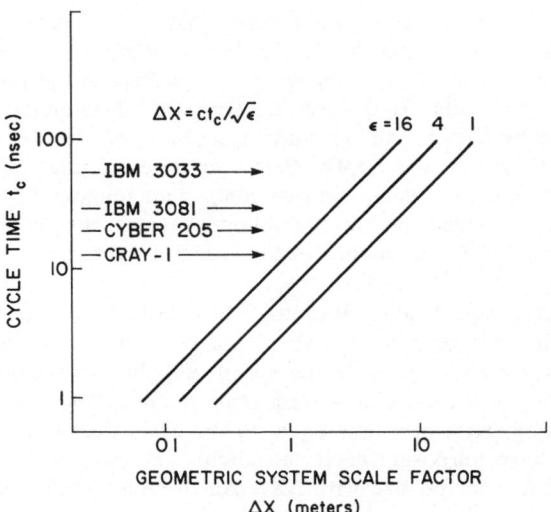

Fig. II-3:   Relationship between the machine cycle time and a geometric scale factor defined as the distance traveled in the time t_c at the speed of light in insulators with dielectric constant ε. The physical size of a machine is coarsely measured in units of ΔX.

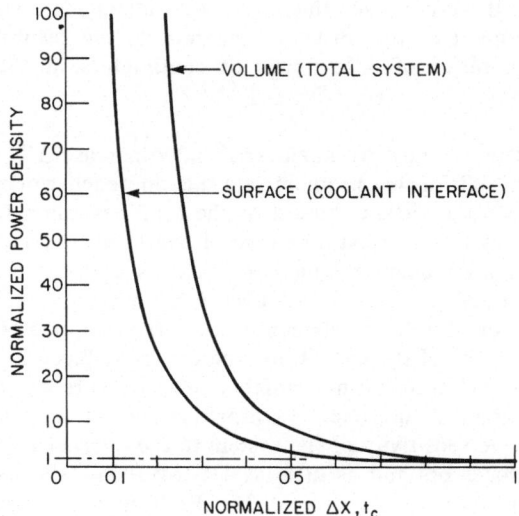

Fig. II-4:   Bounds of the increase in power density resulting from a decrease in cycle time t_c and hence a decrease in physical machine size ΔX.

How else can one effectively increase ΔX? We should note that in nearly all systems, the cycle time is limited by memory access which suggests pipelining and phased memory channels that can be effectively used in vector processors. And, in a sense, a cache memory is a means of circumventing a ΔX problem in the main memory. Thus, such attempts relate to architectural improvements which, except for the general purpose computer, are efficient only if one can tune a machine design to a given problem set. Furthermore, one can always argue that if such systems are implemented in a technology that increases raw speed, then that technology has the advantage over any architectural improvement.

The second question is how, by recognizing the fundamental nature of ΔX, one can contain the largest possible number of circuits within the confines of the space delimited by our scale factor. Clearly, future supercomputers will have to be not only fast but will need to contain at least the same number of circuits as present day machines. This implies a few hundred thousand to one half million circuits in logic alone. The trend is clear. To increase volumetric circuit density by a quantum leap, we must go three-dimensional or, if this is not in today's reach, choose the next best approach, namely a card on board package. In such a structure, chips are connected to cards and these in turn are inserted with minimum spacing into a board. Interestingly enough, the Cray-1 is assembled in a manner that resembles a card-on-board package. This minimizes the overall dimensions of the machine.

It can be argued that ultimately one may not have to face this problem since the circuit density on-chip is constantly increasing. In fact, a chip with nearly one half million *transistors* has been built [13]. This suggests that one will someday be able to place the entire CPU onto a chip. However, one must not forget that a supermachine will require very large amounts of memory and that it will again be memory access that limits the performance of the system.

In itself, the fabrication of a miniaturized package is extremely challenging but it is not a fundamental problem, since a wealth of micro-machining methods are available. The problem lies in the extraction of heat generated by the circuits out of a very small volume between cards. Semiconductor chips now dissipate so much power that special structures are required to extract the heat. In the IBM 3081 system, pistons are pushed onto the chips and the heat is conducted out into a cooling liquid through heat exchangers which increase the effective area of contact with the liquid [14,15]. Such structures are too large to be incorporated into a dense card on board package and instead force the designer to arrange the chips in two dimensions.

Cooling semiconductor chips by immersing them into cryogenic liquids such as nitrogen is also not a real long-term solution to this problem. In contrast to bipolar devices, FET's can operate at low temperatures without performance degradation, but one is limited by the amount of heat that can ultimately be transferred from the chips into the liquid. Initially, as chip power increases, convection increases at the expense of an increasing chip temperature. Then, nucleate boiling sets in and the chip power can be increased considerably with only a relatively small additional temperature increase. However, above a critical power density film boiling occurs. It insulates the chip from the liquid and causes thermal runaway. In liquid nitrogen, the critical power density [16] lies between 10 to 20 W/cm^2. Assuming a design with a safety factor of about 2, one finds a maximum chip power of the order of 1 to 2 W per 5 mm chip. The safety factor is necessary because, if the critical power density is exceeded, the chip may reach temperatures approaching 1000K. In liquids with lower boiling temperatures the situation is generally worse; thus, cooling in liquid helium is out of the question at the power levels we are considering.

It has always been said that as long as one can extract the heat from the chip, semiconductor technology has no problem. Therefore, chip power levels may be increased as cooling methods improve. In this respect, a remarkable result was recently reported [17]. By etching miniaturized channels into the back of a chip and by forcing water through these channels, it was possible to extract up to 800W/cm^2 at a chip temperature not much in excess of the normal operating temperature of semiconductor circuits. Since the channels do not take up additional space and since it is hoped that the plumbing can somehow be integrated, this method suggests the possibility of a very dense semiconductor package. However, we must realize that we are trying to fight nature and indeed the power problem takes on another form. It is now difficult to bring the power into the chip. With 800W/cm^2 and an operating voltage of 2V, a current of 100A must be brought into a 5mm chip through what one would suspect to be a rather large structure. In this context it should be noted that the problem of power distribution is already a significant problem at today's considerably lower power levels.

An interesting question at this point is what the attributes of a new technology should be. Let us normalize ΔX, t_c and the system power density to what is achieved today and let us decrease $\Delta X \sim t_c$. The volumetric system power density will increase as ΔX^{-3}, by three orders of magnitude for a one order of magnitude decrease in t_c. This is shown in Fig. II-4. Volumetric power density is somewhat misleading since the space between the components of a system is still so large that one should consider the surface power density which increases as ΔX^{-2}. The truth will lie somewhere in between these two curves and will probably be quite near to the latter. Such a simple-minded consideration allows us to make interesting observations. First, wherever we start, the power density problem seems innocous enough if we limit ourselves to an incremental decrease in t_c. But the problem grows frightfully fast if one looks sufficiently far ahead into the future. Thus, if we select a new technology that initially has the potential for over one order of magnitude speed improvement, we should like to see a decrease in power dissipation of at least two orders of magnitude. In addition, to insure that it is a true starting point, the extraction of that power should be possible without special structures or forced liquids so that this technology has room for future extension. Coincidentally, Josephson technology promises just that.

C. CHIP AND PACKAGE DESIGN PROBLEMS

When postulating decreases in system size, we assumed that signals were transmitted at the speed of light in the chosen dielectric. This is possible with terminated striplines placed over a groundplane. Such a scheme permits the transmission of information from the sending element to the receiving circuit in one pass at the fastest speed allowed by the dielectric. The problem at room temperature is that lines are resistive, and we are, by definition, talking about thin and narrow lines.

On-chip line resistances are generally so large that one cannot strictly speak of transmission line properties. Instead, the dynamics are rather accurately described by a simple RC network with a total resistance equal to that of the line, shunted to ground by the total capacitance between line and ground. The time constant for a line of length l, width w and thickness d_i, placed on a dielectric with dielectric constant ε and thickness d_o over a groundplane, is given by $RC = \rho \varepsilon \varepsilon_o l^2 / d_i d_o$ where ρ is the resistivity of the conductor and ε_o the permittivity of free space. Interestingly, this time constant remains unchanged if we properly scale to smaller dimensions [18] namely by decreasing in the same proportion length, width, insulator thickness, line spacing and line thickness. The line spacing must be decreased to obtain the desired line density, the insulator thickness to avoid inductive crosstalk and the line thickness to minimize capacitive crosstalk [19].

The independence of the RC time constant to such a scaling may at first seem surprising, since all dimensions, including the line length, decrease. Therefore, the time required to propagate a signal over a given length (l=constant) increases as line dimensions shrink. Yet on-chip performance improvements are made. But note: the speed of the circuits is being increased and, by placing a much larger number of circuits on a chip, one deals with fewer chips and thereby minimizes transmission length through lines in the package.

The problem in the package is worse. Here, to decrease line resistance, one generally uses much thicker and wider lines. These, to minimize crosstalk, are spaced correspondingly further apart. But thick, wide and widely spaced lines lead to a dilemma. Chip contacts must be connected to lines in the package. As the number of circuits per chip N increases, the number of contacts q to the chip will increase as approximately given by Rent's rule [20,21]: $q = AN^B$ where A=1.5 to 3.5 and B=0.55 to 0.75. This requires an increasing line density, for instance in the form of a growing number of wiring planes. Not counting groundplanes and power planes, there are 16 wiring levels [14] in the IBM 3081 chip module. Yet even in this package, chips have to be spaced about twice as far apart than the minimum allowed by their dimensions. The reason is that one requires a certain area around the chips in which the needed wiring channels can be reached. In this area, some space is also provided for structures that permit wiring changes.

To have reasonable line properties requires line lengths with resistances small compared to the line impedance Z_o. A line section that has a resistance of the order of Z_o will attenuate a pulse by nearly a factor of 2. It should be noted here that, because of the relatively small conductivities of the line materials, the skin depths are in most cases of the same order as the line dimensions so that considerations of purely ohmic line resistances lead to valid approximations at present operating frequencies.

As such a package is scaled to smaller dimensions, the problem is compounded. Here too, we must ultimately scale all dimensions. As a result, the ratio of line resistance to line impedance $R/Z_o = \rho l / d_i d_o \sqrt{\mu_o / \varepsilon \varepsilon_o}$ will increase. Thus, in the extreme, if we decrease all dimensions by one order of magnitude, the new line, one tenth of the original length, will have ten times the resistance for the same impedance, or in other words, the length over which it has reasonable line properties is decreased by two orders of magnitude. Only if one finds materials with correspondingly smaller resistivities can one hope to solve this problem.

In very resistive lines, termination is out of the question simply because the line resistance and the terminating resistor form a voltage divider that attenuates the signal in proportion to the line length. But even if the line resistance is small, there is a price to be paid for termination. Assume a fully populated 10,000 circuit chip; Rent's rule for average values for A and B requires about 1000 chip contacts. With line impedances of 50Ω and a driver voltage of 4V, one dissipates 320mW per line. If we assume that signals are present during the entire machine cycle, but that only 10% of the lines are activated at any one time, we find that for a half million circuit CPU, 1.6kW would be dissipated in the terminating resistors of logic lines alone. This explains why termination is used sparingly.

Finally, as one scales down line dimensions, current densities increase and electromigration becomes a design limitation. It depends strongly on the materials used to form the lines and can be decreased significantly only if the operating temperature is decreased.

A very different but nonetheless important problem is the direct result of increased speed. The energy depleted by the circuits must be replenished in a time of the order of the switching time. This is generally done by placing an energy reservoir, such as a capacitor, as near as possible to the chip. Since we are considering a distance of 5mm or more, the available energy is relatively far away resulting in what is called the ΔI pr)lem. It is generally explained in the following way: an inductance exists between the capacitor and the chip. As current is suddenly drawn, the inductance develops a voltage opposing that of the power supply. Thus, the on-chip power dips, preventing some circuits from operating properly. The present solution is to impose design groundrules which limit the number of high power circuits, mostly drivers, that can be switched simultaneously.

But even if that inductance can be largely reduced and power can be brought into the chip through some low impedance line, there remains a speed of light problem. With an insulator having a dielectric constant as low as $\varepsilon=2$ and considering that a few, say two, full reflections back and forth are necessary to smooth out the disturbance, the time it takes to bring energy in over a distance of 5mm is nearly 100ps. One hopes that circuits in a future supermachine will switch faster, and that poses a problem. There are several solutions. One is to minimize load variations using complementary circuits. Another is to regulate the power on-chip, but this will be possible only if the circuit power is small, otherwise it will increase the chip power dissipation significantly.

D.   A DESIGNER'S WISHLIST AND JOSEPHSON TECHNOLOGY

The preceeding discussion was not meant to imply that performance improvements in semiconductor machines have come to a halt. Gains will be made but these are expected to be expensive and more of an incremental nature since the technology is slowly reaching its limits. It is therefore legitimate to ask what kind of technology a hardware designer would like to have for a new start.

First, he would like to have devices with switching speeds about one order of magnitude higher at conventional linewidths, at this time of 2.5 μm, and a promise for further improvements when linewidths shrink to the sub-μm range [22], as experimentally practiced in semiconductors today. Josephson technology has that advantage as shown in Fig. II-5 where semiconductor circuits are compared to experimental Josephson circuits. Gate delays for ring-counters and for realistically loaded circuits are plotted as a function of minimum linewidth.*

Second, since system size has to shrink, the power dissipation must decrease at least two orders of magnitude as suggested by the lower curve in Figure II-4. This will bring us back to the starting point on the normalized power density scale for a system performance improvement of a factor of ten. In other words, the circuit power-delay product must decrease by three orders of magnitude or more and that is again what Josephson technology has achieved on an experimental basis as illustrated in Figure II-6.*

To decrease the power one needs to decrease operating voltages from the order of volts to the order of millivolts, as is the case in Josephson circuits. These, limited by the gap voltage of the superconductors, operate at <3mV. But as switching energies decrease from $\simeq$600meV to $\simeq$3meV, the potential barriers separating the device states decrease and the effect of thermal noise becomes pronounced. Thus, one needs to lower the operating temperature in the same ratio from 350K to a temperature of a few degrees K.] other words, one must operate in liquid He.

Having postulated a fresh start, one would like to be able to carry the heat generated by the new circuits directly into the operating environment, without need for fins, forced fluid flow or other cooling structures. In liquid He at 4.2K this is possible up to power levels of the order of 600 mW/cm^2 at which point film boiling sets in. At this power level temperature differentials between chip and bath are less than 400 mK [23]. Incidentally in superfluid He, the amount of power that can be extracted [24] can be as high as 6.6 W/cm^2.

Operating at <3 mV into impedances of 12Ω and solving the on-chip power regulation problem in a brute force fashion by simply dissipating an order of magnitude more than the basic device, we find that we have the potential for nearly 10^5 devices/cm^2 in a 4.2K bath. This is much more than can presently be accomodated with a 2.5 μm linewidth technology. Thus, even in freely boiling helium at 4.2K, there is a large design space before circuit power must be further reduced or more complex cooling structures become necessary. At that point, the simplest solution is probably to operate instead at $\leq$1.9K in superfluid helium [24]. This increases by one order of magnitude the amount of power the bath can extract directly from the chips.

---

* A more detailed representation of part of such data is given in reference [80]

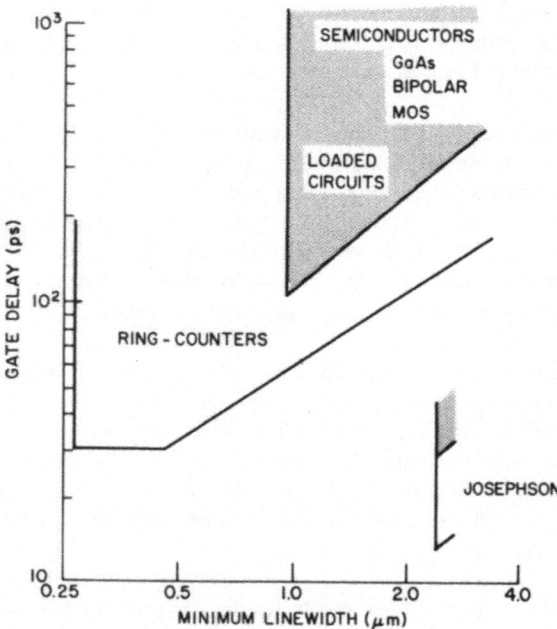

Fig. II-5:   Gate delays as a function of minimum linewidth for semiconductor and
              Josephson circuits. For given linewidth, Josephson circuits have a speed
              advantage of one order of magnitude.

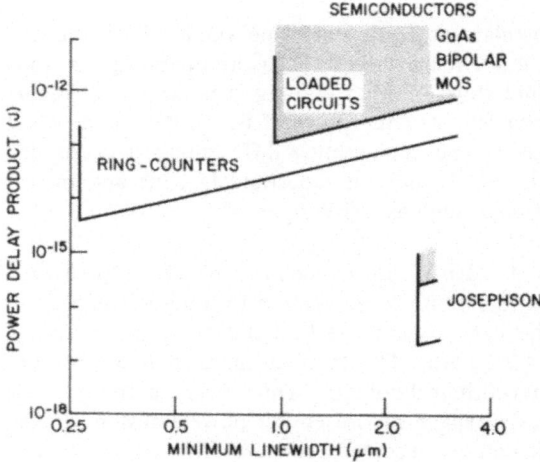

Fig. II-6:   Power delay product of semiconductor and Josephson circuits as a
              function of minimum feature size. Compared to semiconductor circuits
              with the same linewidth Josephson circuits have the required perform-
              ance advantage of over three orders of magnitude.

Lastly, a severe design restriction with semiconductors is that one has to deal with ever shrinking, resistive lines both on chip and in the package. This difficulty is eliminated if one uses superconducting lines. These have zero DC resistance and are near lossless and dispersion-less up to the material dependent gap frequencies of the order of 10^{12} Hz. As a result, very narrow and very thin terminated transmission lines can be used throughout the system. It impacts mainly the package in which large line densities can be achieved. In present experimental Josephson package parts one uses a few 1000Å thick 5 μm lines on 10 μm centers corresponding to a density of 10^3 lines/cm. Such a density is over one order of magnitude larger than semiconductor technology is practicing today in multilevel packages. As a result, a Josephson package can not only be miniaturized but extensively simplified since only two orthogonal wiring levels are required.

No doubt was left of the existence of
a new state of mercury in which its
resistance has practically vanished.

H. Kamerlingh Onnes
Researches Between the Second and
Third International Congress of
Refrigeration

III. ELEMENTS OF A JOSEPHSON COMPUTER

In the following we shall examine the basic elements that are required to build Josephson computer components. Josephson devices have long been reported for their high switching speed and low power dissipation and rightly so. However, equally important is the availability of near-lossless superconducting transmission lines which allow signal transmission without significant distortion from one element to another. For this reason we shall start with a short discussion of line properties before describing the Josephson effect and basic device configurations.

A. *SUPERCONDUCTING TRANSMISSION LINES*

Because superconducting lines have zero DC resistance and very small losses up into the Tera Hertz region, their physical dimensions can be made very small. Yet there are limitations that are worth a more detailed examination [25,26,27,28,29]. A cross-section of a transmission line is shown in Fig. III-1. The structure starts with a superconducting groundplane having a thickness d_g and a London penetration depth λ_g. It is covered with an insulator of thickness d_o and dielectric constant ε. The lines themselves are placed on top having a width w, a thickness d_l and a London penetration depth λ_l. The impedance and the delay per unit length are given by the following equations:

$$Z_o = \frac{d_o}{w} \sqrt{\frac{\mu_o}{\varepsilon_o \varepsilon}\left(1 + \frac{\lambda_g}{d_o}coth\frac{d_g}{\lambda_g} + \frac{\lambda_l}{d_o}coth\frac{d_l}{\lambda_l}\right)} \quad , \qquad (III-1)$$

$$\tau = \sqrt{\mu_o \varepsilon_o \varepsilon \left(1 + \frac{\lambda_g}{d_o}coth\frac{d_g}{\lambda_g} + \frac{\lambda_l}{d_o}coth\frac{d_l}{\lambda_l}\right)} \quad . \qquad (III-2)$$

The equations are valid only if the dielectric thickness is small compared to the linewidth so that fringe fields can be neglected, otherwise correction factors must be added.

Upon inspection we find that both the impedance and the delay are functions of ratios of line dimensions which do not appear in normal transmission line equations. The hyperbolic dependencies of d_l/λ_l and d_g/λ_g result from the fact that the kinetic energy of electron pairs appears as an additional inductance which increases both the impedance and the delay. The multiplying factors of the coth terms are a consequence of the superconducting penetration depths which measure the distance over which a magnetic field can penetrate into the superconductor. It causes an imbalance since the electric field penetration is negligible. Again an additional

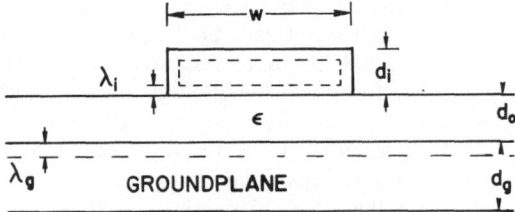

Fig. III-1: Cross-section through a superconducting transmission line situated over a superconducting groundplane. In most cases the insulator encloses the line so that the electric field is confined to a region with dielectric constant ε. The λ's are the London penetration depths.

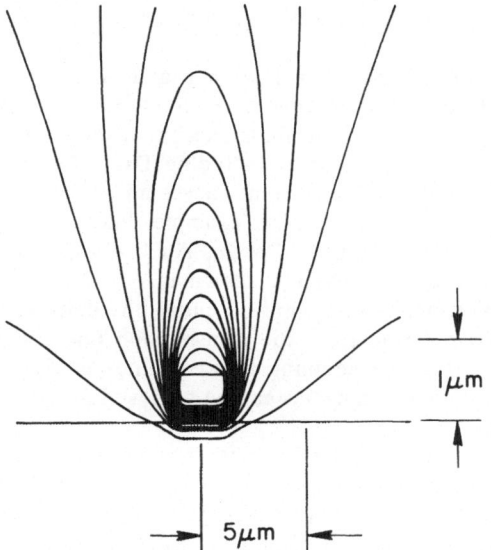

Fig. III-2: Lines of equal vector potential around a current carrying 2.5μm Pb-alloy superconductor placed 0.1μm over a niobium groundplane. Note that the magnetic field which is proportional to the potential line density remains largely confined beneath the conductor. This minimizes cross-stalk between neighboring transmission lines.

additive inductance results with the same overall effect, higher impedances and delays.

In the design of transmission lines, attention must be given to these dependencies although limitations are not significant in the present design space. London penetration depths range from 500Å for pure lead to a few 1000Å for lead alloys such as those presently used in Josephson technology. Since the hyperbolic cotangent increases rapidly as the ratios of dimensions to penetration depths decrease below unity, equation III-2 determines the lower limit of conductor and dielectric thicknesses. Ultimately, the delay in superconducting lines will sharply increase once line thicknesses, linewidths and insulator thicknesses reach dimensions of the order of 1000Å.

There exists one other problem. As equation III-2 indicates, it is undesirable to reduce the impedance of a line by reducing the dielectric thickness much below the penetration depth. Such lines have excessive delays and are totally unusable, except for delay line applications. In this case, the only way to decrease the impedance of a transmission line is to widen it. Such a problem occurs if one attempts to obtain impedances of the order of 1Ω with $2.5\mu m$ wide lines. In contrast, at an impedance level of 12Ω, $2.5\mu m$ lines have delays predominantly determined by ε. As Josephson devices are scaled to smaller dimensions, the required impedances will increase so that no problem is foreseen as linewidths are decreased [22].

With insulator thicknesses of a few thousand Å and lines that are several times wider, coupling to neighboring lines is small. One reason is that in superconducting lines over a superconducting groundplane, the magnetic field remains concentrated below the line at all frequencies including DC. Because of the Meissner effect, the field produced by the line current causes an image current in the groundplane even under static conditions. As an example, lines of equal vector potential around a $2.5\mu m$ wide and $0.5\mu m$ thick Pb, In, Au line ($\lambda_l = 1200$Å) placed 1000Å above a Nb groundplane ($\lambda_g = 850$Å) are shown in Fig. III-2. The magnetic field strength, which is proportional to the equi-potential line density, is seen to decrease very rapidly as one moves away from the transmission line edge. Since conductor thicknesses are small, capacitive coupling between lines is also small. That is the reason why such impressive line densities can be achieved in this technology.

The dynamics of superconducting lines has been extensively studied and a comparison between superconducting niobium and copper lines was made [29]. It shows that almost up to the gap frequency superconducting lines can be very well represented by an ideal lossless and dispersionless transmission line model. If the gap frequency ($\simeq 10^{12}$ Hz) is reached, superconducting lines are only marginally superior to normal metal lines. However, more detailed studies should be made to evaluate dielectric losses. These may not be negligible, as is presently assumed. To measure such losses directly in a frequency range of several hundred Giga-Hertz will require on-chip experiments designed with Josephson devices; nothing else is known to permit a direct transmission line evaluation at such frequencies.

B.   THE JOSEPHSON TUNNEL JUNCTION

The Josephson junction is the active element on which all present superconducting digital circuits are based. Let us start by reviewing its fundamental electrical properties. The boson-like Cooper pairs that carry the supercurrent behave cooperatively and minimize the total energy by locking their phases. As a result, superconductivity can be described as a single entity of pairs that is identified by a Schroedinger-like wave function. If the superconductor carries current, the pairs have a net momentum and, although the wave functions remain locked, there is a gradient of phase [30]. The phase difference ϕ between two points along a superconducting path is an important parameter, related to the current carried by the superconductor.

Assume now that two superconductors, S_1 and S_2, as shown in Fig. III-3, are weakly coupled through a tunneling barrier of thickness $\delta=20\text{Å}$ to 60Å. Here λ_1 and λ_2 represent the superconducting penetration depths. Such a structure is a Josephson junction. A Josephson point-like junction [31] carries a current determined by the phase difference ϕ as

$$I = I_o \sin \phi , \qquad\qquad (III-3)$$

and a voltage across the junction as given by

$$V = K d\phi/dt , \qquad\qquad (III-4)$$

finally,

$$grad\phi = D(\mu_o \overline{H} \times \overline{n})/K , \qquad\qquad (III-5)$$

where I_o is the zero field Josephson threshold current, $K=\hbar/2e$, $D=(\lambda_1+\lambda_2+\delta)$, H the external magnetic field, μ_o the permeability of free space and $\overline{n}$ the unit vector perpendicular to the plane of the junction. These are the Josephson equations [32,33,34].

The current density of a junction $j_1=I_o/a$, where a is the junction area, is a strong exponential function of the barrier thickness. In Pb-PbO-Pb junctions, a variation of 3.5Å in δ causes j_1 to change by one order of magnitude [35]. This indicates the care that must be taken in growing the tunnel barrier for practical devices in which current levels should not change by more than a few tens of percent. In contrast, within any given design range, the junction capacitance C is only a weak function of j_1 and can to first order be related to an assumed constant specific capacitance C_s as $C \simeq C_s a$. The equivalent circuit of a junction is also shown in Fig. III-3. It has a capacitance C, a non-linear quasiparticle tunneling resistance R(V) and a Josephson current element as specified by equation III-3. By adding all the currents one obtains

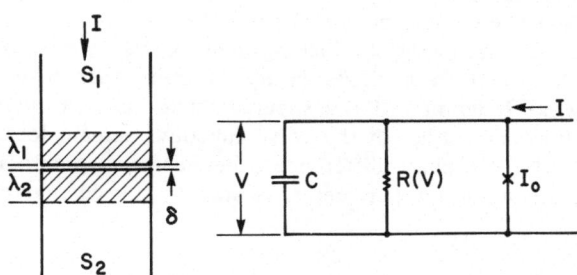

Fig. III-3:   Section through a Josephson tunnel junction formed between two
               superconductors S_1 and S_2. If its dimensions are small compared to the
               Josephson penetration depth, it is a point junction that can be described
               by a simple equivalent circuit as shown.

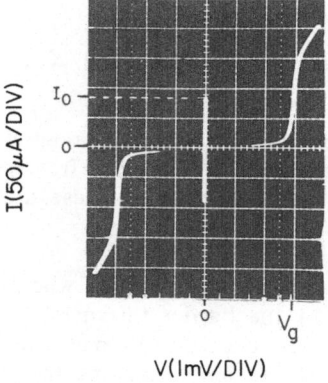

Fig. III-4:   I-V characteristic of a Josephson tunnel junction. It has two states, the
               zero-voltage state up to I_0 and the voltage state as indicated by the
               non-linear single particle tunneling curve. Current and voltage scales
               indicate the low power dissipation of the device.

$$C\frac{dV}{dt} + \frac{V}{R} + I_o \sin \phi = I \ , \qquad (III - 6)$$

and by substituting V as defined in equation III-4

$$CK\frac{d^2\phi}{dt^2} + \frac{Kd\phi}{Rdt} + I_o \sin \phi = I \ . \qquad (III - 7)$$

Equation III-7 is very important. It defines the dynamics of a Josephson point-like junction, a junction smaller than the Josephson penetration depth λ_J, in which magnetic self fields can be neglected. Also, magnetic stray fields which penetrate the junction are assumed much smaller than the magnetic flux quantum $2\pi K = h/2e \equiv \Phi_o = 2.07 \cdot 10^{-15}$ Vs, so that for any practical purpose I_o can be assumed constant. The Josephson penetration depth is given by

$$\lambda_J = [\Phi_o/2\pi\mu_o Dj_1]^{1/2} \ . \qquad (III - 8)$$

Equation III-7 was simultaneously proposed by Stewart [36] and by McCumber [37], who showed its equivalence to the equation of motion of a pendulum. This was extremely fruitful to the field since mental interpretations were possible in terms of a mechanical analog. Another mechanical model is that of the washboard in which a particle moves on a tilted sinusoidal potential. We shall use this latter model later to illustrate a host of dynamic phenomena that one encounters in Josephson devices.

Consider the I-V characteristic of a junction as shown in Fig. III-4. Starting at the origin, the junction can carry a zero voltage current up to I_o. The reason becomes clear by inspecting equations III-3 and III-4. For $I \leq I_o$, III-3 has a static solution with constant $\phi \leq \pi/2$, and therefore V=0. For $I > I_o$, the phase difference increases with time and the junction switches into the voltage state. In the voltage state, the Josephson element produces an AC current that is proportional to the instantaneous voltage. This can be seen from the Josephson equations by setting V=constant. From III-4, we have $\phi = (V/K)t = (2\pi V/\Phi_o)t$ which, by substituting into III-3 yields a sinusoidal current with frequency $f_J = V/\Phi_o$. Numerically, it has the value of 483MHz for every μV across the junction. Being dependent only on V, h and e, the Josephson AC effect is now the basis of the U.S. voltage standard.

In present, practical Josephson junctions, the oscillatory behavior is pronounced only at small voltages and frequencies. With increasing voltage the frequency of the AC current becomes so large that it is almost entirely shunted out by the junction capacitance. Therefore, at higher voltages the dynamic I-V curve is almost identical with the single particle tunneling curve that the junction would exhibit without Josephson effect. It is characterized by a large resistance below and by a sudden onset of current at the material dependent gap voltage $V_g = 2\Delta/e$. Here Δ is the energy gap of the superconductors.

A point-like junction alone is of very limited use. It is a two-terminal device and may be compared to a semiconductor junction. Yet, one characteristic of a computer element is that its input must be isolated with respect to both the power line and the output. To achieve this one has to interconnect more than one junction.

C. BASIC BUILDING BLOCKS

Before we consider multiple junction devices, let us first consider the behavior of a junction shunted by a superconductive inductance L as illustrated in Fig. III-5a. First we must write the current I_L through the inductor in terms of ϕ. This is done by setting the voltage across L equal to that of the junction, equation III-4. We find $L(dI_L/dt)=K(d\phi/dt)$ or $I_L=K\phi/L$. Adding both currents and setting them equal to the external current I, we have in the steady-state (see III-7):

$$I_o \sin \phi \; + \; \frac{K}{L}\phi \; = \; I \, . \qquad\qquad (III-9)$$

This equation is plotted in Fig. III-5b for an arbitrary set of parameter values. First we notice that, as the current I is increased, most of the current is flowing through the junction until its threshold current is reached. At that point, a transition takes place and one flux quantum enters the loop. If the external current is subsequently decreased to zero, the equation has a solution ϕ_1 for I=0 and a quantized current $I_L=K\phi_1/L$ corresponding to one Φ_0 remains trapped in the circuit. Had the current been increased further, two and three, but not more than three flux quanta could have been trapped for the chosen parameters. Only if the inductance and/or I_0 are increased can more flux quanta be stored in the loop. Conversely, with a sufficiently small L or I_0 no circulating current is stable in the absence of an external current.

Being shunted by an inductance, such a device cannot be switched into the voltage state. However, the circuit represents a prototype memory cell which is able to store quantized persistent circulating currents. These currents owe their remarkable stability to flux quantization as shown in Fig. III-5b. Strictly speaking, the parameter that is quantized is the fluxoid. It is distinguished over flux because the energy of the Cooper pairs is only partially associated with the magnetic field, the other part is related to the kinetic energy of the pairs [27].

Let us now add a second junction into the loop and construct a prototype logic device [38,39,40] as shown in cross-section in Fig. III-6a. Its equivalent circuit diagram is illustrated in Fig. III-6b. The device containing the two junctions is sandwiched between the groundplane and at least one overlaid control line. It is called a two-junction interferometer or SQUID for Superconducting QUantum Interference Device. The device is symmetric and the control or input line is arranged such as to give us the desired isolation between the control line and the device. This is the case, since gate current changes (I_g) through the device do not couple into the control line. Only varying circulating loop currents are coupled back into the input.

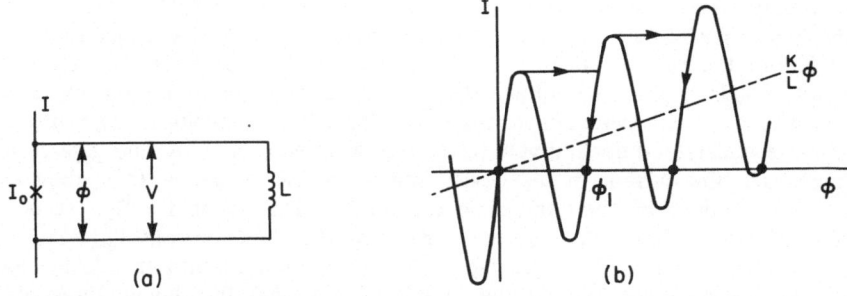

Fig. III-5: (a) Elementary memory loop composed of a superconductive inductance and a Josephson junction. (b) With the proper choice of parameters such a loop can sustain quantized circulating currents in the absence of an external current I.

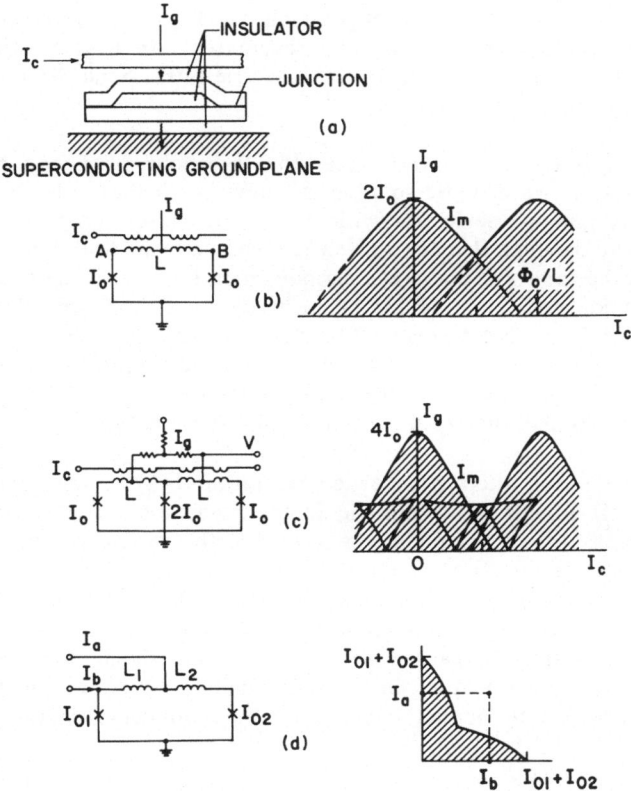

Fig. III-6: (a) Section of a two-junction interferometer. (b), (c) Equivalent
circuits and corresponding threshold curves of a symmetric two- and
three-junction device. (d) An asymmetric two-junction interferometer
with direct current injection. It is used to perform the AND function.

Figure III-6b also shows the so-called threshold curve $I_m(I_c)$ of the device. It is the locus of gate currents I_g and control current values I_c that cause the device to switch into the voltage state. In other words, starting with zero currents, the device will remain in the zero-voltage state as long as the operating point remains in the shaded area. It will switch as it crosses out through the envelope. The threshold curve can be understood qualitatively in the following way. If $I_c=0$ and gate current I_g is increased, the current in the device splits into two equal parts feeding each junction until both reach their threshold current I_0. Thus when $I_g=2I_0$, the device switches into the voltage state at the zero field threshold current $I_{mo}=2I_0$. As control current is increased, a circulating current is induced into the device loop. This current changes the bias of the junctions such that the device threshold is reached for lower values of I_g. At this point only the center lobe exists. As I_c is increased with sufficiently low I_g so that switching does not occur, there comes a point when it is energetically favorable for the device to let one flux quantum enter the loop. That shifts the center lobe to a new position $I_c=\Phi_0/L$. This process repeats itself with a periodicity of Φ_0/L for increasing and for decreasing I_c.

The shape of the threshold curve scales with the LI_0 product. As LI_0 is decreased, the overlap between the lobes decreases. In devices with a built-in asymmetry the threshold curves are also asymmetric [40]. Such devices have found applications in memory design.

For a two-junction interferometer the threshold curve can be calculated analytically [41,42]. No analytic solution is known for higher order interferometers such as the three-junction device shown in Fig. III-6c. It was proposed [40] because it has higher gain, meaning that for a given I_g it can be made to switch with a smaller I_c. Here again the threshold curve has periodic main lobes with the same periodicity as the two-junction device [43,44,45]. But, in addition, there are side lobes placed between the main lobes that correspond to states in which flux quanta entered either the left or the right device loop. Only when the device contains a flux quantum in both loops is it in a quantum state that corresponds to a shift of the center lobe by $I_c=\Phi_0/L$. The zero field threshold current of this device is $I_{mo}=4I_0$.

Another useful interferometer configuration [46] is the injection device shown in Fig. III-6d. It is asymmetric in the inductances and in the Josephson threshold currents. Input currents are injected directly from preceding gates and the interest in this device results from the fact that the threshold curve is wing-shaped. This makes it particularly attractive as an AND gate, where one input alone, I_a or I_b, can vary by relatively large amounts without switching the device, whereas both inputs applied simultaneously cause switching for modest signal values. Interferometer structures are very versatile since within limits they allow one to shape the threshold curve simply by adding junctions or by introducing internal asymmetries [40,47].

The interferometers represented in Fig. III-6 can be extended to contain a large number of point junctions. In the limit they represent a single junction with a spacial extent that can be adjusted to be small or large as compared to the Josephson penetration depth. In the first case, we deal with a short junction that has a threshold curve with the well-known $(\sin x)/x$ form; in the latter case, it represents a long junction or in-line gate. Such gates had been used in Josephson logic and memory applications but, because of their large capacitance and their failure to scale properly to smaller dimensions, they were abandoned in favor of interferometers [40].

D. DEVICE DYNAMICS

Compared to semiconductors, Josephson devices have a different and peculiar dynamic behavior. This stems from the fact that we are dealing with a threshold device which obeys macroscopic quantum effects. Ultimately, we are dealing with interferometers but the analysis can be greatly simplified if we use the equation of motion of a point-like junction, as given in Equation III-7, and assume that the threshold I_o can be altered by a control current. This simplifying assumption is valid because the LI_o product of most practical devices is small and as a result the junctions are relatively tightly coupled. With the exception of *internal* device dynamics, Equation III-7 yields excellent approximations.

Dynamic effects are best explained by analogy with a mechanical equivalent, the so-called washboard model as illustrated in Fig. III-7. Equation III-7 not only describes a Josephson point junction but also the dynamics of a massive particle in sinusoidal potential wells with an amplitude proportional to the Josephson threshold current I_m. This is easily seen by comparing the equation of motion of such a particle with equation III-6. The behavior of the particle is described by $dp/dt + \eta p + dU(x)/dx = 0$ where m is its mass, $p = mdx/dt$ its momentum, η represents the losses and $U(x)$ the potential.

We find that the velocity of the particle corresponds to the device voltage, that the mass is equivalent to the total junction capacitance and that losses incurred by the particle represent losses due to the parallel combination of junction resistance and external load. Current fed into the junction appears as an additive straight line with a slope proportional to the gate current I_g.

1. Plasma Oscillations

Consider the particle trapped in a potential well as in point A or B of Figure III-7 and assume it is slightly disturbed. If damping is light, the particle will oscillate in the well and undergo so-called plasma oscillations. In a Josephson device this correponds to a resonance between the junction capacitance and the non-linear inductive Josephson current. The equivalent Josephson inductance L_J can be derived from Lentz's law and we have
$V = L_J(dI/dt) = (d\phi/dt)L_J I_m \cos \phi = (VL_J I_m/K) \cos \phi$, or $L_J = K/I_m \cos \phi$.
Thus, for small oscillations the device is a resonator composed of L_J and the total device capacitance C with a resonant or plasma frequency given by

$$\omega_p = 1/\sqrt{KC/I_m \cos \phi}. \qquad (III - 10)$$

After any disturbance, a device will exhibit damped oscillations at that frequency. These decrease exponentially in amplitude. Conversely, a biased device which receives a signal at that frequency may switch into the voltage state.

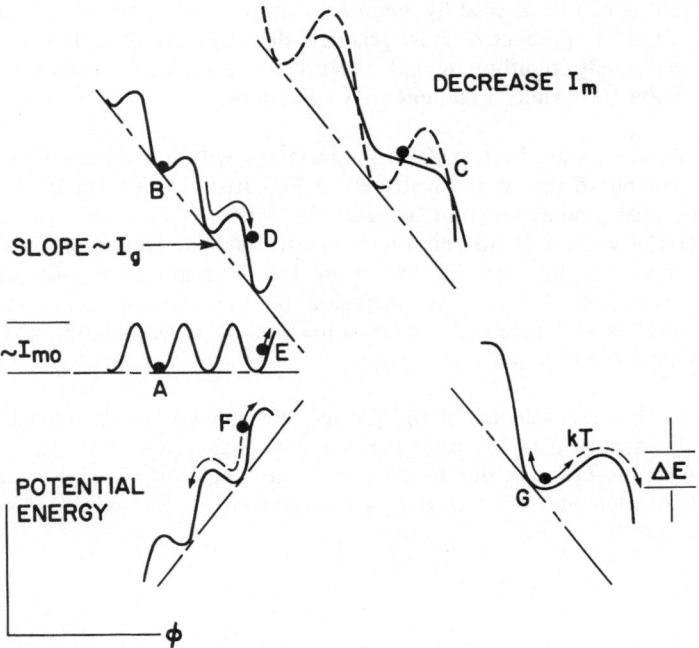

Fig. III-7: Potential energy of a single Josephson point junction plotted as a
            function of the phase difference ϕ. The slope of this periodic function
            is proportional to the applied current I_g.

2. Switching into the Voltage State and Turn-On Delay

Assume the gate current I_g is gradually increased. That increases the overall slope of the potential wells, point B in Fig. III-7. Unless I_g exceeds the zero field threshold current which in symmetric devices is given by $I_{mo} = \Sigma I_{ot}$, the device remains in the zero-voltage state, or in the model the particle remains in its well. If I_m is decreased, there comes a point when the wells open and become flat, point C. At this point, the threshold is reached, the particle escapes and accelerates downhill. The particle having a non-zero velocity means that the junction has a non-zero voltage. If the threshold is only slightly exceeded, the slope on which the particle dwells is very small, so that the initial acceleration is also small. This causes the so-called turn-on delay t_o which can be very large (ns) for control currents near the threshold. But if the control current overdrives the threshold by 50% or more, it is only of the order of a few ps in present devices. The turn-on delay corresponds to the first Josephson oscillation. It can be accurately derived from simulations and excellent analytic approximations have been derived for junctions as well as for interferometers [48,49]. Significantly t_o scales as

$$t_o \sim (C/I_{mo})^{1/2} \qquad\qquad (III - 11)$$

This means that it decreases as one decreases linewidths. The reason is that the device capacitance C (junction size) decreases as w^2 whereas, with increasing line impedance Z_o, current levels are expected to decrease as w [22], see equation III-1.

In the same way that the particle accelerates toward its final average velocity, as determined by the viscosity of the medium, the junction voltage increases exponentially up to the value determined by its total resistive load R_T [50]. In the subgap region, the device has a large near-linear resistance R_j so that for any practical purpose, R_T is determined by the parallel combinaton of R_j and any externally applied resistance, as long as the operating voltage remains below the gap voltage. The risetime of the signal is usually defined as one time constant of the exponential rise. Thus,

$$t_r = CR_T \qquad\qquad (III - 12)$$

After switching the particle will periodically change its velocity as it traverses the wells. This means that in a Josephson device we do not only build up a DC voltage but have superimposed an AC voltage directly proportional to V. It is the Josephson AC effect as discussed in Section III-B.

3. Resonances

A long Josephson junction is a resonator composed of an open ended section of tranmission line. Similarly, an interferometer has the ability to resonate by exchanging energy between the junction capacitances and the inductances. In present designs the resonant frequency lies below half the gap frequency. As the Josephson frequency increases with increasing signal voltage, it will at one point coincide with the resonant frequency of the device and build up energy in the resonator. This energy shines back onto the junctions. In our model, which represents only one of these junctions, it corresponds to having the particle roll down hill while it is accelerated and decelerated in synchronism with the crossing of each potential well. The particle will lock into that mode in such a manner that, as the slope or I_g is increased, it will not accelerate substantially but instead increase the amplitude of the oscillation. This means that as one enters a resonance the current can be increased at nearly constant voltage and a device resonance appears as a current step in the I-V characteristic [51,52]. There are as many resonance steps as there are loops in the device. Switching into a resonance must be avoided in digital circuits [53,54,55,56]. This is achieved by decreasing the step height through resistors connected across the interferometer inductances [54].

4. Latching and Non-Latching Operation

If, after the particle is accelerated, the control current is removed so that the potential barriers between the wells are reestablished, either of two things can happen. In the case of viscous damping, the particle will simply stop in one of the potential wells. In contrast, with light damping it will not (point D in Fig. III-7). It will gain enough kinetic energy during the fall into each well to be able to escape from it into the next. Obviously, the situation depends not only on damping but also on the mass of the particle and the height of the potential barriers. A Josephson device behaves the same way. Upon removal of the control current, it will switch back into the zero-voltage state if $\beta = CR^2 I_m / K \leq 1$. In contrast if $\beta > 1$, it will latch into the voltage state and must be reset by decreasing the gate current [36,37]. Resetting occurs at a characteristic voltage [57] which for large β can be approximated by $V_{min} \simeq (2I_m K/C)^{1/2}$.

Non-latching devices behave somewhat like transistors. They switch on and off with the input signal. However, for reasons explained in the next chapter, latching devices are preferred.

5. Punchthrough

Punchthrough in latching Josephson devices is very different from punchthrough in semiconductors. In the former case it pertains to the fact that, as the gate current is decreased, there is a non-zero probability that the device will reach zero velocity at a particular well only for a moment. Then, having stored internal energy, it will switch again after the gate current is being brought back to its original value [58,59,60,114]. To explain this phenomenon consider the particle moving through the wells, point D, while I_g is decreased through zero to a negative value. If this inversion of I_g proceeds reasonably slowly the particle will, in most of the cases, come to rest in a well near $I_g=0$. Sometimes however it will have enough energy near point E to climb the top of the potential barrier, point F, where it can remain for arbitrarily long times. Once I_g has become negative, the particle falls back and rolls down the potential hill. The same can happen if I_g is brought to zero and then increased back to a positive value. Random noise has no influence on this effect since, although it will disturb the particle we just considered, it can place a particle that would normally not punchthrough on top of the barrier.

In terms of a Josephson device this means that resetting cannot be guaranteed with certainty. However, the probability of punchthrough can be made arbitrarily small. At present, Josephson devices are reset by inverting I_g from a constant positive value $I_g = \gamma I_{mo}$ to a constant negative value $I_g = -\gamma I_{mo}$, and back. If we denote the transition time as t_p, then the probability of punchthrough p_o during the inversion of current can, for a single device, be approximated as [60]

$$p_o \simeq (t_p/2\pi\gamma R_T C) \exp (\sqrt{\beta} - 4t_p/\pi\gamma R_T C) \ . \qquad (III - 13)$$

In circuit design, low device capacitances, small loads and sufficiently large t_p must be chosen to ensure that p_o is sufficiently small so that punchthrough occurs, for instance, only once per year in 10^6 devices. It then appears as an occasional soft failure with a frequency that can be made arbitrarily small.

Experiments [61,62] have shown that for a single device, the punchthrough probability falls off exponentially as qualitatively predicted by equation III-13. But more work is required to check the theory quantitatively and to evaluate circuit punchthrough.

For a given t_p, the punchthrough probability can be drastically decreased [60] if the gate current waveform is shaped such that it dwells for a short moment at $I_g = 0$. Unfortunately, that complicates the AC power supply design. However, more design work should be done in this area since t_p is a wasteful overhead in the cycle time of a Josephson machine.

6. Noise

Erroneous switching due to noise must be avoided. Yet, as with punchthrough, one cannot strictly eliminate the possibility of such an event [63,64,65,66]. Again the probability of occurrence must be sufficiently decreased. Investigations of various noise factors such as shot noise, 1/f noise, quantum noise, etc. indicate that thermal noise is the predominant source [67].

The situation is again best understood in terms of the washboard model. Assume that we increase I_g to a value below the zero field threshold current I_{mo}, point G in Fig. III-7. The junction is powered up but must remain in the zero-voltage state. With $I_g \neq 0$, the barrier height ΔE is decreased and thermal noise can cause the particle to escape. The rate of escape for moderately light damping [68] is given by a rate equation $r = (\omega_p/2\pi) \exp (-\Delta E/kT)$ where k is Boltzmann's constant and T the effective device temperature. For a given device and a given temperature, the only adjustable parameter is ΔE. This means that one must choose I_g sufficiently below I_{mo} to avoid switching. By expressing ΔE in terms of the circuit parameters (equation III-7), one can relate the safety margin of gate current $\Delta I = I_{mo} - I_g$ to a chosen rate r by [67]

$$\Delta I/I_{mo} \simeq [(kT/2KI_{mo}) \ln (\omega_p/2\pi r)]^{2/3} \ . \qquad (III - 14)$$

Experiments have shown that this relation is a good approximation and that it can be applied to interferometers with sufficiently low LI_o [69]. Care must be taken to determine the effective noise temperature of the circuit which in most practical cases is only slightly higher than the substrate temperature.

> Since Boole showed that logics can
> be reduced to very simple algebraic
> systems -- it was possible for Bab-
> bage and his successors to design or-
> gans for a computer that could per-
> form the necessary logical tasks.
>
> *H. H. Goldstine Ref. [3]*

IV. JOSEPHSON CIRCUITS

A. LOGIC

The Josephson effect is found in many structures in which two superconductors are weakly coupled. Besides the tunnel junction we discussed here there are
weak links, point contacts, and junctions in which the barrier is a thick metal or a
thick degenerate semiconductor [70,71,72,73,74].

From these devices, the tunnel junction was chosen for digital applications.
The reason is that it has a very large subgap resistance R_J as shown in Fig. IV-1a.
With a gate current $I_g < I_{mo}$ and an external load R_1 which ensures operation just
below the gap voltage, we find that, once switched, the current I_g initially flowing
through the junction is to the largest part transferred into the load. Here it is
available to switch other devices. For this reason, the junction quality was expressed
in terms of a characteristic voltage V_m that is related to R_J as $V_m = j_1 a R_J$. V_m
increases with increasing quality of the junction. Resonance current steps are kept
small by damping the interferometer inductances as previously discussed [53,54].

Operation above V_{min} leads to latching operation. If the load R_2 is made
small enough to intersect the voltage axis at or below V_{min}, non-latching operation
can be obtained. In Fig. IV-1b, the experimental I-V curve of a non-latching
three-junction interferometer with a load of 0.15Ω is shown for different control
currents. Such a device can be powered with a DC current at nW power levels. The
problem is that the signal in the load has to be carried to the next device via terminated transmission lines and, as discussed in section III-A, it is difficult to design fast
low impedance lines. They are either narrow and situated above a very thin dielectric, which results in an extremely large delay (equation III-2), or fast with a thick
insulator, but then the linewidth becomes large (equation III-1). This impacts line
and circuit density, and for this reason, latching circuits were chosen as the work-
horse of the technology.

1. Operating Principles

An elementary logic circuit is illustrated in Fig. IV-2. It shows the simplest
member of the family, a 2-input OR [40]. The circle represents a symmetric three-
junction interferometer with two control lines, each of which can independently
switch the device. Nominal threshold currents are $I_{mo} = 200\mu A$. The circuit is
powered from a power bus that distributes a trapezoidal voltage V_p and generates a
gate current I_g through a power resistor R_p. The gate operates with positive as well
as negative gate currents and it is reset during the transition time t_p. A transmission

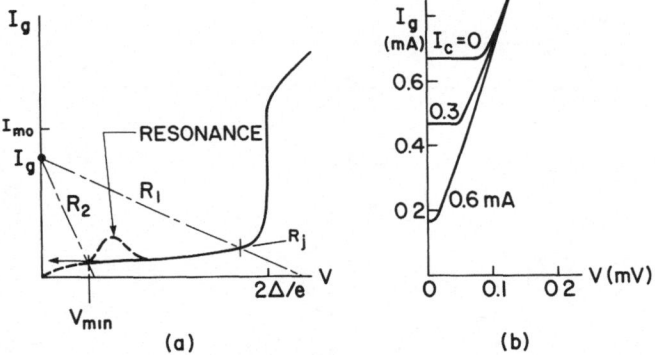

Fig. IV-1: (a) I-V curve of a Josephson interferometer. Good quality implies a high sub-gap resistance R_j or high $V_m=aj_1R_j$. A load R_1 intersecting above a characteristic voltage V_{min} leads to hysteretic behavior. A small load R_2 intersecting at or below V_{min} permits reversible switching as shown in (b).

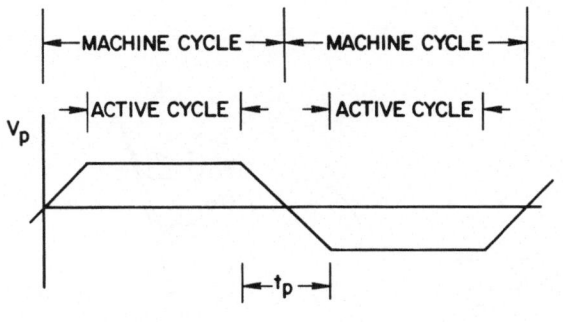

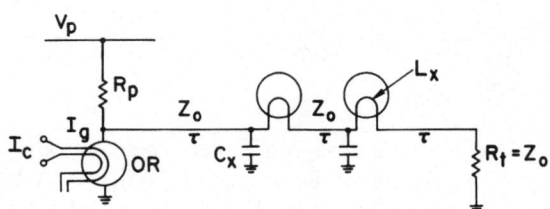

Fig. IV-2:   Elementary latching logic circuit with serial fanout using terminated
superconducting transmission lines.   The discontinuities at a fanout are
matched to Z_0.   Trapezoidal power permits circuit resetting at the end of
the machine cycle.   Note that the circuits operate for supply currents of
both polarities.

line with impedance $Z_0 = 12\Omega$ is connected across the interferometer. It forms the load that is carrying the control signal to the next device. Other sections of transmission lines continue the path and string in series all the devices that are controlled by that output. At the end, the line is terminated into a resistance $R_t = Z_0$ so that reflections are avoided. In that manner the signal is transmitted to the fanout devices in one pass at the speed $\bar{c}$ as determined by the dielectric.

In contrast to semiconductor circuits, the output signal is distributed serially and fanout is essentially unlimited since additional fanout gates do not load the driving gate. At each fanout there exists a discontinuity because a control line has the character of an inductance L_x. To avoid reflections, one can add padding capacitors C_x which match the impedance of the control line sections to the transmission line. If one restricts the risetime to be $t_r \geq 3L_x/Z_0$, ringing is avoided in the L_x, C_x circuit and the padding capacitors then permit operation with a fully matched line [40,75,76].

2. Delay Components

At the beginning of the machine cycle, with the information stored in latches, the AC power supply voltage increases to its plateau and gate current biases all devices below I_{mo}. The onset of the plateau defines the beginning of the active cycle. At that time, under the control of the decoded instruction set, the information from some of the latches is released. This information switches a number of interconnected logic levels asynchronously.

We shall here define the gate delay from the moment the input signal is at threshold to the moment at which the output signal reaches threshold. This yields the following delay components. The turn-on delay t_o is followed by the exponentially rising output voltage with risetime t_r as defined by equation III-12. The signal is transmitted in the transmission line sections of length l_i with a delay per unit length τ as given by equation III-2. At each crossing of a fanout control line, the output signal is delayed by $t_x = L_x/Z_0$ if padding capacitors are used. For a fanout of F, the total nominal circuit delay to the last stage is therefore simply the sum of all of these components.

In most cases, one is interested in the average gate delay t_g that one may encounter in a machine. It is based on an average wiring distance $\bar{l}$ and an average fanout $\bar{F}$. It is given by

$$t_g = t_o + t_r + \tau\bar{l} + \bar{F}t_x . \qquad (IV-1)$$

There is a fundamental difference between the fanout characteristic of semiconductor circuits and Josephson circuits. In the former case, output lines are fanned out in parallel and the signal will rise as determined by the total line capacitance. This means that if we connect to a nearby and to a far-away gate, the longer line will load down the shorter. In Josephson circuits, the nearby gate will switch first regardless of the distance to the other fanout circuits. Thus, one may take better advantage of the on-chip placement of gates in critical paths.

The active cycle is usually determined by a worst case path that involves about 10 logic gate levels and an instruction cache access. It therefore involves not only gate delays but also package and memory delays. At the end of the cycle, with information again stored in latches, the power waveform crosses over to the negative side to reset all logic circuits. The shortest transition time that one can obtain with present power supply designs is about 20% of the machine cycle, but generally, circuit punchthrough is the limiting factor (equation III-13).

3.   Current Injection Logic (CIL) Family

The simplest member of the CIL family is the 2-input OR shown in Fig. IV-2. It can be extended to a 4-input OR simply by interconnecting two three-junction interferometers through a small resistance. In this case, if one switches, the other will also always switch. The AND function takes advantage of the injection interferometer shown in Fig. III-6d. To ensure circuit isolation it is driven by two 2-input OR gates as illustrated in Fig. IV-3. Note that with such an arrangement one always obtains a free 2-input OR function at every AND input. The 2-input AND can also be extended to a 4-input AND by using two such gates with their output feeding into a third injection interferometer. Experimental 2.5 μm CIL gates have been investigated [77,78,79,80] and were found to operate with an average logic gate delay of 44 ps at a power level of 5 μW. Most of this power is being dissipated in the power resistors R_p. The fastest operation of the fastest member, the 2-input OR, gave a delay of 13 ps at 2.6 μW.

Designing a machine with OR's and AND's is not possible unless one also provides an INVERTER. This poses a problem with latching logic. An INVERTER must give an output signal until an input signal arrives. But with latching logic, once the output signal is present, the driven gates will switch, latch and disregard the subsequent disappearance of the INVERTER output. For this reason, the INVERTER must be timed and must be made to switch only if no input is received at a time at which one is sure that the signal to be inverted should have arrived. Timed INVERTERS were designed and experimentally investigated. The timing aspect is unfortunately a complication one would like to avoid. Therefore, it was decided to use INVERTERS sparingly and to perform the inversion at the output of the latches. These are automatically timed at the beginning of the active logic cycle. At the expense of 20 to 30% more circuits, complement signals can be carried into the logic wherever an inversion is necessary.

A latch contains two storage elements, a master and a slave. The reason is that the information entered into the latch must appear at the output only at the beginning of the next cycle. The master of the CIL latch is based on a superconducting storage loop [81] as shown in Fig. III-5. Here the junction was replaced by a three-junction interferometer, as seen in Fig. IV-4, for a data latch [82,83,84]. Input currents for that gate are obtained from an appropriate set of control circuits. If a "zero" is written, the storage loop contains $0 \pm 1\Phi_o$ and if a "one" is received, a current equivalent to $6 \pm 1\Phi_o$ is stored in the loop. The switching time of the latch is typically 100 ps. Part of the loop inductance forms the control lines of two sense gates in the slave circuit. If the latch is updated in a given cycle, the slave output remains unchanged until after it has been reset during t_p and is being powered up

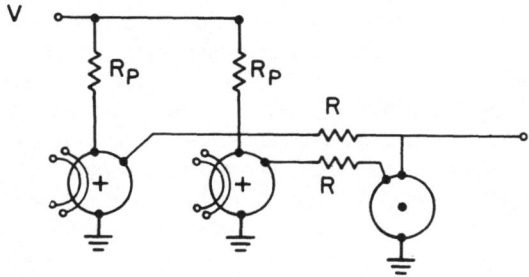

Fig. IV-3: Two-input AND circuit. Circuit isolation is provided through 2 two-input OR devices. These yield a free OR function at each input of the circuit. The AND function is performed through direct injection into a two-junction interferometer as shown in Fig. III-6d.

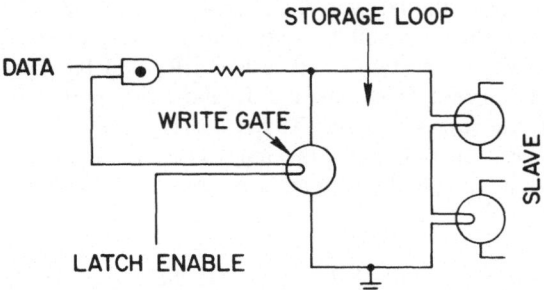

Fig. IV-4: Simplified representation of a Data Latch. Information is stored in the form of quantized persistent circulating currents in a superconducting loop.

again at the beginning of the next cycle. The slave provides both true and complement outputs. CIL latches, both masters and slaves, have been designed and were
successfully operated [82,83].

The remaining members of the CIL family are long line drivers and receivers
[85]. Although we are dealing with a system based on 12Ω terminated transmission
lines, drivers are required. The reason is that in the package there are inductive
discontinuities and small unshielded sections at which crosstalk can occur. The signal
risetimes must be increased so that these padded discontinuities behave as transmission line sections. At the same time, crosstalk is minimized. Receivers at the end of
long lines are needed because it takes the same time to discharge a transmission line
as it takes to send a signal to its end. Therefore, the part near the terminating
resistor of a long line which received its signal at the end of the active logic cycle can
hold that signal over into the next cycle and erroneously switch logic gates at that
time. This is avoided by designing a receiver which compares the polarity of the
signal with that of the power supply. Only if the signal has the same polarity as the
power will the receiver switch. Driver and receiver designs have been experimentally
evaluated and were found to perform as predicted by the models [85].

4.   *Direct Coupled Logic*

Alternate approaches to interferometer logic have been proposed and are
being explored. They dispense with the relatively large inductance and the control
lines of the interferometer and produce isolation by using the fact that a junction
switches from a zero DC resistance to a very large resistance as long as the operating
voltage remains below the gap. Two circuits are shown in Fig. IV-5. The Direct
Coupled Isolation (DCI) device is illustrated in Fig. IV-5a. It forms a bridge with
two junctions and two resistors [86]. Input current I_c is injected through terminating
resistors (not shown) from the left and the load R_t is connected to the right. Initially, the junctions are in the zero voltage state and the gate current divides in the two
branches as determined by the two small resistors R_1 and R_2. If an input signal is
received, it subtracts current from junction J_1 and adds to that flowing through J_2
which switches into the resistive state. That transfers most of the gate current into
J_1 which also switches. In that state, the input is isolated from the output through
the large resistance of J_1 and the output is disconnected from ground through the
high resistance of J_2.

The first proposed circuit of this class was the Josephson Atto Weber Switch
(JAWS) as shown in Fig IV-5b [87]. Its operation is very similar. A second
negative gate current (I_b) is used to improve operating margins. The input signal
again switches first J_2, then J_1, disconnecting the input from the device and the
output from ground. DCI devices of varying complexity have been proposed
[88,89,90,91,92,93,94]. Not having the isolation of a control line, such devices
cannot be serially connected into a single output line; instead, parallel fanout is
required. Also, it is difficult to perform the AND function with sufficiently large
margins. This was remedied by using, as in CIL, the injection interferometer as the
basic AND element. A flip-flop circuit such as a direct coupled logic latch, the
"Huffle" [95], was also reported.

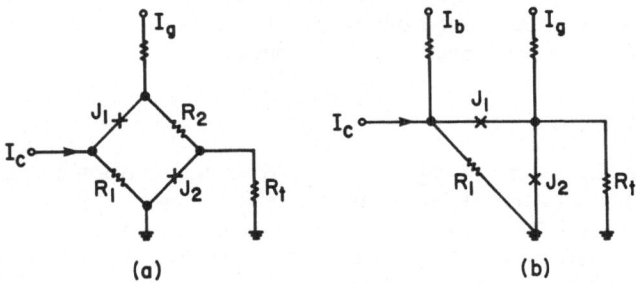

Fig. IV-5: Direct coupled logic (a) DCI and (b) JAWS. In such circuits, gate isolation is achieved through switched high resistance point junctions. J_1 disconnects the input I_c and J_2 disconnects the output load from ground. [From reference [80], © 1980 IEEE]

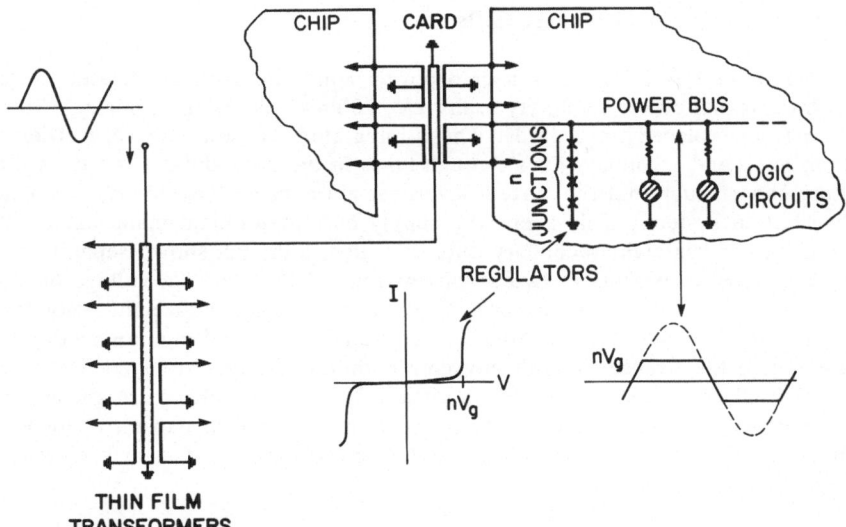

Fig. IV-6: Power supply system for latching logic circuits. Sinusoidal current is distributed to 16 chips through thin film transformers that fan out in a star-like manner. This minimizes timing skew. On-chip the sinewave is clipped with serially connected junctions (regulators) to obtain the desired trapezoidal power waveform.

Direct coupled circuits have the advantage of a simpler vertical structure, because the control layer is not needed. The problem is that compared to interferometer logic the operating margins are smaller. The basic underlying reason is that the output signal must be fanned out in parallel, each line competing for output current, in a device that has limited gain at the outset.

5. *Power Supply*

Latching circuits are chosen because of their high performance on a systems level. This necessitates the design of an AC power supply [96] which must fulfill several requirements [97]. First, it must periodically reset the circuits. Second, during the active part of the cycle, the gate current must be held as constant as possible. Third, by virtue of the anticipated speed of the circuits, the power must be regulated on chip so as to avoid the equivalent of the ΔI problem encountered in semiconductor technology. Fourth, the AC power must be distributed with a minimum of timing skew throughout the system, so that circuits everywhere will be powered up and reset at the same time. This last requirement, if satisfied, has the advantage that the machine clock is available to any circuit with small timing skew. Finally, the system power currents must be kept sufficiently small to permit the use of integrated lines operating far from their critical current. As will be explained shortly, all these conditions can be satisfied if one bases the design on the distribution of sinusoidal power through a balanced network of transformers arranged in a star-like fashion to minimize phase shifts.

The operating principle of such a power supply network is illustrated in Fig. IV-6. An external power signal (typically 2V, 200mA) enters the primary of a thin film transformer placed on a card. These integrated transformers very efficiently couple primary and secondary lines through holes in the groundplane and the voltage ratios are chosen such that the current levels remain constant ($\simeq$200mA) throughout the distribution system. This keeps the supply current problem manageable. The first transformer has eight secondary outputs which, although shown separately, are in fact connected in parallel in two groups of four. These carry a voltage of about 200mV and are connected to the primary of a transformer with eight secondaries, operational at a voltage of about 20mV. The transformers and interconnecting lines are laid out in the package before any consideration is given to signal lines. This ensures that all power waveforms are in phase. If wiring problems are encountered, it is in principle possible to adjust the phase with inductive or capacitive loads. A chip is powered by four secondaries of the last transformer so that such a system is able to service 16 chips.

On-chip, the supply signal is again distributed in a star-like manner to regulators formed by n series-connected Josephson junctions in which the Josephson current is suppressed with a DC control bias. Regulation is achieved through the voltage limiting current jump at n times the junction gap voltage. A trapezoidal voltage waveform is produced and distributed in a low impedance power bus that feeds a few tens of logic circuits [98,99,100]. As previously discussed, the bipolar nature of the power waveform is of no consequence since the considered circuits operate with both polarities. In fact, with such an arrangement one has the advantage of obtaining two machine cycles per period of the power sinewave.

The power voltage is distributed to the logic gates through individual power resistors R_p. These isolate the circuits and minimize disturbs in the power bus. For a 12mV supply with four lead-alloy junction regulators, the power resistors are of the order of 80Ω.

The design trade-offs in such a supply are numerous and shall not be discussed here. One should note however that, as in logic circuits, the regulator junctions must have a high subgap resistance R_j or, in other words, a large V_m. One also requires a sharp onset of the gap to obtain as large a ratio of active to total cycle as possible. Furthermore, the gap must be well-defined, reproducible, and must exhibit a large slope (dI/dV) to permit good control of the gate current.

Supply circuits able to power 16 chips in systems with cycle times of a few ns have been designed and were successfully operated [99,100].

B. MEMORY

A general purpose Josephson computer must contain a cryogenic memory. Clearly, to take advantage of a high-speed CPU, at least a fast cache memory is needed. Its access time must be commensurate with the machine cycle. Whether a Josephson main memory is necessary in a first machine is open to debate. However, as progress is made, the technology will require a main memory or an additional member of the memory hierarchy, in the form of a second, much larger and somewhat slower cache.

As discussed in Section III-C, memory design in Josephson technology has been based on the storage of persistent quantized circulating currents in superconducting loops. Cache memory loop cells with non-destructive readout (NDRO), as well as cache memory circuit components, have been investigated [101,102,103,104]. Also, the design of a main memory with destructive readout (DRO), in which circulating currents are stored in two-junction interferometers, Fig. III-6a, was explored [105,106,107]. It was found that, for such a main memory, the operating margins are too small because of our present inability to control the device threshold currents adequately. For this reason, we shall concentrate on cache memory in this discussion.

1. Cell Operating Principle

The memory cell we are considering here [108,109] is the product of a rather long evolution that has been extensively reviewed in Ref. [110]. It is composed of a superconducting loop containing a three-junction interferometer. A "1" is stored in the form of a circulating current and a "0" is defined by the absence of a stored current. Sensing is performed with a two-junction interferometer placed beneath the loop. The current flowing in the loop acts as a control current to that sense gate.

A memory cell must have properties which allow it to be written and read by the coincidence of orthogonal currents. This permits the selection of a single cell (or bit) in the array if the memory is bit-organized, or of a row of cells if it is word-organized. Here we shall be concerned with bit-organized memories only.

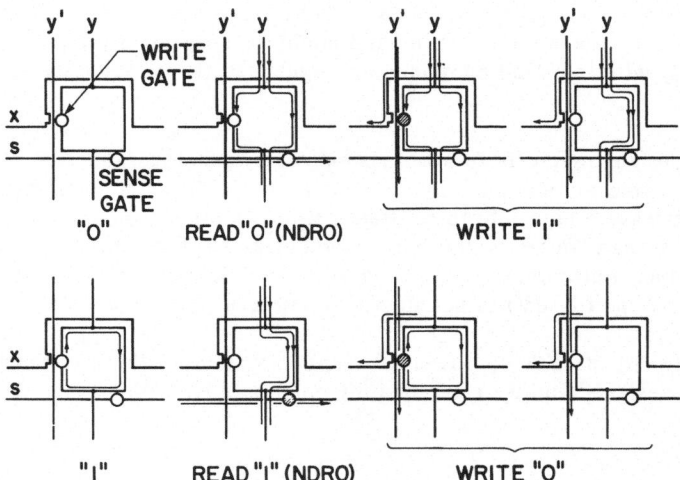

Fig. IV-7:   The cache memory cell stores information as a quantized circulating
current for a "1" and no current for a "0". The write gate momentarily
opens the storage loop during switching.   Non-destructive reading
(NDRO) is obtained through the switching of an inductively coupled
sense gate. All possible cell functions are indicated.

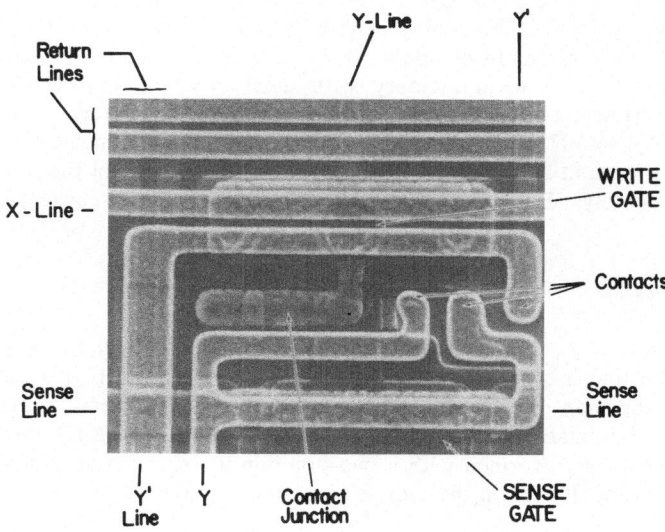

Fig IV-8:   Cache memory cell designed in 2.5μm technology. The cell has a size of
58μmx48μm. Cell switching time is about 30ps. [From reference [109],
© 1979 IEEE]

Cell operation is illustrated in Fig. IV-7. Let us start with an empty cell representing a "0". To read this "0", current is simultaneously applied to the y line and to the orthogonal s line. The sense gate has full gate current s but the y current, fed into the cell, splits into two equal parts because the inductances of the two loop branches are equal. Thus, half the y current acts as control to the sense gate. It is designed to switch only with the full y current as control. The lack of switching indicates the stored "0".

The writing of a "1" requires activation of the write gate which has two control lines, one y' in the Y direction, the other x in the X direction. Again, y current is applied and if there is coincidence of y' and x, the write gate switches transferring the entire y current into the right hand branch of the loop. Upon removal of first y' and x, followed by the removal of y, a circulating current remains stored. This is best understood by imagining that the y current is removed through the application of a negative y current which will cancel half of the current on the right hand side and add half of the applied negative current in the left hand branch.

Sensing of a "1" proceeds in the same manner as before except that now the presence of the circulating current ensures that the full y current acts as control to the sense gate which switches. The writing of a "0" necessitates only a y' and an x current in coincidence. If a "1" is stored, the write gate switches and dissipates the circulating current in the loop.

To form the array, cells of the type shown in Fig. IV-7 are interconnected. In the Y direction the loops are strung into y and y' lines and in the X direction the sense gates are connected into sense lines and the write gate controls into x lines. As a result, 2N lines in the Y direction and 2N lines in the X direction permit the selection of one cell out of N^2.

Cells of this type, designed with 2.5μm linewidths, have been experimentally operated with stored currents equivalent to two flux quanta [109]. A micrograph is shown in Fig. IV-8. In trying to increase bit density, efforts were made to integrate the elements as close together as possible. As a result, details are difficult to discern. This cell has a size of 58 x 48 μm^2 and it switches from one state to another in about 30 ps.

2. Auxiliary Circuits

Auxiliary circuits are required to provide currents to the cells and to channel the information into and out of the memory chip. A typical memory unit [103] is shown in Fig. IV-9. Signals are received from logic. Since the signal levels of logic and memory may differ, it is necessary to provide interface circuits which, in the case of a Josephson cache chip, accept the logic levels and produce pulses of appropriate amplitudes. For an array containing N^2 cells, the interface accepts $2\log_2 N$ addresses. In addition, it receives a function signal that determines whether one reads or writes and a data signal that determines whether the written information is a "0" or a "1". This information is entered and latched into two decoders [111] which act as address, data and function registers. Each decoder has $\log_2 N + 1$ stages. The last stage of the X decoder decides whether to feed current into the x or the s line of the

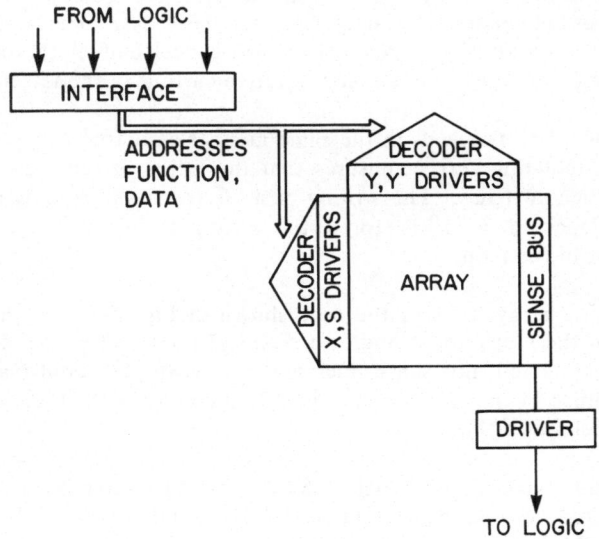

Fig. IV-9:   Basic diagram of a bit-organized memory chip. The cells form an array.
Also shown are the auxiliary circuits which provide the coincident
currents needed for the selection of a single cell. [From reference [103],
© 1980 IBM J. Res. Develop.]

selected row, whereas the last stage of the Y decoder provides to the selected column a y and y' current to write a "1", or a y current alone to sense, or a y' current alone to write a "0". Each decoder switches one of 2N drivers that direct current into the selected array lines. During a read cycle, the information sensed by the sense gate is collected by a sense bus, an N input OR circuit, and is channeled to an off-chip driver that reconverts the memory sense signal back into a logic level.

In present cache design, all logic functions are performed with DC powered loop logic. The basic circuit is illustrated in Fig. IV-10a. It is essentially a large memory cell (see Fig. III-5), which contains a second so-called reset gate. Thus the operation of all circuits, including the cell, are based on the same operating principle. DC gate current is fed through a set or driver gate. Many such gates can be strung serially into the power line in which crosstalk is minimized through isolating power resistors R_p. If the set gate is switched, the gate current I_g, initially flowing through the gate, is completely transferred into the loop. The loop may be an x, s or y' line or may contain memory cells to form a y line or interconnect sense gates into a sense line. At the end of the cycle, the reset gate is switched and the current is transferred out of the loop back into the set gate. In the sense line, the current is transferred back before resetting occurs if a "1" switches a sense gate. All sense lines are transformer coupled to the sense bus [112]. It detects the early disappearance of the sense loop current as a "1".

Decoders [103,111] are designed with a similar set of loops. These form through the interaction of serially interconnected devices, a tree with 2N branches. A detailed design of a Josephson cache chip has been documented and should be referred to for further study [103].

3. Circuit Dynamics

In loop logic, only one of the gates in the loop switches at any one time; the others remain in the zero-voltage state and can be treated as shorts. As a result, the dynamics of superconducting loops can be described by the equivalent circuit shown in Fig. IV-10b. The equation of motion of this circuit is obtained by summing all currents, see equation III-7 and III-9,

$$CK\frac{d^2\phi}{dt^2} + \frac{K}{R}\frac{d\phi}{dt} + \frac{K}{L}\phi + I_m \sin \phi = I \ . \qquad (IV-2)$$

If the LI_{mo} product of the circuit is much larger than Φ_o, quantum effects can be neglected and the Josephson term acts to first order as a switch which opens when the gate current exceeds I_m. In other words, in the present range of operating frequencies, the Josephson AC current tends to average out and can be neglected. In this case, equation IV-2 becomes a complete linear second order differential equation that describes a parallel resonant circuit.

In loop logic, one would like to transfer the entire gate current out of the set gate into the loop and back. This is achieved by an appropriate choice of damping resistance. First, since the subgap tunnel resistance is generally difficult to control,

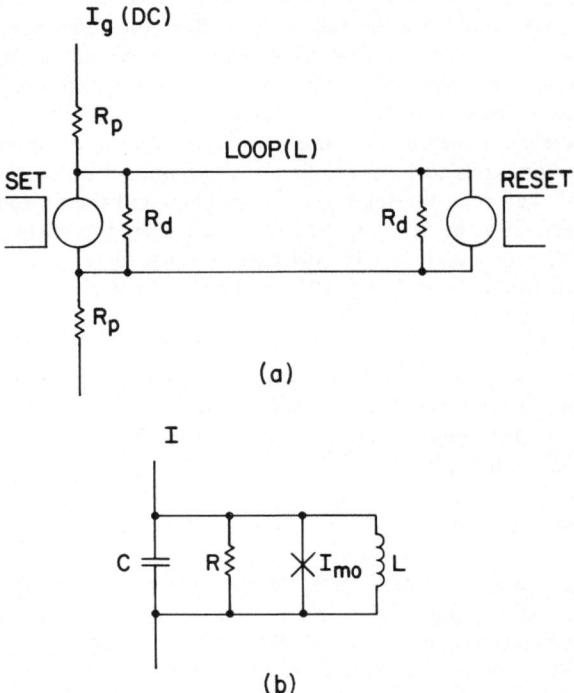

Fig. IV-10: Memory design is based mainly on loop logic. It is performed with serially DC powered loops as shown in (a). R_d are external damping resistors which adjust circuit dynamics. A simplified equivalent circuit is shown in (b).

one chooses junctions with high V_m. In this case, a well-controlled external damping resistor R_d, with a much lower resistance, can be connected across the gate (see Fig. IV-10). It will dominate the damping conditions and is chosen such that the circuit is slightly underdamped. An approximate value for the total damping resistance, $R = R_d // R_J$ is $R \simeq 0.8(L/C)^{1/2}$. The value of the damping resistor must be chosen such that the current, which in this case has the tendency to overshoot, is zero in the gate when the voltage reaches V_{min}. This means that the gate resets when the entire gate current is transferred into the loop.

In most array loops, the lines are relatively long and behave as transmission line sections. However, the current transfer time into such structures is still long as compared to the time of flight so that a number of reflections back and forth occur during the risetime. This means that these loops still have a predominantly inductive character. However, in a final design, detailed simulations are necessary and additional damping resistors must be connected across the center of the loops to attenuate the strength of the reflections [103]. A different complication occurs in a cell which stores only a few flux quanta. Here, simulations are necessary because quantum effects invalidate equation IV-2.

Delays in these circuits are measured in units of $(LC)^{1/2}$. This means that, except for time of flight delays in address lines, the chip access time will scale in the same manner. As gates are miniaturized, performance improvements are obtained by reducing the junction capacitance. But the scaling also indicates that, to obtain speed, one should reduce the loop inductance. This means that a design involving several small arrays is faster than one based on a single large array. In present designs, a typical current transfer time into a 2mm long loop is on the order of 150 ps. This permits one to expect 4K bit cache memory chips with sub-nanosecond access times.

C. *MODELS AND SIMULATIONS*

By and large, Josephson circuits can be modelled with excellent accuracy. There are several reasons. One is that the junction is essentially a two-dimensional device described by a set of highly non-linear, but simple, equations. In its I-V characteristic, the main features scale with the normal tunneling resistance as determined from the tunneling probability across the barrier. However, the junction current level, calculated from barrier thickness and barrier height, is far from being predictable [35,113]. Also, the value of the subgap tunneling resistance is presently beyond the reach of accurate theoretical prediction. Further studies are needed in this area.

Of course, this is of minor concern to the circuit designer once a controllable junction process is available. He is in a position to simulate quite accurately the static and dynamic behavior of his networks. In most cases, these can be modelled with ideal elements, capacitors, resistors, inductances and transmission lines.

The junctions are accurately described by the Stewart-McCumber model [36,37] (equation III-7). Limitations should be expected at frequencies approaching the gap frequency, however, in present devices the junction capacitance tends to

shunt out such effects. For some time now, this model has been successfully extended in describing the non-linear tunneling resistance by a piece-wise linear approximation and more complex models have been described [115,116,117,118]. Inductances
that enclose the junctions into an interferometer can be calculated with existing
programs which simultaneously solve both the London and Maxwell equations in two
dimensions [119,120,121,122]. Penetration depths are determined from measurements based on very accurate experimental methods [123]. Our understanding of the
kinetic inductance in presently used superconductors has been experimentally verified
[27] and the effect of increasing field penetration into electrodes that become thin as
compared to the penetration depths is well-understood [124]. Thus, we can determine all inductive device components and mutual inductances [79,125] necessary to
calculate threshold curves with existing programs. Most are based on the following
algorithm: first, the effective control currents flowing into the device are determined.
Then, the gate current I_g is set equal to the sum of the junction currents
$I_g = \Sigma I_{oi} \sin \phi_i$ and the quantum conditions $\Sigma \phi_i = 2\pi n$, where n=0, ± 1, ± 2...,
are imposed around each loop. Currents through junctions and inductances, ex-
pressed in terms of the ϕ_i, are summed at nodes and, by using Lagrange's multipliers,
the ϕ_i corresponding to the maximum I_g are determined. That determines the
threshold curve [43]. Various computational methods have been explored to solve
this problem, however, the formulation has essentially remained unchanged.

Internal resonances in symmetric interferometers can be predicted with
experimentally verified theories [53,54,55]. A better understanding of resonances in
asymmetric devices is still required. Also, the onset of device switching at mode
boundaries below the threshold curve is not theoretically understood at this point
[126].

Circuit dynamics are simulated by using programs such as ASTAP [127] in
which the network is topologically described. Boundary value problems, as one may
encounter in the description of a long junction, are generally reduced to initial value
problems by starting the calculation with currents and voltages at zero. In circuit
simulations, transmission lines are treated as ideal which, for short structures or low
frequencies, is an excellent approximation. In fact, the static behavior of strip lines
is well understood. Detailed investigations are needed to evaluate high frequency
losses in long lines since experimental results have indicated that, even at modest
frequencies on the order of 10^9Hz, losses are larger than expected from theory.
These are suspected to stem from the dielectrics rather than the superconductors
[128].

Thus, despite some shortcomings, a wealth of design and simulation tools
have become available and are found to predict circuit behavior very accurately.
This is an important asset of Josephson technology since it minimizes the number of
steps required for experimental verification as designs progress.

D. GAIN, MARGINS AND DESIGN-LIMITED YIELD

Compared to semiconductors, Josephson devices have relatively low gain.
No doubt, digital computer elements, strictly speaking, have a gain of unity since
signals, propagating through a large number of logic circuits, remain essentially at the

same voltage amplitude. Yet, gain in general can often be translated into improved operating margins.

For Josephson devices, let us define gain as the ratio between the current available at the output and the current required to switch the next device with sufficient overdrive. First, we have to remind ourselves that Josephson interferometer circuits have a fundamental advantage over semiconductor circuits in that the fanout gates can be serially connected without loading effect. That means that, if a circuit has enough gain to switch one device, it is able to switch as many devices as one may wish, as long as one takes into account that each additional fanout gate will be switched with an additional delay. This lessens the burden on gain. Still, one would like to have more sensitive devices or larger output signals.

The situation is illustrated in Fig. IV-11. We plot gate current versus control current. Three threshold curves are shown, one nominal at I_{mo}, the others representing threshold current variations to higher and lower values. Contained in these spreads must not only be on-chip variations, but also the run-to-run reproducibility, threshold variations caused by fluctuations in chip temperature and long-term instabilities that can be expected between thresholds as fabricated and those found at the end of life of the machine.

The maximum gate current I_g must remain below the minimum threshold by a safety margin determined from the thermal noise that one expects in that circuit. This noise margin is chosen for an acceptable low error frequency, see section III-D-6. It is added worst case. Note that, compared to semiconductor circuits, Josephson device energies are lower by about a factor of 200 (from $\simeq 600$meV to 3meV) whereas the operating temperature has been lowered by less than two orders of magnitude from 350K to 4K. As a consequence, the relative importance of thermal noise is somewhat larger.

The minimum gate current is determined by the sidelobe of the highest threshold curve. This bounds the allowed variations in gate current. The effective output or control current I_c is that part of the gate current which is not shunted away by the relatively high subgap resistance R_j ($<10\%$ with good quality junctions). Also subtracted are losses incurred because of less than perfect coupling between the control lines and the devices. In general, the control current is $I_c \simeq 0.8I_g$. The linear relationship $I_c = 0.8I_g$ is drawn in Fig. IV-11. It determines the available range of gate currents which produce an I_c that exceeds all thresholds. To illustrate the various limitations a designer has to take into account, the threshold curves are drawn here in such a way that I_c and not the sidelobe is limiting I_g. Clearly the device is not optimized; it should have a slightly larger LI_o product.

The problem is that one cannot arbitrarily increase threshold variations, since they impose decreasing gate currents which in turn result in decreasing output currents. That is where gain could effectively be translated into margins. Larger output currents would permit the use of low LI_o product interferometers whose threshold curves have lower sidelobes. But additional gain can be obtained in various ways. One can use double control windings, which effectively double the control current as suggested in the figure by the $I_c = 1.6I_g$ line. Also, buffer devices

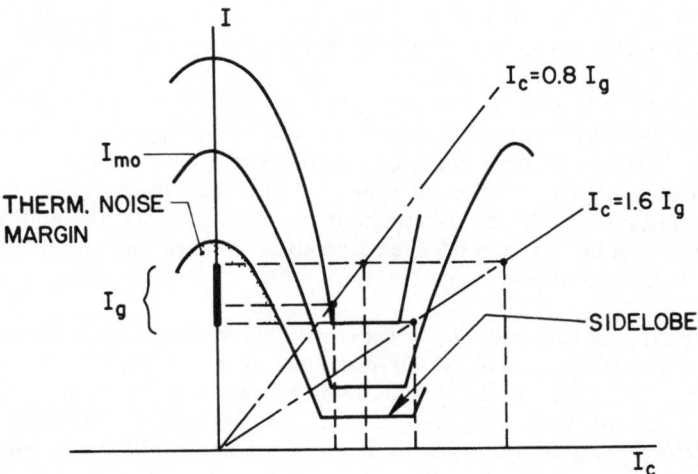

Fig. IV-11: Operating margins of a three-junction interferometer. Variations in threshold current I_{mo} limit the operating region. The control current I_c is a linear function of the gate current I_g. Available control currents for a single and a double ($I_c \simeq 1.6 I_g$) control line are shown.

can be connected to the circuits to increase the available output current [86]. Note that the gain of actual logic circuits is larger than suggested by the figure. Nominally, control currents exceed the threshold by about a factor of two.

It goes without saying that the example we just considered addresses only one of the many aspects that a full design must take into account. One must produce chip designs such that the spreads of all design parameters within established distributions permit the fabrication of fully functioning chips with a very high probability of success . In other words, LSI designs are statistical in nature.

Let there be n design parameters P_1 with known spread distribution, so that the failure probability p_i of being outside of the design range is established. The yield probability Y_1 of being within range in all N circuits of a chip is then $Y_1 = (1-p_1)^N$. If we now assume, for reasons of simplicity, that all design parameters are independent, the design-limited yield Y_D is given by

$$Y_D = \prod_i^n Y_i \qquad\qquad (IV-3)$$

Y_D determines the fraction of chips that will be functional with a perfect process, rejects being caused only as a result of extreme parameter variations within the specified distributions. Although process-limited yield is generally very low, ranging from a few percent to a few tens of percent, the design-limited yield should be 0.9 or higher.

There are two reasons for this requirement. First, design-limited yield has generally the shape of an n dimensional mesa, relatively flat on top but with rather abrupt cliffs where the parameters reach the limits of the design range. Therefore, setting Y_D to 0.5 would mean that very slight changes in process control can yield a very good product or no product at all. The second reason relates to the fact that functional chips should be fabricated in relatively few runs. With $Y_D = 0.5$, one would think that only twice as many runs are necessary. This is not the case, as one may experimentally confirm by tossing a coin.

A large design-limited yield requires very small individual failure probabilities, especially if the number of circuits per chip is large. To evaluate the Y_1, one can simulate the circuit with a large number of Monte Carlo runs, testing each time whether operation is satisfactory or not. However, many hundreds of thousands of runs are required if one insists on a high confidence level. The problem can be simplified if one deals with normal distributions. In such cases, it is reasonably straightforward to evaluate failure probabilities analytically, simply from the overlap of the tails of the distributions.

E.   OTHER DEVICE AND CIRCUIT APPROACHES

Up to now, Josephson digital circuits were exclusively based on tunnel junctions. One reason is that a good quality junction has a large sub-gap resistance R_j, the other reason is that it can be integrated on-chip. In these applications, the junction is essentially used as a relay contact that can change its impedance from zero to a high value. In fact, the interferometer forms a kind of relay and its latching nature is reminiscent of latching electromechanical relays that were used in early business machines. A Josephson junction, however, because of its inherent non-linearities, exhibits other phenomena that may be usefully employed in digital applications.

Assume a junction with a length that exceeds the Josephson penetration depth λ_J by a factor of six or more. It can be conceptually represented by an infinite set of point junctions interconnected through differential inductances, or macroscopically by a large set of parallel two-junction interferometers as the one shown in Fig. III-6a. External fields will be shielded from the junction by screening currents confined to distances of the order of $2\lambda_J$ at both ends. This means that the point junctions at one end are almost completely decoupled from the point junctions at the other end, or, more precisely, a change in phase difference ϕ on one side causes a negligibly small phase change at the other. Assume now that an impulse causes ϕ to change by 2π at one junction edge. A flux quantum is injected but the corresponding circulating currents are spatially confined to a distance $2\pi\lambda_J$ and no significant change in ϕ occurs elsewhere. A soliton was formed. If such a junction is uniformly biased with gate current $I_g<I_{mo}$, the soliton will be accelerated along the junction by the Lorentz force induced through the magnetic field of the bias current [129,130,131].

Because of the moving localized phase change of 2π, a voltage pulse (equation III-4) and a corresponding displacement current move along the junction [132]. The direction of movement depends on the polarity of the bias current. Under the influence of the Lorentz force, the soliton accelerates and contracts relativistically as it approaches the speed of light $\bar{c}$ in the junction. The ultimate velocity, however, will be limited by losses. In accelerating, the soliton gains energy as evidenced by the increasing amplitude of the voltage pulse. This means that the device has substantial power gain (of the order of 40) [133] which, by an appropriate change in device impedance, a gradually widening or narrowing of the junction along its length, can be transformed into voltage or current gain.

Digital switching devices based on solitons were proposed [134,135,136] and experimentally investigated [136]. In one approach [136], the output of the junction is interconnected to two devices through resistors that match the junction impedance. The two devices, forming a fork, are biased, one with the same polarity as the device that forms the handle, the other in the opposite direction. A soliton generated in the handle will trigger a moving soliton in only one of the fork devices. In the other, the polarity of the bias is such that it pushes the incurred change in ϕ back toward the branch point where it is dissipated.

The basic problem with this device is that the input, the bias current I_g, is a current level whereas the output is a pulse. But applications can be found in regularly structured circuits such as decoders, etc.

The device can be extended to function as a flux flow device in a manner similar to that of a transistor [137,138]. In this case, a control current is applied and the device is switched into the voltage state. The control field produces solitons that move along the junction and accelerate. An increase in gate current is translated into an increased acceleration at a velocity that gradually approaches $\bar{c}$. Thus a current step develops at a voltage determined by the time averaged voltage of the pulses associated with the moving solitons. As the control field is increased, more solitons are generated per unit of time and the voltage at which the current step occurs shifts gradually to higher values. Since voltage and current in an intersecting load are given by the position of the step, one can design inverting analog amplifiers [138]. The problem with these devices is presently their requirement for a low output impedance.

Shift register applications based on the phased movement of vortices along long junctions [139] or linear arrays of interferometers [140] are being studied and extended to the design of counters [141] and phased logic networks [142,143]. Unfortunately, up to now, experimental verification of the high speed potential of these circuits is lacking.

Logic circuit designs have been mostly confined to the class of AC powered switches previously described, and, in memory, DC powered loop logic is used. Such circuits can be extended to perform specialized logic applications or can be used as building blocks for Programmable Logic Arrays, or PLAs. In a PLA, logic is performed not through the interconnection of a seemingly random assembly of ORs, ANDs, etc., but in ordered, memory-like arrays. Typically, information is released from latches into a set of decoders that feed the inputs of an array of wide AND circuits which in turn feed the inputs of a structured array of OR circuits.

The advantage of the PLA is that non-personalized chips can be fabricated in numbers. The logic to be performed is then determined by personalizing the last level which specifies the connection points to the circuits. This greatly reduces turnaround time in making masks for a (usually large) number of levels and, most importantly, saves considerable time when errors are detected and changes have to be made in a given chip design.

Josephson PLAs have been proposed [144]. These use arrays of a read only memory (ROM) design which has been experimentally tested in part [145,146]. The circuit functions are performed with loop logic similar to the one described in the context of the Josephson cache design.

A promising novel device is the so-called QUITERON [215]. It is composed of three thin superconducting electrodes made with materials that have different gap voltages. These form a sandwich of two tunnel junctions. Heavy injection of quasiparticles by means of a current fed into one junction causes the gap of the center electrode to vanish if a given critical power density is reached. In this non-equilibrium state the I-V characteristic of the second junction changes and an inverting transistor-like switch can in principle be built. Significantly, the QUITER-ON is composed of Josephson junctions, but its operation is not based on the Josephson effect.

> No man would start to do anything, if
> he did not expect to reach some end.
> *Aristotle, Metaphysics*

V. PROCESS ASPECTS AND CHIP FABRICATION TECHNOLOGIES

A detailed review of material issues and fabrication techniques in Josephson technology is available [147]. We shall therefore only highlight basic aspects, but will include recent accomplishments on the chip level. This work was based on lead alloys [148,149,150], despite continuous efforts by technologists to introduce a harder and more reliable niobium-based technology. Reluctance to do so came from the circuit designers because of the relatively large specific capacitance of niobium junctions. However, recently proposed edge structures with small area junctions make niobium circuits attractive from a circuit and system point of view.

A. LEAD ALLOYS

1. The Vertical Structure

A typical cross-section through the vertical structure surrounding a junction is shown in Fig. V-1. The structure is placed onto a silicon wafer with a thin surface layer of SiO_2. The first metal layer is a $0.3\mu m$ thick Nb groundplane ($\lambda \simeq 850$Å) which is anodized to form a 350Å thick insulating film of Nb_2O_5 ($\varepsilon=30$). Except in areas where contact must be made to the groundplane, the structure is covered with an evaporated film of SiO ($\varepsilon=5.8$). Its thickness is $0.145\mu m$ and $0.275\mu m$ for logic and memory designs, respectively. This is followed by a 300Å thick resistor layer of $AuIn_2$ ($2\Omega/\square$) [151,152]. All the SiO and Pb-alloy layers are patterned with a photoresist stencil lift-off technique while conventional lithography is used for Nb and Nb_2O_5.

The junction base electrode, also used as the first wiring level, is a $0.2\mu m$ thick deposited superconductor composed of $Pb_{0.84}In_{0.12}Au_{0.04}$ [148,153] ($\lambda \simeq 1400$Å). It is followed by the evaporation of a contact layer of the same composition and is covered with $0.275\mu m$ of SiO that defines the junction area. In memory, a second identical SiO layer, used on top, permits the increase of interferometer inductances. The approximately 60Å thick tunneling barrier is grown in an rf or plasma discharge [154] in which, as a result of a developing DC potential, ions are accelerated toward the sample and sputter-etch the oxide surface during oxide growth [155,156]. In simple systems, such as pure Pb, a final time independent oxide thickness is obtained when the growth rate and the etching rate become equal.

The counterelectrode is $0.4\mu m$ thick ($\lambda \simeq 2000$Å) and composed of a $Pb_{0.71}Bi_{0.29}$ alloy [148,157]. It is followed by a $0.1\mu m$ thick counterelectrode protection layer and a $0.5\mu m$ thick SiO layer over which the control line layer is evaporated. The control line layer has the same composition as the base electrode. Its thickness is $0.8\mu m$ ($\lambda \simeq 1200$Å) and it also serves the function of a second wiring level. Finally, the finished chip is protected with a $2\mu m$ SiO layer.

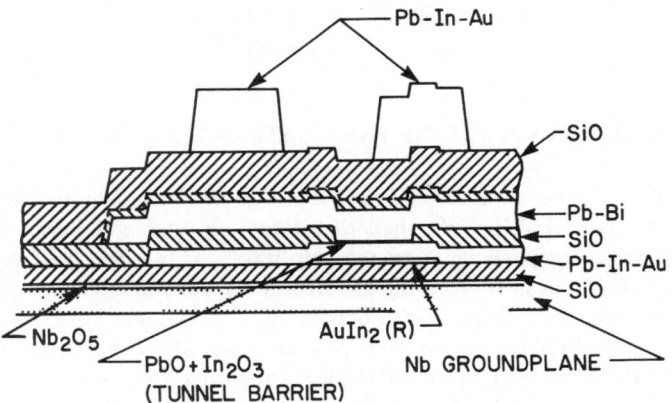

Fig. V-1:   Cut through the vertical structure of a lead alloy device. The groundplane is made of niobium and insulators are evaporated SiO. Photoresist
lift-off stencils are used to define the shapes of the various layers.
[From reference [148], © 1980 IBM J. Res. Develop.]

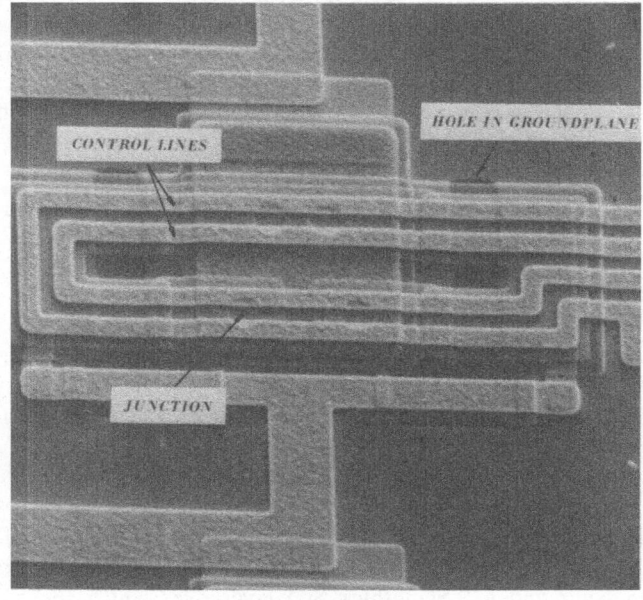

Fig. V-2:   Planar three-junction logic interferometer. Most of the device inductance is obtained through holes in the groundplane. Resistors to damp
resonances are connected between the outer junctions and the double
center junction.

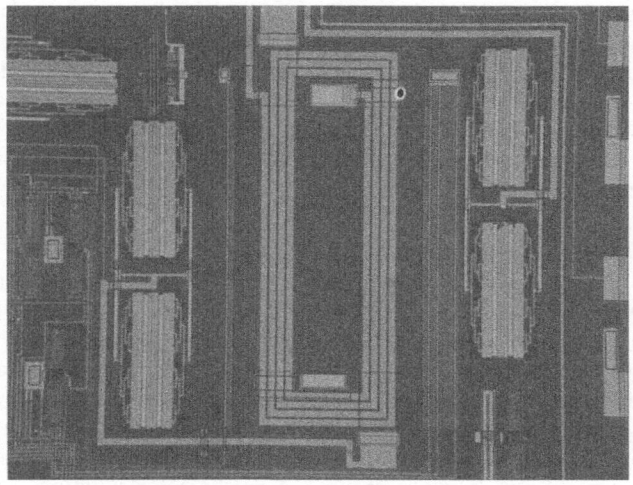

Fig. V-3:   Thin film power transformer. The primary and secondary lines are very efficiently coupled through a hole in the groundplane.

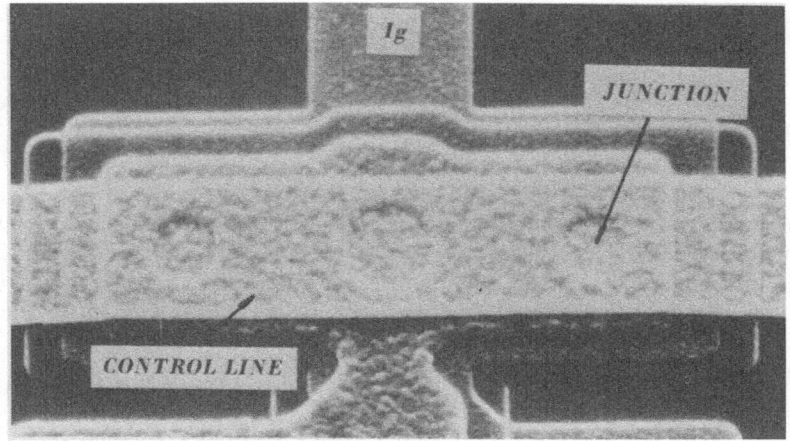

Fig. V-4:   Bridge type three-junction memory interferometer. Because of density requirements in memory, these undamped devices are preferred. Resonances are accommodated in the design of these DC powered circuits.

The basic reason why lead-alloys were chosen was that good quality junctions ($V_m>25$mV) could be made with reasonable control [158]. The technology was developed, became a standard, and so far has remained the basis of all circuit work primarily because of the small specific capacitance of its junctions. Junction failures occur during thermal cycling between room temperature and 4K because of a rupture of the tunnel barrier oxide [159,160,161,162]. Such failures were initially frequent but significant improvements have been made over the years [163]. Cycling of large area junctions, equivalent in junction area to about one half million three-junction logic interferometers, indicated an onset of first failures only after 400 thermal cycles [164,147].

2. Some Device Configurations

A photomicrograph of a three-junction logic interferometer is given in Fig. V-2. As mentioned in section III-D, logic interferometers require damping resistors across the device inductances to reduce resonance step amplitude [54]. This is difficult to achieve in a bridge-type interferometer as shown in Fig. III-6a. For this reason, the interferometer loop, instead of being closed over the base electrode, is looped back on the side, over the groundplane. In these devices, the large center junction is made out of two adjacent 2.5μm diameter junctions so that all junctions have the same dimensions. To decrease the device size, the loop inductance is predominantly obtained through holes in the groundplane placed beneath the wings of the interferometer. Current is fed into the structure through a resistive divider at the midpoint of the inductances as shown in Fig. III-6c. Minimum linewidths are 2.5μm.

Holes in the groundplane are also used to build integrated AC power supply transformers as depicted in Fig. V-3. This particular transformer is the first stage of the power distribution circuit that was discussed in section IV-A-5. As indicated in Fig. IV-6, such transformers will ultimately be placed on the module from where the power is fanned out in a star-like manner, first to the second stage transformers and then to the chips.

In memory applications, resonances play a less significant role. Switching into resonances is allowed in the write gate of the cell and resonance amplitudes are sufficiently small in asymmetric interferometers [165] such as the sense gate. The photomicrograph of a three-junction bridge type write gate is shown in Fig. V-4. However, some auxiliary circuits use damped planar logic-type interferometers.

3. A Sub-Nanosecond Cycle Time Chip

To test the capability of 2.5μm linewidth Josephson logic on a chip level, an encoding circuit with 10 levels of shallow logic and 12 latches was designed with CIL circuits [166]. The latches of the type described in section IV-A-3 provide true and complement signals to perform the predominantly required EXOR function in the logic. The maximum number of logic levels, including the gating functions in the latches, is five. The circuits are powered with a regulated AC supply. Key signals that determine the encoding function are applied quasi-statically from the outside.

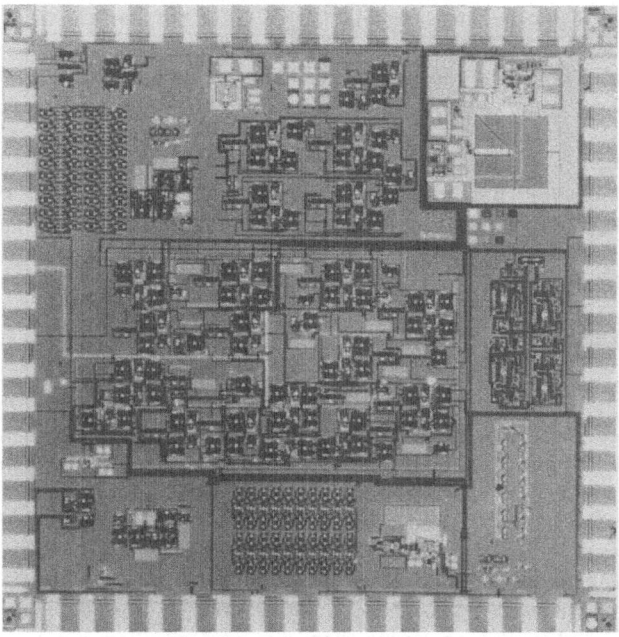

Fig. V-5: A sub-nanosecond cycle time enscription circuit designed with $\simeq 100$ gates is shown in the center of the chip. It is surrounded by other exploratory circuits. The ability of 2.5μm linewidth Josephson logic circuits to perform at cycle times as low as 670ps was demonstrated at a chip power dissipation of 350μW. [Reprinted from the ISSCC 82 Digest of Technical Papers, © 1982 IEEE and with permission of the publisher]

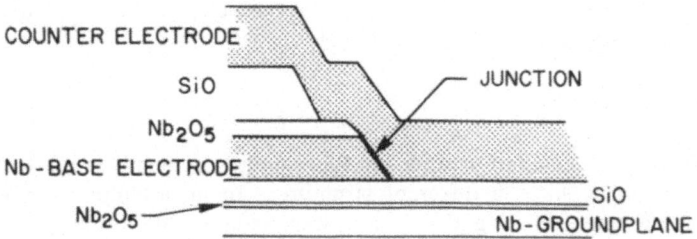

Fig. V-6: Cross-section through an edge junction. Such devices have a very small junction area. They permit the use of a niobium based technology without the performance penalty that would otherwise result from the increased specific junction capacitance of niobium.

The chip is shown in Fig. V-5. Only the portion near the center, approximately one quadrant of the chip, is the encoding circuit itself. Other unrelated test circuits are placed at the periphery. The encoding circuit contains 102 logic gates designed with 177 devices. It performed the proper logic function at a cycle time as low as 665ps with a total chip power dissipation of 350μW. One should note that this is the total cycle time which includes the power transition time t_p. It is a remarkable achievement and demonstrates the potential for sub-nanosecond chip cycle times with circuits designed in a 2.5μm technology. To the best of our knowledge, such a cycle, corresponding to a bit rate of 1.5 Gbits/sec, has never been achieved before.

B. NIOBIUM

Niobium is a much stronger metal than lead and a niobium junction hardly ever failed during thermal cycling. But being a high melting point material, it has drawbacks. For instance, it is a strong getter and high quality junctions are difficult to fabricate [167,168,169]. Its biggest drawback, however, is the high dielectric constant ($\varepsilon \simeq 30$) of its most interesting oxide Nb_2O_5. If used as an insulator in transmission lines, time of flight delays become over a factor of two larger than in the same lines over SiO. However, the real problem centers around the specific junction capacitance [170] which, considering the spread in the data, could be as high as a factor of four larger than that of lead-alloy junctions. This results in a corresponding increase of the circuit risetime and translates in present 2.5μm logic circuits into a gate delay increase by roughly a factor of two. Memory circuits have delays that scale with $(LC)^{1/2}$. Consequently, they will also operate at about half speed. In addition, a large delay is incurred in the lengthening of the power transition time, to keep the punchthrough probability sufficiently low.

If one takes device stray capacitances into account and considers a Pb-alloy system in which the circuit delay, the package delay and the transition time are in the ratio of 0.4:0.4:0.2, an increase in junction capacitance by a factor of four leads to an increase in total cycle time by approximately a factor of 1.7. A power supply with dwell, see section III-D-5, will improve the situation but even then, the system's performance will be degraded by over 30% and the design of a power system with dwell is presently a cumbersome and costly proposition. One way to remedy this problem is to consider very small area junctions which can be fabricated without further taxing the established 2.5μm technology.

1. Niobium Edge Junctions

To decrease the junction area by four implies a diameter of 1.25μm in a circular junction or requires a different structure. In an attempt to fabricate extremely small normal tunneling devices, it was demonstrated that one can form tunnel junctions [171,172] and ultimately Josephson junctions [173,174,175] at the edge of the base electrode. A cross-section of a Josephson edge device is illustrated in Fig. V-6. Since the thickness of a 0.2μm thick metal film can be very well controlled, a 2.5μm wide edge junction device holds the promise of a nearly sevenfold decrease in active junction area. This means that the junction width can exceed the minimum linewidth and that, therefore, the use of niobium at the same performance level as Pb-alloys leads in principle to a better junction area control.

Edge junctions with niobium base and lead counterelectrodes have been made with reasonably high V_m [174]. A blanket Nb base electrode is deposited. The Nb is wet anodized except where one needs to fabricate contacts. Then the base electrode shapes are delineated with photoresist and the structure is plasma etched. Since the etch rate of photoresist is larger than the etch rates of either the Nb or the Nb_2O_5, a slanted edge is formed. The tunnel barrier is grown on this edge and the junction is completed with a Pb or Pb-alloy counterelectrode. However, a considerable effort is needed if one wants to bring Nb-edge junction technology to the same technological level as is now achieved in Pb-alloy technology.

2. Alternate Barriers

Alternate low capacitance barriers that permit the use of planar junctions are being investigated. An interesting process is based in a deposited Si barrier. The Selective Niobium Anodization, or SNAP, process consists of a blanket deposition of a Nb base electrode followed by blanket depositions of a Si barrier and a Nb counterelectrode [176,177]. This means that a large junction extending over the entire wafer is made first. It can be tested and only those wafers that have good quality, high V_m junctions near the design value of the threshold current may be processed further. The junctions are then delineated by oxidizing the top electrode through its thickness. Simple direct current injection logic devices have been successfully made with this process. Measurements indicate a specific junction capacitance smaller than that of Pb-alloy junctions.

Similar methods involving the deposition of subsequently oxidized aluminum films as tunnel barriers are also being studied [178]. Here too, low junction capacitances can be expected. In both methods, the crucial question will be whether junctions with sufficient threshold control and high V_m can be realized with high yield.

> The image of it gives me content already; and I trust it will grow to a most prosperous perfection.
> *Shakespeare, Measure for Measure*

VI. PACKAGE TECHNOLOGY

A. BASIC REQUIREMENTS

In developing the technology for a future supercomputer, it is necessary that package delays be decreased in the same proportion as the circuit delays [10,179,180,181]. In a planar package, for a given circuit density and chip size, such a decrease is only possible until the chips touch. For any further improvement, we need to exploit the third dimension. Present $2.5\mu m$ superconducting lines over SiO propagate signals at a velocity of about $c/3$. Therefore, once we consider cycle times of the order of a nanosecond, the longest path length traversed in one cycle cannot exceed a distance of the order of 10cm.

To develop a miniaturized three-dimensional packaging technology is a challenge that equals that of developing new high speed logic and memory circuits [182,183,184,185]. The package parts must not only be very small three-dimensional structures, but must be designed such that they can be taken apart and reassembled to permit system bring-up, engineering changes and servicing. A schematic sketch of the present approach is shown in Fig. VI-1. Chips are attached to one side of cards which carry two orthogonal levels of transmission lines for interchip connections. Some of these lines are carried downward to a set of microscopic pins that interconnect with lines on a wiring module to exchange signals between cards.

Let us now evaluate the advantage of such a card-on-board approach over that of a planar package. The relevant dimensions are the card length D_w and height D_h and the card interdistance D_c. If there are M cards, the total area available for chip placement will be MD_wD_h and the maximum length a signal can travel (from corner to opposite corner) in a two-dimensional package with the same number of chips is $2(MD_wD_h)^{1/2}$. Note that the square two-dimensional package is always assumed optimized.

We shall not be concerned with the granularity resulting from a finite number of chips and assume continuous functions. But we must take into account that in present semiconductor systems, such as the 3081, the chips are about a factor of two further apart than necessary from simple size considerations. As a result, we define the maximum two-dimensional path as $X_2(Jos)$ for Josephson chips and $X_2(Sem) \simeq 2X_2(Jos)$ for semiconductor chips. The maximum path length in the card-on-board package must be taken from point A to point B. We have, therefore, $X_3 = (M-1)D_c + 2D_h + D_w$. Ratios of X_2/X_3 are shown in Fig. VI-2 for various relative sizes of package parts. The parameter $g = D_c/D_h$.

Let us first concentrate on the curves drawn in full. These correspond to the single sided board with single sided cards as shown in Fig. VI-1. No attempt was

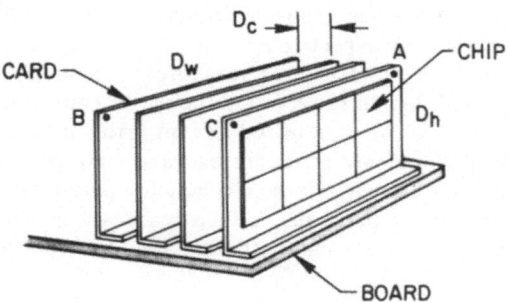

Fig. VI-1:   Concept of a Josephson card-on-board package. Because of the small
power dissipation of the circuits the cards can be tightly spaced. Cards
are interconnected through wiring modules beneath the board.

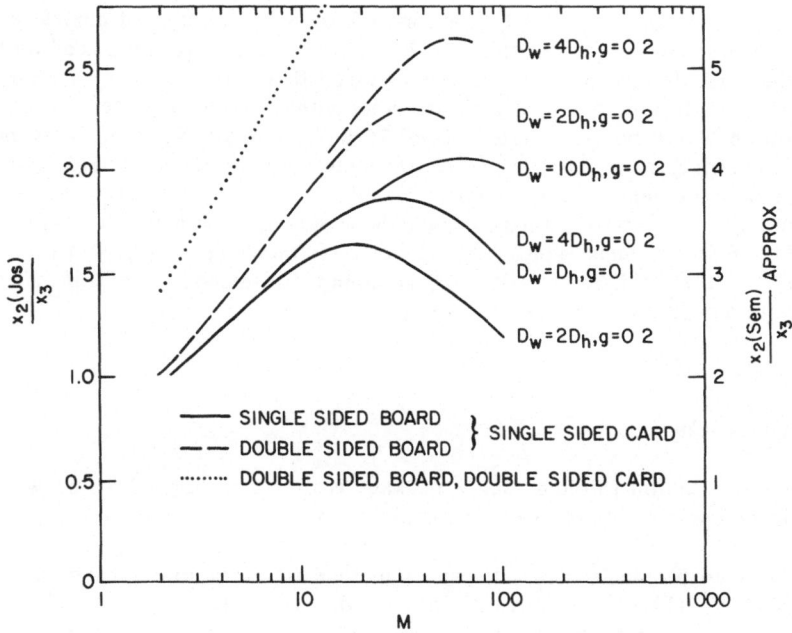

Fig. VI-2:   Time of flight delay advantage of a three-dimensional Josephson
package X_3 over a two-dimensional Josephson package $X_2(Jos)$ and a
two-dimensional semiconductor package $X_2(Sem)$. In the latter, because
of other design constraints, chips must be kept further apart than is
necessary from simple geometric considerations.

made to optimize the three-dimensional package. The reason is that the relative sizes will be dictated by other design considerations such as pin densities, etc. For a single card (M=1), one should expect that both X_2 and X_3 be equal since we are dealing in both cases with two-dimensional structures. This is not the case because of our definition of the path length for X_3. On the single card, it is taken as the long path from A to C. For small M, this introduces an error in favor of X_2. As M is increased, the advantage of the card-on-board package increases by a factor of three to four over comparable path lengths in semiconductor packages and up to a factor of two over a planar Josephson package. It then reaches a maximum beyond which the card-on-board package becomes simply too long and far from optimum.

From these results, one would tend to conclude that the performance advantage of the proposed three-dimensional structure over a two-dimensional Josephson package is too small to warrant the effort. This is especially true for a package with a small number of cards or in a large package for path lengths involving small circuit islands.

Although such time of flight improvements are far from being negligible, one must realize that the structure we just described is only the first step toward the ultimate need of this technology. The situation is improved by placing the cards on both sides of the board, top and bottom, as indicated by the dashed curves in Fig. VI-2. An even larger improvement (by $\sqrt{2}$) is obtained if chips are placed on both sides of the card, dotted curve. Here it is assumed that chips communicate through vias in the card. In a moderately large assembly with 20 such cards, ten on top, ten on the bottom of the board, $X_2(Jos)/X_3=3.25$ and $X_2(Sem)/X_3 \simeq 6.5$. This means that as we approach a true 3-D structure, we tend nearer to the postulated order of magnitude improvement. More importantly, if we set $D_w=3cm$, $D_h=1.5cm$ and $D_c=0.3cm$, the maximum distance through the assembly is <9cm. It could easily contain 320, 6.35mm chips, which with 1K logic circuits/chip represent a processor of the size of the IBM 3033 [180], or, assuming 4K memory bits/chip, a cache memory with a capacity in excess of 128Kbyte.

B. PACKAGE DESIGN

1. Mechanical Construction

A cross-section through such a package is drawn in Fig. VI-3 and we shall describe its components by following a signal line through the various parts.

To avoid problems with differences in thermal expansion, all substrates are made of silicon. The chip is provided with In, Bi, Sn solder contacts on the circuit side. Matching solder contacts exist on the card. To attach the chip, it is flipped over and placed onto the card and the solder is reflown [186]. On the card, such contacts connect to 5μm wide Pb-alloy ($Z_o=12\Omega$) transmission lines on 10μm centers. Because of the high wiring density, only two orthogonal wiring levels are required. The groundplane is Nb.

Signal lines that must communicate with another card are brought to the bottom, where a right angle connection is made to a line on the foot [187]. This

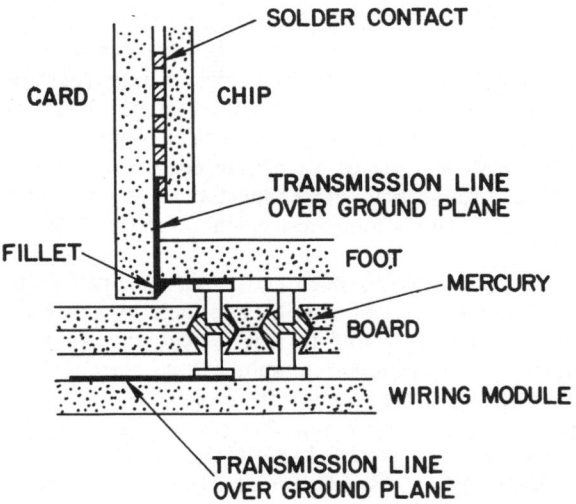

Fig. VI-3: Cross-section of a Josephson package. Chips are soldered to cards and cards are interconnected to wiring modules through terminated 12Ω transmission lines via microscopic Pt pins. Mercury is used as a low temperature solder to ensure pin contacts that are pluggable at room temperature.

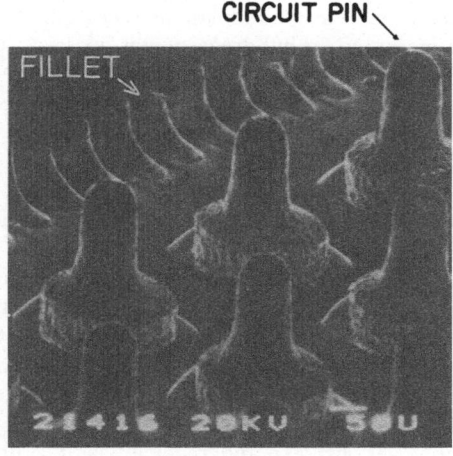

Fig. VI-4: Microphotograph of contact pins on the foot of a card. The pins have a height of 0.2mm and are spaced 0.3mm apart. [From reference [185], © 1982 IEEE]

connection is called a fillet; it is a In, Sn solder joint that also provides the mechanical attachment between card and foot. Similar electrically inactive fillets (not shown) are provided on top of the foot. The fillets are on $76\mu m$ centers and alternately connect to signal lines and ground. On the foot, the lines connect to Pt pins that are batch fabricated with a metal mask using electro-discharge machining, EDM. These pins are 0.2mm high and placed on 0.3mm centers as shown in the photomicrograph of Fig. VI-4. Active lines are connected to active pins and groundplane connections through ground fillets to nearest neighbor groundpins.

The pins form a pluggable connector to a non-personalized board. It is made of two pieces of Si into which pyramidal holes are made by anisotropic etching. The two pieces are fused together so as to form open ended cavities which are filled with a drop of mercury. Similar pins on wiring modules that carry 12Ω transmission lines to other cards are inserted into the mercury from the other side. The concept of the wiring module permits engineering wiring changes without having to replace large package parts. As this package is cooled to 4K, the mercury freezes and effectively solders the pins together. Yet at room temperature, the entire structure can again be disassembled. The thickness of the silicon parts in present experimental assemblies ranges from 0.2mm for the card to 0.6mm for foot and wiring module.

2. *Electrical Properties*

Signal lines throughout the package are matched 12Ω transmission lines, except when crossing from the card to the wiring module. The solder connection between the chip and the card are able to carry Josephson logic signals without significant distortion or delay. But there are unshielded regions through the pins and at the fillets under which the groundplane is recessed. Despite the fact that the parts have dimensions measured in hundreds of μm, or less, sufficiently large inductances are formed to impede the propagation of signals with risetimes as those found on the chip [188,189]. Experiments performed on early package parts were compared with three-dimensional inductance calculations made on these structures [188]. Good agreement was found. By applying such calculations to the present connectors, one finds that the inductances of fillets and pins range between 100 and 200 pH respectively, even though care is taken to carry ground connections in nearest neighbors.

To avoid reflections, padding capacitors are placed between these inductances and ground. As a result, the discontinuities are matched to the 12Ω lines. To avoid ringing, this necessitates that the signal risetime $t_r \geq 3L_x/Z_0$ as mentioned in section IV-A-1. With $L_x = 200pH$ and $Z_0 = 12\Omega$ it becomes necessary to slow the risetime to about 50ps. This is achieved through matched filters in the off-card drivers. The delay through the matched sections, given by the L/Z_0 time constant, is not excessive, however, drivers and receivers add a noticeable delay.[85] Consequently, in terms of speed of light distances, the cards are further away from the wiring module than indicated by the package dimensions. However, the relatively long risetimes required for card crossings have the advantage that crosstalk between fillets and between pins is not a concern [190].

> Grau, teurer Freund, ist alle Theorie.
>
> *Goethe, Faust*

VII. A SYSTEM LEVEL EXPERIMENT

A. DESCRIPTION OF THE EXPERIMENT

The aim of the experiment [191] was to demonstrate that the existing circuit technology can operate in a package as described in the previous section.   The circuits mapped a single bit of a worst case path of a signal processor with a one cycle cache access [179].

The path, illustrated in Fig. VII-1, starts at a register (in this case a single latch), traverses ten levels of logic, activates a driver to send the signal to a cache memory support chip that receives it and drives the information to a cache chip. Since an operating cache memory was not yet available, the cache chip simulates the access time with a 700ps long delay line and sends it back to the support chip. From there it is sent to the logic chip, received and stored in a register, again a single latch.

All CIL circuits required for this experiment were fitted into one quadrant of a 6.35mm chip and were copied into all four quadrants. This gave sufficient redundancy to be able to select the three required chip functions among four chips placed onto two modules.   The personalization that selected operating quadrants was realized with an appropriately chosen wiring module.   The circuits were powered with an AC supply as described in section IV-A-5, although some shedding of the simulated full load was necessary because of accidental problems with the current carrying capability of superconducting contacts. A card is shown in Fig. VII-2 and the full experiment with its two cards in Fig. VII-3.   The cards have a size of only 15x14mm.

The logic delay through the entire path was measured to be 2.6ns which, with a power supply transition time of 1.1ns, gave a total cycle time of 3.7ns. This at a chip power dissipation of <5mW per quadrant.   The logic circuits included all members of the CIL family described in section IV-A-3. These were fabricated in lead alloy technology as discussed in section V-A.

B. CRYOINSERT, I/O, AND MAGNETIC SHIELDING

To communicate with the outside world, the two cards, the board and the wiring module of the system level experiment, were placed onto a 2.54cm diameter circular silicon adapter. Contact was made through the board to pins on the adapter situated to the left and right of the cards. A rectangular hole in the adapter permitted the insertion of the wiring module. The assembly was placed into a cryoinsert [192], constructed with specially selected materials to minimize magnetic stray fields. Inputs and outputs were carried over 50Ω copper strip lines, 0.18mm wide on 0.51mm centers, placed onto a thin polyimide with a copper groundplane on its underside. At the lower end, this I/O cable was split and configured in such a way that signals could be brought in and out from both sides of the adapter as shown in Fig. VII-3. The I/O lines with a bandwidth of DC to $\simeq$1GHz were terminated on

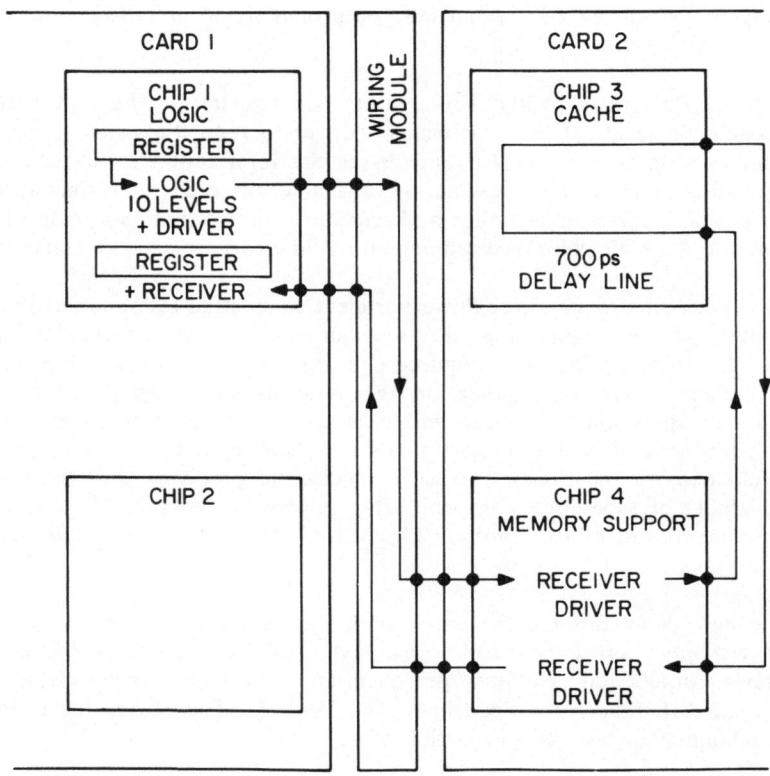

Fig. VII-1: Block diagram of the systems test vehicle. Three of four chips on two
cards interconnected through a wiring module demonstrate the operation
in Josephson technology of a one bit wide worst case path through a
signal processor.

Fig. VII-2: Photograph of a card used in the Josephson systems test vehicle. Pins and solder pads for chip bonding can be clearly seen. [From reference [185], © 1982 IEEE]

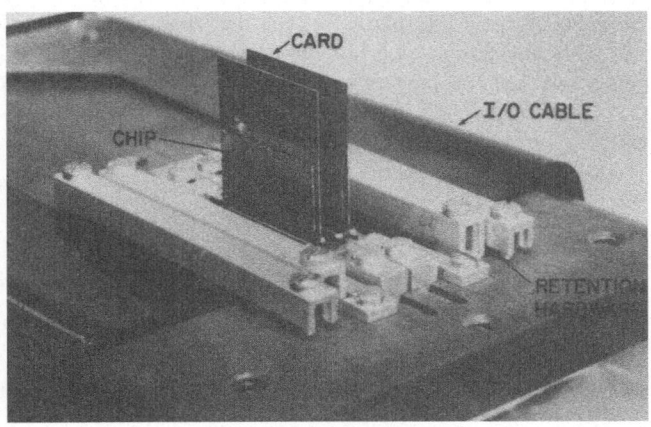

Fig. VII-3: Assembly of the Josephson systems test vehicle on the tip of its cryoinsert. I/O strip lines on a polyimide cable are connected from both sides to the adapter (not shown) on which the assembly rests. The experiment demonstrated that in Josephson technology the chosen worst case path is compatible with a system cycle time of 3.7ns. [From reference [185], © 1982 IEEE]

the adapter. A signal with a risetime of 200ps traverses such cables with 15% attenuation and a risetime degradation to 400ps. The heat loss is 1.7mW per line [192]. To support the assembly, the experiment and the retention hardware were attached to a pyrex support tube and inserted into a vacuum tight pyrex housing.

Care must be taken to shield the earth's magnetic field. For this reason, the experiment was placed into a set of four μ-metal shields containing compensating coils on the inside. This permitted operation in an environment with a remnant magnetic field of about $3\mu G$. In addition, the devices on-chip were surrounded by so-called moats, slots in the groundplane, which collect moving trapped fluxoids [193] and prevent them from reaching the devices. The problem is that superconductors in general, and superconducting groundplanes in particular, are rather imperfect diamagnets. If cooled down to 4K in a magnetic field, they tend to trap quantized circulating currents which produce localized magnetic fields. In fact, even in zero external field, if the cooling rate is too large, electric currents caused by temperature differences in the conductors can contribute rather large magnetic fields. Further studies are required to find adequate methods for shielding and cooling a large Josephson computer.

The system level experiment was performed by ultimately immersing the package into liquid helium in an open dewar. A Josephson computer could not realistically function in such an environment. A closed cycle refrigerator will be needed. Refrigerators are being studied for this purpose [194]. In the range of a few watts of dissipation in helium, an engine is likely to contain a compressor feeding two expanders which cool the gas first from room temperature to about 60K, then to 15K. Finally, a Joule-Thomson valve reliquifies the helium which evaporates at a rate 1.4 liters/hour for every watt dissipated in the dewar. Such systems would contain a reservoir of liquid helium of a few tens of liters, so that in case of a power failure the machine stops, but the content of Josephson memories is retained [195].

Sane sicut lux seipsam et tenebras
manifestat, sic veritas norma sui et
falsi est.

Spinoza, Ethics

VIII.   CONCLUSION

Josephson technology has come a long way during a decade of determined effort [196,197,101,102,10,198,191] and yet considerably more work is needed before one can expect to see superconductive computers compete with and overtake semiconductor systems.   The achievements will have to proceed at a steady pace; ultimately any new technology has a time window.

Superconductivity is not new; in fact, its discovery preceded semiconductors by three decades.   It was slow in establishing itself mostly because it took nearly fifty years before the necessary theoretical understanding of its microscopic origin was finally reached.   In the second half of this century, everything came together: the BCS theory, the tunnel junction, the Josephson effect, all Nobel prizes for increasing the fundamental understanding of nature.   These were shortly followed by proposals of circuit applications and a realization of the advantages of a Josephson computer technology [199].   However, the technology needs liquid helium to operate its circuits.   As discussed in section II-D, the low operating temperature and the ability to use superconductivity give this approach many fundamental advantages over room temperature electronics.   Yet it is often felt that the need for a cryogenic environment constitutes a disadvantage.

First, there is a concern about the safety aspects related to the use of liquid helium.   It is perceived to be dangerous as a result of its low boiling temperature. Such fears are of course subjective and will have to be alleviated by pointing out that the gas itself is inert and that its use in liquid form is based on an established technology that has developed simple and effective safety mechanisms.   Nevertheless, it may be of advantage to select initial customers among those that show little concern about liquid helium on their premises to prove that a closed cycle liquid helium refrigerator is a perfectly acceptable machine.

A second objection often heard is that, because of the need for a refrigerator, a Josephson computing system will be invariably expensive.   This is not necessarily the case.   The price of a 10W helium refrigerator lies today between $100K and $150K, and costs are expected to decrease as such machines are mass produced. A 10W Josephson machine promises to be a very powerful computer [11] as compared to present semiconductor systems.   In such a machine, the expense of the refrigerator will be traded against savings incurred from the lack of large power supplies and of chip cooling gear.   In semiconductor systems, the cost of power supplies scales linearly with system power and represents 10 to 20% of the cost of the machine [200].   In addition, the hardware necessary to distribute air or liquid coolants to the chips and the equipment required to extract the heat from the system, are not negligible expenses.

It has also been said [201] that Josephson technology simply transposes the power problem and that the effective power required for the circuits is considerably higher than claimed. This is correct, yet totally irrelevant. It is true that a 10W refrigerator requires about 2kW for every Watt dissipated in the cryostat. However, as we have amply shown in the foregoing, to achieve ultra-fast computing with large systems necessitates low power dissipation in the machine itself. The total system power is of little concern to the hardware designer of a CPU or of a cache memory. But even with respect to total power, Josephson technology is competitive with semiconductors. The hypothetical 10W machine would at the power outlet require an estimated 20kW, just about the same as the 23kW needed for the IBM 3081-D16 processor.

There is one problem that results from the need for liquid helium. It stems from the fact that distances, and therefore delays, into and out of the dewar are large. For this reason, it proved difficult to find applications which permit the inclusion of a few Josephson chips into a semiconductor system to improve its performance. Instead, it rapidly became clear that one had to develop all aspects of an entire system without the ability of taking advantage of existing hardware, except in the tools required to fabricate Josephson parts. From the start, one had to think of new approaches in every respect.

It was necessary to invent an entire family of logic circuits, memory circuits, RAMs, ROMs, and a totally new package. Work was required on dewars, magnetic shields, input-output lines and circuits that interface with the rest of the world. Basic materials had to be selected and these had to yield working devices rapidly or else the circuit designers could not make progress. Lead was chosen because one was able to build circuits that could operate at 4K in open helium dewars and although these early circuits often failed during thermal cycling, circuit design could proceed.

Progress was made, the thermal cyclability was improved and early fears centering around the necessity of growing a stable tunnel oxide, only a few tens of Angstroms thick, have now turned into optimism. In fact, all major circuit and system demonstrations were made in lead alloy technology, both at IBM and elsewhere. With the advent of edge junctions, niobium has the potential of producing high performance circuits. The problem of its large specific capacitance is circumvented by using a small junction area. But it is not a simple substitution. Such a technology must first be developed and must prove to be at least the equal of lead in all other aspects. Investigations are under way.

With the successful operation of the system test vehicle, the technology has reached a certain maturity in the laboratory, but not from a manufacturing point of view. One needs to prove that chips and package parts can be made again and again, with some acceptable yield. Preparations are underway at IBM to attempt this in a development line situated in East Fishkill. The problem centers mainly around the reproducibility of the device threshold, but other questions are still to be resolved as well.

To substantiate the promises of the technology, the full operation of a Josephson prototype machine is required. It must contain all aspects of a large system, but be neither too large nor too small. In the first case, it may be overly difficult to achieve and in the latter it may not prove the point. The construction of such a prototype is planned at IBM [179], notwithstanding the fact that by prototype one does not imply the first of a series of machines. It is simply a vehicle to demonstrate the technology.

We mentioned that much work still has to be done. Even a company as large and determined as IBM finds it very difficult to revolutionize the electronics field with such a new approach. A higher degree of involvement at universities and other industries is needed. The problem is not unlike that encountered by the transistor in the early 1950s. A quote [6] is appropriate: "Thus the transistor had something of a mixed reception ranging from wild enthusiasm to open hostility. The enthusiastic were largely those who did not appreciate the complexity of the problems. Neither, probably, did the hostile, but they were also unaware of the vast potential of the device. In the middle was a silent minority which realized how much work needed to be done and also had some inkling of the possible rewards." In 1952, Bell held a symposium at which it revealed the state of the art and it did its utmost to stimulate others to play a part in the development of semiconductor devices. With its open publication policy which has now been practiced for many years, IBM has done the same in Josephson technology.

Is it because one must design every aspect of an entire system that widespread activities are so difficult to seed in this technology? Probably not; it applies only to computing systems. Small scale Josephson applications are waiting to be exploited. SQUID magnetometers are being developed for a variety of applications [202]; Josephson samplers with resolutions of a few picoseconds have been realized [203,204,205] and await development into a product, A/D converters are being studied [206,207,208,209,210,211], and detectors and mixers are being investigated [212]. These need not be encumbered with large closed cycle refrigerators. Instead, such studies should be coupled with ongoing efforts to design small refrigerators [213]. In fact, one may ultimately be able to build a refrigerator into a chip [214].

IBM, and presumably a number of Japanese companies, are interested in Josephson computing systems, and that is where the largest challenge lies. Whether Josephson technology will be a technical and commercial success is still an open question. But we must try to succeed, since there is no other approach known that promises a quantum leap in performance over a semiconductor technology that is starting to feel its limits.

ACKNOWLEDGEMENTS

I would like to thank J. H. Greiner, E. P. Harris, C. J. Kircher, S. Lahiri, J. C. Logue, M. J. Marcus, E. Shapiro and B. van der Hoeven for their suggestions concerning the content of this paper. Special thanks go to J. Meyer and B. J. Juliano for their help in assembling and organizing the material.

REFERENCES:

1.   R. D. Levine, *Scientific American* **246** #1, 118 (1982).
2.   D. Schmandt-Besserat, *Scientific American* **238** #6, 50 (1978).
3.   H. H. Goldstine, The Computer from Pascal to von Neumann, Princeton University Press (1972).
4.   L. M. Branscomb, *Science* **215**, 755 (1982).
5.   M. S. Pittler, D. M. Powers, D. L. Schnabel, *IBM J. Res. Develop.* **26**, 2 (1982).
6.   E. Braun, S. MacDonald, Revolution in Miniature, Cambridge University Press, Cambridge (1978).
7.   J. Worlton, A Philosophy of Supercomputing, Los Alamos Scientific Laboratory Report LA-8849MS (1981).
8.   J. Matisoo, *Appl. Phys. Lett.* **9**, 167 (1966).
9.   J. Matisoo, *Proc. IEEE* **55**, 2052 (1967).
10.   W. Anacker, *IEEE Spectrum*, **16**, 26 (1979).
11.   W. Anacker, *IBM J. Res. Develop.*, **24** 107 (1980).
12.   E. Shapiro, Digital Technology Status and Trends, **12**, 199, R. Oldenbourg Verlag, Munchen, Wien (1981).
13.   J. W. Beyers et al., IEEE International Solid-State Circuits Conference Digest of Technical Papers (1981).
14.   A. J. Blodgett, D. R. Barbour, *IBM J. Res. Develop.*, **26**, 30 (1982).
15.   R. C. Chu, U. P. Hwang, R. E. Simons, *IBM J. Res. Develop.*, **26**, 45 (1982).
16.   E. G. Brentari, P. J. Giarratano, R. V. Smith, Boiling Heat Transfer for Oxygen, Nitrogen, Hydrogen and Helium, NBS Technical Note No. 317 (1965).
17.   D. B. Tuckerman, R. F. W. Pease, *IEEE Electr. Dev. Lett.* **EDL-2**, 126 (1981).
18.   R. H. Dennard et al., *IEEE J. of Solid-State Circuits* **SC-9**, 256 (1974).
19.   J. T. Wallmark, *IEEE Trans. Electr. Dev.* **ED-29**, 451 (1982).
20.   B. S. Landman, R. L. Russo, *IEEE Trans. on Computers* **G20**, 1469 (1971).
21.   W. E. Donath, *IBM J. Res. Develop.* **18**, 401 (1974).
22.   H. H. Zappe, "Proceedings NSF Workshop on Opportunities for Microstructure Devices," Airlie, VA, Nov. 19-22 (1978).
23.   E. Flint, J. Van Cleve, *Advances in Cryogenic Engineering* **27**, to be published.
24.   D. N. Lyon, "International Advances in Cryogenic Engineering," Plenum Press, NY, **T-5**, 371 (1965).
25.   J. C. Swihart, *J. Appl. Phys.* **32**, 461 (1961).
26.   D. C. Mattis, J. Bardeen, *Phys. Rev.* **111**, 412 (1958).
27.   W. H. Henkels, C. J. Kircher, *IEEE Trans. Magn.* **MAG-13**, 63 (1977).
28.   R. L. Kautz, *J. Appl. Phys.* **49**, 308 (1978).
29.   R. L. Kautz, *J. of Research of the National Bureau of Standards* **84**, 247 (1979).
30.   T. Van Duzer, C. W. Turner, Principles of Superconductive Devices and Circuits, Elsevier, North Holland, Inc., New York (1981).
31.   For convenience, equation III-3 is written in terms of current rather than current density. This is valid since we are here assuming a point-like junction as defined later in the text.

32. B. D. Josephson, *Phys. Lett.* **1**, 251 (1962).
33. J. M. Rowell, *Phys. Rev. Lett.* **11**, 200 (1963).
34. S. Shapiro, *Phys. Rev. Lett.* **11**, 80 (1963).
35. S. Basavaiah, J. M. Eldridge, J. Matisoo, *J. Appl. Phys.* **45**, 457 (1974).
36. W. C. Stewart, *Appl. Phys. Lett.*, **12**, 277, (1968).
37. D. E. McCumber, *J. Appl. Phys.*, **39**, 3113 (1968).
38. R. C. Jacklevic, J. Lambe, J. E. Mercereau, and A. H. Silver, *Phys. Rev.*, **140**, A1628, (1965).
39. H. H. Zappe, *Appl. Phys. Lett.* **27**, 432 (1975).
40. H. H. Zappe, *IEEE Trans. Magn.*, **MAG-13**, 41 (1977).
41. W. Tsang, T. Van Duzer, *J. Appl. Phys.* **46**, 4573 (1975).
42. P. Wolf, Proceedings International Conference on Superconducting Quantum Devices, Berlin 1976 Walter de Gruyter, Berlin (1977).
43. B. S. Landman, *IEEE Trans. Magn.* **MAG-13**, 871 (1977).
44. E. O. Schultz-Dubois, P. Wolf, *Appl. Phys.* **16**, 317 (1978).
45. R. L. Peterson, C. A. Hamilton, *J. Appl. Phys.* **50**, 8135 (1979).
46. T. Gheewala, *Appl. Phys. Lett.* **33**, 781 (1978).
47. H. Beha, *Electron Lett.* **13**, 596 (1977).
48. E. P. Harris, *IEEE Trans. Magn.* **MAG-15**, 562 (1979).
49. R. L. Peterson, SQUID'80, eds. Hans-Dieter Hahlbohm, Heinz Lubbig, Walter de Gruyter & Co., Berlin-New York, 685 (1980).
50. H. H. Zappe, K. R. Grebe, *J. Appl. Phys.* **44**, 865 (1973).
51. N. R. Werthamer, *Phys. Rev.* **147**, 255 (1966).
52. I. O. Kulik, *Sov. Phys.-Tech. Phys.* **12**, 111 (1967).
53. H. H. Zappe, B. S. Landman, *J. Appl. Phys.* **49**, 344 (1978).
54. H. H. Zappe, B. S. Landman, *J. Appl. Phys.* **49**, 4149 (1978).
55. D. B. Tuckerman, J. H. Magerlein, *Appl. Phys. Lett.* **37**, 241 (1980).
56. S. Faris, E. A. Valsamakis, *J. Appl. Phys.* **52**, 915 (1981).
57. H. H. Zappe, *J. Appl. Phys.*, **44**, 1371, (1973).
58. T. A. Fulton, R. C. Dynes, *Solid-State Comm.* **9**, 1069 (1971).
59. R. E. Jewett, T. Van Duzer, *IEEE Trans. Magn.* **MAG-17**, 599 (1981).
60. E. P. Harris, W. H. Chang, *IEEE Trans. Magn.* **MAG-17**, 603 (1981).
61. A. Mukherjee, T. Gheewala, *IEDM* **5.6**, 122 (1981).
62. M. B. Ketchen, C. J. Anderson, *Appl. Phys. Lett.* **40**, 272 (1982).
63. N. Raver, private communication.
64. V. Ambegaokar, B. I. Halperin, *Phys. Rev. Lett.* **22**, 1364 (1969).
65. T. A. Fulton, L. N. Dunkleberger, *Phys. Rev.* **B9**, 4760 (1974).
66. P. A. Lee, *J. Appl. Phys.* **42**, 325 (1971).
67. E. P. Harris, private communication.
68. M. Buttiker, E. P. Harris, R. Landauer, *Bull. Am. Phys. Soc.* **27**, 267 (1982).
69. M. Klein, A. Mukherjee, *Appl. Phys. Lett.* **40**, 744 (1982).
70. K. K. Likharev, *Rev. Mod. Phys.* **51**, 101 (1979).
71. W. Y. Lum, T. Van Duzer, *J. Appl. Phys.* **48**, 1693 (1977).
72. J. Seto, T. Van Duzer, LT-13, Plenum Publishing Corp. **3**, 328.
73. C. L. Huang, T. Van Duzer, *Appl. Phys. Lett.* **25**, 753 (1974).
74. R. Ruby, T. Van Duzer, *IEEE Trans. Electr. Devices* **ED-28**, 1394 (1981).
75. D. J. Herrell, *IEEE J. Solid-State Circuits* **SC-9**, 277, (1974).
76. M. Klein, D. J. Herrell, *IEEE Trans. Solid-State Circuits* **SC-13**, 577 (1978).
77. T. Gheewala, *Appl. Phys. Lett.*, **34**, 670 (1979).

78.   T. R. Gheewala, *IEEE J. Solid-State Circuits* **SC-14**, 787 (1979).
79.   T. Gheewala, *IBM J. of Res. Develop.* **24**, 130 (1980).
80.   T. R. Gheewala, *IEEE Trans. Electr. Dev.* **ED-27**, 1857 (1980).
81.   A. Davidson, *IEEE J. Solid State Circuits* **SC-13**, 583 (1978).
82.   H. C. Jones, T. R. Gheewala, *Internat. Electr. Dev. Meeting* **CH1511**, 884 (1980).
83.   H. C. Jones, T. R. Gheewala, *IEEE J. Solid State Circuits* to be published.
84.   S. Dhong and T. Gheewala, *Appl. Phys. Lett.* **38**, 936 (1981).
85.   M. Klein, to be published in *IEEE J. Solid-State Circuits*, **SC-17**, (1982).
86.   T. Gheewala, A. Mukherjee, *IEDM Tech. Dig.* 482, (1979).
87.   T. A. Fulton, J. H. Magerlein, L. N. Dunkleberger, *IEEE Trans. Magn.* **MAG-13**, 56 (1977).
88.   J. H. Magerlein, L. N. Dunkleberger, *IEEE Trans. Magn.* **MAG-13**, 585 (1977).
89.   J. H. Magerlein, L. N. Dunkleberger, T. A. Fulton, *AIP Conference Proceedings 44* **ISSN**, 459 (1978).
90.   T. A. Fulton, S. S. Pei, L. N. Dunkleberger, *Appl. Phys. Lett.* **34**, 709 (1979).
91.   T. C. Wang, R. M. Josephs, B. F. Stein, P. L. Young, W. E. Flannery, *IEDM81* **CH1708**, 118 (1981).
92.   T. C. Wang, R. M. Josephs, B. F. Stein, P. L. Young, W. E. Flannery, *IEEE Trans. Electr. Dev.* **ED-29**, 414 (1982).
93.   S. Takada, S. Kosaka, H. Hayakawa, *Jap. J. Appl. Phys.* **19**, 607 (1980).
94.   J. Sone, T. Yoshida, H. Abe, *Appl. Phys. Lett.* **40**, 741 (1982).
95.   A. F. Hebard, S. S. Pei, L. N. Dunkleberger, T. A. Fulton, *IEEE Trans. Magn.* **MAG-15**, 408 (1979).
96.   K. Lofstrom, T. Van Duzer, *IEEE Trans. Magn.* **MAG-13**, 599 (1977).
97.   P. C. Arnett, D. J. Herrell, *IEEE Trans. Magn.* **MAG-15**, 544 (1979).
98.   M. B. Ketchen, *IEDM79* **CH1504**, 489 (1979).
99.   M. B. Ketchen, *IEEE Int. Conf. Circuits and Computers* (1980).
100.   C. J. Anderson, M. B. Ketchen, *IEEE Trans. Magn.* **MAG-17** 595 (1981).
101.   H. H. Zappe, *IEEE J. Solid State Circuits* **SC-10**, 12 (1975).
102.   W. H. Henkels, H. H. Zappe, *IEEE J. Solid-State Circuits* **SC-13**, 591 (1978).
103.   S. M. Faris, W. H. Henkels, E. A. Valsamakis, H. H. Zappe, *IBM J. of Res. Develop.* **24**, 143(1980).
104.   W. H. Henkels, *J. Appl. Phys.* **50**, 8143 (1979).
105.   P. Gueret, Th. O. Mohr, P. Wolf, *IEEE Trans. Magn.* **MAG-13** 52 (1977).
106.   R. F. Broom, P. Gueret, W. Kotyczka, Th. O. Mohr, A. Moser, A. Oosenbrug, P. Wolf, *1977 IEEE Int. Solid-State Circuits Conf., Dig. Tech. Papers* **60**, (1977).
107.   P. Gueret, A. Moser, P. Wolf, *IBM J. Res. Develop.* **24**, 155 (1980).
108.   P. Wolf, *IBM Tech. Discl. Bull.* **16**, 214 (1973).
109.   W. H. Henkels, J. H. Greiner, *IEEE J. Solid-State Circuits* **SC-14**, 794 (1979).
110.   H. H. Zappe, *IEEE Trans. Electr. Dev.* **ED-27**, 1870 (1980).
111.   S. M. Faris, *IEEE J. Solid-State Circuits* **SC-14**, 699 (1979).
112.   S. M. Faris, A. Davidson, *IEEE Trans. Magn.* **MAG-15**, 416 (1979).
113.   J. W. Matthews, C. J. Kircher, R. E. Drake, *Thin Solid Films* **47**, 95 (1977).
114.   M. Koyanagi, T. Endo, A. Nakamma, *Jap. J. Appl. Phys.* **20**, L901 (1981).

115. D. G. McDonald, E. G. Johnson, R. E. Harris, *Phys. Rev. B* **13**, 1028 (1976).
116. R. E. Harris, *Phys. Rev. B* **11**, 3329 (1975).
117. R. E. Harris, R. C. Dynes, D. M. Ginsberg, *Phys. Rev. B* **14**, 993 (1976).
118. R. E. Harris, *J. Appl. Phys.* **48**, 5188 (1977).
119. L. E. Alsop, A. S. Goodman, F. G. Gustavson, W. L. Miranker, *J. Comp. Phys.* **31**, 216 (1979).
120. W. H. Chang, *J. Appl. Phys.* **50**, 8129 (1979).
121. W. H. Chang, *IEEE Trans. Magn.* **MAG-17**, 764 (1981).
122. W. H. Chang, *J. Appl. Phys.* **52**, 1417 (1981).
123. W. H. Henkels, *Appl. Phys. Lett.* **32**, 829 (1978).
124. M. Klein, *IEEE Trans. Magn.* **MAG-13**, 59 (1977).
125. L. M. Geppert, J. H. Greiner, D. J. Herrell, S. Klepner, *IEEE Trans. Magn.* **MAG-15**, 412 (1979).
126. H. H. Zappe, *Appl. Phys. Lett.* **25**, 424 (1974).
127. Advanced Statistical Analysis Program (ASTAP) Program Reference Manual, IBM Corporation, White Plains, NY, 1973.
128. W. H. Henkels, private communication.
129. E. BenJacob, Y. Imry, *Journal de Physique Colloque C6* Suppl. au n°8, **39**, 569 (1978).
130. T. V. Rajeevakumar, L. M. Geppert, J. T. Chen, *J. Appl. Phys.* **51**, 2744 (1980).
131. L. Gunther, Y. Imry, *Phys. Rev. Lett.* **44**, 1225 (1980).
132. P. M. Marcus, Y. Imry, *Solid-State Comm.* **33**, 345 (1980).
133. T. V. Rajeevakumar, private communication.
134. K. Nakajima, Y. Onodera, *J. Appl. Phys.* **47**, 1620 (1976).
135. A. Matsuda, H. Yoshikiyo, *J. Appl. Phys.* **52**, 5727 (1981).
136. T. V. Rajeevakumar, *IEEE Trans. Magn.* **MAG-17**, 1 (1981).
137. K. Enpuku, K. Yoshida, F. Irie, K. Hamasaki, *IEEE Trans. Electron Devices* **Ed-27**, 1973 (1980).
138. T. V. Rajeevakumar, *Appl. Phys. Lett.* **39**, 439 (1981).
139. T. A. Fulton, L. N. Dunkleberger, *Appl. Phys. Lett.* **22**, 232 (1973).
140. H. Beha, W. Jutzi, G. Mischke, *IEEE Trans. Electron Devices* **ED-27**, 1882 (1980).
141. J. P. Hurrell, D. C. Pridmore-Braun, A. H. Silver, *IEEE Trans. Electron Devices* **ED-27**, 1887 (1980).
142. K. K. Likarev, *IEEE Trans. Magn.* **MAG-13**, 242 (1977).
143. H. Tamura, Y. Okabe, T. Sugano, *IEEE Trans. Electron Devices* **ED-27**, 2035 (1980).
144. S. M. Faris, *IBM Tech. Discl. Bull.* **21**, 3384 (1979).
145. S. M. Faris, Hardware and Software Concepts in VLSI, Chapter 9, Van Nostrand Reinhold Company, (1982).
146. S. M. Faris, *IEEE Circuits and Systems Magazine* **3**, 2 (1981).
147. M. R. Beasley, C. J. Kircher, Superconductor Materials Science - Metallurgy, Fabrication and Applications, Chapter 9, eds. S. Foner and B. B. Schwartz, Plenum Press, New York, London.
148. J. H. Greiner et al., *IBM J. Res. Develop.* **24**, 195 (1980).
149. J. H. Greiner, S. P. Klepner, *J. Vac. Sci. Technol.* **18**, 262 (1981).
150. C. Y. Fu, T. Van Duzer, *IEEE Trans. Magn.* **MAG-17**, 290 (1981).

151.   S. K. Lahiri, *Thin Solid Film* **41**, 209 (1977).
152.   C. J. Kircher, S. K. Lahiri, *IBM J. Res. Develop.* **24**, 235 (1980).
153.   S. K. Lahiri, S. Basavaiah, *J. Appl. Phys.* **49**, 2880 (1978).
154.   J. H. Greiner, *J. Appl. Phys.* **42**, 5151 (1971); **45**, 32 (1974).
155.   J. M. Baker, J. H. Magerlein, R. W. Johnson, *J. Vac. Sci. Technol.* **20**, 175 (1982).
156.   G. B. Donaldson, H. Faghihi-Nejad, *IEEE Trans. Electron Devices* **ED-27**, 1988 (1980).
157.   S. K. Lahiri, S. Basavaiah, C. J. Kircher, *Appl. Phys. Lett.* **36**, 334 (1980).
158.   S. Basavaiah, J. H. Greiner, H. H. Zappe, S. J. Singer, *J. Appl. Phys.* **51**, 1702 (1980).
159.   S. K. Lahiri, O. C. Wells, *Appl. Phys. Lett.* **15**, 234 (1969).
160.   S. K. Lahiri, *J. Appl. Phys.* **41**, 3172 (1970).
161.   M. Murakami, *Thin Solid Films* **55**, 101 (1978).
162.   C. Y. Fu, T. Van Duzer, *J. Vac. Sci. Technol.* **17**, 752 (1980).
163.   S. Basavaiah, J. H. Greiner, *J. Appl. Phys.* **48**, 4630 (1977).
164.   H.-C. W. Huang, et al., *IEEE Trans. Electron Devices* **ED-27**, 1979 (1980).
165.   D. B. Tuckerman, private communication.
166.   A. Mukherjee, *IEEE Electron Device Lett.* **EDL-3**, 29 (1982).
167.   S. I. Raider, R. W. Johnson, R. E. Drake, R. A. Pollak, presented at the Electrochemical Society Meeting, Hollywood, FL (1980).
168.   R. F. Broom, S. I. Raider, A. Oosenbrug, R. Drake, W. Walter, *IEEE Trans. Electron Devices* **ED-27**, 1998 (1980).
169.   S. I. Raider, R. W. Johnson, T. S. Kuan, R. E. Drake, R. A. Pollak, submitted to the Applied Superconductivity Conference, Knoxville, TN (1982).
170.   J. H. Magerlein, *IEEE Trans. Magn.* **MAG-17**, 286 (1981).
171.   M. Heilblum, S. Wang, J. R. Whinnery, T. K. Gustavson, *IEEE J. Quantum Electronics*, **QE-14**, 159 (1978).
172.   R. H. Havemann, *J. Vac. Sci. Technol.*, **15**, 389 (1978).
173.   R. E. Howard, E. L. Hu, L. D. Jackel, L. A. Fetter, R. H. Bosworth, *Appl. Phys. Lett.* **35**, 879 (1979).
174.   R. F. Broom, A. Oosenbrug, W. Walter, *Appl. Phys. Lett.* **37**, 237 (1980).
175.   A. W. Kleinsasser, R. A. Buhrman, *Appl. Phys. Lett.* **37**, 841 (1980).
176.   H. Kroger, L. N. Smith, D. W. Jillie, *Appl. Phys. Lett.* **39**, 280 (1981).
177.   D. W. Jillie, L. N. Smith, H. Kroger, IEDM Technical Digest, 701 (1981).
178.   J. M. Rowell, M. Gurvitch, J. Geerk, *Phys. Rev. B*, **24**, 2278 (1981).
179.   F. F. Tsui, *IBM J. Res. Develop.* **24**, 243 (1980).
180.   M. J. Marcus, *Proc. IEEE* **69**, 404 (1981).
181.   J. Matisoo, *Scientific American* **242**, 50 (1980).
182.   A. V. Brown, *IBM J. Res. Develop.* **24**, 167 (1980).
183.   B. van der Hoeven, *J. Vac. Sci. Technol.* **18**, 841 (1981).
184.   S. K. Lahiri, P. Geldermans, G. Kolb, J. Sokolowski, M. J. Palmer, *IEEE Trans. Components Hybrids Manuf. Technol.* **CHMT-5**, 166 (1982).
185.   S. K. Lahiri, et al., to be published in *IEEE Trans. Components Hybrids Manuf. Technol.* (June 1982).
186.   C. Y. Ting, K. R. Grebe, D. P. Waldman, 157th Electrochem. Soc. Meeting **80-1**, 210 (1980).
187.   K. R. Grebe, C. Y. Ting. D. P. Waldman, Extended Abstracts, 157th Electrochem. Soc. Meeting **80-1**, 213 (1980).

188. H. C. Jones, D. J. Herrell, *IBM J. Res. Develop.* **24**, 172 (1980).
189. J. Temmyo, H. Yoshikiyo, *IEEE Trans. Microwave Theory, Techniques* **MTT-30**, 27 (1982).
190. C. J. Anderson, M. Klein, M. B. Ketchen, submitted to the Applied Superconductivity Conference, Knoxville, TN (1982).
191. M. B. Ketchen, et al., *IEEE Electron Device Lett.* **EDL-2**, 262 (1981).
192. P. A. Moskowitz, R. W. Guernsey, J. W. Stasiak, to be published in Advances in Crygenic Engineering, New York: Plenum Press **27** (1982).
193. S. Bermon, T. Gheewala, submitted to the Applied Superconductivity Conference, Knoxville, TN (1982).
194. E. B. Flint, L. C. Jenkins, R. W. Guernsey, Refrigeration for Cryogenic Sensors and Electronic Systems, eds. J. E. Zimmerman and S. E. McCarthy, Conference held at NBS, Boulder (1980), Proceedings (page 93), issued 1981.
195. R. W. Guernsey, E. B. Flint, Refrigeration for Cryogenic Sensors and Electronic Systems, eds. J. E. Zimmerman and S. E. McCarthy, Conference held at NBS, Boulder (1980), Proceedings (page 15), issued 1981.
196. D. J. Herrell, *IEEE Trans. Magn.* **MAG-10**, 864 (1974).
197. D. J. Herrell, *IEEE J. Solid-State Circuits* **SC-10**, 360 (1975).
198. Two special issues on Josephson technology: *IBM J. Res. Develop.* **24** (1980), and *IEEE Trans. Electron Devices* **ED-27** (1980), highlight details of recent achievements.
199. W. Anacker, *IEEE Trans. Magn.* **MAG-5**, 968 (1969).
200. J. C. Logue, private communication.
201. C. Mead, L. Conway, Introduction to VLSI Systems, Section 9.9, Addison-Wesley, Series in Computer Science (1980).
202. J. Clarke, *IEEE Trans. Electron Devices* **ED-27**, 1896 (1980).
203. C. A. Hamilton, F. L. Lloyd, R. L. Peterson, J. R. Andrew, *Appl. Phys. Lett.* **35**, 718 (1979).
204. S. M. Faris, *Appl. Phys. Lett.* **36**, 1005 (1980).
205. D. B. Tuckerman, *Appl. Phys. Lett.* **36**, 1008 (1980).
206. M. Klein, ISSCC **20**, 202 (1977).
207. R. E. Harris, C. A. Hamilton, F. L. Lloyd, *Appl. Phys. Lett.* **35**, 720 (1979).
208. R. L. Peterson, *J. Appl. Phys.* **50**, 4231 (1979).
209. C. A. Hamilton, F. L. Lloyd, *IEEE Electron Device Lett.* **EDL-1**, 92 (1980).
210. C. A. Hamilton, F. L. Lloyd, R. L. Kautz, *IEEE Trans. Magn.* **MAG-17**, 577 (1981).
211. C. A. Hamilton, F. L. Lloyd, *IEEE Trans. Magn.* **MAG-17**, 3414 (1981).
212. P. Richards, T. Shen, *IEEE Trans. Electron Devices* **ED-27**, 1909 (1980).
213. D. B. Sullivan, J. E. Zimmerman, J. T. Ives, Proc. NBS Cryocooler Conference, eds. J. E. Zimmerman, D. B. Sullivan, S. E. McCarthy, NBS Special Publication **607**, 186 (1981).
214. W. A. Little, Proc. NBS Cryocooler Conference, eds. J. E. Zimmerman, T. M. Flynn, NBS Special Publication **508**, (1978).
215. S. M. Faris, S. I. Raider, J. H. Greiner, R. E. Drake, A. J. Warnecke, *Bull. Amer. Phys. Soc.*, **26**, 306 (1981).

ORGANIC SUPERCONDUCTIVITY

D. Jérome

Laboratoire de Physique des Solides
Université Paris-Sud, 91405 Orsay (France)

ABSTRACT

Superconductivity which is observed in the series of Quasi-One-Dimensional Organic Conductors $(TMTSF)_2X$ is characterized by experimental features which can be related to the strong anisotropy of the band structure of these materials. We shall emphasize the interplay between magnetically ordered and superconducting ground states and the existence of a field dependent depression of the density of states at the Fermi level below 30 K or so. Finally, we point out briefly that the quasi mean-field behaviour or the superconducting transition at 1.2 K in $(TMTSF)_2ClO_4$ can be easily reconciled with a broad one dimensional fluctuation regime within the framework of a quasi one dimensional theory which takes into account the quantum phase fluctuations of the order parameter and a band structure anisotropy not larger than 10.

INTRODUCTION AND HISTORICAL PERSPECTIVE

The discovery of superconductivity in organic substances[1] is the achievement of an intense investigation of the electronic properties of molecular conductors which began about 10 years ago with the preparation of the first stable highly conducting organic solids, namely those belonging to the TTF-TCNQ series[2,3]. It has been quickly recognized that the unusual properties of these conductors; existence of phase transitions towards non-magnetic insulating states at low temperature and lattice precursor signs of these transitions observed at temperatures as high as about 3 times the actual phase transition temperature by X-ray diffuse scattering experiments[4], required the use of One-Dimensional Physics

129

    The concept of One-Dimension is relevant to Organic Conductors
as a result of the planar shape of the constituent molecules and of
their particular arrangement in the crystal, namely a packing of
the molecules in stacks along a preferred direction allowing strong
and weak overlap between the π-orbitals of near neighbour molecules
belonging to the same stacks and to adjacent stacks respectively.

    In actual conducting molecular crystals the tunnelling of
electrons from one chain to its neighbours cannot be disregarded.
However, as long as the interchain tunnelling matrix element remains
small, the main concepts of One Dimensional Physics still apply.
Being more specific, the quantities which must be compared are
(i) the thermal broadening $\hbar/\tau_{\shortparallel}$ of the 1-D Fermi surface coming
from a finite intrachain electron scattering time $\tau_{\shortparallel}$ and (ii) the
warping of the Fermi surface due to interchain coupling which
amounts to $\approx 2\pi t_{\perp}$, where $t_{\perp}$ is the interchain overlap integral.When
the broadening is smaller than the warping ($\hbar/\tau_{\shortparallel} < 2\pi t_{\perp}$), the interchain motion of the electrons proceeds coherently and a plasma edge
in the reflectance properties becomes visible with light polarized
along a transverse direction[5].

    Whenever $\hbar/\tau_{\shortparallel} > 2\pi t_{\perp}$ the transverse electron motion is diffusive.
However even for the situation of transverse coherent motion, the
Fermi surface of a quasi-one-dimensional (Q-1-D) single-chain conductor such as $(TMTSF)_2X$ does not present closed orbits; 1-D physics
is still applicable. In several highly conducting organic charge-
transfer complexes or salts a transverse diffusive to coherent cross
over occurs below 100 K.

    The Peierls-Fröhlich nature of the phase transition discovered
in TTF-TCNQ[4] is only one of the particular theoretical features
exhibited by a 1-D electron-gas, which is predicted to be unstable
at low temperature. The nature of the instability depends on the
interactions between the electrons within a given stack of molecules: direct coulombic interactions, or indirect electron-electron
interactions mediated by phonons or possibly by the electric polarization of highly polarizable organic molecules. In the presence
of interactions the ground state of a 1-D electron gas is unstable
against the formation of charge (or spin) density waves. This instability competes with an other instability, namely superconductivity
which arises in the 1-D electron gas as well as in a usual 3-D gas.

    The interplay between spin (charge) density waves and super-
conducting instabilities of a 1-D electron gas becomes clear if
the electron-hole (CDW-SDW channel) or the electron-electron (Cooper
channel) response functions are calculated perturbatively. In the
expansion of the superconducting response of a 1-D electron gas divergent contributions from density fluctuations occur and vice-ver-
sa, i.e. the two types of instability become coupled[6]. Another

problem in 1-D systems is the destruction of long-range order by
thermal fluctuations[7] at any non-zero temperature i.e. the absence
of phase transition. Hence, a treatment of phase transitions via a
self consistent field acting on the order parameter which neglects
possible deviations from its mean value (the so-called mean-field
approximation) is inappropriate for a 1-D conductor. More elaborate
treatments (beyond mean-field approximation) are thus required.

Keeping the two specific features of a 1-D electron gas in mind
(i) interplay between magnetism and superconductivity, (ii) possible
existence of a strong fluctuation regime, we shall summarize here
some of the fascinating properties of the organic superconductors.

For a more thorough review of the properties of organic conductors and superconductors we refer the reader to other recent
articles on the subject[8,9].

The charge transfer compound TMTSF-DMTCNQ[10] has been the first
organic conductor in which a high conductivity state has been stabilized under pressure down to zero temperature[11].

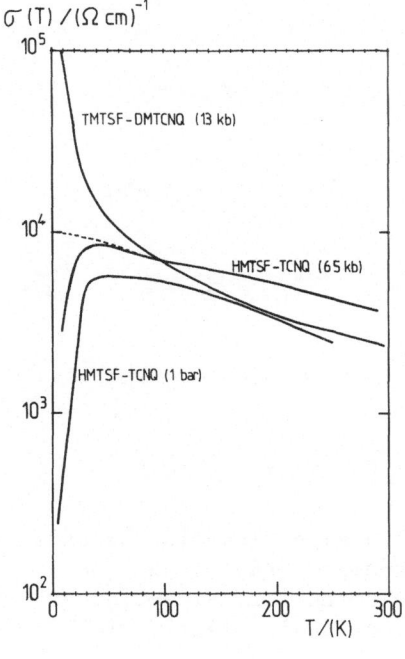

Fig. 1. Temperature dependence
of the conductivity of
TMTSF-DMTCNQ under 13 Kbar
and HMTSF-TCNQ under
ambient pressure and
6.5 Kbar. The dashed-line
shows the wiping out of
the Peierls transition
by crystalline disorder
in some early HMTSF-TCNQ
samples.

The application of a hydrostatic pressure in excess of $\sim$ 10 Kbar
suppresses the low temperature insulating state observed below 42 K
under ambient pressure. Above 10 Kbar, $\sigma_{\shortparallel}$ reaches and even overwhelms
$10^5 (\Omega cm)^{-1}$ at 1.2 K, figure (1). Moreover, an enormous magnetoresistance is found for fields perpendicular to the molecular stacking
axis.

Admittedly it is tempting to explain the highly conducting
state of TMTSF-DMTCNQ at low temperature in terms of a semimetallic
Fermi surface similar to the model which has been proposed for the
interpretation of early experimental data of an other two-chain
conductor HMTSF-TCNQ under pressure[12]. In that picture the diffusive to coherent transverse motion cross-over at low temperature is
accompanied by a Fermi surface shape changing from planar at high
temperature to semimetallic-like at low temperature.

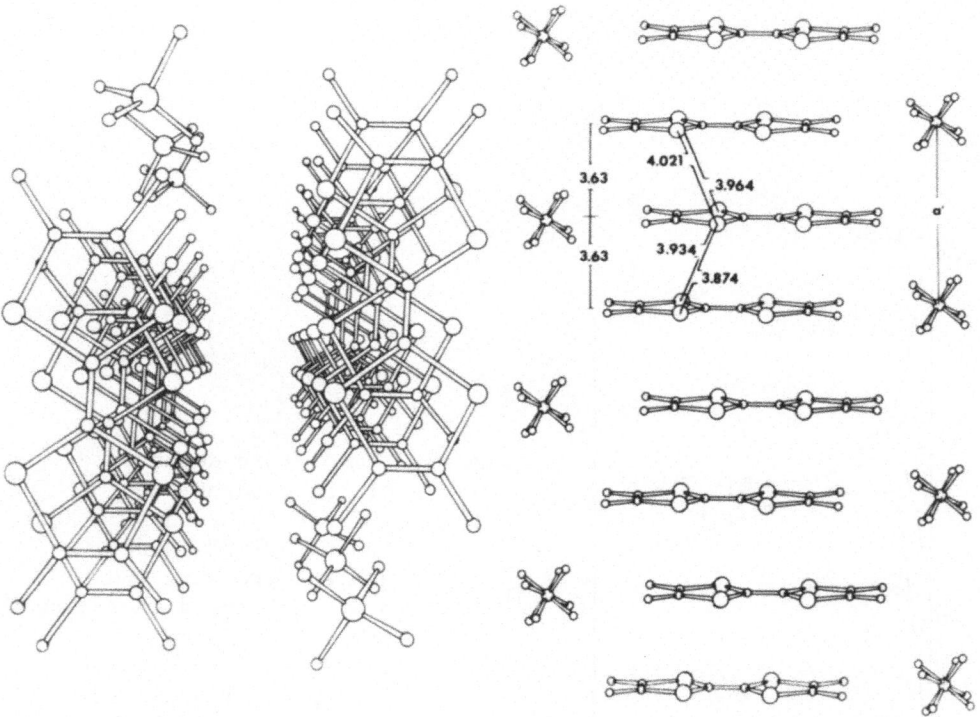

Fig. 2. View of $(TMTSF)_2ReO_4$ along the stacking direction, a-axis
(left). The ReO_4^- ions show alternating order along the
a-axis (T < 180 K) leading to a 2a potential. Side-view of
$(TMTSF)_2ClO_4$ (right). The dimerization is clearly visible
in the Se-Se intermolecular contacts.

The conductivity is not severely affected by the Fermi surface
dimensionality cross-over since both the density of carriers and
the carrier mobility vary in about the same proportion through the
transition region[13]. The comparison of TMTSF-DMTCNQ and HMTSF-TCNQ
data, figure (1) reveals striking differences between these two systems, in particular in the low temperatrue domain the conductivity of TMTSF-DMTCNQ begins the increase sharply below 100 K
whereas the conductivity of some HMTSF-TCNQ samples exhibits a saturation. TMTSF-DMTCNQ under pressure is thus the first system in
which the existence of superconducting short-range pairing contributing to the conductivity has been claimed[11]. Furthermore, superconducting fluctuations may coexist with semimetallic-like Fermi
surfaces at low temperature.

The only molecular conductors exhibiting superconductivity so
far belong to the $(TMTSF)_2X$ series of organic conductors[14] where
TMTSF are quasi planar molecules stacked in a zig-zag pattern along
the a-axis of a triclinic crystal (figure 2), X is an inorganic
negatively charged anion in the salt and which plays an essential
role in allowing a partial filling of the outer shell of the organic molecule TMTSF, but otherwise does not contribute directly to
electron transport. Anions do, however, affect the electronic properties of the $(TMTSF)_2X$ series via possible molecular ordering
arising at low temperature[15]. Despite the 2:1 stoichiometry of the
$(TMTSF)_2X$ series the conduction band derived from the overlap of
π-molecular orbitals is half-filled instead of quarter-filled as
it can be inferred from the stoechiometry of the salt. There is an
alternation of the band overlap integral $t_{''}$ along the a-axis which
is due to a slight dimerization of the TMTSF molecule as shown by
the 3-D crystalline structure.

MAGNETISM VERSUS SUPERCONDUCTIVITY

In the simplest theoretical treatment of the 1-D electron gas
problem the nature of the low temperature divergent response depends
on the sign and spatial shape of the eldctron-electron interactions
[16]. Approximating the Fourier transform of the electron-electron
interaction by two constants g_1 and g_2 related to scattering at
wave vectors $q = 2k_F$ and $q = 0$ respectively allows a derivation of
a phase diagram in the $g_1 g_2$ plane establishing in which regions of
the plane the correlation function of a given type is most strongly
divergent. First order renormalization treatment provides a power
law divergence of the static and uniform correlation function
(q=ω=0) spin (charge) density waves or superconductivity at low
temperature. As the inequality $g_1 < 2g_2$ is satisfied the spin (charge)
density wave divergence is predominant over the superconducting
channel whereas the opposite becomes true when $g_1 > 2g_2$. In addition,
for $g_1 > 0$ triplet superconductivity (TS) and spin density wave(SDW)

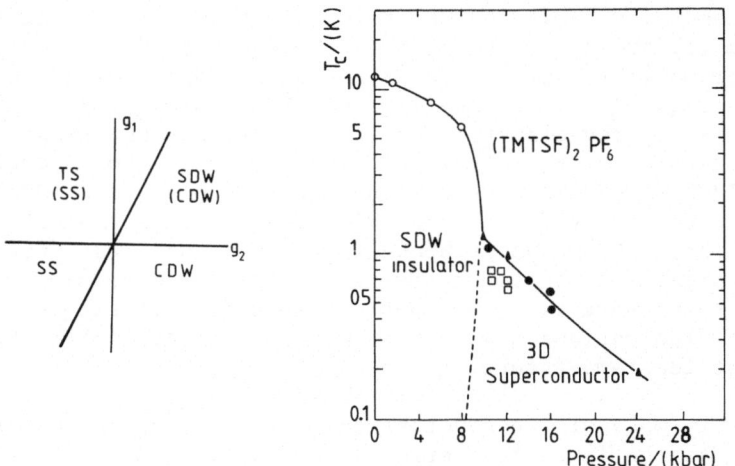

Fig. 3.   Most divergent response functions of the 1-D electron
          gas within the g_1/g_2 model (left).
          Experimental phase diagram of the $(TMTSF)_2PF_6$ salt (right)

divergencies are logarithmically stronger than SS and CDW
divergencies respectively (figure 3 a).

     Long range order is however possible at low temperature when
coupling between adjacent chains is taken into account. Using a
mean-field approximation for the interchain coupling and exact results for the single chain problem [17], the transition temperature
is given by $1 - \alpha_\perp \chi_{1D}(0)\big|_{T=T_c} = 0$ where $\chi_{1D}(q)$ is the order para-
meter susceptibility of a chain and $\alpha_\perp$ the transverse coupling.
Therefore the establishment of long-range order does not occur at
a temperature which is of the order of the intrachain interaction
$\alpha_{\shortparallel}$, as for the mean-field approximation but at a temperature which
can be significantly lower, depending on the anisotropy of the interaction $\alpha_\perp/\alpha_{\shortparallel}$.

     For an order parameter with two components, the analogy with
the ordering of a Heisenberg two-dimensional (XY) ferromagnet leads
to the phase transition temperature.

$$T_c \underset{\sim}{\sim} (\alpha_\perp \quad \alpha_{\shortparallel})^{1/2} \qquad\qquad [1]$$

in the weak interchain coupling limit, i.e. $\alpha_\perp \ll \alpha_{\shortparallel}$. Equation [1]
provides two important results: (i) the actual transition tempera-
ture of a quasi-one-dimensional (Q-1-D) conductor at T_c is lower
than the transition which would be derived using a mean-field
treatment of the 1-D problem in the ratio $(\alpha_\perp/\alpha_{\shortparallel})^{1/2} < 1$. (ii) the
strengths of the inter and intrachain couplings have no necessary

direct relation in any of the four instabilities (TS, SS, SDW, CDW) of a Q-1-D conductor.

The following situation may also be encountered

$$\alpha_{//}^A > \alpha_{//}^B \quad \text{and} \quad \alpha_{\perp}^A < \alpha_{\perp}^B \qquad [2]$$

where A and B are two divergent channels. In the situation exemplified by equation [2] the instability B may appear below a 3-D ordering temperature called T_c. But at temperatures well above T_c the 1-D divergence of the A-channel will dominate. We think this is the relevant situation for $(TMTSF)_2PF_6$ at ambient pressure even though the ground state stable below 12 K shows clearly the onset of magnetism[18]. Interplays between different ground states on the one hand and competition between different possible long-range ordered states and fluctuating regimes on the other hand are peculiarities of Q-1-D conductors. In the rest of this article we briefly summarize some experimental data of the $(TMTSF)_2X$ series supporting this picture.

Pressure dependence

Figure(3b) displays the phase diagram of $(TMTSF)_2X$-like salts when X is an anion of octahedral symmetry (PF_6, AsF_6, TaF_6,...).

At low pressure an itinerant antiferromagnetic ground state has been firmly established by magnetic measurements: anisotropic magnetic susceptibility [18], proton and selenium NMR [19, 20] and antiferromagnetic resonance [21]. The occurence of a spin structure (SDW) is likely to double the periodicity of the electron potential along the a-direction and therefore triggers the opening of a gap at the Fermi level driven by exchange interactions. The stabilization of the SDW state below 12 K at ambient pressure is accompanied by a sharp drop of the conductivity. However, the unperfect nesting of the Q-1-D Fermi surface is probably responsible for a semimetallic nature of the low temperature magnetic state, exhibiting a large and anisotropic magnetoresistance[22,23]. Above 8 Kbar or so, the ground state of $(TMTSF)_2PF_6$ and related compounds is superconducting as demonstrated by resistive[1] and Meissner effect data[24]. In the vicinity of 8 Kbar the phase diagram is rather intricate: a reentrance of the superconducting ground state is observed below the magnetic state, as shown by a pronounced upturn of the resistance around 4-5 K on cooling, followed by a subsequent transition towards a superconducting state around 1 K [25,26]. There is not yet any experimental evidence regarding the possible persistance of antiferromagnetic order in the superconducting state when the pressure is such that the reentrance phenomenon is observed.

The compounds $(TMTSF)_2X$ for which the anions exhibit a tetra-

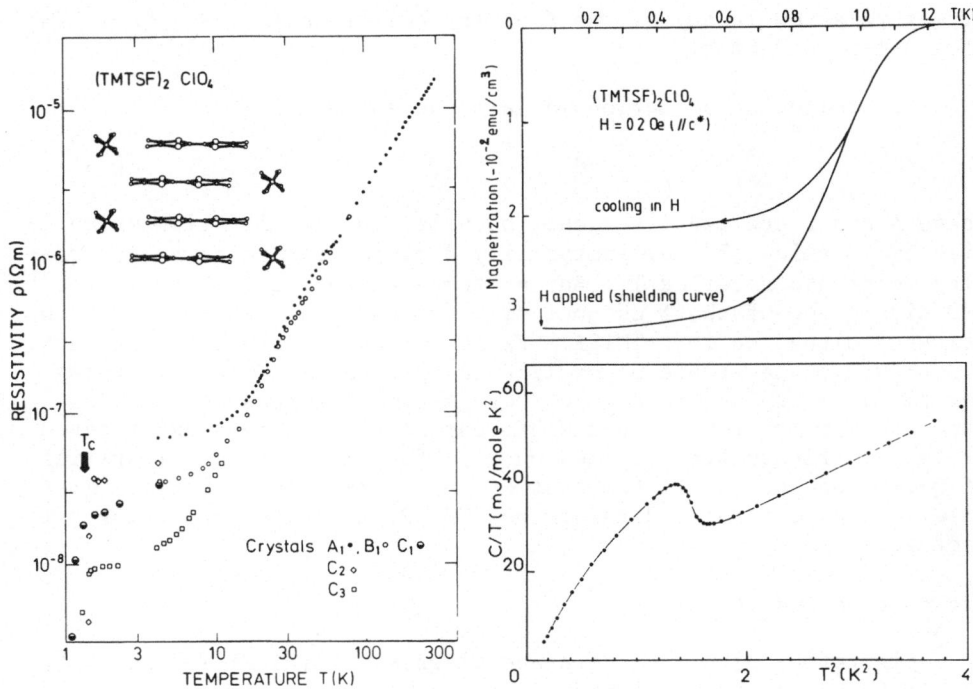

Fig. 4. Observation of organic superconductivity in $(TMTSF)_2ClO_4$ via different techniques: resistive transition of several crystals, Meissner effect and molar specific heat for the R-state.

hedral symmetry (X = ReO_4, ClO_4) do show similar behaviours, namely insulating or superconducting ground states depending on the applied pressure[27]. However, as far as $(TMTSF)_2ReO4$ is concerned, the insulating state stable below 180 K at ambient pressure is suppressed by a pressure of about 8 Kbar. It has also been shown to be non-magnetic[28] and due to an alternating order of the $ReO4$ doubling the lattice periodicity along the a-direction[29]. $(TMTSF)_2ClO4$ is also an interesting system since superconductivity can be achieved in that compound without the need of and external pressure[20]. Figure 4 summarizes the superconducting transition of $(TMTSF)_2ClO4$ observed by resistive[30], DC magnetization[31] and specific heat techniques[32]. The Meissner effect was measured on one single-crystal cooled through the transition in a small magnetic field of 0.2 Oe aligned with the c^*-axis. Furthermore it has been found, by comparison with a tin reference sample that shielding of an external vanishing field, namely H $\approx$ 0.01 Oe, is complete for all three crystallographic axis after $(TMTSF)_2ClO4$ has been cooled to 50 mK in a "zero" applied field.

The Meissner signals which are 80% and 55% of the shielding for
the $c^{\times}$ and $b^{\times}$ directions respectively in low field and at low temperature indicate the bulk and 3-D nature of the superconducting sta-
te. However, as far as the Meissner effect along the a-direction is
concerned, only 1% of the shielding signal is observed. Following
the Meissner experiments one can expect qualitatively that the lon-
gitudinal overlap $t_{\prime\prime}$ is larger than both transverse integrals $t_{\perp}^{c^{\times}}$
and $t_{\perp}^{b^{\times}}$. Although the exact ratios between the electronic coupling
along the main crystallographic directions are still somewhat uncertain we believe that a fair compromise using, optical data[5],
H_{c2} critical field anisotropy [33], [34] and conductivity anisotropy [14] is
given by 10:1:1/15 for the sequence $t_a : t_{b^{\times}} : t_{c^{\times}}$. Considering the
above determination of the band parameters, the Fermi surface of
the conducting phase of $(TMTSF)_2X$ salts is open, although some warping is expected from finite transverse overlaps[35]. Hence, the
magnetoresistance of that phase should not present any de Haas-
Shubnikov oscillations. This Q-1-D character by no means prevents
the transverse electron motion from being coherent (band-like) at
low temperature with the concomittant existence of a transverse
plasma edge in the optical reflectance spectrum[5].

Additional interesting informations related to the superconducting state below 1.2 K are given by the interpretation of the specific heat anomaly. The condensation energy and the thermodynamical
critical field amount to 2×10^{16} eV/mole and 44 Oe respectively[32].
The approximate determination of the gap at the Fermi level below
T_c is $2\delta = 0.35$ meV. Notice the ratio $2\delta/T_c$ equals approximately
the BCS weak coupling value. Heat capacity measurements[36] have also
revealed a behaviour of the electronic properties at low temperature which seems up to now, specific to $(TMTSF)_2ClO_4$.

A fast-cooled crystal of $(TMTSF)_2ClO_4$ below 40 K undergoes a
superconducting transition at 0.9 K instead of 1.2 K when the sample is cooled slowly at a rate of about 0.1 K/minute. This behaviour
is due to the onset of a phase transition at 24 K, as shown by resistive[37],[38], EPR[38] and NMR[39] data. Furthermore, X-ray investigations[40] have shown that below 24 K a 3-D superlattice structure
is observed with no change of the periodicity along a and c axis
but a doubling occuring along b. A careful study of the dynamics
of this structural transition shows that it corresponds probably
to the ClO_4 ions ordering ferroelectrically along the a-axis below
24 K. While slow cooling results in quasi perfect ordering at low
temperature (called the relaxed R-state), rapid cooling below 30 K
($\gtrsim$ 10 K/minute) leads to a partially ordered (quenched-Q state) exhibiting an excess of entropy in the superconducting state with
respect to the R-state[36]. This excess entropy has been attributed
to frozen-in lattice (ClO_4 ions) disorder. It is remarkable however
to notice that in spite of a lowering of T_c in the Q-state no marked
change of $N(E_F)$ in the temperature domain 1.2 - 2 K can be detected

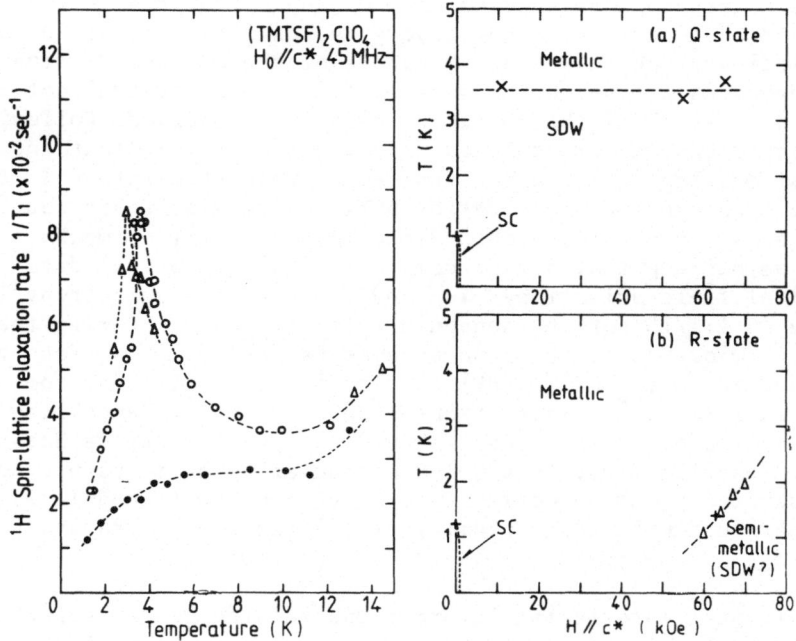

Fig. 5. Proton spin-lattice relaxation rate of (TMTSF)$_2$ClO$_4$ ver-
         sus temperature in the Q-state (O), partially annealed
         (Δ) and R-state (o)(left). Proposed phase diagram of
         (TMTSF)$_2$ClO$_4$: Q-state (a) and R-state (b). The data
         of specific heat under high field (+ reference 32) and
         threshold of de Haas-Shubnikov oscillations (Δ reference
         45) are also included.

via C_V data. Therefore, unless an important reorganization of the
phonon spectrum is associated with the 24 K structural transition
the lowering of T_c can be attributed to the residual (non magnetic)
disorder of the Q-state (see section II.3).

Effects of high magnetic field

     Proton and Selenium NMR[39] and high field C_V data[32] performed
on the R-state of (TMTSF)$_2$ClO$_4$ allow to map the T-H phase diagram
(figure 5). As far as the R-state is concerned, superconductivity
is observed below 1.2 K in zero field and up to a critical field
$H_{c2} \perp$ of $\sim$1 kOe at T = 0 K for H//c*. C_V experiments under high-field
have revealed the existence of a transition at $T_c \simeq 1.4$ K (H = 63 kOe)
between a high $N(E_F)$ state at T> 1.4 K and a small $N(E_F)$ state at
lower temperatures (possibly a semimetallic state). Furthermore,
^{77}Se NMR data strongly support the occurence of SDW's in the high
field induced semimetallic state of (TMTSF)$_2$ClO$_4$.

     The phase line between the metallic and the high-field induced

semimetallic state of $(TMTSF)_2ClO_4$ is in remarkable agreement with
the threshold field necessary for the observation de Haas-Shubnikov oscillations, a fact which proves the two-dimensional
nature of the Fermi surface (closed orbits) at fields larger than
the threshold field[41]. Recent de Haas-Shubnikov data showing a
threshold field of $\sim$ 50 kOe at 0.47 K[42] and $\sim$ 43 kOe at 25 mK[43]
for H//c* , tend to support the phase diagram of figure (5 b) and
could suggest that a threshold field of about 40 kOe is required
for the stabilization of the semimetallic (magnetic phase of
$(TMTSF)_2ClO_4$ even at zero temperature.

We may emphasize that a similar high field restoration of a
semimetallic [44](magnetic)[45] state is also observed in $(TMTSF)_2PF_6$
under pressure.

Effects of lattice disorder in the Q-state of $(TMTSF)_2ClO_4$

The onset of a phase transition around 3.5 K in the Q-state of
$(TMTSF)_2ClO_4$ is best demonstrated by a sharp peaking of the spin-
lattice (^1H and ^{77}Se) relaxation rates at that temperature[39]. Below
3.5 K an increase of the NMR linewidth is also observed. Consequently,
a SDW state is stabilized below 3.5 K in agreement with the observation of an AFMR signal at helium temperature[46].

In the same temperature domain a significant drop of the conductivity is noticed at low temperature, whereas $N(E_F)$ measured by specific heat does not seem to be strongly affected by the onset of magnetism[36]. Superconductivity arises at a temperature lower than in
the R-state, namely T_c < 0.9 K. Annealing the Q-state at a temperature T>30 K, followed by a slow cooling allows the restoration of
the R-state not showing any onset of magnetism[38,39] prior to the
superconducting transition at 1.2 K.

A tentative interpretation of the Q-state behaviour can be proposed following the theory on the effect of disorder on phase transitions in Q-1-D conductors[47]. In the framework on this theory weak
impurity potentials may induce independent (and non-coherent) phase
shifts of the electron wavefunction on each filament resulting in a
decrease of the amplitude for hopping of a Cooper pair from chain to
chain. It follows therefore that with an increase of the concentration of impurity scattering potentials T_c decreases but the 1-D divergence of the superconducting response function is only more weakly
affected.

As far as $(TMTSF)_2ClO_4$ is concerned, the smearing of the 1-D SC
divergent channel in the Q-state may be large enough to allow the
establishment of a SDW on the nested portions of the Fermi surfaces
below 3.5 K or so. We may notice that intermediate situations between

figure 5 a and b can be achieved varying the annealing time or(and) the cooling rate below 30 K. However we believe that figure (5b) represents the behaviour of a R-state in which the only external parameter governing the balance between the 1-D divergences is the magnetic field.

Since band parameters are obviously unchanged by the magnetic field, the competition between superconductivity and magnetism at low temperature in $(TMTSF)_2ClO_4$ (and in $(TMTSF)_2PF_6$ under pressure) cannot be attributed to a variation of the interchain tunnelling coupling, as it could be inferred from the pressure dependent phase diagram of $(TMTSF)_2PF_6$ in zero field. We suggest instead an interpretation relying on the 1-D character of the electronic properties: after an easy removal of the 3-D ordered superconducting state by a small magnetic field ($H_{c_2}^{c*} \sim 1$ kOe at T = 0), the 1-D superconducting divergence is weakened by larger fields and finally as the field reaches the threshold value ($H//c* \sim 63$ kOe at 1.4 K) the balance between 1-D superconducting divergence and 3-D SDW instability turns in favour of the second one which condenses and gives rise to a semimetallic state accompagnied by quantum oscillations of the magnetoresistance.

THE FLUCTUATING REGIME

This section is devoted to a brief survey of experimental results suggesting that the superconducting channel is the dominant divergence in the 1-D regime (transverse coherence length < interchain distance) up to high temperatures, 30 K or so, for the $(TMTSF)_2X$ series. All the results of this section refer to the low temperature behaviour of either $(TMTSF)_2ClO_4$ (R-state) under ambient pressure or $(TMTSF)_2PF_6$ - AsF_6 at $P > P_c$ when the T domain lower than 25 K is concerned.

Electrical conductivity

The longitudinal DC conductivity becomes larger than $10^4 (\Omega cm)^{-1}$ at low temperature[14]. It reaches $10^5 - 10^6 (\Omega cm)^{-1}$ at helium temperatures[30] (figure 4), a value which in terms of a single-particle interpretation suggests an electron mean free path of 1000 Å or more. σ_{DC} is also greatly diminished by the application of a magnetic field (the magnetoresistance is strongest when the field is aligned with the c* direction[23]. However optical reflectance data point towards a drastic frequency dependence of the conductivity in the very far infrared domain[48]. In both $(TMTSF)_2ClO_4$ and $(TMTSF)_2PF_6$ at ambient pressure the optical electron life time increases by a factor 3 at most between 300 K and low temperature[5,49] ($T \sim 25$ K) (figure 6), whereas the corresponding increase of the DC conductivity exceeds a factor 50. Furthermore the FIR conductivity of $(TMTSF)_2PF_6$ is about 400 $(\Omega cm)^{-1}$ at T = 25 K, $\omega = 10$ cm^{-1}, i.e. $\sim$ 100 times smaller

than DC or microwave conductivities[48]. Therefore, FIR conductivity
data imply the existence of a long life time collective mode at 25 K
($10^{-12}s < \tau_c < 2.10^{-11}s$).

Thermal conductivity of $(TMTSF)_2ClO_4$

The study of the thermal conductivity of $(TMTSF)_2ClO_4$ and
$(TMTSF)_2PF_6$ (P = 12 Kbar) shows a significant drop below 50 K
(figure 6). Such a behaviour is very striking for two reasons: i)
In the same T-domain σ_{DC} is strongly T-dependent and therefore one
could expect heat to be carried by the electrons when σ_{DC} reaches
10^5 $(\Omega cm)^{-1}$, following the Wiedemann-Franz proportionality relation
between κ_e and σ. ii) The drop of κ is greatly suppressed, especially below 25 K, by the application of a magnetic field. This latter
behaviour is again in contradiction with the Wiedemann-Franz law

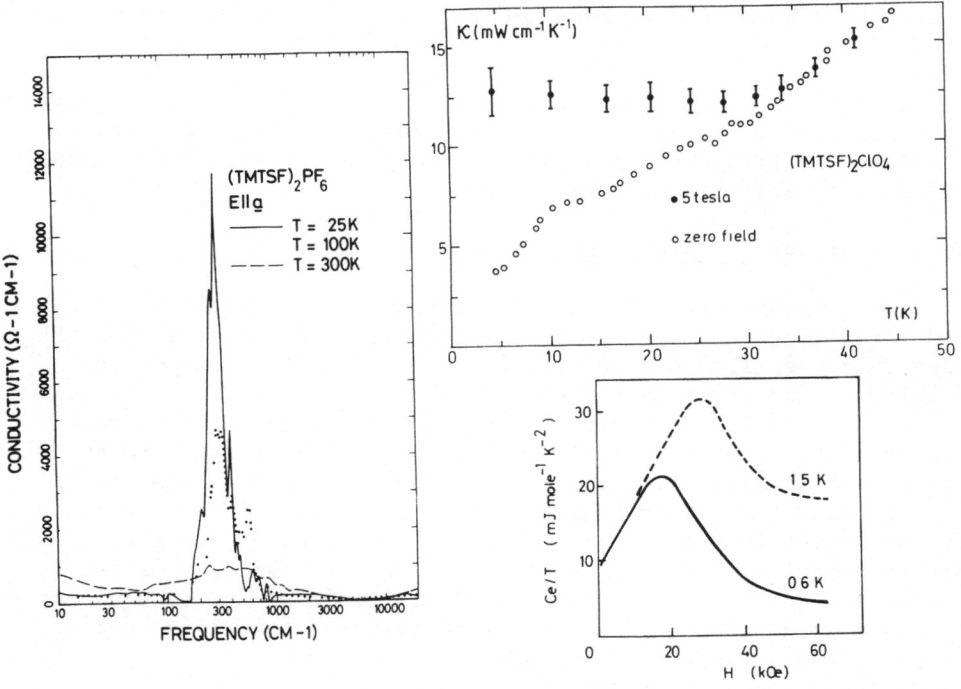

Fig. 6. Frequency dependence of the conductivity of $(TMTSF)_2PF_6$
at 25 K under atmospheric pressure (left).
Temperature dependence of the thermal conductivity of
$(TMTSF)_2ClO_4$ and influence of a magnetic field (top right).
Field dependence of the electronic specific heat of
$(TMTSF)_2ClO_4$ at 0.6 and 1.5 K (H//c*), (bottom right).

which would predict a decrease of κ (and not an increase) under magnetic field, according to the large positive magnetoresistance observed at low temperature. The drop of κ at low temperature must be attributed to a change in the electronic structure (typically a decrease of $N(E_F)$ at low temperature) which is sensitive to magnetic field.

Field dependence of the specific heat

At low temperature (T < 3 K) the specific heat of $(TMTSF)_2ClO_4$ is well described by the typical law $C_v = \gamma T + \beta T^3$ where γ and β are related to $N(E_F)$ and the phonon spectrum respectively . However a striking field dependence of $N(E_F)$ is observed[51] at low temperature (figure 6) up to 20 kOe or so after the suppression of the superconducting state above $H_{C_2}^{C^*} \lesssim$ 1 kOe. The 70 % increase of $N(E_F)$ in a field of 20 kOe at helium temperature is a remarkable effect which agrees qualitatively with the field dependence of the thermal conductivity and which suggests a field-induced restoration of $N(E_F)$.The study of the field dependence of $N(E_F)$ has not yet been performed above 3 K since at higher temperatures the electronic contribution becomes very much smaller than the lattice contribution.

Summarizing : (i) the density of states at the Fermi level is depressed at low temperature below its value at high temperature (T $\gtrsim$ 40 K) and (ii) the depression is partly suppressed by the application of a large magnetic field.

Energy width of the pseudo-gap

Two different techniques, electron quantum tunnelling and FIR absorption experiments performed on $(TMTSF)_2ClO_4$ and in $(TMTSF)_2PF_6$ under pressure allow a determination of the energy width at the vicinity of the Fermi energy over which the density of states is depressed at low temperature. The energy dependence of the transition probability of electrons which obey Fermi statistics to tunnel between two metallic electrodes separated by a thin insulating barrier can reproduce to some extent the energy dependence of the density of states in the vicinity of the Fermi level of the electrodes. This technique is largely used to derive quantities such as energy gaps in the quasi-particle density of states if one or both electrodes are superconductors. Schottky-type electron tunnelling using N-doped evaporated GaSb on $(TMTSF)_2PF_6$[52] or $(TMTSF)_2ClO_4$[53] performed under 11 Kbar and ambient pressure for the two organic compounds respectively supports the existence of a depression of $N(E_F)$ in the organic superconductor over an energy width $2\Delta \sim$ 3.6 meV centered at the Fermi energy (figure 7a). The FIR reflectance of $(TMTSF)_2ClO_4$ studied at helium temperature reveals a drop of 5% around 3.8 meV. As shown by magneto-absorption experiments[54] (figure 7b), this optical absorption threshold can be observed without

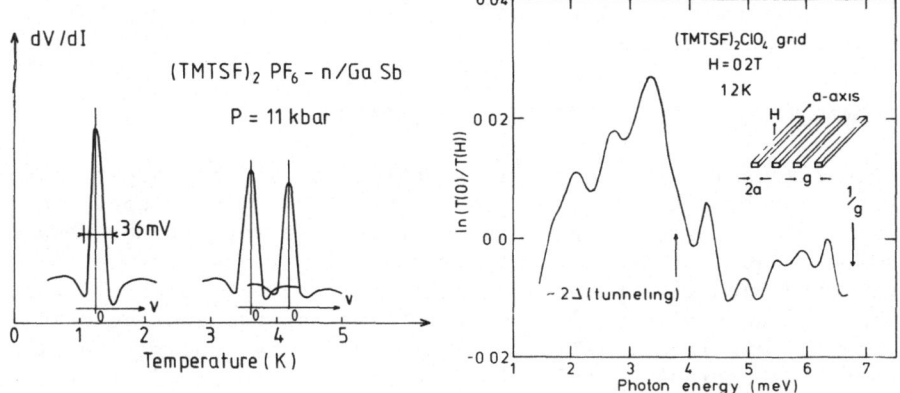

Fig. 7. a) Temperature dependence of the Schottky tunnelling
            characteristics of N-doped GaSb/(TMTSF)$_2$PF$_6$ junctions
            under 11 Kbar.
         b) Magnetoabsorption of a (TMTSF)$_2$ClO$_4$ grid. There is a
            pronounced onset of magnetoabsorption below $\approx$ 3.6 meV.
            The electromagnetic radiation is polarized along the
            a-axis and the magnetic field is normal to this axis.

significant shift of its energy up to 15 K or so. The vanishing
of the absorption threshold occurs between 20 and 50 K [55].

    All experimental results presented in this section suggest
the existence of a pseudo-gap of width 3.6-3.8 meV at the Fermi
level. This pseudo-gap is suppressed by a large magnetic field
and remains visible up to temperatures which are about ten times
the temperature for the onset of long range superconducting order.
Since organic conductors are Q-1-D conductors, precursor signs
of the low temperature instabilities are expected to occur at
temperatures larger than the 3-D ordering temperature. Thus, we
must consider precursor effects of two different kinds: SDW or
Superconductivity.

    We may rule out the SDW origin of the pseudo-gap for several
reasons: (i) no sign of magnetism has been detected (via NMR experiments) up to the temperature domain in which the pseudo-gap is
observed[39], (ii) whenever a SDW gap is observed, it is stabilized
and not suppressed by a magnetic field[39,18], (iii) there is no
SDW state stable at low temperature in (TMTSF)$_2$PF$_6$ under pressure[45],
(iv) electron tunnelling characteristics related to a SDW gap which
have been observed in (TMTSF)$_2$PF$_6$ below 12 K at ambient pressure
do not show the typical resistance minima on both sides of the zero

bias[56], (v) commensurability would prevent the fluctuating SDW
from contributing to the DC conduction[57].

SUPERCONDUCTING FLUCTUATIONS AND CONCLUSION

The discussion in the previous section shows that superconductivity is a very reasonable origin to the strongly developed precursor regime[58]. However, how can one reconcile a precursor domain
extending to 30 K or so with a superconducting transition which
behaves very much like phase transition within the mean-field theory
(figure 4). The small critical width of the superconducting transition seen by specific ($\Delta T/T_c \lesssim 10\%$ can by no means support the existence of fluctuations up to $T \sim 10 T_c$ unless the point of view of
phase transitions in Q-1-D conductors is taken. In such a case for
a two degrees of freedom order parameter (amplitude and phase) some
decoupling of the two components should occur at $T \gg T_c$ [8]. The spatial correlation function of the order parameter is given by:
$\langle \Delta(x)\ \Delta(0) \rangle = |\Delta|^2 e^{-x/\xi(T)}$ where $\xi(T) \sim a/k_B T$. Consequently, at $T > 0$,
the short range order which arises with a coherence length $\xi(T)$,
digs a pseudo-gap of width $2\Delta \sim a_{\shortparallel}$ at the Fermi level. The amplitude of the order parameter reaches a significant value ($\sim a_{\shortparallel}$) already
below the 1-D mean-field temperature, $T_1 \sim a_{\shortparallel}$, whereas its phase is
still free to take any value until T_c is attained. At that temperature an inter-chain locking of the phases takes place. If the zero point motion of the phase (quantum fluctuations) is taken into
account[59] a true gap develops below T_c. Its value at $T = 0$ amounts
to a real gap $2\Delta\ (T = 0) = 3.5 T_c$, a value much smaller than the amplitude of the pseudo-gap. For strong quantum fluctuations Schulz
and Bourbonnais[59] have derived $2\ \Delta(T=0) = 3.5\ T_c$; i.e. the mean-
field behaviour is recovered for the phase locking transition
(onset of 3-D order). Furthermore with reasonable values of the
intrachain coupling ($g_1 - 2g_2 \sim 0.6$) and of the band structure anisotropy ($t_{\shortparallel}/t_\perp = 10$), the same authors have calculated, using $T_c = 1.2K$,
that the superconducting precursor regime should extend up to about
30 K.

In conclusion organic superconductors of the $(TMTSF)_2X$ series
exhibit several characteristic features of the Q-1-D electron gas,
namely competition between magnetic and superconducting ground
state, existence of a very broad temperature domain in which 1-D
superconducting seeds are strongly developed. Furthermore, the narrow 3-D ordering transition satisfying approximately the mean-field
model agrees fairly well with the existence of significant quantum
phase fluctuations at low temperature.

Admittedly superconductivity of organic conductors although
existing is probably not yet optimized with the $(TMTSF)_2X$ series.
One could take advantage of the strong 1-D divergence at high temperature. An increase of the interchain coupling could allow therefore a substantial increase of the critical temperature.

Acknowledgments

The work at Orsay has benefited from the constant cooperation of K.Bechgaard and H.J.Schulz. I also wish to acknowledge the efficient collaboration of all my colleagues at the Laboratoire de Physique des Solides.

REFERENCES

1. D.Jérome, A.Mazaud, M.Ribault and K.Bechgaard, J.Phys.Lett.41, L-95 (1980).
2. J.Ferraris, D.O.Cowan, V.Walatka and J.H.Perlstein, J.Am.Chem. Soc.95, 948 (1973).
3. L.B.Coleman, M.J.Cohen, D.J.Sandman, F.G.Yamagishi, A.F.Garito and A.J.Heeger, Solid State Commun.12, 1125 (1973).
4. F.Denoyer, R.Comès, A.F.Garito, A.J.Heeger, Phys.Rev.Lett.35, 445 (1975).
5. C.S.Jacobsen, D.B.Tanner, K.Bechgaard, Phys.Rev.Lett.46, 1142 (1981).
6. Y.A.Byschkov, L.P.Gorkov and I.E.Dzyaloshinskii, Sov.Phys.JETP, 23, 499 (19).
7. D.J.Scalapino, M.Sears and R.A.Ferrel, Phys.Rev.B6, 3409 (1972).
8. D.Jérome and H.J.Schulz, Adv.in Physics 31, 000 (1982).
9. J.Friedel and D.Jérome, Contemp.Phys.23,000, (1982).
10. C.S.Jacobsen, K.Mortensen, J.R.Andersen, K.Bechgaard, Phys.Rev. B18, 905,(1978).
11. A.Andrieux, P.M.Chaikin, C.Weyl, K.Bechgaard, J.R.Andersen, J.Physique 40, 1199,(1979).
12. J.R.Cooper, M.Weger, D.Jérome, D.Le Fur, K.Bechgaard, A.N.Bloch and D.O.Cowan, Solid State Comm.19, 749 (1976).
13. M.Weger, Solid State Comm 19, 1149 (1976).
14. K.Bechgaard, C.S.Jacobsen, K.Mortensen, H.J.Pedersen, N.Thorup, Solid State Commun.33, 1119 (1980).
15. J.P.Pouget, R.Moret, R.Comès, K.Bechgaard, J.M.Fabre, L.Giral, Mol.Cryst.Liq.Cryst.79, 129 (1982).
16. J.Solyom, Adv.Phys.28, 201 (1979).
17. D.J.Scalapino, Y.Imry, P.Pincus, Phys.Rev.B11, 2042 (1975).
18. K.Mortensen, Y.Tomkiewicz, T.D.Schultz, E.M.Engler, Phys.Rev. Lett.46, 1234 (1981).
19. A.Andrieux, D.Jérome, K.Bechgaard, J.Phys.Lett 42, L-871 (1981).
20. J.C.Scott, H.J.Pedersen, K.Bechgaard, Phys.Rev.B24, 475 (1981).
21. J.B.Torrance, H.J.Pedersen and K.Bechgaard, to be published.
22. P.M.Chaikin, P.Haen, E.M.Engler, R.L.Greene, Phys.Rev.B24, 7155 (1981).
23. C.S.Jacobsen, K.Mortensen, M.Weger and K.Bechgaard, Solid State Commun.38, 423 (1981).
24. K.Andres, F.Wudl, D.B.McWhan, G.A.Thomas, D.Nalewajek and A.L.Stevens, Phys.Rev.Lett.45, 1449 (1980).
25. R.L.Greene and E.M.Engler, Phys.Rev.Lett.45, 1587 (1980).
26. R.Brusetti, M.Ribault, D.Jérome and K.Bechgaard, J.Physique 43, 801 (1982).

27. S.S.P.Parkin, M.Ribault, D.Jérome and K.Bechgaard, J.Phys.C.
     Solid State 14, 5305 (1981).
28. K.Bechgaard, Mol.Cryst.Liq.Cryst.79, 1 (1982).
29. J.P.Pouget, R.Moret, R.Comès and K.Bechgaard, J.Physique Lett.
     42, L-543 (1982).
30. K.Bechgaard, K.Carneiro, F.B.Rasmussen, M.Olsen and C..Jacobsen
     Phys.Rev.Lett.46, 852 (1981).
31. D.Mailly, M.Ribault, K.Bechgaard, J.M.Fabre and L.Giral,
     J.Physique Lett., to be published (1982).
32. P.Garoche, R.Brusetti, D.Jérome, K.Bechgaard, J.Physique Lett.
     43, L-147 (1982).
33. D.Jérome, Mol.Cryst.Liq.Cryst.79, 155 (1982).
34. R.L.Greene, P.Haen, S.Z.Huang, E.M.Engler, M.Y.Choi and P.M.
     Chaikin, Mol.Cryst.Liq.Cryst.79, 183 (1982).
35. B.Horovitz, H.Gutfreund, M.Weger, Phys.Rev.B12, 3174 (1975).
36. P.Garoche, R.Brusetti and K.Bechgaard, preprint (1982).
37. D.U.Gubser, W.W.Fuller, T.O.Poehler, J.Stokes, D.O.Cowan, M.Lee
     and A.N.Bloch, Mol.Cryst.Liq.Cryst.79, 225 (1982).
38. S.Tomic, D.Jérome, P.Monod and K.echgaard, J.Physique Lett.
     to be published (1982).
39. T.Takahashi, D.Jérome and K.Bechgaard , J.Physique Lett.43,
     L-565 (1982).
40. J.P.Pouget, G.Shirane, K.Bechgaard and J.M.Fabre, preprint(1982).
41. J.F.Kwak, J.E.Schirber, R.L.Greene and E.M.Engler, Mol.Cryst.
     Liq.Cryst. 79, 111 (1982).
42. H.Bando, K.Oshima, M.Suzuki, H.Kobayashi, G.Saito, preprint
     (1982).
43. T.Ishiguro, private communication.
44. J.F.Kwak, J.E.Schirber, R.L.Greene and E.M.Engler, Phys.Rev.
     Lett.46, 1296 (1981).
45. L.J.Azevedo, J.E.Schirber, R.L.Greene, E.M.Engler, Physica 108B,
     11831 (1981).
46. W.M.Walsh, F.Wudl, E.Aharon-Shalom, L.W.Rupp, J.Vandenberg,
     K.Andres, J.B.Torrance, preprint 1982.
47. A.I.Larkin and V.I.Melnikov, Sov.Phys.JETP 44, 1159 (1976).
48. C.S.Jacobsen, D.B.Tanner and K.Bechgaard, Mol.Cryst.Liq.
     Cryst. 79, 25 (1982).
49. K.Kikuchi, I.Ikemoto, K.Yakushi, H.Kuroda, K.Kobayashi, Solid
     State Commun.42, 433 (1982).
50. D.Djurek, M.Prester, D.Jérome and K.Bechgaard, J.Phys.C Lett.
     21, L-669 (1982).
51. R.Brusetti, P.Garoche and K.Bechgaard, preprint (1982).
52. C.More, G.Roger, J.P.Sorbier, D.Jérome, M.Ribault, K.Bechgaard,
     J.Physique Lett.42, L-313 (1981).
53. A.Fournel, C.More, G.Roger, J.P.Sorbier, J.M.Delrieu, D.Jérome
     M.Ribault, K.Bechgaard, J.M.Fabre and L.Giral, Mol.Cryst.Liq.
     Cryst.79, 261 (1982).
54. H.K.Ng, T.Timusk, J.M.Delrieu, D.Jérome, K.Bechgaard, J.M.
     Fabre, J.Physique Lett.43, L-513 (1982).

55. T.Timusk, private communication.
56. D.Jérome.Proceedings of the IVth Conference on Superconductivity in d and f-band metals, Karlsruhe (1982).
57. P.Lee, T.M.Rice and P.W.Anderson, Solid State Commun.14,703 (1974).
58. H.J.Schulz, D.Jérome, A.Mazaud, M.Ribault and K.Bechgaard, J.Physique, 42, 991 (1981).
59. H.J.Schulz and C.Bourbonnais, preprint (1982).

SOLITONS IN LONG JOSEPHSON JUNCTIONS

N. F. Pedersen

Physics Laboratory I
The Technical University of Denmark
DK-2800 Lyngby, Denmark

INTRODUCTION

The Josephson junction transmission line (JTL) is approximately described by the sine-Gordon equation, and hence it is a convenient experimental solid state system for the study of solitons. In the JTL the physical manifestation of a soliton is a (propagating) fluxon, i.e. a magnetic flux quantum, $\Phi_o = h/2e = 2,064 \times 10^{-15}$ V·s. This and the corresponding relation to simple observable quantities (zero field steps) was noted first in a pioneering paper by Fulton and Dynes, 1973. The field has since attracted a large number of researchers and considerable progress has been made both theoretically and experimentally. A number of applications have been suggested; they include a soliton microwave generator, a vortex transistor, and the use of a fluxon as the basic bit in data processing circuits. The spectacular advances in fabrication technology connected with programmes on a Josephson junction computer implies that such ideas are not a priori unrealistic. In general there have been four methods of investigating perturbed sine-Gordon solitons on the JTL: (i) analytical or perturbation methods, (ii) numerical simulation, (iii) measurements on mechanical models, and (iv) experiments on real Josephson junctions. Much of the progress has taken place through interaction and stimulation between researchers working in those four directions. As a result we are moving in a direction where large Josephson junctions with solitons are understood almost as well as small junctions with spatial homogeneity.

The field of solitons has grown exponentially in recent years. Even specializing to sine-Gordon solitons in superconductors makes it impossible to cover the subject in a limited space. Hence the present paper does not pretend to be balanced. First of all, in

investigating non-linear wave phenomena in Josephson junctions two
schools appear to exist. One originates in a theory (Kulik 1967)
based on standing electromagnetic waves and is formally valid only
when the length of the junction is much less than the Josephson
penetration depth. The other is based on non-linear running waves
(solitons) and is applicable in the opposite limit. Although there
is a region where both models may be applicable, only the soliton
approach will be used here. With respect to applications only rf
applications will be discussed. Digital electronic applications
such as the flux shuttle (Fulton et al. 1973), JTL computer circuits
(Rajeevakumar 1981), and the vortex transistor (Likharev et al.
1979) will not be discussed.

Finally most of the work described here is centered around
activities that have taken place in an informal cooperation between
Physics Laboratory I and the Laboratory of Applied Mathematical
Physics at the Technical University of Denmark, the University of
Salerno, Italy, the PTB Institute in Berlin, and the National
Research Laboratory in Los Alamos, U.S.A. With these limitations
in mind, the paper is organized as follows: Section I discusses the
perturbed sine-Gordon model. In section II this is used to calculate
the IV-curve of zero field steps for inline and overlap junctions,
and some experimental examples are shown. Section III treats aspects
of the magnetic field dependence within the perturbed sine-Gordon
model. Finally section IV treats questions connected with microwave properties of long junctions.

I.   THE PERTURBED SINE-GORDON EQUATION

In a small Josephson junction (Fig. 1a) characterized by a
maximum pair current I_o, a resistance R, and a capacitance C the
Josephson equation in dimensionless units may be expressed as

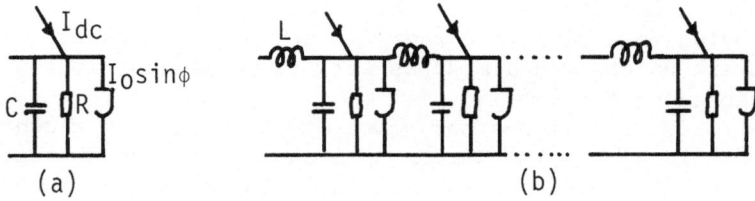

                (a)   (b)

Fig. 1 Equivalent diagram for (a) a small Josephson junction, and
       (b) a large Josephson junction.

$$\phi_{tt} + \alpha\phi_t + \sin\phi = \eta \tag{1}$$

where $\phi(t)$ is the phase difference between the two superconducting films and time t is measured in units of $1/\omega_o$ where $\omega_o = (2eI_o/\hbar C)^{\frac{1}{2}}$. η is the normalized dc bias current, $\eta_1 = I_{dc}/I_o$, and α (a damping parameter) is given by $\alpha = (\hbar/2eR^2 I_o C)^{\frac{1}{2}}$. $-e$ is the electronic charge, and $2\pi\hbar$ is Plancks constant. Eq. 1 does not have a general analytic solution, however a vast literature on the qualitative behaviour and approximated solutions exist.

A Josephson junction transmission line (JTL) is long in one direction (for example for the x direction). In that case the phase difference becomes space dependent and $\phi(x,t)$ obeys an equation of the form

$$-\phi_{xx} + \phi_{tt} + \alpha\phi_t + \sin\phi = \eta \tag{2}$$

where the space dimension x is measured in units of the Josephson penetration depth $\lambda_J = (\hbar/2 \mu_o edJ)^{\frac{1}{2}}$. Here d is the magnetic thickness of the barrier $(d = 2\lambda_L + t_o)$ and J is the maximum pair current density. λ_L is the London penetration depth, and t_o the oxide thickness. A simple way to derive Eq. 2 is to consider the nonlinear transmission line in Fig. 1b as an equivalent circuit for the JTL. Kirchoffs equations are

$$\frac{\partial V}{\partial X} = - L' \frac{\partial I}{\partial t}$$

$$\frac{\partial I}{\partial X} = - C' \frac{\partial V}{\partial t} - J'\sin\phi + J'_{dc} \qquad ; \frac{\partial \phi}{\partial t} = \frac{2eV}{\hbar} \tag{3}$$

In these equations L', C', J' and J'_{dc} are inductance, capacitance, maximum pair current and bias current per unit length i.e.

$$L' = \frac{\mu_o \cdot d}{W}, \ C' = \frac{W \epsilon_r \epsilon_o}{t_o}, \ J' = J \cdot W, \ J'_{dc} = \frac{I_{dc}}{L} \tag{4}$$

L and W are the length and width of the junction, and ϵ_r is the relative dielectric constant of the oxide layer. With time and space normalizations introduced in connection with Eqs. 2,3 the above equations yield Eq. 2. Also we find from Eqs. 3,4 :

$$\phi_t = v = V/(\hbar\omega_o/2e) \tag{5}$$

i.e. that ϕ_t is the voltage across the junction normalized to $\hbar\omega_o/2e$.

$$\phi_x = -i = -I/J \cdot W \cdot \lambda_J \tag{6}$$

i.e. ϕ_x is the (normalized) surface current flowing in the x-direction.

With these normalizations velocities are measured in units of

$$\bar{c} = \frac{1}{\sqrt{L'C'}} = c \cdot \sqrt{\frac{t_o}{\varepsilon_r (2\lambda_L + t_o)}} \tag{7}$$

Finally the characteristic impedance of the transmission line becomes

$$Z_o = \sqrt{\frac{L'}{C'}} = \frac{1}{W} \sqrt{\frac{(2\lambda_L + t_o) t_o}{\varepsilon_o \varepsilon_r}} = \frac{(2\lambda_L + t_o)}{W} \frac{\bar{c}}{c} \cdot 377\Omega \tag{8}$$

Eq. 2. may conveniently be written in the form

$$\phi_{tt} - \phi_{xx} + \sin\phi = \varepsilon f \; ; \; 0 \le \varepsilon \le 1 \tag{9}$$

where we for small values of α and η consider the right hand side as a perturbation.

With $\varepsilon f = 0$ Eq. 9 is the ordinary sine-Gordon equation which has the Hamiltonian (Barone et.al. 1971, McLaughlin and Scott 1978)

$$H = \int_{-\infty}^{\infty} (\tfrac{1}{2}\phi_x^2 + \tfrac{1}{2}\phi_t^2 + 1 - \cos\phi) \; dx \quad . \tag{10}$$

Soliton and antisoliton solutions to the sine-Gordon equation are given by

$$\phi_o^{\pm}(x,t) = 4 \tan^{-1}\exp[\pm\gamma(u)(x-ut-x_o)]; \; |u| \le 1 \tag{11}$$

with upper and lower signs respectively. Here x_o is the initial position of the soliton with velocity u and $\gamma(u)$ is the Lorentz factor

$$\gamma(u) = (1-u^2)^{-\tfrac{1}{2}} \tag{12}$$

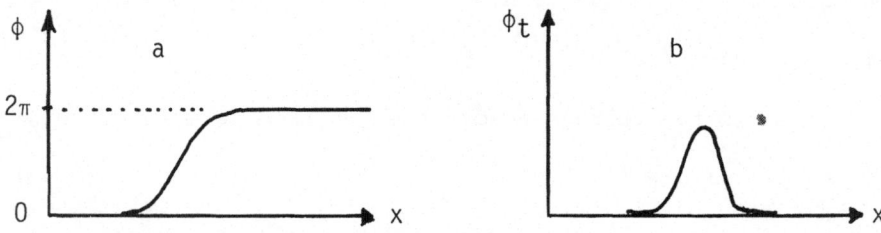

Fig. 2 Schematic drawing of (a) the phase, and (b) the time derivate of the phase for a sine-Gordon soliton.

Fig. 2 shows qualitatively ϕ_o and $(\phi_o)_t$ for a sine-Gordon soliton; inserting Eq. 11 in Eq. 10 yields the energy of a single soliton

$$H_s = 8 \cdot \gamma(u) \qquad (13)$$

With $\epsilon f = \eta - \alpha\phi_t$ (corresponding to Eq. 2) small we may look for solutions close to the soliton solution, Eq. 11. Assuming that the effect of the perturbation on the soliton is to modulate its velocity, we obtain (McLaughlin and Scott, 1978)

$$\phi(x,t) = 4 \tan^{-1} \exp (\gamma(u(t)))(x-X(t)) + \arcsin \eta \qquad (14)$$

where the location of the soliton is given by

$$X(t) = \int_o^t u(t)dt + x_o \qquad (15)$$

By combining Eqs. 2, 10 and 13 it may be shown that the velocity $u(t)$ approximately satisfy the differential equation

$$\frac{d}{dt}(u\gamma(u)) = \gamma(u)^3 \cdot \frac{du}{dt} = -\alpha u\gamma(u) + \frac{\pi}{4} \cdot \eta \qquad (16)$$

From Eq. 16 with $du/dt = 0$ it follows that the stationary value, u_∞, of u is given by

$$u_\infty = 1/(1+4\alpha/\pi\eta)^2)^{\frac{1}{2}} \qquad (17)$$

Hence the soliton velocity is determined as a balance between the bias current η and the losses represented by α. The qualitative behaviour of $u(\eta)$ is shown in Fig. 3.

Here the perturbation technique was introduced in the simplest possible way to treat the propagation of a soliton on an infinite

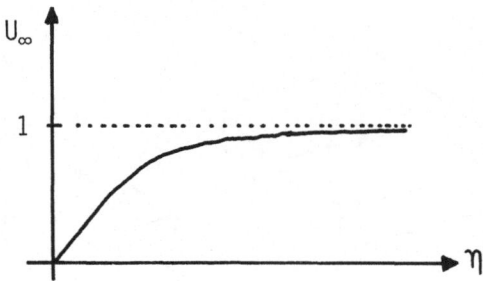

Fig. 3 Qualitative behaviour of the velocity, u_∞, of a soliton on the perturbed sine-Gordon line.

line with loss and bias. Much more complicated problems, such as the
collision of a soliton with an impurity, the collision of a soliton
and an antisoliton etc. have been calculated using this method
(McLaughlin and Scott 1978, Christiansen and Olsen, 1982). In the
following we will use the method to calculate the dc IV curve of
junctions of various geometries.

II.   THE dc IV-CURVE

IIa. PERTURBATION CALCULATION OF THE IV-CURVE OF LONG JUNCTIONS

In section I the single fluxon motion on an infinite line
with loss and bias was discussed in the framework of the perturbed
sine-Gordon model. In a Josephson junction of finite length the
boundary conditions become crucial to the fluxon dynamics. Two
geometries - the overlap geometry shown in Fig. 4a, and the inline
shown in Fig. 4b are of particular interest. In both cases the
long side of the junction defines the x-direction and the short
the y-direction. The film thickness is assumed to be much larger
than the London penetration layer of depth λ_L where all the surface
currents flow. We assume further that $\ell = L/\lambda_J >> 1$ and $w = W/\lambda_J << 1$
so that spatial variations of the phase occur only in the x-direction.
From Eq. 6 the surface current in the London penetration layer is
given by

$$I_x = W\lambda_J J\phi_x \qquad\qquad (18)$$

and thus the magnetic field in the oxide layer is

$$H_y = I_x/W = \lambda_J J\phi_x \qquad\qquad (19)$$

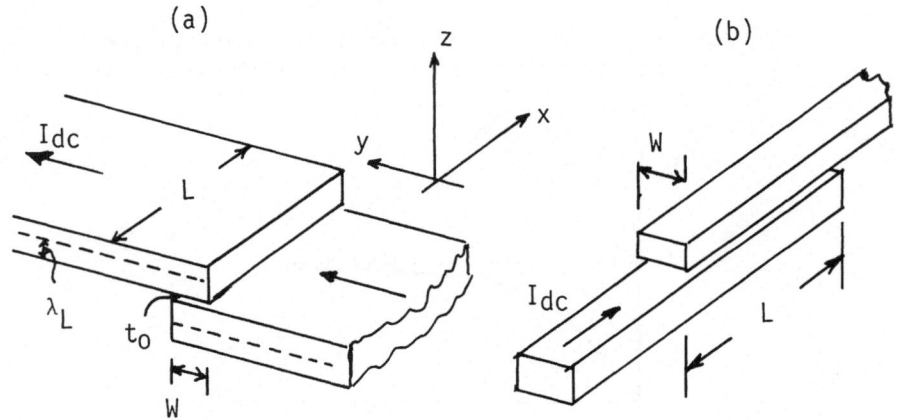

Fig. 4 Geometry of (a) the overlap junction and (b) the inline
       junction.

These considerations are common to both geometries. We now introduce the bias current I_{dc} and the boundary conditions separately for the two cases: (Levring et al., 1982a).

In the overlap junction (Fig. 4a) the bias current I_{dc} is uniformly distributed and enters through the term η in Eq. 2, which is then given by

$$\eta^{ov} = \frac{I_{dc}}{J \cdot W \cdot L} \tag{20}$$

i.e. the critical current I_c^{ov} is given by $I_c^{ov} = JW \cdot L$. Since no currents flow out at the junction ends the boundary conditions are (Levring et al., 1982)

$$\phi_x (0,t) = \phi_x(\ell,t) = 0 \tag{21}$$

since - because of the geometry - there are no selffields in the y direction. With an external magnetic field in the y-direction we get (Eq. 19)

$$\phi_x (0,t) = \phi_x(\ell,t) = \frac{H_{ext}}{\lambda_J \cdot J} \equiv \kappa_{ext} \tag{22}$$

In the inline junction (Fig. 4b) we obtain because of the special bias current distribution (Scott and Johnson, 1969)

$$\eta^{in} = 0 \tag{23}$$

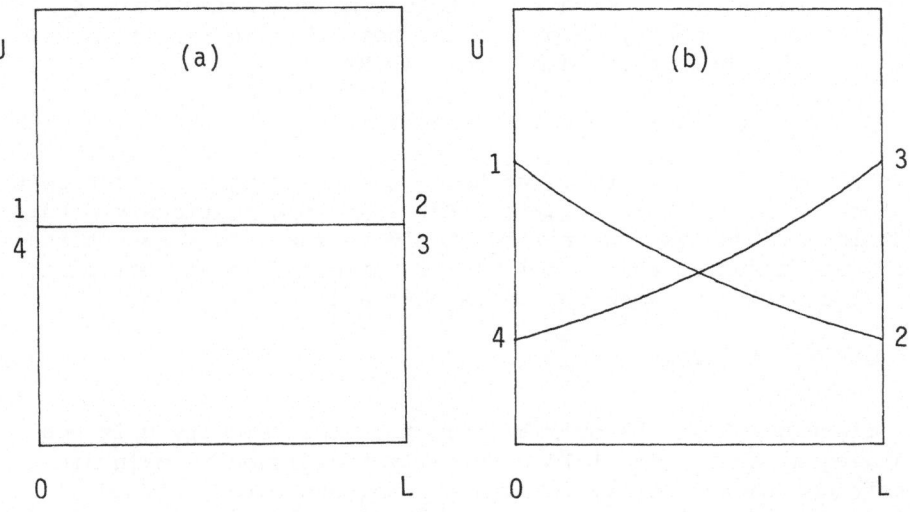

Fig. 5 Trajectories of a soliton in (a) the overlap junction, and (b) the inline junction. The numbers give the sequence of the trajectories.

in Eq. 2, and the bias current enters only through the magnetic
field it creates at the ends of the junction through the boundary
conditions (Barone et al., 1975)

$$\phi_x(0,t) = -\phi_x(\ell,t) = \frac{I_{dc}}{2\lambda_J \cdot W \cdot J} = \kappa^{in} \tag{24}$$

For this case the critical current I_c^{in} is given by $I_c^{in} = 4\lambda_J \cdot W \cdot J$
which is smaller than in the overlap case. In our notation the
maximum supercurrent correspond to $\kappa^{in} = 2$.

With an external magnetic field in the y-direction the
boundary conditions become

$$\left.\begin{array}{c} \phi_x(0,t) \\[6pt] \phi_x(\ell,t) \end{array}\right\} = \pm\,\kappa^{in} + \kappa_{ext} = \frac{1}{\lambda_J \cdot J}\left\{\pm\frac{I_{dc}}{2W} + H_{ext}\right\} \tag{25}$$

If (for both geometries) one fluxon moves back and forth with
average velocity u, a dc voltage v_{dc} (normalized to $\hbar\omega_9/2e$) is
developed across the junction. Here $v_{dc} = 2\cdot2\pi/T = 4\pi u/2\ell$ where T
is the (normalized) period. In order to determine this average
velocity we write the total (normalized) energy on the junction
(Eq. 10)

$$H = \int_0^\ell (\tfrac{1}{2}\phi_x^2 + \tfrac{1}{2}\phi_t^2 + (1-\cos\phi))\,dx \quad . \tag{26}$$

Differentiating Eq. 26 with respect to time and using η and κ for
the particular geometry we obtain the perturbation result (Levring
et al., 1982a, McLaughlin and Scott, 1978)

$$\frac{dH}{dt} = -\kappa[\phi_t(\ell,t) + \phi_t(0,t)] - 8u^2\cdot\gamma(u)\alpha + 2\pi\eta u \quad . \tag{27}$$

To obtain the last two terms we have used the single soliton solution to the sine-Gordon equation (Eq. 14). In a stationary condition
dH/dt integrated over one period should be zero. In zero external
field, and assuming that a soliton is reflected as an antisoliton,
the phase shift over one period is -4π, and Eq. 27 gives

$$\alpha 8u^2\cdot\gamma(u) = 2\pi\eta u + \kappa\cdot\frac{8}{T}\pi \quad . \tag{28}$$

Since $\kappa = 0$ in the overlap geometry, the velocity u is constant, equal to u_∞ (Eq. 17). Physically the fluxon is maintained
at this constant velocity through the Lorentz force from the
uniformly distributed bias current. The dynamic picture is
shown in Fig. 5a.

For the inline geometry $\eta = 0$, (Eq. 23) and the soliton

velocity decreases towards $u_\infty = 0$ along the Josephson line (Fig. 5b).
When it reaches the boundary it gets an energy input of $4\pi\kappa$ from
the boundary condition and continues its steady state motion. If
for the inline case the velocity change along the junction is small
(i.e. $\alpha\ell \ll 1$) we may insert $T = 2\ell/u$ in Eq. 28 to determine u.

The (normalized) voltage across the junction, v_{dc}, is then
given by $v_{dc} = 2\pi u/\ell$, where u is determined from Eq. 28 for the two
geometries separately. The result is (Levring et al. 1982a)

$$u\gamma(u) = \begin{cases} \dfrac{\pi\eta}{4\alpha} = \dfrac{I_{dc}}{4\alpha JWL} & ; \quad \text{all } \alpha\ell \qquad \text{(overlap junction) (29a)} \\[3em] \dfrac{\pi\kappa}{2\alpha\ell} = \dfrac{I_{dc}}{4\alpha JWL} & ; \quad \alpha\ell \ll 1 \qquad \text{(inline junction) (29b)} \end{cases}$$

(Eq. 29a is identical to Eq. 17). Hence in the limit $\alpha\ell \ll 1$ results
for the two geometries are identical and the so-called zero field
steps from overlap junctions also exist in inline junctions.

The common result for the two geometries is shown in Fig. 6.
The only difference is that the critical current for the inline
junction, I_c^{in}, is smaller than that for the overlap junction, I_c^{ov},
as discussed earlier. Higher order zero field steps may be obtained
by scaling the voltage with the number of fluxons.

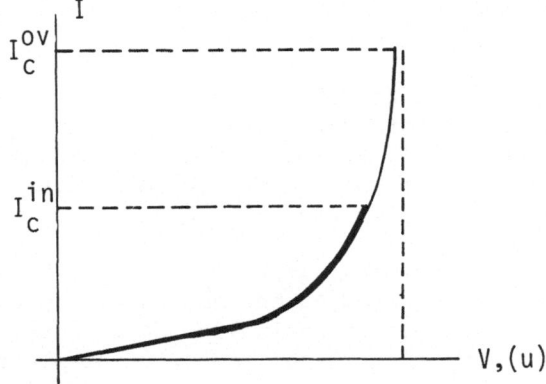

Fig. 6   Qualitative drawing of the IV curve of a zero field step
         in the inline and overlap junctions. (After Levring et
         al., 1982a).

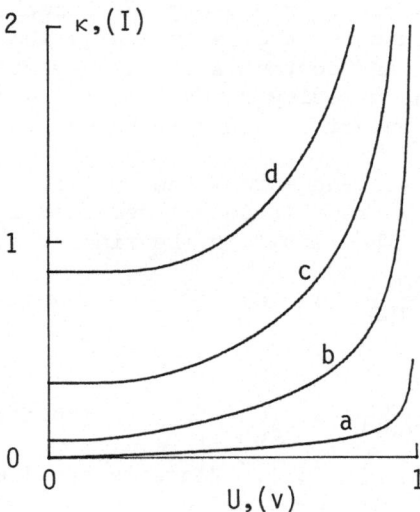

Fig. 7   Calculation of the zero field step of an inline junction.
$\alpha\ell = 0.1$ (a), 0.5 (b), 1.0 (c), and 1.5 (d).
(After Levring et al. 1982b).

If in the inline geometry the condition $\alpha\ell \ll 1$ is not satisfied
integration is necessary to obtain T in Eq. 28. For the average
velocity, u_{av}, the analytical expression is (Levring et al. 1982b)

$$\kappa^{in} = \frac{4}{\pi} \frac{\sinh(\frac{\alpha\ell}{2})}{\sqrt{\left[\frac{\tanh(\frac{\alpha\ell}{2u_{av}})}{\tanh(\frac{\alpha\ell}{2})}\right]^2 - 1}} \; ; \quad \text{all } \alpha\ell \quad \text{(inline junction)} \quad (30)$$

This expression approaches Eq. 29b in the limit of $\alpha\ell \ll 1$. For
other values of $\alpha\ell$ the relation between $I_{dc} \propto \kappa$ and $v_{dc} \propto u_{av}$ is
shown in Fig. 7. The cut-off at low $\kappa \sim I_{dc}$ is the main qualitative
difference from the overlap case, where for all $\alpha\ell$ the zero field
steps start in the origin (Eq. 29a).

If, in the inline geometry, $\alpha\ell \gg 1$ and $I_{dc} > I_c^{in}$ ($\kappa > 2$)
fluxons are continously created in one end of the junction and
antifluxons in the other. They annihilate each other in the center
of the junction and give rise to the so-called displaced linear
branch as discussed by Scott and Johnson, 1969. This displaced
linear branch may in unnormalized units be expressed as

$$V = \frac{1}{2} \sqrt{\frac{L'}{C'}} (I_{dc} - I_c^{in}) \tag{31}$$

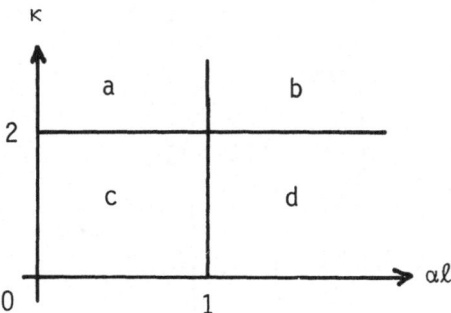

Fig. 8   Regions of qualitatively different behaviour for the inline
         junction: (a) continuous generation of fluxons and anti-
         fluxons with reflections in both ends. (b) displaced
         linear branch. (c) zero field steps, and (d) static fluxon
         solutions (zero voltage).

        The regions of existence of the various modes in inline
junctions is shown in Fig. 8. It is not yet clear whether there
may be coexistence of zero field steps and a displaced linear branch
in the same junction in the vicinity of $\alpha\ell \sim 1$.

        It should be mentioned that there have been direct analytic
approaches to calculating the IV-curve of long overlap junctions.
Parmentier 1978, Costabile et al. 1978, and DeLeonardis et al.,
1980, treat soliton dynamics in finite junctions based on the
pure sine-Gordon equation. The former authors are able to
calculate analytically the zero field steps by assuming a quadratic
loss term ($\sim \Gamma\phi_t^2$ instead of $\alpha\phi_t$ in Eq. 2) and periodic boundary
conditions. Marcus and Imry, 1980, transform Eq. 2 to an ordinary
differential equation and assume periodic boundary conditions.

IIb. dc EXPERIMENTS ON LONG JUNCTIONS

        Experimentally zero field steps were first reported by Chen et
al., 1969. Fig. 9 shows an example from their work (Chen et al.
1974). (Note that their numbering of steps is twice that used here).
To illustrate the previous section we have chosen some results
recently obtained in our laboratory because - for the purpose of
comparison - inline and overlap junctions with equal dimensions
(L × W = 1000 μm × 20 μm) were fabricated on the same substrate.
The junctions are Nb-Nboxide-Pb junctions with $\ell \simeq 6$. The IV-curves
are shown in Fig. 10. We note the existence of zero field steps at
almost the same voltages in inline and overlap junctions. The
junctions had some excess current probably due to some NbO instead of
Nb_2O_5 in the oxide layer; (this excess current gives rise to α-
values in the range 0.1 - 0.5 which are convenient for numerical
simulations).

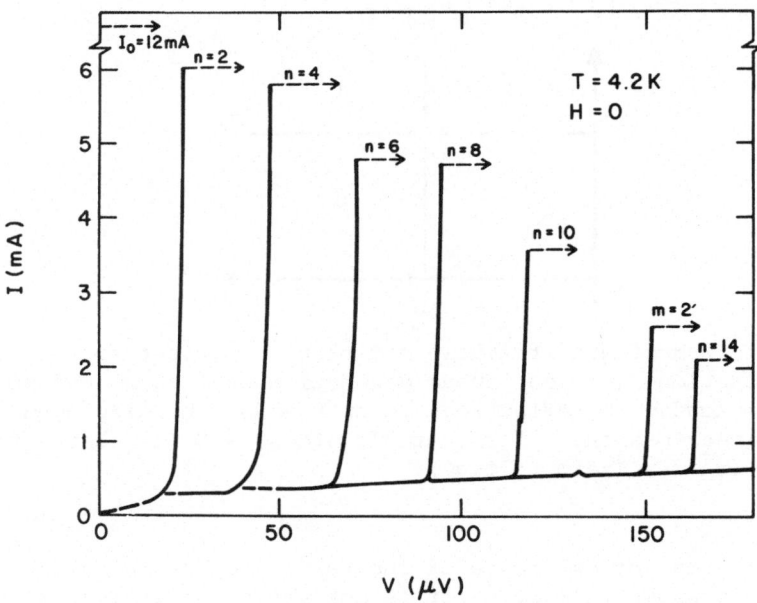

Fig. 9   Zero field steps in a Pb-Pboxide Pb junction (after Chen
         and Langenberg, 1974).

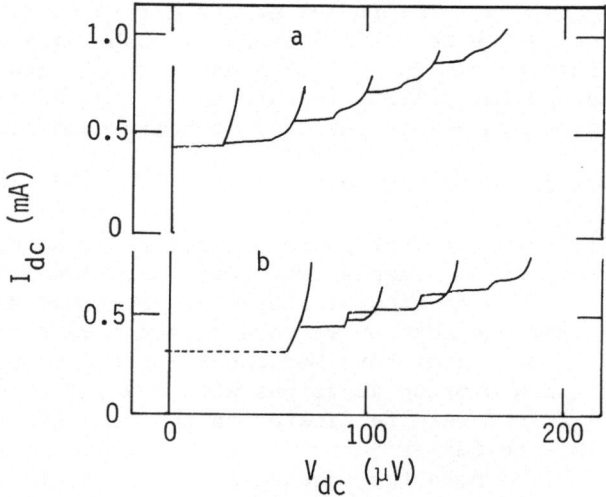

Fig. 10   Zero field steps of (a) overlap junction and (b) inline
          junction. For both curves: L = 1000 μm, W = 20 μm.
          Nb-Nboxide-Pb junctions, T = 4.2 K. There is excess
          background current due to NbO contamination of the barrier.

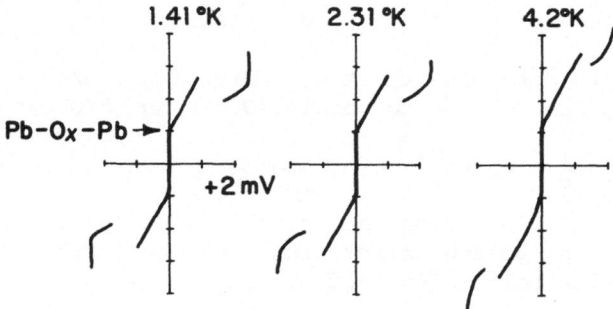

Fig. 11   Displaced linear branch in an inline Pb-Pboxide-Pb-junction:
          Vertical 500 mA/div. horizontal 2 mV/div. (after Scott and
          Johnson, 1969).

An example of the displaced linear branch corresponding to the
vortex flow situation in inline junctions (region b in Fig. 8) is
shown in Fig. 11 (after Scott and Johnson, 1969). Although the
present paper is concentrated on resonant fluxon propagation it
should be mentioned that active research on the flux flow regime in
inline junctions (Yoshida et al. 1978, Hamasaki et al. 1980, and
Rajeevakumar et al. 1980) and overlap junctions (Nakajima et al.
1978) is continuing.

III. APPLICATION OF A MAGNETIC FIELD

IIIa MAGNETIC FIELD TUNING OF THE ZERO FIELD STEPS

In the possible microwave applications magnetic field tuning
of the zero field step is very important. Again to investigate that
problem we have to rely on the four methods - analytical, numerical,
analog, and experimental - and unfortunately, at present those
methods do not always agree.

Recently Levring et al. 1982b obtained analytical solutions
for magnetic field tuning of inline - and overlap zero field steps
in the framework of the perturbed sine-Gordon model described in
sections I and II. Following their derivation we introduce for
convenience the quantities a and z defined by

$$u = \tanh a = (z^2/(1+z^2))^{\frac{1}{2}} \tag{32}$$

$$z = u \cdot \gamma(u) = P_f/8$$

where P_f is the momentum of a sine-Gordon soliton. Eq. 32 implies
that

$$\gamma(u) = \cosh a, \quad z = \sinh a, \quad u_\infty = \tanh a_\infty, \quad z_\infty = \frac{\pi\eta}{4\alpha} \tag{33}$$

In order to determine the trajectory of a fluxon we integrate Eq. 16 and use Eqs. 32, 33 to obtain (Olsen and Samuelsen, 1982)

$$z = z_\infty + (z_o - z_\infty)e^{-\alpha t} \tag{34}$$

where $z_o = u_o \cdot \gamma(u_o)$, u_o being the initial velocity of the fluxon. Finding u from Eq. 32 and integrating one step further (Eq. 15) we obtain (valid for both inline and overlap cases)

$$x(t) = x_o + u_\infty t - \frac{1}{\alpha}\ln\frac{z+\sqrt{z^2+1}}{z_o+\sqrt{z_o^2+1}} - \frac{u_\infty}{\alpha}\ln\frac{1+z_\infty z_o + \sqrt{z^2+1}\sqrt{z_o^2+1}}{1+z_\infty z_o + \sqrt{z_\infty^2+1}\sqrt{z^2+1}} \tag{35}$$

Physically the main effect of the magnetic field is the following. Since the current flowing in the x-direction is given by ϕ_x and the voltage across the junction is given by ϕ_t, and since there is a phaseshift of 4π connected with the fluxon reflected as an anti-fluxon we get an energy increase at $x = 0$ of $-4\pi \cdot \phi_x(0,t)$ and at $x = \ell$ of $4\pi \cdot \phi_x(\ell,t)$.

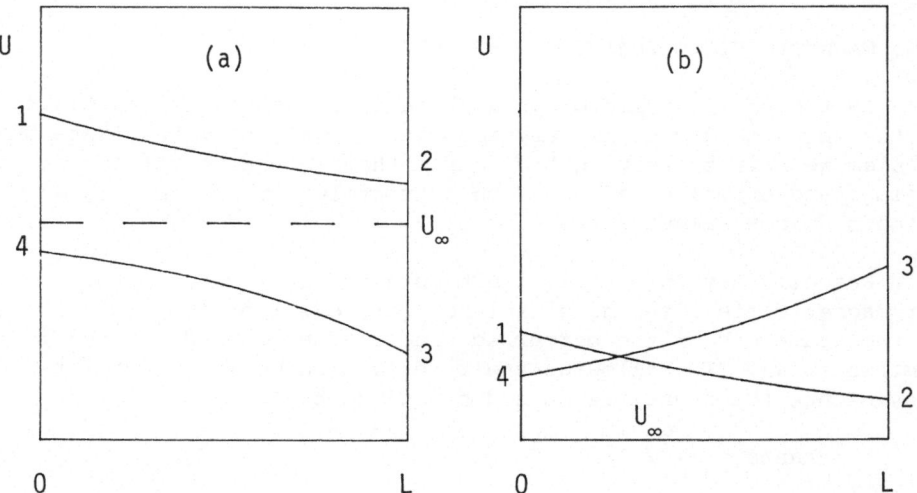

Fig. 12   Trajectories of a soliton in an external magnetic field.
(a) overlap junction (b) inline junction. The numbers
give the sequence of the trajectories.

This is illustrated in Figs. 12a, b which show trajectories for the two geometries. For the overlap junction we get with the boundary conditions Eq. 22 and using Eqs. 13, 33

$$\pi \frac{\kappa_{ext}}{2} = \cosh a_i - \cosh a_j \; ; \quad i,j = \begin{cases} 2,3 \\ 1,4 \end{cases} \tag{36}$$

where i, j refer to the end points of the trajectories in Fig. 12a. The times of flight t_{12} and t_{34} are determined from Eq. 34

$$\alpha t_{ij} = \ln \frac{z_i - z_\infty}{z_j - z_\infty} = \ln \frac{\sinh a_i - \sinh a_\infty}{\sinh a_j - \sinh a_\infty} \; ; \quad i,j = \begin{cases} 1,2 \\ 3,4 \end{cases} \tag{37}$$

and finally Eq. 35 gives

$$\frac{\alpha\ell}{u_\infty} = \alpha t_{ij} - \frac{a_j - a_i}{u_\infty} - 2 \ln \frac{\cosh\frac{a_i + a_\infty}{2}}{\cosh\frac{a_j + a_\infty}{2}} \; ; \quad i,j = \begin{cases} 1,2 \\ 3,4 \end{cases} \tag{38}$$

The six equations Eqs. 36, 37, 38 determine a_1, a_2, a_3, a_4, t_{12} and t_{34}, and the voltage of the zero field step is given by the average velocity $u_{av} = 2\ell/(t_{12} + t_{34})$. We have not been able to find a closed analytical expression for this quantity based on the above equations, however, an expansion valid to second order in the magnetic field is

$$v_{dc}(H_{ext}) = \frac{2\pi u_{av}}{\ell} = v_{dc}(0) \left\{ 1 - \frac{\pi^2}{8} \frac{\tanh\frac{\alpha\ell}{2u_\infty}}{\frac{\alpha\ell}{2u_\infty}} \frac{(2+u_\infty^2)(1-u_\infty^2)^2}{u_\infty^4} \kappa_{ext}^2 \right\} \tag{39}$$

where $v_{dc}(0) = 2\pi u_\infty/\ell$ with u_∞ given by Eq. 17. We note from Eq. 39 that the voltage of the zero field step can only decrease, and that the effect is largest for small η. Fig. 13 shows a calculation of the first zero field step for various values of the magnetic field.

For the inline junction the trajectories are shown in Fig. 12b; the boundary conditions Eq. 25 give the energy inputs at $x = 0$ and $x = \ell$

$$\frac{\pi}{2} (\kappa + \kappa_{ext}) = \cosh a_1 - \cosh a_4$$
$$\frac{\pi}{2} (\kappa - \kappa_{ext}) = \cosh a_3 - \cosh a_2 \tag{40}$$

For the times of flight t_{12} and t_{34} we obtain from Eq. 37

$$e^{\alpha t_{ij}} = \frac{\sinh a_i}{\sinh a_j} \; ; \quad i,j = \begin{cases} 1,2 \\ 3,4 \end{cases} \tag{41a}$$

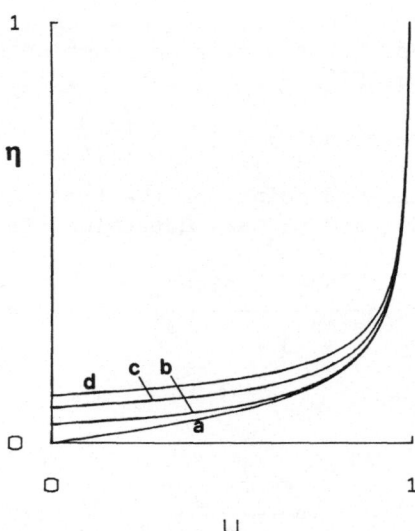

Fig. 13   Magnetic field tuning of the zero field step of an overlap
          junction. $\alpha = 0.1$, $\ell = 10$, $\kappa_{ext} = 0$ (a), 0.1 (b) 0.3 (c)
          and 0.5 (d).

$$e^{\alpha(t_{12}+t_{34})} = e^{2\alpha\ell/u_{av}} = \frac{\sinh a_1 \cdot \sinh a_3}{\sinh a_2 \cdot \sinh a_4} \tag{41b}$$

Finally Eq. 35 (with $u_\infty = 0$ and $z_\infty = 0$) gives

$$\alpha\ell = \ln\frac{\sinh a_i + \cosh a_i}{\sinh a_j + \cosh a_j} \quad ; \qquad i,j = \begin{cases} 1,2 \\ 3,4 \end{cases} \tag{42}$$

which implies $a_1 - a_2 = a_3 - a_4$. Again combining the five equations
Eqs. 40, 41b, and 42 determine a_1, a_2, a_3, a_4, and u_{av} and hence the
voltage of the zero field step. If we for convenience define

$$A = \left(\frac{\pi\kappa}{4\sinh\frac{\alpha\ell}{2}}\right)^2, \qquad B = \left(\frac{\pi\kappa_{ext}}{4\cosh\frac{\alpha\ell}{2}}\right)^2 \tag{43}$$

we find after some rather lengthy calculations that the voltage of
the zero field step is given by (Levring et al., 1982b)

$$v_{dc} = \frac{2\pi u_{av}}{\ell} = 4\pi\alpha\left\{\ln\frac{[(A-B)^{\frac{1}{2}}\cosh\frac{\alpha\ell}{2} + (1+A-B)^{\frac{1}{2}}\sinh\frac{\alpha\ell}{2}]^2 - \frac{B}{A-B}}{[(A-B)^{\frac{1}{2}}\cosh\frac{\alpha\ell}{2} - (1+A-B)^{\frac{1}{2}}\sinh\frac{\alpha\ell}{2}]^2 - \frac{B}{A-B}}\right\}^{-1} \tag{44}$$

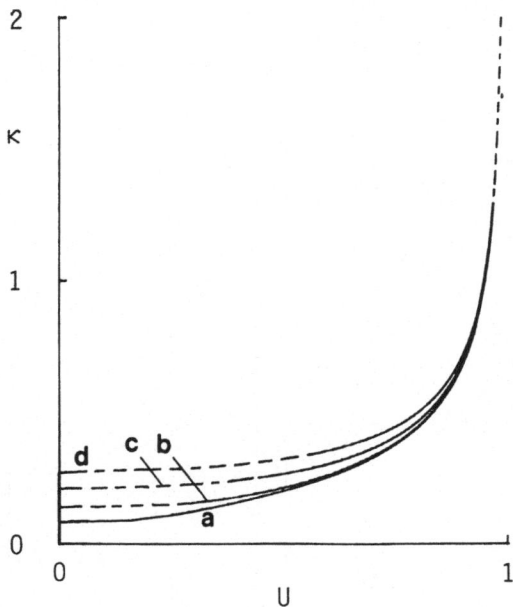

Fig. 14   Magnetic field tuning of the zero field step of an inline
          junction $\alpha l = 0.5$, $\kappa_{ext} = 0$ (a), 0.1 (b), 0.3 (c), and
          0.5 (d).

from which the zero field steps as a function of bias current and
magnetic field may be calculated directly. Eq. 44 yields Eq. 30
and Eq. 29b in the limits of $\kappa_{ext} \to 0$ and $\kappa_{ext} \to 0$, $\alpha l \to 0$,
respectively. Fig. 14 shows a calculation of the inline zero field
step for various values of the magnetic field. Note that the
qualitative behaviour is the same as for the overlap case; in
particular the voltage can only decrease and the magnetic tuning
is largest for small values of the bias current. An important
limitation on this perturbation calculation is the assumption that
a soliton is reflected as an antisoliton without radiation losses.
The validity of this assumption was numerically investigated by
Olsen and Samuelsen 1981 for the pure sine-Gordon equation. They
find that, depending on the parameters, a soliton may also be
reflected as a soliton, several anti-solitons or even annihilated.
The dashed parts in Fig. 14 indicate approximately regions where
such restrictions occur.

As mentioned in the beginning of this section the four methods
do not always agree. Recently a numerical simulation of magnetic
field tuning for both inline and overlap junctions was done with
$l = 5$ and $\alpha = 0.25$ (Erné et al., 1982a).

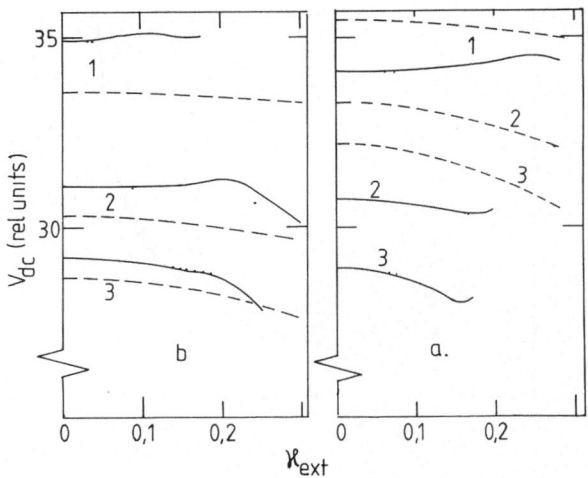

Fig. 15 Comparison of the perturbation calculation with a numerical
        calculation by Erné et al. 1983. (a) overlap case, η = 0.65
        (1), 0.50 (2), and 0.45 (3).   (b) inline case, κ = 1.625 (1),
        1.25 (2), and 1.125 (3). α = 0.25, ℓ = 5 for all curves.
        Full curves: Numerical calculation. Dashed curves:
        Perturbation result. Dotted curves: Perturbation result
        shifted to coincide at κ_{ext} = 0. Note the broken scale at
        the voltage axis.

Although our perturbation method is only valid for α << 1, ℓ >> 1,
and hence not applicable to those parameters we dare a comparison
in Figs. 15a,b for the overlap and inline geometries respectively.
Comparing the perturbation result with the numerical simulation we
note the following; even in zero magnetic field there are differences
(∿3%). This may possibly be ascribed to the fact that a phaseshift
in connection with the reflection is not included in the
perturbation calculation (with the rather short junction, ℓ = 5,
this may be important). With an applied magnetic field the
qualitative behaviour deviates (since the numerical calculation

gives a frequency increase) for the largest values of the bias
current, whereas the agreement is much better for lower values of
the bias current. This difference is to be expected in as much as
the use of the formulas connected with the pure sine-Gordon equation
is less appropriate the higher the average velocity. In any case the
comparison made in Figs. 15a,b is not a priori feasible because the
conditions for the perturbation calculation are not satisfied. .

A very important feature of the IV-curve when a magnetic field
is applied is the appearence of so-called Fiske steps (Coon and
Fiske 1965), which appear in between the zero field steps, i.e.,
at voltages $V_n = n(2\pi u_{av}/\ell)$ where $n = \frac{1}{2}$, $3/2$, $5/2$, etc. In the
soliton models so far, there have not appeared realistic numerical
or analytical calculations of those Fiske steps; however, some work
has been done within the framework of standing electromagnetic waves
(Kulik 1967) valid only for $\ell \lesssim 1$.

For the $n = \frac{1}{2}$ step the soliton model predicts on the average
one half soliton in the junction. This can be understood in a
picture (Dueholm et al. 1981b) where a soliton propagates from $x = \ell$,
is annihilated at $x = 0$ due to the energy decrease (discussed in
connection with Eqs. 36, 40), a plasma wave propagates to $x = \ell$
where it reappears as a soliton due to the energy increase there.
That this mechanism is indeed possible was demonstrated in numerical
experiments (Dueholm et al. 1981b).

A final remark in connection with the magnetic field dependence
of zero field steps is that it still remains to calculate the height
of the zero field steps within the framework of perturbed sine-
Gordon model.

IIIb EXPERIMENTS ON MAGNETIC FIELD TUNING

It would be natural to compare the calculations in the previous
section to experiments on real junctions, however very limited
experimental data exist in the literature. For overlap junctions
Costabile et al. 1980 report a frequency increase in contradiction
to the perturbation result; however, a comparison is not really
fair since the normalized length of their junction was only about 2.
For inline junctions Finnegan et al. 1975 report a frequency decrease,
although no further details were given. Very recently Erné et al.
1982b have done more detailed measurements, an example of which
is shown in Fig. 16 for a junction with $L/\lambda_J \simeq 14$. Note that the
maximum in frequency is shifted (probably due to trapped flux) but
that the overall picture is a frequency decrease with magnetic
field, in agreement with the perturbation result.

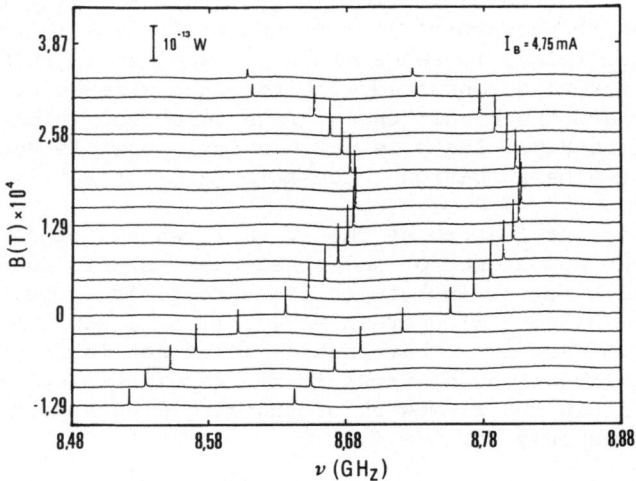

Fig. 16   Emitted power from the first zero field step for an inline
          junction with $L/\lambda_J \simeq 14$. The two peaks correspond to the
          two sidebands from the microwave receiver (from Erné et
          al. 1982b).

IV.   rf-PROPERTIES

IVa   INTERNAL MICROWAVE GENERATION: THEORY

     As mentioned previously it is possible to use the long Josephson
junction as a microwave generator when it is biased on one of the
zero field steps. If the average velocity of one soliton is u_{av} then
it will reach one end of the junction with a frequency $f_1 = u_{av}/2\ell$.
If parts of the soliton energy is coupled out one may detect an rf
signal at the frequency f_1 in an external circuit. When biased on
the second zero field step two solitons are present. In the original
picture (Fulton and Dynes, 1973) it was assumed that a soliton and
an antisoliton would move in opposite directions in a symmetric
motion as shown in the upper part of Fig. 17b. We will denote this
the $F + \bar{F}$ mode. This is natural to expect since two solitons on an
infinite line repel each other. The frequency of the radiation in
that symmetric case would be $f_2^s = 2f_1$. However, another configuration
where two solitons move together (FF-mode) as shown in the lower
part of Fig. 17b is in principle possible on the second zero field
step. For this antisymmetric configuration the frequency of the
radiation would be $f_2^a = u'_{av}/2\ell$ where u'_{av} is the average velocity
of the two soliton bunch. Assuming $u'_{av} \simeq u_{av}$ we get $f_2^a \simeq f_1$, i.e.
half the frequency of the symmetric mode.

     On the third zero field step (3 solitons) three modes could
in principle exist (Fig. 17c); (i) three single solitons (anti-
solitons) $(F + F + \bar{F}$ - mode) giving a frequency $\sim 3f_1$; (ii) a three

soliton bunch with frequency $\sim f_1$ (FFF-mode) and (iii) a two soliton
bunch and a single (anti) soliton moving in a roughly symmetric way
and giving a frequency $\sim 2f_1$ (FF + $\bar{\text{F}}$ - mode). From the point of view
of the sine-Gordon equation on the infinite line the bunching as
discussed here cannot take place. This is - however - a shortcoming
of the model, because already in 1974 it was demonstrated by
numerical calculations that complicated bunching of solitons on a
line with loss and bias may take place (Nakajima et al., 1974a,
Nakajima et al., 1975). The same effect was discovered in mechanical
models (Nakajima et al., 1974b, Cirillo et al., 1981) and recently
in an experiment on real junctions (Dueholm et al., 1981a). Later,
within the framework of the perturbed sine-Gordon model Karpman et
al., 1981, investigated the bunched two soliton problem both on an
infinite line and in an overlap junction of finite length. On the
infinite line they find that two solitons started very close together
approach an equilibrium distance that is roughly the dimension of
a soliton ($\sim \lambda_J$), and stay there for very long time. As time goes
to infinity they diverge slowly (logarithmically). On a junction
of finite length they find that the reflections at the boundaries
does not change that picture, and the bunched two-soliton mode is
observed. The effect of boundary conditions and other parameters
on the bunching phenomenon is still a matter of dispute (T. H.
Soerensen et al., 1982).

A full numerical simulation of Eq. 2 on the second zero field
step of an overlap junction was performed by Erné and Parmentier,
1981a. They made the following conclusion based on their numerical
studies: For junctions of small length, $\ell \lesssim 6$ only the "traditional"
symmetric F + $\bar{\text{F}}$ mode exists. For lengths $\ell \gtrsim 6$ both modes may exist.
The general rule was that the F + $\bar{\text{F}}$ mode exist for small values of
the bias current and that the bunched FF mode exists for larger
values. For intermediate bias values there is a region where both

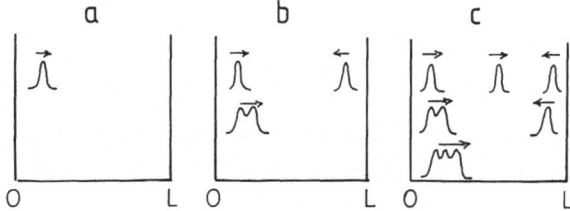

Fig. 17 Schematic drawing of possible soliton configurations on (a)
        the n = 1 step, (b) the n = 2 step, and (c) the n = 3 step.

modes are stable. Connected with a change from one mode to the
other there is a small change (less than a percent) in the voltage
of the zero field step and - of course - a factor of two change in
the frequency of emitted radiation as discussed previously. Such
mode transitions are speculated to be the cause of fine structure
often observed in zero field steps (Chen and Langenberg, 1974).
These observations have been confirmed in a series of measurements
(Cirillo et al., 1981) on a mechanical analog of a long overlap
junction. For completeness it should be mentioned that the addition
of surface impedance damping term (a term $\beta\phi_{xxt}$ in Eq. 2) may help
create bunching on the infinite line. (Johnson 1968, Nakajima et
al., 1974a). At this point it is tempting to make the following
parallel. A sine-Gordon soliton is often treated - and have
properties as a particle. Perturbing the sine-Gordon soliton with
external forces ($\sim$bias) and various losses ($\sim\alpha\phi_t$, $\sim\beta\phi_{xxt}$) and
boundary conditions may create n-soliton (n-particle) bound states;
we may talk about binding forces and "molecular theory" of
solitons on the perturbed sine-Gordon line.

A question important for the possible use of the soliton microwave generator is how much power can be coupled to an external
circuit. For a simple estimate we assume that our JTL with
characteristic impedance Z_o (Eq. 8) is loaded with a frequency independent load resistor R_L. The power dissipated in the load
resistance P_L is approximately

$$P_L \simeq \frac{4Z_o R_L}{(R_L + Z_o)^2} \cdot P_{sol} \tag{45}$$

where P_{sol} is the power in the soliton. Assuming $(2\lambda_L + t_o) \simeq 0.15$ μm,
$\bar{c}/c \simeq 0.04$ and $W = 15$ μm, we find from Eq. 8, $Z_o \simeq 0.15$ Ω. The
parameters used here are representative of Nb-Nboxide-Pb junctions,
and give the typical impedance level of the JTL. Note that Z_o is
approximately independent of the current density, but depends
critically on the width. In practice it is difficult to obtain such
low impedance levels for R_L in microstrip circuits, i.e. the normal
experimental situation is $R_L \gg Z_o$. The power in the soliton may
be estimated to

$$P_{sol} \sim J \cdot W \cdot \lambda_J \cdot V_n \tag{46}$$

where V_n is the voltage of n'th zero field step (determined largely
by the length L of the junction). Thus in the limit $R_L \gg Z_o$ we find

$$P_L \sim \frac{4Z_o}{R_L} \cdot JW \cdot \lambda_J V_n \tag{47}$$

Using Eq. 8 for Z_o we note that P_L is independent of W and increases as the square root of the current density. For a junction with $J \simeq 200$ A/cm^2, $\lambda_J \simeq 30$ μm, W = 15 μm biased on the second zero field step at ≈140 μV, we find at 35 GHz (FF-mode) $P_{sol} \simeq 10^{-7}$ W. With a load resistor $R_L \simeq 1.5\Omega$ we estimate $P_L \simeq 4 \times 10^{-8}$ W.   This crude estimate is of course optimistic. Firstly the power was assumed to be only dissipated at the fundamental soliton frequency, secondly the assumption of a frequency independent load resistor of 1.5 Ω is unrealistic, and thirdly the system was treated as if it was linear. Nevertheless the example illustrates the power levels in question. A fundamental problem is how hard the JTL can be loaded without destroying the soliton reflection. This was investigated numerically by Erné and Parmentier, 1981. As a guideline they found that the simple soliton picture survives if $R_L \gtrsim 5Z_o$. A potential application of the soliton microwave generator is as pump source for Josephson junction mixers (or SIS mixers) or parametric amplifiers. For such applications the power level is typically nanowatts; thus even with allowance for an optimistic estimate, the soliton microwave generator appears to be a realistic device.

    Apart from the emitted power the linewidth is important for applications. Thus single microbridges have also been considered for microwave oscillators; for those the emitted power may be comparable to that of long Josephson tunnel junctions, however the linewidths are typically tens of Megahertz - too large for most applications. For long Josephson junctions analytical calculations of the linewidth within the framework of the perturbed sine-Gordon model was recently performed (Salerno and Scott, 1982, Joergensen et al., 1982). A particularly simple expression was derived by Joergensen et al., 1982 who finds for the overlap junction

$$\Delta f = \left(\frac{2e}{h}\right)^2 \cdot \frac{\pi^2}{8\ell} \cdot \frac{4\pi kRT}{\gamma(u_\infty)^5} = \left(\frac{2e}{h}\right)^2 \cdot \pi kT \cdot \frac{R_d^2}{R_s} \; ; \qquad (48)$$

here $R_s = V_{dc}/I_{dc}$ taken at the bias point on the zero field step and R_d is the dynamic resistance in the same point. Note that surprisingly this expression is the same as that used for microbridges and small tunneljunctions although the model (solitons in the perturbed sine-Gordon equation) is completely different. The details of this calculation will be published shortly (Joergensen et al., 1982). Because the dynamic resistance approaches zero at the top of a zero field step Eq. 48 predicts linewidths as small as a few kilohertz with realistic junction parameters; this is a very important advantage over other Josephson type microwave oscillators.

IVb. EXPERIMENTS ON MICROWAVE GENERATION

    The first experiment on detecting the emitted microwave power

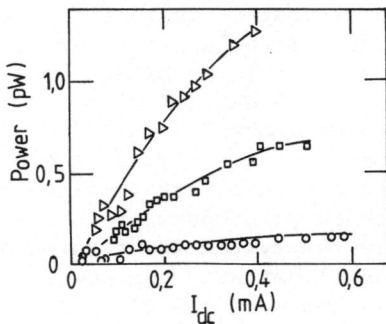

Fig. 18 Emitted power at $f_1 \sim 10$ GHz from the n = 1 step (circles),
         n = 2 step (squares), and n = 3 (triangles). Smooth curves
         have been fitted to the data points.

and relating it to fine structure on zero field steps in long
Josephson junctions was done by Chen and Langenberg, 1974; a more
recent one, which is compared with numerical calculations is Erné
et al., 1981 b. Here I want to concentrate on some recent
experiments done by the Lyngby group, because that work - supple-
mented with very detailed numerical calculations - revealed many
details of the various soliton modes discussed in the previous
section. In that series of experiments (Dueholm et al., 1981a,
Dueholm et al., 1981b) Nb-Nboxide-Pb junctions with $2 \lesssim L/\lambda_J \lesssim 22$
were investigated. The lengths L were such that the fundamental
one soliton frequency f_1 was either ~ 5 GHz or ~ 10 GHz; the
emitted microwaves were detected with a receiver at ~ 10 GHz. The
experiment led to the following conclusions: (i) In the short
samples ($f_1 \sim 10$ GHz) microwave radiation at 10 GHz was observed on
all zero field steps implying bunched 2 soliton propagation on the
n = 2 zero field step, bunched 3 soliton propagation on the n = 3
step etc. Fig. 18 shows the emitted power from the n = 1, 2, and 3
steps. Note that qualitatively the power increases with step
number (V_n in Eq. 47) and bias current as can be expected from
simple considerations. (ii) In the long sample ($f_1 \sim 5$ GHz) a mode
transition on the second zero field step from the symmetric $F + \bar{F}$
(~ 10 GHz) to the bunched FF mode (~ 5 GHz) was clearly identified
at a value of η in qualitative agreement with numerical calculations
(Erné et al., 1981a). (iii) On higher order steps complicated fine
structure with corresponding changes in the frequency and power of
emitted microwaves was observed. Fig. 19 shows an example of this
more complicated behaviour with mode transitions on an n = 5 step.
In this latter case there are clearly numerous combinations of
bunching with 2, 3, 4, and 5 solitons. (iv) The linewidth of the
emitted radiation was typically of order a few kilohertz at the
top of a zero field step in qualitative agreement with Eq. 48.
(v) When a magnetic field was applied, the n = ½, 3/2, ...
Fiske steps appeared, all with radiation at the fundamental soliton

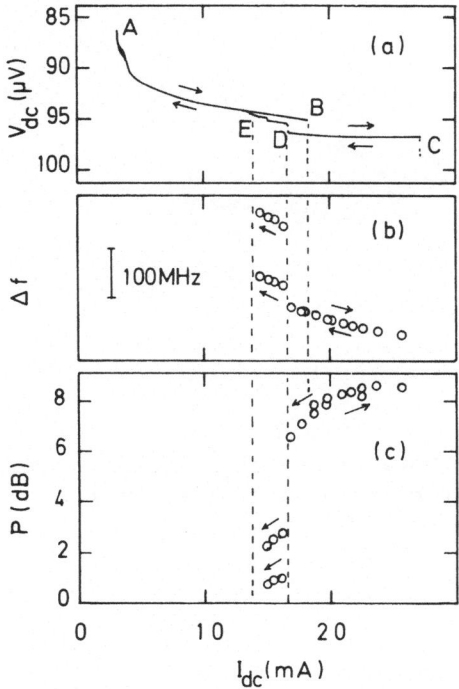

Fig. 19 Measurements at 10 GHz ($f_1 \sim$ 5 GHz) on an n = 5 step. (a)
the zero field step (b) and (c): frequency and power of
the emitted microwave radiation. Note the region with
two simultaneous signals. (After Dueholm et al., 1981a).

frequency. Thus in this case bunching still occurs, however, the
picture is further complicated by the annihilation and creation
of one or more solitons at the boundaries as discussed in section
III. Because these experiments spanned from the simple, intuitively
understandable picture (n = 1 step) to very complicated multi-
soliton propagation with mode transitions and annihilation/creation
of solitons a project was undertaken to see how far a full numerical
investigation with realistic experimental junction parameters could
reproduce the experimental observations. This is described in the
next section.

IVc. DETAILED COMPARISON

     This work on a detailed numerical comparison with experiments
was recently carried out in Lyngby, and is described in a series
of papers: Soerensen et al., 1981, Lomdahl et al., 1981,
Christiansen et al., 1981a, Lomdahl et al., 1982. Since it is
not possible to cover the details here, the reader is urged to
consult those papers. The calculations assume parameters derived
from the experiments, and a surface impedance term, $\beta\phi_{xxt}$, is
included. Fig. 20 shows a calculation of the first three zero

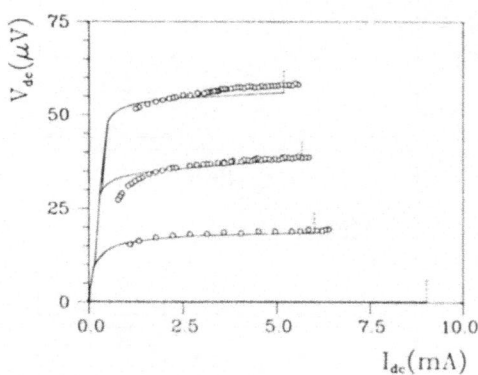

Fig. 20 Comparison between experimental (full curves) and
         numerically calculated zero field steps (after Lomdahl et
         al., 1981).

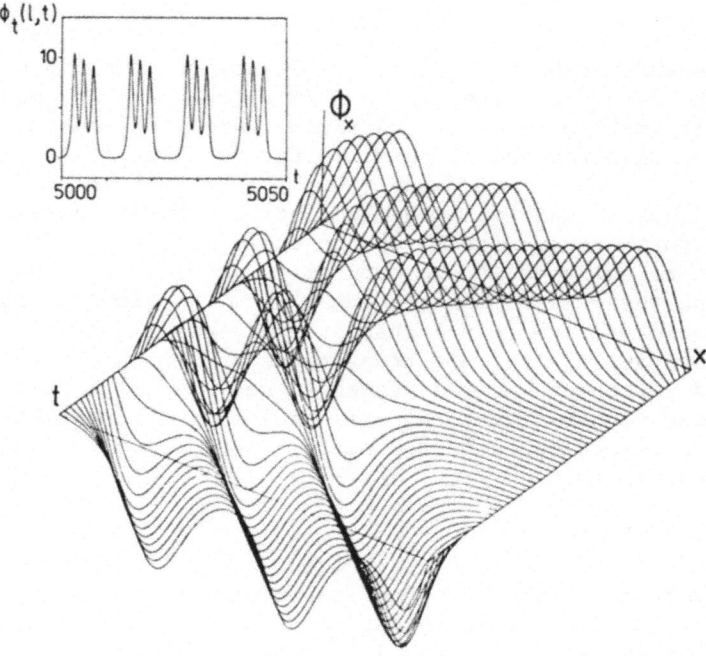

Fig. 21 Numerical calculation of the FFF-mode (after Lomdahl et
          al., 1982).

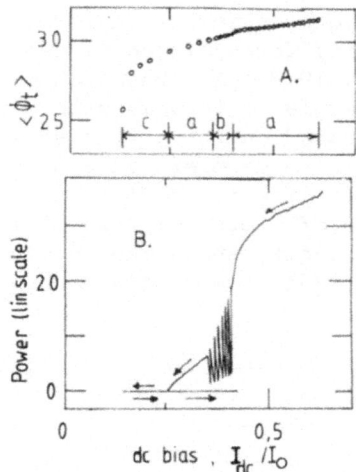

Fig. 22 A. Numerical calculation of the n = 3 step (a) FFF-mode,
(b) FF+F̄-mode, and (c) F+F+F̄-mode.
B. Numerical calculation of the emitted power at f_1
(after Lomdahl et al., 1981 and Lomdahl et al.,
1982).

field steps together with the measured ones. (A similar type of
calculation was recently reported by Erné et al., 1981b). On the
second zero field step both the ranges of the FF and the F + F̄
modes were found in good agreement with the measurements. Here we
concentrate on the third zero field step with 3 solitons.
Numerically was found both the FFF mode, the F+F+F̄ mode, and the
FF+F̄ mode. Fig. 21 show $\phi_x(x,t)$ for the FFF mode.
Fig. 22A shows the details of the third zero field step, with the
bias ranges of the different modes. Fig. 22B shows the calculated
power at the fundamental soliton frequency $\sim f_1$; note in particular
the rapid oscillations connected with the FF + F̄ mode. These are
due to a small difference in the velocity between a 2 soliton bunch
and a single soliton, and correspond to the simultaneous generation
of two frequencies very close together. (Fig. 19 for the n = 5 step
is such an experimental example). Concluding, this detailed
numerical simulation reproduced almost all the experimental
observations and gave details that are far beyond what is possible
within the framework of the perturbed sine-Gordon model.

IVd. OTHER MICROWAVE APPLICATIONS

It deserves mentioning that other schemes than the long inline or overlap junctions have been suggested for microwave applications. An interesting possibility for a microwave generator was suggested by McLaughlin and Scott, 1978. In their scheme a circular JTL is partly interrupted at regular intervals with microshorts or weak links. The analysis - based on the perturbed sine-Gordon equation - shows that as a soliton is decelerated and accelerated at the passage of the microshort it emits radiation. The average frequency in the passage of microshorts is $\omega^* = 2\pi u_m/a$, where u_m is the average fluxon velocity and a the spacing between the microshorts. They find that radiation at the (doppler shifted) frequencies

$$\omega^{\pm} = \frac{-\omega^* \pm u_m\sqrt{\omega^{*2} + u_m^2 - 1}}{1 - u_m^2} \tag{49}$$

appears in the JTL. Another circular microwave oscillator was recently suggested by Christiansen et al., 1981b.

A quite different application was suggested by Yoshida and Irie, 1975. In this scheme a signal enters in one end of a long junction and is amplified through a parametric interaction with moving fluxons, i.e. a Josephson junction traveling wave amplifier. All these ideas remain to be explored experimentally.

IVe. rf PROPERTIES: APPLIED rf

Although a lot of progress has taken place in understanding the long Josephson junction both mathematically and experimentally, only very limited experience exists when an external rf signal is applied. In the small junction this has led to the development of Josephson mixers, SIS mixers, parametric amplifiers, volt standard etc. In the perturbed sine-Gordon equation application of an external microwave signal consists in adding a term $\eta_1 \sin\Omega t$ to the right hand side of Eq. 2. Here η_1 is the (normalized) rf current and $\Omega = \omega_{rf}/\omega_o$, where ω_{rf} is the frequency of the external signal. Recently Mineev and Shmidt, 1980, published an analytical calculation in the framework of the perturbed sine-Gordon equation for the limit where $\eta_1 \ll 1$. Their analytical results may be summarized as follows. They find that instead of having traditional sum and difference frequencies of the applied frequency and some soliton frequencies, the frequency spectrum in the long junction is given by Eq. 49, if Ω is substituted for ω^* and u substituted for u_m. This is a rather surprising result not to be expected from simple intuition. Further for $\Omega < 1$, they find a resonance in the

dc IV-curve of the zero field step at a voltage $v_{res} \simeq 2\pi \sqrt{1-\Omega^2/\ell}$.
Neither the spectrum corresponding to Eq. 49 (with proper sub-
stitutions) nor the resonance in the zero field step has yet been
observed experimentally. Indeed no experimental work in such directions has been reported.

In the limit of large applied rf signals ($\eta_1 \gtrsim 0.5$ and $\Omega \simeq 1$)
numerical experiments was recently reported. (Eilbeck et al., 1981,
Bennett et al., 1982). They show that - as in the case of a small
junction - chaos may appear in the long Josephson junction. This
is a very new field of research and few conclusions can be made
yet. However, the interplay between the very orderly and coherent
motion of solitons and the chaos introduced in the nonlinear system
by a large amplitude rf signal is very intriguing and a research
subject of large current interest. On the experimental side
investigations of chaos should be possible by measuring a broadband
noise rise in the frequency spectrum of the junction. Staying on
the subject of coherence or chaos in the perturbed sine-Gordon
system it should be mentioned that chaos has been predicted under
certain circumstances even in the absence of applied rf signals.
(Maginu, 1982).

CONCLUSION

A main line through this paper has been to demonstrate the
enormous developments that have taken place in recent years in
understanding analytically the dynamics of soliton motion in the
long Josephson junction. This progress has taken place mainly
through the application of the perturbed sine-Gordon equation
formalism. Examples of successes are numerous, however, we hope
also to have demonstrated that in some cases - even if an analytical
calculation can be made - it does not agree with numerical
simulation. Thus a deep respect for the complexity of soliton
dynamics must be preserved, and analytic results should be critically
compared with numerical and experimental data before being trusted.
An example where such unresolved disagreements exist was the
magnetic field tuning of the zero field steps. On the experimental
side there are still a lot of remaining good experiments to do,
particularly in areas of applied magnetic fields and applied microwave signals. Research aiming at the use of the soliton properties
of the Josephson transmission line in microwave and computer
circuits also appears very promising.

ACKNOWLEDGEMENT

Discussions and contributions from P. L. Christiansen, M.
Cirillo, B. Dueholm, S. Erné, T. F. Finnegan, E. Joergensen, V.
Karpman, O. A. Levring, K. K. Likharev, R. Monaco, J. Mygind, O. H.
Olsen, R. D. Parmentier, M. R. Samuelsen, A. C. Scott, and O. H.
Soerensen are gratefully acknowledged. This research was financed
in part by the Danish Natural Science Research Council.

REFERENCES

Barone, A., Esposito, F., Magee, C. J., and Scott, A. C.,
    1971, Theory and application of the sine-Gordon
    equation, Riv. d. Nuovo Cim., 1:227.
Barone, A., Johnson, W. J., and Vaglio, R., 1975, Current flow
    in large Josephson junctions, J. Appl. Phys., 46:3628.
Bennett, D., Bishop, A. R., and Trullinger, S. E., 1982,
    Coherence and chaos in the driven, damped sine-Gordon
    chain, Preprint.
Chen, J. T., Finnegan, T. F., and Langenberg, D. N., 1969,
    Anomalous dc-current singularities in Josephson tunnel
    junctions, Proc. Int. Conf. on Superconductivity,
    August 1969:413, (North-Holland, Amsterdam 1971).
Chen, J. T. and Langenberg, D. N., 1974, Fine Structure in the
    anomalous dc current singularities of a Josephson tunnel
    junction, Low Temp. Phys., LT 13, vol. 3:289 (Plenum,
    New York 1974).
Christiansen, P. L., Lomdahl, P. S., Scott, A. C., Soerensen,
    O. H., and Eilbeck, J. C., 1981a, Internal dynamics
    of long Josephson junction oscillators, Appl. Phys. Lett.
    39:108.
Christiansen, P. L., Lomdahl, P. S., Zabusky, N. J., 1981b,
    Tunable oscillator using pulsons on large area lossy
    Josephson junctions, Appl. Phys. Lett. 39:170.
Christiansen, P. L. and Olsen, O. H., 1982, Propagation of
    fluxons on Josephson lines with impurities, Wave Motion.
    4:163.
Cirillo, M., Parmentier, R. D., and Savo, B., 1981, Mechanical
    analog studies of a perturbed sine-Gordon equation,
    Physica, 3D:565.
Coon, D. D. and Fiske, M. D., 1965, Josephson ac and step
    structure in the supercurrent tunneling characteristic,
    Phys. Rev., 138:A744.
Costabile, G., Parmentier, R. D., Savo, B., McLaughlin, D. W.,
    and Scott, A. C., 1978, Exact solutions of the sine-
    Gordon equation describing oscillations in a long (but
    finite) Josephson junction, Appl. Phys. Lett., 32:587.
Costabile, G., Cucolo, A. M., Pace, S., Parmentier, R. D.,
    Savo, B., and Vaglio, R., 1980, Magnetic field
    behaviour of zero field steps in long Josephson
    junctions, SQUID'80:147 (W. de Gruyter, Berlin 1980).
DeLeonardis, R. M. and Trullinger, S. E., 1980, Theory of
    boundary effects on sine-Gordon solitons, J. Appl.
    Phys. 51:1211.
Dueholm, B., Levring, O. A., Mygind, J., Pedersen, N. F.,
    Soerensen, O. H., and Cirillo, M., 1981a, Multisoliton
    excitations in long Josephson junctions, Phys. Rev.
    Lett., 46:1299.

Dueholm, B., Joergensen, E., Levring, O. A., Mygind, J.,
    Pedersen, N. F., Samuelsen, M. R., Olsen, O. H., and
    Cirillo, M., 1981b, Dynamic fluxon model for Fiske
    steps in long Josephson junctions, Physica, 108B:1303.
Eilbeck, J. C., Lomdahl, P. S., and Newell, A. C., 1981,
    Chaos in the inhomogeneously driven sine-Gordon equation,
    Phys. Lett., 87A:1.
Erné, S. N. and Parmentier, R. N., 1981, Loading effects on
    Josephson junction fluxon oscillators,J.Appl.Phys.52:1608.
Erné, S. N. and Parmentier, R. D., 1981a, Fluxon propagation
    zero field steps, and microwave radiation in very long
    Josephson tunnel junctions, J. Appl. Phys., 52:1091.
Erné, S. N., Ferrigno, A., Finnegan, T. F., and Vaglio, R.,
    1981b, Fluxon propagation in long overlap Josephson
    junctions, Physica, 108B:1299.
Erné, S. N., Ferrigno, A., DiGenova, S., Parmentier, R. D.,
    1982a, Frequency tuning by magnetic field in Josephson
    junction fluxon oscillators, Preprint.
Erné, S. N., Ferrigno, A., Finnegan, T. F., and Savo, B., 1982b,
    Experiments on the fluxon dynamics of long Josephson
    junctions, Verhandl. DPG (VI), 17:1001, and Emitted
    microwave radiation and fluxon dynamics of long
    Josephson junctions, to be published.
Finnegan, T. F., Toots, J., and Wilson, J., 1975, Frequency-
    pulling and coherent-locking in thin film Josephson
    oscillators, Low Temp. Phys., LT 14, 4:184.
Fulton, T. A. and Dynes, R. C., 1973, Single vortex propagation
    in Josephson tunnel junctions, Sol. St. Comm., 12:57.
Fulton, T. A., Dynes, R. C., and Anderson, P. W., 1973, The
    flux shuttle - Josephson junction shift register
    employing single flux quanta, Proc. IEEE, 61 No. 1:28.
Hamasaki, K., Yoshida, K., Irie, F., Enpuku, K., and Inoue, M.,
    1980, Microwave-induced vortex motion in long Josephson
    junctions, Jap. Journ. of Appl. Phys., 19, No. 1:191.
Joergensen, E., Koshelets, V. P., Monaco, R., Mygind, J.,
    Samuelsen, M. R., and Salerno, M., 1982, Thermal
    fluctuations in resonant motion of fluxons on a
    Josephson transmission line; theory and experiment,
    Preprint.
Johnson, W. J., 1968, Nonlinear wave propagation on super-
    conducting tunneling junctions, Thesis, Univ. of
    Wisconsin (unpublished).
Karpman, V. I., Ryabova, N. A., and Solov'ev, V. V., 1981,
    Congelation of fluxons in a long Josephson junction,
    Phys. Lett. A, 85A:251.
Kulik, I. O., 1967, Zh. Tek. Fiz., 37:157, (Sov. Phys. - Tech.
    Phys., 12:111).
Levring, O. A., Pedersen, N. F., and Samuelsen, M. R., 1982a,
    On fluxon motion in long overlap and inline Josephson
    junctions, Appl. Phys. Lett., May 1982 to appear.

Levring, O. A., Pedersen, N. F., and Samuelsen, M. R., 1982b,
    Perturbation calculation of magnetic field dependence
    of fluxon dynamics in long inline and overlap Josephson
    junctions, Preprint.

Likharev, K. K., Semenov, V. K., Snigirev, O. V., and Todorov,
    B. M., 1979, Josephson junction with lateral injection
    as a vortex transistor, IEEE MAG-15:420.

Lomdahl, P. S., Soerensen, O. H., Christiansen, P. L., Scott,
    A. C., and Eilneck, J. C., 1981 , Multiple frequency
    generation by bunched solitons in Josephson tunnel
    junctions, Phys. Rev. B, 24:7460.

Lomdahl, P. S., Soerensen, O. H., and Christiansen, P. L.,
    1982, Soliton excitations in Josephson tunnel junctions,
    Phys. Rev. B, 25: 5737.

Marcus, P. M. and Imry, Y., 1980, Steady oscillatory states
    of a finite Josephson junction, Sol. Stat. Commun.,
    33:345.

Maginu, K., 1982, Spatially homogenous and inhomogeneous
    oscillations and chaotic motion in the active Josephson
    junction line, Preprint.

McLaughlin, D. W. and Scott, A. C., 1978, Perturbation analysis
    of fluxon dynamics, Phys. Rev. A, 18:1672.

Mineev, M. B. and Shmidt, V. V., 1980, Radiation from a vortex
    in a long Josephson junction placed in an alternating
    electromagnetic field, Sov. Phys. JETP, 52:453.

Nakajima, K., Onodera, Y., Nakamura, T., and Sato, R., 1974a,
    Numerical analysis of vortex motion on Josephson
    structures, J. Appl. Phys., 45:4095.

Nakajima, K., Yamashita, T., and Onodera, Y., 1974b, Mechanical
    analogue of active Josephson transmission line,
    J. Appl. Phys. 45:3141.

Nakajima, K., Sawada, Y., and Onodera, Y., 1975. Nonequilibrium
    stationary coupling of solitons, J. Appl. Phys., 46:5272.

Nakajima, K., Ichimura, H., Onodera, Y., 1978, Dynamic vortex
    motion in long Josephson junctions, J. Appl. Phys.
    49:4881.

Olsen, O. H. and Samuelsen, M. R., 1981, Fluxon propagation
    in long Josephson junctions with external magnetic
    field, J. Appl. Phys., 52:6247.

Olsen, O. H. and Samuelsen, M. R., 1982, Influence of the $\cos\phi$
    conductance on fluxons propagating in long Josephson
    junctions, Phys. Rev. B, 25:3181.

Parmentier, R. D., 1978, Fluxons in long Josephson junctions,
    Solitons in action:173 (Academic Press 1978, New York).

Rajeevakumar, T. V., Geppert, L. M., and Chen, J. T., 1980,
    Critical conditions of flux flow in long Josephson
    tunnel junctions, J. Appl. Phys. 51:2744.

Rajeevakumar, T. V., 1981, A Josephson vortex flow device,
        Appl. Phys. Lett., 39:439.
Salerno, M. and Scott, A. C., 1982, Line width for fluxon
        oscillators, Preprint.
Scott, A. C. and Johnson, W. E., 1969, Internal flux motion
        in large Josephson junctions, Appl. Phys. Lett., 14,
        No.10:316.
Soerensen, O. H., Lomdahl, P. S., and Christiansen, P. L.,
        1981, Fluxon dynamics in intermediate length Josephson
        tunnel junctions, Physica 108B:1299.
Soerensen, T. H., Christiansen, P. L., and Lomdahl, P. S.,
        1982, An approximate theory for bunched solitons on
        finite Josephson tunnel junctions, Phys. Lett. A,
        89A: 308.
Yoshida, K. and Irie, F., 1975, Frequency conversion in a
        long Josephson junction with a moving vortex array,
        Appl. Phys. Lett. 27:469.
Yoshida, K., Irie, F., and Hamasaki, K., 1978, Flux-flow
        characteristics of a large Josephson junction,
        J. Appl. Phys. 49:4468.

JOSEPHSON EFFECTS: BASIC CONCEPTS

Antonio Barone* and Gianfranco Paternò**

*Istituto di Cibernetica del C.N.R.
Arco Felice, Napoli, Italy
Istituto di Fisica, Università di Napoli
Napoli, Italy

**Associazione EURATOM-ENEA sulla Fusione
Centro Ricerche Energia Frascati
C.P. 65, 00044 Frascati, Rome, Italy

1. INTRODUCTION

Obviously, at the level of a simple overview, it would be unrealistic to attempt to cover all the various aspects of the Josephson effect[1] in one lecture. Rather, we confine ourselves to a selection of ideas just to give the "flavor" of[2] the subject and provide the basic tool for the following lectures.

Probably, even the far-reaching vision of Bryan Josephson would have found it difficult to predict in far-off 1962 the fantastic growth of interest which still places this subject twenty years later into the limelight of the scientific world, both for the underlying physics involved and for the varied applications it has given rise to.

2. SIMPLE THEORETICAL APPROACH

An introduction to the theory of the Josephson effect is a hard task. Although several approaches can be devised a choice of a rather comprehensive approach is not consistent with the requirements of simplicity and brevity.

Let us start by considering the archetype Josephson junction,

namely a tunneling structure made by two superconducting layers
divided by a thin dielectric barrier. As indicated by Giaever,[3]
when two superconductors are taken at a distance of say about 50 Å
quantomechanical tunneling leads, as a macroscopic result, to a net
tunneling current I whose voltage dependence is depicted by the
finite voltage branch of the V-I characteristic in Fig. 1. The
simplest model to account for this effect, in which quasiparticles
are treated as normal electrons, is based on a mere application of
the "golden rule"

$$I = \frac{2\pi}{\hbar} |T|^2 \int_{-\alpha}^{+\infty} N_L(E) N_R(E + eV) [f_L(E) - f_R(E + eV)] dE$$

$|T|$ is the tunneling matrix element assumed energy independent and
N and f represent the density of states and the Fermi factor respectively; L and R label left and right superconductors whose
Fermi energy levels are shifted by eV. At a temperature T, above
the critical also the N's can be assumed to be constant leading to
an ohmic I vs V dependence. At $T < T_c$ the B.C.S. densities of
states are considered and a numerical solution of the integral
gives the V-I characteristics of Fig. 1.

If the tunneling barrier thickness is further reduced down to
say 10-20 Å then a supercurrent flows through the structure.
Roughly speaking we can say that in this case the two macroscopic
wave functions describing the L and R superconductors

$$\psi_{L(R)} = \rho^{\frac{1}{2}} e^{i\phi_{L(R)}}$$

have a significant overlap which results in a "correlation" of the
two superconductors. The phases in the two electrodes cannot be
altered independently as it was for the two "isolated" superconductors. Indeed is the relative phase $\phi = \phi_L - \phi_R$ which becomes the
most significant parameter which appears in observable quantities.
In this case in fact ϕ is related to current and voltage as
follows:

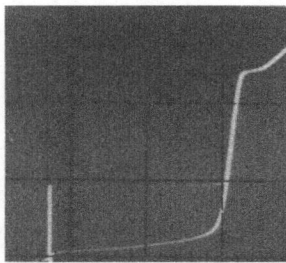

Fig. 1.   Voltage current characteristic of a   Nb-NbO$_x$-Pb tunnel-
           ing junction. Horizontal scale 1 mV/div, vertical scale
           1 mA/div.

$$I = I_1 \sin\phi \qquad ; \qquad \frac{\partial\phi}{\partial t} = \frac{2e}{\hbar} V \qquad\qquad (1)$$

which represent the basic Josephson relations. These clearly indicate that a finite supercurrent of maximum value I_1 can flow with zero voltage drop (d.c. Josephson effect) indicating that the junction behaves, as a whole, like a single superconductor although involving lower values of the critical current, field etc. ("weak superconductivity"). A constant applied voltage will lead to an alternating current of frequency $v/V = 486.6$ MHz/V (ac Josephson effect).

Josephson relations 1 can be obtained by resorting to over-simplified approaches. One possibility is based on the analogy with a linear atomic chain in the "tight binding" approximation. Within such an analogy the eigenfunctions are given by:

$$\Phi_\alpha = \sum_{n=-\infty}^{+\infty} \psi_n \, e^{in\alpha}$$

where the cristal momentum $\hbar\alpha$ and n are conjugate variables. The corresponding eigenvalues are:

$$E_\alpha = E_o - 2M \cos\alpha$$

where $-M$ is the matrix element for the transfer of a pair. Thus Hamilton equations apply leading to the Josephson equations (1).

Both the approach given by Giaever to describe single particle tunneling and these models describing pairs tunneling are surprisingly simple with respect to the goal they reach. However, in spite of their value and interest, they cannot provide a comprehensive picture of the tunneling of both quasi particle and Cooper pairs in a unified view. This objective necessarily requires a microscopic analysis which we shall develope in a rather simplified form.

3. TUNNELING HAMILTONIAN FORMALISM

Tunneling processes in superconducting junctions can be described by the following Hamiltonian:

$$H = H_L + H_R + H_T$$

where H_L (H_R) is the complete Hamiltonian of the left (right) electrode. H_T is responsible for the transfer of electrons from one metal to the other. The expression of H_T is

$$H_T = \sum_{kq\sigma}[T_{kq}c^+_{k\sigma} d_{q\sigma} + T^*_{kq} d^+_{q\sigma} c_{k\sigma}] \qquad (2)$$

$c_{q\sigma}$ ($d_{q\sigma}$) is the destruction operator of one particle state of momentum k and spin σ of the unperturbed left (right) Hamiltonian H_L (H_R). T_{kq} is a matrix element related to the transition probability of one electron from the k-state in the left metal to a q-state in the right one. H_T describes instantaneous transfer of electrons without spin flip or energy absorption in the barrier. Furthermore time reversal symmetry implies:

$$T^*_{-k-q} = T_{kq}$$

A detailed discussion of the motivation, results and limits of the tunneling Hamiltonian method is given in the excellent paper by Feuchtwang.[4] Let's also observe that a direct formulation of the tunneling process which avoids transfer Hamiltonian formalism is discussed in the above quoted article and references reported therein.

The tunneling current I(V,T) at a finite temperature and for a voltage V applied to the junction can be found from the expectation value of the time derivative of the number of particles operator of the right metal:

$$I(V,T) = -e\langle\dot{N}_R\rangle$$

where

$$N_R = \sum_{q,\sigma} d^+_{q\sigma} d_{q\sigma}$$

From the equation of motion for N_R, since it commutes with H_L and H_R we have:

$$\dot{N}_R = \frac{i}{\hbar}[H_T, N_R] = \frac{i}{\hbar}\sum_{kq}(T_{kq} c^+_{k\sigma} d_{q\sigma} - T^*_{kq} d^+_{kq} c_{k\sigma})$$

where the anticommutation relations for $c_{k\sigma}$ and $d_{q\sigma}$ have been used.

By using the relation:

$$T^*_{k\sigma}\langle d^+_{q\sigma} c_{k\sigma}\rangle = \{T_{kq}\langle c^+_{k\sigma} d_{q\sigma}\rangle\}^*$$

we get for the current the expression:

$$I(V,T) = \frac{2e}{\hbar} \text{ Im } \{\sum_{kq\sigma} T_{kq} \langle c^+_{k\sigma} d_{q\sigma}\rangle\}$$

This can be evaluated in the framework of the linear response theory. The interaction term H_T is considered as a perturbation. The system evolution is described in the interaction representation in which the time dependence of the operators is determined by the unperturbed hamiltonian $H_L + H_R$, whereas the time evolution of the eigenstates is determined by the perturbation term H_T.

To the first order in H_T we get:

$$I = - \frac{2e}{\hbar} \text{ Re } \sum_{kq\sigma} T_{kq} \int_{-\infty}^{t} d\tau \ e^{\eta\tau} <[c_{k\sigma}^{+}(t) \ d_{q\sigma}(t), H_T(\tau)]> \qquad (3)$$

where the expectation value refers to the unperturbed Hamiltonian. The factor $e^{\eta\tau}$ implies that the perturbation is "turned on" adiabatically from $t = -\infty$. Due to the applied voltage $V(t)$ the Fermi levels of the two electrodes μ_L, and μ_R will be relatively shifted by $\mu_L - \mu_R = - eV$. This situation can be described assuming that:

$$H_L(V) = H_L(0) - eV \ N_L$$

where N_L is the number of particles operator of the left side.[5] Therefore the expression for the destruction operator of electrons in the left metal will be

$$c_{k\sigma}(t) = e^{i\Phi(t)/2} \ \bar{c}_{k\sigma}(t) \qquad (4a)$$

where

$$\frac{d\Phi(t)}{dt} = \frac{2e}{\hbar} \ V(T) \qquad (4b)$$

$\bar{c}_{k\sigma}(t)$ is the time dependent operator for $V = 0$. Thus inserting (2) in (3) we have:

$$I(V,T) = - \frac{2e}{\hbar} \text{ Re } \sum_{\substack{kq\sigma \\ k'q'\sigma'}} T_{kq} \int_{-\infty}^{t} d\tau \ e^{\eta\tau} \{T_{k'q'} \ e^{-i[\Phi(t)+\Phi(\tau)]/2}$$

$$<[\bar{c}_{k\sigma}^{+}(t)d_{q\sigma}(t), \ \bar{c}_{k'\sigma'}^{+}(\tau)d_{q'\sigma'}(\tau)]>$$

$$+ T_{k'q'}^{*} \ e^{-i[\Phi(t)-\Phi(\tau)]/2} \ <[\bar{c}_{k\sigma}^{+}(t)d_{q\sigma}(t),$$

$$d_{q'\sigma'}^{+}(\tau)\bar{c}_{k'\sigma'}(\tau)]>\}$$

Let us introduce the two functions $S(t)$ and $R(t)$ defined as follows:

$$S(t-\tau) = -\frac{2ie}{\hbar^2} \Theta(t-\tau) \sum_{\substack{kq\sigma \\ k'q'\sigma'}} T_{kq} T^*_{k'q'} <[c^{-+}_{k\sigma}(t)d_{q\sigma}(t),$$

$$d^+_{q'\sigma'}(\tau)\bar{c}_{k'\sigma'}(\tau)]> \tag{5a}$$

$$R(t-\tau) = -\frac{2ie}{\hbar^2} \Theta(t-\tau) \sum_{\substack{kq\sigma \\ k'q'\sigma'}} T_{kq} T_{k'q'} <[c^{-+}_{k\sigma}(t)d_{q\sigma}(t),$$

$$\bar{c}^{-+}_{k\sigma}(\tau)d_{q'\sigma'}(\tau)]> \tag{5b}$$

where $\Theta(x)$ is the step function defined as $\Theta(x) = 0$ for $x < 0$, $\Theta(x) = 1$ for $x \geq 0$.

$S(t)$ contains only terms like:

$$c^+_{k\sigma}(t)d_{q\sigma}(t)d^+_{q'\sigma'}(\tau)c_{k'\sigma'}(\tau)$$

in which at a time t an electron tunnels from the right to the left and at a time τ another one tunnels from the right to the left metal. Therefore it accounts for quasiparticle tunneling process. $R(t)$ contains only terms like:

$$c^+_{k\sigma}(t)d_{q\sigma}(t)c^+_{k'\sigma'}(\tau)d_{q'\sigma'}(\tau)$$

which describes a process in which two electrons are destroyed in the right metal and two are created in the left metal. Therefore it is related to the tunneling of pairs.

By using (5a) and (5b) we have

$$I(V,T) = -\operatorname{Im} \int_{-\infty}^{+\infty} d\tau e^{\eta\tau} \{e^{-i[\Phi(t)+\Phi(\tau)]/2} R(t-\tau)$$

$$+e^{-i[\Phi(t)-\Phi(\tau)]/2} S(t-\tau)\} \tag{6}$$

If $V(t) = V = $ constant, from (4b) it follows

$$\Phi(t) = \omega_F t \qquad \text{where} \qquad \omega_F = \frac{2e}{\hbar} V$$

and therefore expression (6) becomes

$$I = \operatorname{Im} \{e^{-i\omega_F t} \int_{-\infty}^{+\infty} dt' \, e^{i(\omega_F/2 + i\eta)t'} R(t')$$

$$+ \int_{-\infty}^{+\infty} dt' e^{i(i\eta-\omega_F/2)t'} S(t')\}$$

where the change of variable $t' = t - \tau$ has been made. $S(t)$ and $R(t)$ can be expressed in terms of spectral functions $A_i(k,\omega)$ and $B_i(k,\omega)$ $(i = L; R)$ of the Green's functions of the superconducting electrodes:

$$S(t) = - i\frac{4e}{\hbar^2} \Theta(t) \sum_{kq} |T_{kq}|^2 \{\iint_{-\infty}^{+\infty} \frac{d\omega d\omega'}{(2\pi)^2} e^{i(\omega-\omega')t} A_L(k,\omega)A_R(q,\omega')$$

$$[f(\omega) - f(\omega')]\}$$

$$R(t) = e^{-i(\phi_L-\phi_R)} R'(t)$$

where:

$$R'(t) = i \frac{4e}{\hbar^2} \Theta(t) \sum_{kq} |T_{kq}|^2 \{\int_{-\infty}^{+\infty} \frac{d\omega d\omega'}{(2\pi)^2} e^{-i(\omega-\omega')t} B_L(k,\omega)B_R(q,\omega')$$

$$[f(\omega') - f(\omega)]$$

and $f(\omega) = [e^{\omega/k_B T} + 1]^{-1}$ is the Fermi function. The phase factor $e^{-i(\phi_L-\phi_R)}$ comes out from the factorization of the phase factors of the spectral functions B_R and B_L. These functions are in fact proportional to the complex superconducting order parameter Δ_R on the right metal and Δ_L^* on the left.

Introducing the Fourier transforms of $S(t)$ and $R'(t)$:

$$S(\omega) = \int_{-\infty}^{+\infty} dt\, e^{i\omega t} S(t) \quad ; \quad R'(\omega) = \int_{-\infty}^{+\infty} dt\, e^{i\omega t} R'(t)$$

we get for the current at a constant voltage V the expression:

$$I(V,T) = \operatorname*{Im}_{\eta\to 0^+}\{S(i\eta - \frac{\omega_F}{2}) + e^{-i\phi(t)}R'(i\eta + \frac{\omega_F}{2})\} \tag{7}$$

where $\phi(t) = \phi_L - \phi_R + \omega_F t$ identifies with the relative phase between the superconductors already introduced in the phenomenological approach. Defining the quantities:

$$I_{qp}(\omega) = \operatorname*{Im}_{\eta\to 0^+} S(i\eta - \omega)$$

$$I_{qp1}(\omega) = \operatorname*{Re}_{\eta\to 0^+} S(i\eta - \omega)$$

$$I_{j2}(\omega) = \operatorname*{Im}_{\eta\to 0^+} R'(i\eta + \omega)$$

$$I_{j1}(\omega) = -\operatorname{Re}_{\eta \to 0^+} R'(i\eta + \omega)$$

(7) becomes:

$$I(V,t) = I_{qp}(V,T) + I_{j1}(V,T)\sin\phi(t) + I_{j2}(V,T)\cos\phi(t) \tag{8}$$

This is the expression for the total current in a tunnel junction at constant voltage bias, derived the first time by Josephson in 1962.[1] The first and second terms are the quasi particle and Josephson currents respectively the third term is the phase dependent current arising from the interference of pairs and quasi particles. The real and imaginary part of $S(\omega)$ and $R'(\omega)$ are related by the Kramers-Kronig or dispersion relations[6]

$$\operatorname{Re} S(\omega) = \frac{1}{\pi} P \int_{-\infty}^{+\infty} d\omega' \frac{\operatorname{Im} S(\omega')}{\omega' - \omega}$$

$$\operatorname{Im} S(\omega) = -\frac{1}{\pi} P \int_{-\infty}^{+\infty} d\omega' \frac{\operatorname{Re} S(\omega')}{\omega' - \omega}$$

Where P indicate the principal part of the integral. Analogous relations are valid for $\operatorname{Re} R'(\omega)$ and $\operatorname{Im} R'(\omega)$. As it can be seen from (8) the real part of $S(\omega)$ does not give contribution. As shown by Larkin and Ovchinnikov[7] this result holds even for a slowly varying voltage. This term becomes significant for a rapidly varying time dependent voltage applied to the junction.

4. TOTAL CURRENT IN THE B.C.S. APPROXIMATION

By computing the Fourier transform of $S(t)$ and $R't)$ the following expressions are derived:[6,7]

$$I_{qp}(V,T) = \frac{\hbar}{eR_N} \int_{-\infty}^{+\infty} d\omega \, n_L(\omega) n_R(\omega-\omega_0)[f(\omega-\omega_0) - f(\omega)]$$

$$I_{qp1}(V,T) = \frac{\hbar}{\pi eR_N} P \int_{-\infty}^{+\infty} d\omega \int_{-\infty}^{+\infty} d\omega' \frac{n_L(\omega) n_R(\omega')}{\omega-\omega'-\omega_0}[f(\omega)-f(\omega')]$$

$$\tag{9}$$

$$I_{j2}(V,T) = \frac{\hbar}{eR_N} \int_{-\infty}^{+\infty} d\omega \, p_L(\omega) \, p_R(\omega-\omega_0)[f(\omega-\omega_0) - f(\omega)]$$

$$I_{j1}(V,T) = \frac{\hbar}{\pi eR_N} P \int_{-\infty}^{+\infty} d\omega \int_{-\infty}^{+\infty} d\omega' \frac{p_L(\omega) p_R(\omega')}{\omega-\omega'-\omega_0}[f(\omega) - f(\omega')]$$

where ω_o = eV/$\hbar$. R_N is a constant that can be interpreted as the junction resistance when both metals are in the normal state. The energies are referred to the Fermi levels and are measured in units of $\hbar$.

In the B.C.S. approximation:

$$n_i(\omega) = \frac{\omega}{\sqrt{\omega^2-\Delta_i^2}} \, \Theta(|\omega| - |\Delta_i|)$$

$$p_i(\omega) = \frac{\Delta_i}{\sqrt{\omega^2-\Delta_i^2}} \, sng(\omega)\Theta(|\omega| - |\Delta_i|) \qquad\qquad (i = L,R)$$

Δ_i is the energy gap of the corresponding superconducting electrode and the two functions $\Theta(x)$ and $sng(x)$ are defined as follows:

$\Theta(x) = 1$; $sng(x) = 1$ for $x \geq 0$

$\Theta(x) = 0$; $sng(x) = -1$ for $x < 0$

Let us observe that the expression I_{qp} closely reproduces that obtained by Giaever in the oversimplified model previously outlined.

For $T = 0$ analytical expressions have been obtained by Werthamer[6] in terms of elliptic integrals. These expressions are shown in Fig. 2. All the curves are normalized to the maximum Josephson current at zero voltage whose expression is:

$$I_1 = I_{j1}(0,0) = 2 \frac{\Delta_L \Delta_R}{R_N(\Delta_L+\Delta_R)} \, K(\frac{|\Delta_L-\Delta_R|}{\Delta_L + \Delta_R})$$

where $K(x)$ is the complete elliptic integral of the first kind. The curves refer to the case $\Delta_L = 2\Delta_R$. Let us observe that instead of $I_{qp1}(V,0)$ we have reported the quantity $I_{qp1}(V,0) - I_{qp1}(0,0)$ in order to remove the singularity at $V \to 0$.

As it is clear from the figure, I_{j1} exhibits a singularity at eV = $\Delta_L + \Delta_R$. This singularity ("Riedel peak") is related to the singularity of the density of states in the superconductor at eV = Δ. The expressions of current terms at finite temperature have been numerically computed by Harris[8,9].

Figure 3 gives the ratio of the zero voltage current $I_{j1}(0,0)$ to the finite current discontinuity $\Delta I = I_{qp}(V,0)$ at eV = $\Delta_L + \Delta_R$ as function of $\delta = |\Delta_L - \Delta_R|/(\Delta_L + \Delta_R)$. We observe that when $\Delta_L = \Delta_R$ (i.e. $\delta = 0$), $I_{j1} = \Delta I$ and when one of the gap tends to

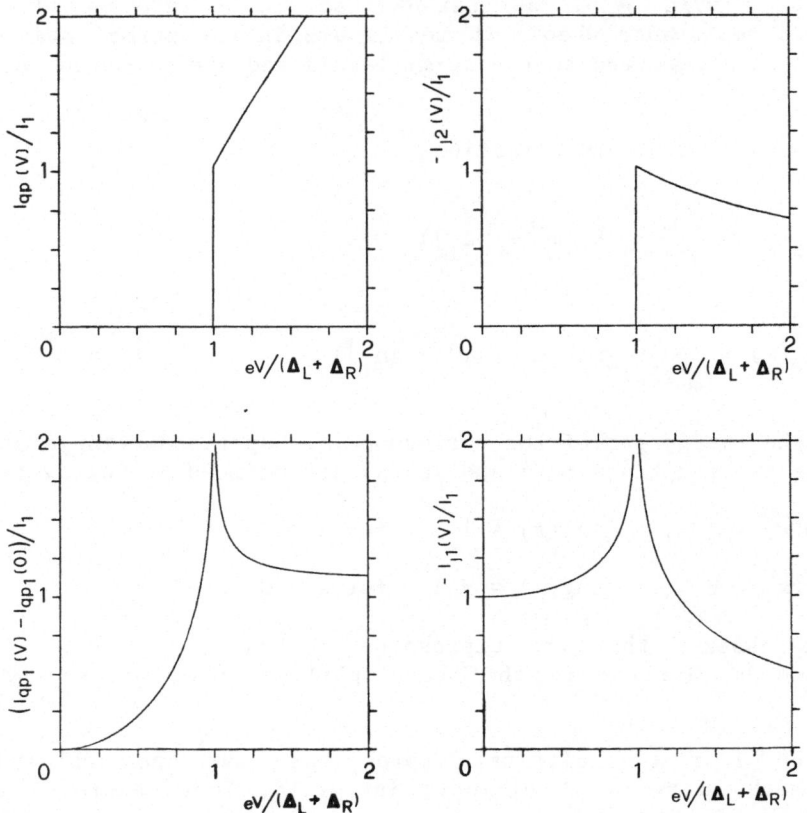

Fig. 2.   Real and imaginary parts of $R'(\omega)$ and $S(\omega)$ for $T = 0K$ and
$\Delta_L = 2\Delta_R$.

zero ($\delta = 1$), $I_{j1}(0,0)$ tends to zero as well since it corresponds
to a junction with an electrode going in the normal state.

Expression (8) can be written as

$$I(V,T) = I_{j1}(V,T) \sin\phi + I_{qp}(V,T)[1 + \varepsilon(V,T) \cos\phi] \qquad (10)$$

where $\varepsilon(V,T) = I_{j2}(V,T)/I_{qp}(V,T)$. The microscopic theory predicts
for $V \to 0$, $T \to 0$, the value $\varepsilon(0,0) = 1$ whereas from experiments the
value $\varepsilon(0,0) = -1$ has been found. Such a discrepancy appears to be
accounted by the introduction of a finite width in the Riedel
peak[10,11].

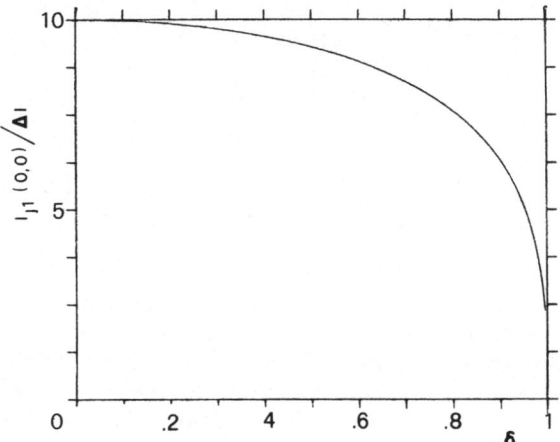

Fig. 3.   The ratio of the zero voltage current $I_{j1}(0,0)$ to the
finite current discontinuity $\Delta I = I_{qp}(V,0)$ at eV =
$= (\Delta_L + \Delta_R)/e$, as a function of $\delta = |\Delta_L - \Delta_R|/(\Delta_L + \Delta_R)$, for
T = 0K.

Usually equation (10) is well approximated by the following

$$I = I_1 \sin\phi + \sigma_o(V)V \tag{11}$$

where the $\cos\phi$ term has been neglected and the quasi particle
current has been written as $\sigma_o(V)V$.

5.   ASPECTS OF JUNCTION ELECTRODYNAMICS

Let us consider now the junction in a magnetic field H. It can
be easily shown that in this case a space modulation of the phase
occurs, that is

$$\frac{\partial\phi}{\partial x} = \frac{2e}{\hbar c} H_y d \quad ; \quad \frac{\partial\phi}{\partial y} = -\frac{2e}{\hbar c} H_x d$$

with $d = (2\lambda_L + t)$ where λ_L is the London depth and t the physical
barrier thickness. These relations together with relations 1 can be
combined with the Maxwell equation

$$\frac{\partial H_y}{\partial x} - \frac{\partial H_x}{\partial y} = \frac{4\pi}{c} J_z + \frac{1}{c}\frac{\partial D_z}{\partial t}$$

giving

$$\frac{\partial^2 \phi}{\partial x^2} + \frac{\partial^2 \phi}{\partial y^2} - \frac{1}{\bar{c}^2} \frac{\partial^2 \phi}{\partial t^2} = \frac{1}{\lambda_j^2} \sin\phi \tag{12}$$

where

$$\bar{c} = c\left(\frac{1}{4\pi Cd}\right)^{\frac{1}{2}} \qquad \text{and} \qquad \lambda_j = \left(\frac{\hbar c^2}{8\pi e d J_1}\right)^{\frac{1}{2}}$$

where $C = \varepsilon_r/4\pi\,t$ is the junction capacitance and λ_j is the Josephson penetration depth which gives a measure of the distance within which Josephson currents are confined at the edges of the junction as a result of the current screening effect due to the self generated magnetic field. To account for the dissipative term in (11) equation (12) can be written in the general form

$$\frac{\partial^2 \phi}{\partial x^2} + \frac{\partial^2 \phi}{\partial y^2} - \frac{\partial^2 \phi}{\partial t^2} - \frac{\beta}{c^2} \frac{\partial \phi}{\partial t} = \frac{1}{\lambda_j^2} \sin\phi \tag{13}$$

where $\beta = \sigma_o/C$.

This equation is a very complex one and has been widely considered in a number of cases using different approximations. A lot of information on the propagation phenomena have been obtained by investigating the Sine-Gordon equation[12]

$$\Box\phi = \sin\phi$$

where $\Box$ is the D'Alambert operator and $\lambda_j = 1$. Also equation

$$\frac{\partial^2 \phi}{\partial x^2} - \frac{\beta}{c^2} \frac{\partial \phi}{\partial t} = \frac{1}{\lambda_j^2} \sin\phi$$

has been investigated in the context of the flux-flow in metal barrier junctions where the energy storage associated with the capacitance can be neglected with respect to dissipation.[13] In the third lecture on the Josephson effect the time-independent case of equation (13) for $\beta = 0$ will be considered both in one and two spatial dimensions.

REFERENCES

1. B.D. Josephson, "Possible New Effects in Superconductive Tunneling", Phys. Lett. 1, 251 (1962).
2. For a general account of the Josephson effect and extensive updated bibliography see: A. Barone, G. Paternò, "Physics and

Applications of the Josephson Effect", Wiley, New York (1982).

3. I. Giaever, "Energy Gap in Superconductors Measured by Electron Tunneling", Phys. Rev. Lett. 5, 147 (1960).

4. T.E. Feuchtwang, "Tunneling Theory without the Transfer-Hamiltonian Formalism I", Phys. Rev. 10, 4121 (1974).

5. G. Rickaizen, "Theory of Superconductivity", Wiley, New York (1965), Chap. 10.

6. N.R. Werthamer, "Nonlinear Self-coupling of Josephson Radiation in Superconducting Tunnel Junctions", Phys. Rev. 147, 255 (1966).

7. A.I. Larkin and Yu.N. Ovchinnikov, "Tunnel Effect between Superconductors in an Alternating Field", Sov. Phys. J.E.T.P. 24, 1035 (1967).

8. R.E. Harris, "Cosine and Other Terms in the Josephson Tunneling Current", Phys. Rev. B10, 84 (1974).

9. R.E. Harris, "Josephson Tunneling Current in the Presence of a Time-dependent Voltage", Phys. Rev. B11, 3329 (1975).

10. M.R. Samuelsen, unpublished (1978).

11. A.B. Zorin, I.O. Kulik, K.K. Likharev, J.R. Schrieffer," The Sign of the Interference Current Component in Superconducting Tunnel Junctions", Sov. J. Low Temp. Phys. 5, 537 (1979).

12. A. Barone, F. Esposito, C.J. Magee, A.C. Scott, "Theory and Application of the Sine-Gordon Equation", Nuovo Cimento 1, 227 (1971).

13. J.R. Waldram, A.B. Pippard, J. Clarke, "Theory of Current Voltage Characteristics of SNS Junctions and Other Superconducting Weak Links", Phil. Trans. Roy. Soc. A268, 265 (1970).

INVESTIGATIONS ON THE JOSEPHSON EFFECT BY

PHOTOSENSITIVE TUNNEL STRUCTURES

A. Barone and M. Russo

Istituto di Cibernetica del Consiglio Nazionale

delle Ricerche, 80072 Arco Felice, Italy

INTRODUCTION

Within the framework of the semiconducting barrier tunnel junctions, a particular interest has been recognized in light-sensitive junctions[1,2]. A light-sensitive structure consists of two superconducting layers separated by a thin (50 ÷ 1000 Å) photoconducting film such as CdS, CdSe, CdTe, etc. We can devise three main reasons of interest in the study of such junctions. The first lies in the possibility of performing investigations on the underlying physics of superconductive tunneling as well as on the properties of the semiconductor itself. A second aspect of interest is related to potential applications of these structures as optically controlled devices[3]. Finally, due to the possibility of a continuous adjustability of the tunneling barrier by light exposure, these junctions can exhibit properties simulating in a single sample a variety of situations occurring in conventional junctions.

In the following we shall use this last aspect to describe several aspects of the Josephson effect. The light-sensitive structures discussed here employ CdS barriers. The choice of this photoconductor is suggested among other reasons by its "storage-like" behavior. The light-induced modifications persist, at low temperature, after removal of the light input. This feature is also exhibited by other semiconductors (e.g. CdSe, CdTe). The different vapour pressure of cadmium and sulphur which would appear as a technological drawback indeed allows the possibility of obtaining a large variety of CdS's by a suitable control during the preparation. This represents a useful degree of freedom for planning structures of different characteristics.

CURRENT-VOLTAGE CHARACTERISTICS

In the following we refer to symmetrical (Pb-Pb and In-In) and asymmetrical (Pb-In) structures with CdS barriers. A typical current-voltage, I-V, characteristic of a Pb-Pb junction at a temperature above the critical one is shown in Fig. 1.a. The linear behavior at low bias and the very rapid increase of the current at high applied voltage (V > 25 mV) typical of a tunneling structure is clearly exhibited. A dV/dI vs. V experimental curve is also shown, Fig. 1.b. We observe that, in Pb-Pb samples, the minimum of the conductance is slightly shifted from zero bias to V_o, though this effect is not observable in In-In structures. From the data reported, a voltage dependent conductance $G(V') \equiv G(V-V_o)$, such as $G(V') = \alpha(1+3\ \gamma\ V'^2)$ is found. Assuming an electron effective

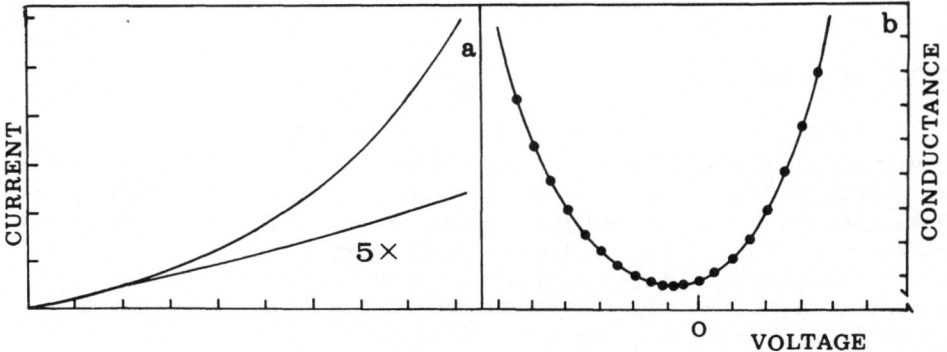

Fig.1 - (a) current-voltage characteristic of a Pb-CdS-Pb light sensitive tunnel junction in dark conditions. Horizontal scale: 10 mv/div; vertical scale: 10 mA/div.; T=77K. (c) Conductance versus voltage for the same sample. Horizontal scale: 2 mV/div.; vertical scale: 0.01 mho/div.; T=77K.

mass ratio of unity, the parameters α and γ are related to the height, $\overline{\varphi}$, and width, s, of the rectangular equivalent average barrier by the equations [4]:

$$\alpha \equiv G(O) = 3.16\ 10^{10}\ \overline{\varphi}^{1/2}\ s^{-1} exp(-1.025\ s\ \overline{\varphi}^{1/2})$$

$$\gamma = 1.15\ 10^{-2}\ s^2\overline{\varphi}^{-1} - 3.15\ 10^{-2}\ s\ \overline{\varphi}^{-3/2}$$

where s is in Ångstroms, $\overline{\varphi}$ in volt and image forces are not included. Typically the results give low barrier heights and quite large widths. For the date of Fig. 1 it is $\overline{\varphi}$ = 28 mV and s=71 Å in dark conditions. In Fig. 2, the I-V characteristics of a Pb-Pb

junction at T=4.2K are reported.   The quasi-particle tunneling
branch of the junction in dark conditions is shown in Fig. 2.a.
The I-V curves at different light exposure times are reported in
Fig. 2.b.   The main feature is a continuous decrease of the tunneling resistance with the light exposure time (a variation up to a factor 10^3 can be found).   The low resistance states are stable and the
initial conditions can be recovered either by warming the sample
up to about 100 K or by injection through the junction of a suitable
current pulse. The quenching of photoconductivity by infrared irradiation was also proposed but this method is still matter of experimental investigations in tunnel junctions. The Fig. 2.c shows a
light-induced dc Josephson current previously absent (Fig. 2.a,b).
Thus from these results we observe that a single photosensitive
junction can provide a variety of behaviors: absence of tunneling,
quasi-particle and pair tunneling regimes.

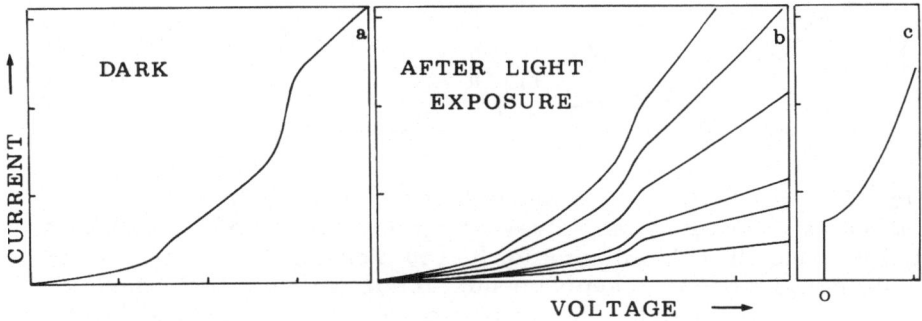

Fig. 2 - Current-voltage characteristics of a Pb-CdS-Pb light-
sensitive tunnel junction at T=4.2K. The first characteristic
(a) refers to the junction in dark conditions.   The curves
shown in (b) refer to increasing values of illumination time
on the same sample; the tunneling resistance decrease with
the light exposure.   In (c) it is shown the light-induced Josephson current.   Horizontal scale: 1 mV/div.; vertical scale:
0.1 mA/div (a,c), 1 mA/div.   (b).

TEMPERATURE DEPENDENCE OF THE JOSEPHSON CURRENT

When measurements of the critical dc Josephson current versus temperature are performed on semiconducting barrier junctions
one finds results which, in general, are quite different from the
predictions of Ambegaokar and Baratoff calculations for oxide barrier junctions. In particular, results concerning a Pb-CdS-In structure is reported in Fig. 3.a.   In this case, the peculiar differences
from the conventional oxide barrier junctions can be ascribed to
the occurrence of proximity effects at the semiconductor/supercon-

ductor boundary. This circumstance is also reflected in the dependence of the junction behavior on the barrier film thickness and the
nature of the electrodes. For instance, in Pb-In junctions, light-
induced Josephson current, I_J^L, has been observed even for a barrier film thickness up to several hundreds Ångstroms.   In this
case, a confinement of the barrier at the Pb-CdS interface can be
assumed whereas the In-CdS interface produces an ohmic contact.
Typically, the semiconducting film thickness, t, which allows light
induced Josephson current is   t ≈ 200 Å for Pb-Pb junctions, t ≈ 400
Å for Pb-In junctions and t ≈ 800 Å for In-In junctions.

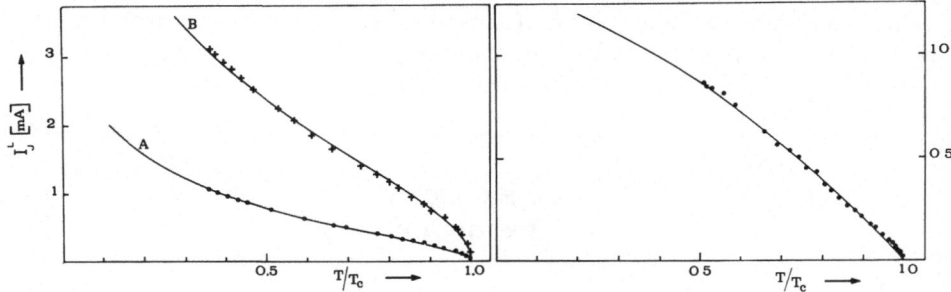

Fig. 3 – (a) Temperature dependence of the dc Josephson
current at two different values of light-induced critical current
in the same Pb-CdS-In junction.   (b) Temperature dependence
of light-induced dc Josephson current in an In-CdS-In sample.
The experimental data are compared with the theoretical behavior (solid curves) calculated in the framework of proximity
effect.

On the basis of the whole phenomenological picture, the temperature dependences of the light-sensitive structures considered
here have been interpreted in terms of a proximized S-I-N-S
sandwich.   The results reported in Fig. 3.a refer to two different
levels of light-induced Josephson current in the same sample. The
experimental data are compared with the theoretical behavior based
on a simple model of the proximity effect [5,6]. The effect of the
light input appears to be observable only in the critical current level and in the value obtained for the ratio between the thickness
of the normal layer and its coherence length, d_N/ξ_{NO}. The same
approach applies also to a symmetrical configuration S-N-I-N-S
(In-CdS-In); see Fig. 3.b.   A different approach to explain the
phenomenology of semiconducting barrier junctions has been recently proposed by Aslamazov and Fistul'[7] in the framework of the microscopic theory. To date, however, the experimental results and
the theory are not yet suitable for a straightforward comparison.

MAGNETIC FIELD DEPENDENCE OF THE LIGHT-INDUCED JOSEPHSON CURRENT.

Let us refer to the simple junction configuration of Fig.4 and let us assume an external magnetic field $H_e \equiv H_y$ and negligible self-field effects (small junction, L, $W < \lambda_J$). It can be easily shown that

$$I_J(k) = \left| \int_{-\infty}^{\infty} \zeta(x) \; exp \; ikx \; dx \right| \tag{1}$$

with $k = \dfrac{2 \pi d}{\Phi_0} H_y$, ($d = \lambda_1 + \lambda_2 + t$ and Φ_0 is the flux quantum).

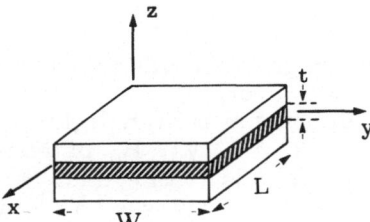

Fig. 4 - Junction sketch and relative framework.

Thus the maximum Josephson current at a given applied magnetic field is given by the modulus of the Fourier transform of the linear current density:

$$\zeta(x) = \int J_1(x,y) \, dy \tag{2}$$

where J_1 is the maximum current density.

In the case J_1 = constant te I_J vs H_e gives the classical Fraunhofer-like pattern. This situation (J_1= const.) occurs for a uniform illumination of the junction barrier. By increasing the light exposure the critical current increase as well. As a consequence λ_J decreases and the junction can no longer be considered as a small one. In Fig. 5.a,b are reported the two situations of a small and large junction respectively, realized in the same sample. In Fig. 5.c it is also shown the maxima of various I_J^L vs. H_e patterns corresponding to different illumination levels.

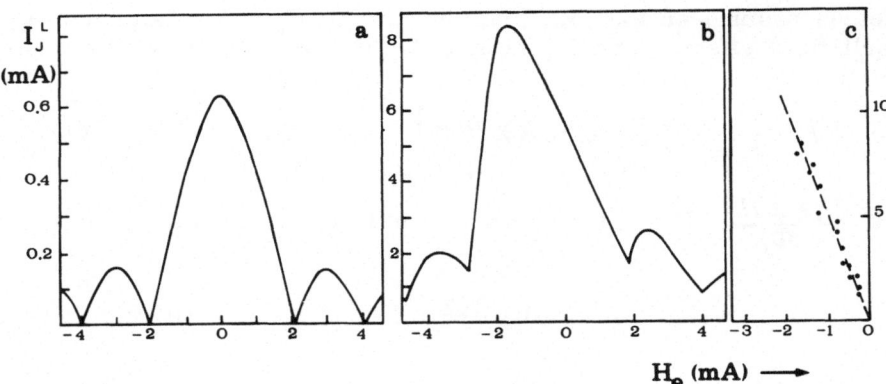

Fig. 5 - (a) Applied magnetic field dependence of a low light-induced critical current level.
(b) I_J^L vs. H_e dependence for a high light-induced critical current level.   (c) Maxima of I_J^L vs. H_e patterns corresponding to different illumination levels.   The data refer to the same Pb-CdS-In junction. H_e is expressed in units of current in a Helmotz coil.

       Let us consider now what happens when J_1 is no longer constant. In this case a suitable current density profile can be considered in $\zeta(x)$. It should be remarked, however, that the possibility of recovering such a profile from the critical current versus magnetic field pattern is not univocal [8]. Ambiguity can be removed by additional assumptions based on plausible physical circumstances.[9] For instance, in the case of light-sensitive structures an enhancement of the current at the junction edges (those not covered by a "thick" counterelectrode) is very likely to occur due to the higher local illumination. A great consistency has been found in a variety of situations by a proper "shading" of the samples. In general the I_J^L vs H_e dependence is the result of a combination, of diffraction and interferential patterns, see Fig. 6. One limiting situation (diffraction) is that previously considered of J_1 = const.; the other (interference) is that of a high current density at the edges simulating a two-junctions system. A simple model to take into account the junction behavior can be based on a current distribution such as that of Fig. 7.c:

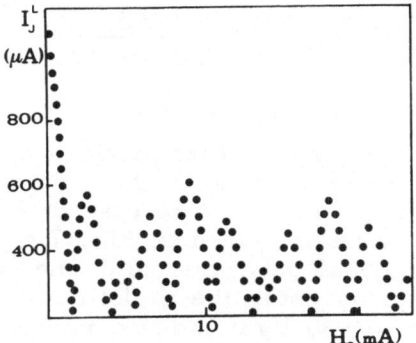

Fig. 6 - Magnetic field dependence of the light-induced Josephson current in a In-CdS-In junction. Due to a peaking of the
current density profile at the junction edges, the dependence
is a combination of diffraction and interference patterns.

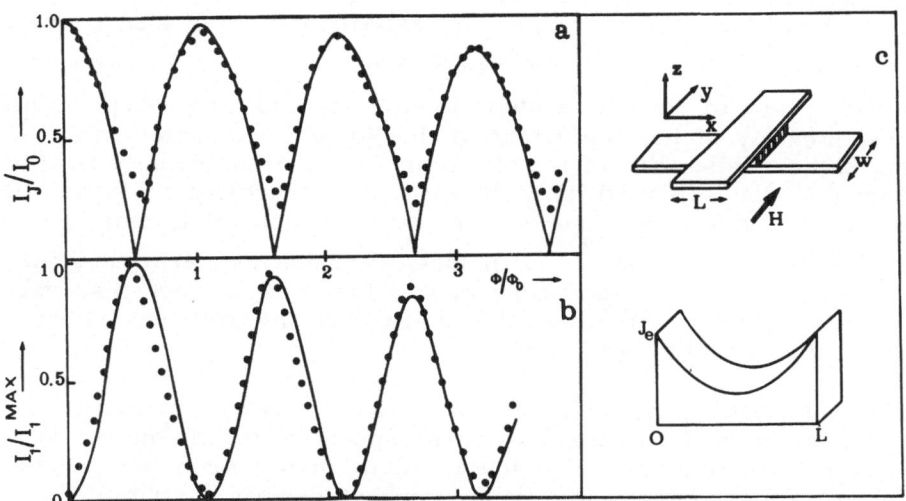

Fig. 7 - (a) Magnetic field dependence of the light-induced maximum zero-voltage Josephson current of an In-CdS-In junction. The
experimental data (full circles) are normalized to the maximum value of the current. The solid line is the theoretical behavior for a
nonuniform current density profile described by (3) with χ =18.
(b) Magnetic field dependence, for the same junction of the light-
induced first Fiske step amplitude. The experimental data are normalized to the maximum current value. The theoretical behavior
is calculated for a nonuniform current density profile with χ =18.
(c) Sketch of the junction and of the assumed shape for the current density distribution. The shown profile corresponds to $\chi \simeq 2$.

$$J_1(x,y) = J_e \frac{\cosh \chi (1 - 2x/L)}{\cosh \chi} \qquad (3)$$

where J_e is the value of the supercurrent at the edges and χ is a parameter which shapes the current profile i.e. the current distribution is more and more confined at the junction edges as far as the χ value goes from 0 to ∞ . The case of a current density highly peaked at the edges is shown in Fig. 7.a. Full circles correspond to the experimental data; the solid line represents the theoretical fitting obtained by using the model outlined above with $\chi = 18$. This approach to light-induced non-uniformities has been employed successfully in several problems. In particular, by using the properties of light-sensitive junctions, the occurrence of self-resonant modes have been studied in junctions exhibiting a non-uniform current density profile [10]. In Fig. 7, together with the I_J^L vs H_e pattern (a), it is shown the experimental magnetic field dependence of the first "Fiske" step, I_1 , (b). The data are compared with theoretical calculation (solid line) performed in the low Q limit [11] and assuming for the current density profile the value deduced by the critical current behavior. Besides the dependence of $I_n(H_e)$ on the J_1 shape, the experiments have also confirmed theoretical previsions such as a $\max(I_{nmax}/I_{Jmax}) \leq 0.68$ and the occurrence of zero field steps having an amplitude proportional to the intensity of the coefficient of the Fourier expansion of J_1 [12]. Besides the availability of a well controlled current density profile, a relevant rôle is played by light-sensitive structures in these studies as well as in experiments concerning structural fluctuations in the tunneling barrier[12,13]. In fact, experiments performed only on light-induced Josephson current and Fiske steps (not observable in dark conditions) guarantee the absence of any spurious effect such as the occurrence of shorts.

In conclusion, with the aid of light sensitive tunneling structures, we have outlined various aspects of the phenomenology of a Josephson junction. Particular attention has been devoted to important effects related to various current density distributions created by "nonuniform" barriers. The next lecture will be focused on the problem of the geometrical configuration of the junction electrodes.

REFERENCES

1.   I. Giaever, Photosentsitive tunneling and superconductivity, Phys. Rev. Lett. 20: 1286 (1968).

2.   A. Barone, M. Russo, and P. Rissman, Effect of preparation parameters on light sensitivity in superconductive tunnel junctions, Rev. Phys. Appl.: 9: 73 (1974).

3.   S.M. Faris, The role of VLSI Superconducting Technologies in Future Computers, in: "Hardware and Software Concepts in VLSI", G. Rabbat ed. Van Nostrand, (1982).

4.   J. G. Simmons, Generalized formula for the electric tunnel effect between similar electrodes separated by a thin insulating film, J. Appl. Phys. 34: 1793 (1963)

5.   N.L. Rowell, and H.J.T. Smith, Investigation of the superconducting proximity effect by Josephson tunneling, Can. J. Phys. 54: 223 (1976)

6.   F. Andreozzi, A. Barone, M. Russo, G. Paternò, and R. Vaglio, Measurements of the dc Josephson current in light-sensitive junctions, Phys. Rev. B18: 6035 (1978)

7.   L.G. Aslamazov, and M.V. Fistul', Temperature dependence of the critical current of superconductor-semiconductor-superconductor junctions, Sov. Phys. JETP 54: 206 (1981)

8.   H.H. Zappe, Evaluation of tunnel junction barriers using the magnetic field dependence of the dc Josephson current, Phys. Rev. B11: 2535 (1975)

9.   A. Barone, G. Paternò, M. Russo, and R. Vaglio, Diffraction and interference phenomena in single Josephson junctions, Phys. Stat. Sol. A41: 393 (1977)

10.   M. Russo, and R. Vaglio, Self-resonant modes in Josephson junctions exhibiting a nonuniform maximum current distribution, Phys. Rev. B17: 2171 (1978)

11.   I.O. Kulik, Theory of resonance phenomena in superconducting tunneling, Sov. Phys.-Tech. Phys. 12: 111 (1967)

12.   C. Camerlingo, M. Russo, and R. Vaglio, Zero-field "Fiske" steps in small Josephson junctions, J. Appl. Phys. (in press).

13.   A. Barone, G. Paternò, M.Russo and R.Vaglio, Experimental results and analysis of structural fluctuations in light-sensitive Josephson junctions, Sov. Phys. JETP 47: 776 (1978)

THE PROBLEM OF GEOMETRICAL CONFIGURATION

IN JOSEPHSON JUNCTIONS

Antonio Barone* and Ruggero Vaglio**

*Istituto di Cibernetica
del Consiglio Nazionale delle Ricerche
80072 Arco Felice, Italy

**Istituto di Fisica, Università di Salerno
84100 Salerno, Italy

INTRODUCTION

In dealing with the actual Josephson junctions it is
of great importance to take into account geometrical factors. As we shall see in detail the junction configuration
geometry plays a fundamental role in the behavior of a
Josephson structure. The effects are particularly evident
when the response of the junction in magnetic field is
investigated via measurements of the critical supercurrent I_J vs. applied magnetic field H_e.
For a given shape of the junction the I_J vs. H_e patterns are rather insensitive with respect to the various
electrodes configuration as long as the junctions dimensions are smaller than a charcteristic lenghth λ_J (Josephson penetration depth).

For "large" (compared to λ_J) junctions effects related to the presence of self-induced field are present
leading to a variety of behaviors whose interest does
concern not only the importance of the physical interpretation of the junction electrodynamics but also important

207

features for device applications. We shall treat this problem of junction geometry in some detail beginning with an oversimplified analysis based on a linear current-phase relationship which provides, in spite of its crudeness, important informations which can be used at a first glance. More detailed procedures and exact analysis will be then developed which lead, we hope, to a rather exhaustive picture of the topic.

EQUATIONS AND BOUNDARY CONDITIONS

The behavior of a Josephson tunnel junction, in the stationary case, can be obtained in principle by solving, with the appropriate boundary conditions, the equation[1]:

$$\frac{\partial^2 \phi}{\partial x^2} + \frac{\partial^2 \phi}{\partial y^2} = \frac{1}{\lambda_J^2} \sin\phi \qquad (1a)$$

where the junction lies in the x-y plane, ϕ is the phase difference across the superconductors forming the junction and λ_J is the Josephson penetration depth. The current density in the barrier is given by:

$$J(x,y) = J_1 \sin\phi \qquad (1b)$$

so that J_1 is the maximum allowed supercurrent density. The effective field $\vec{H}$ is related to the phase ϕ by the relation

$$\lambda_J \vec{\nabla}\phi = \frac{1}{H_0} (\vec{z} \times \vec{H}) \qquad (1c)$$

where $H_0 = \frac{4\pi}{c} \lambda_J J_1$ and $\vec{z}$ is the unit vector normal to the junction plane.

The boundary conditions for Eq. 1a are obtained from Eq. 1c and can be written as

$$\lambda_J \frac{\partial \phi}{\partial n}\bigg|_\Gamma = \frac{1}{H_0} (\vec{z} \times \vec{H})_n\bigg|_\Gamma \qquad (2)$$

where n is the outgoing normal to the boundary Γ of the tunneling region. $\vec{H}$ can be expressed as the sum of an externally applied field $\vec{H}_e$ and the field $\vec{H}_s$ produced by the current I circulating in the superconducting films forming the junction itself. The exact computation of the field distribution along the boundaries and in particular of $\vec{H}_s$ is, in most cases, laborious (see final section). Let us restrict ourselves, for the moment, to the special case of rectangular boundaries for the tunneling region with an uniform field H_e applied in the y direction (see fig.1). The boundary conditions 2 can be rewritten as:

$$\lambda_J \left. \frac{\partial \phi}{\partial x} \right|_{x=0} = - \alpha(y) \frac{L}{\lambda_J} \frac{I}{I_1} + \frac{H_e}{H_0}$$

$$\lambda_J \left. \frac{\partial \phi}{\partial x} \right|_{x=L} = \beta(y) \frac{L}{\lambda_J} \frac{I}{I_1} + \frac{H_e}{H_0}$$

(3a)

$$\lambda_J \left. \frac{\partial \phi}{\partial y} \right|_{y=0} = - \gamma(x) \frac{W}{\lambda_J} \frac{I}{I_1}$$

$$\lambda_J \left. \frac{\partial \phi}{\partial y} \right|_{y=W} = \delta(x) \frac{W}{\lambda_J} \frac{I}{I_1}$$

(3b)

where $I = WLJ_1$.

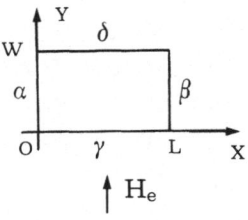

Fig.1. Junction contour in the x-y plane.

The functions $\alpha(y)$, $\beta(y)$, $\gamma(x)$, $\delta(x)$, depend on the specific electrode configuration (or "junction geometry") and are determined by the field distribution along the corresponding edge as determined by the current flow in the superconducting films ($\alpha(y)$ depends on the value of $\vec{H}_s$ along the line x=0 between 0 and W and so on).

In many papers it is assumed, for the sake of simplicity, $\alpha,\beta,\gamma,\delta$ = constant. In this case, from the application of the Ampere's law to the contour depicted in fig.1, it follows:

$$\alpha + \beta + \gamma + \delta = 1 \tag{3c}$$

in the general case, condition 3c has to be replaced by an integral relation.

First of all, as it is evident from Eqs. 3a,b, if the junction is small compared to the Josephson penetration depth, i.e. if $\dfrac{L}{\lambda_J}$, $\dfrac{W}{\lambda_J}$ << 1 (zero dimensional junctions), the problem is independent of the functions $\alpha,\beta,\gamma,\delta$, and so of the specific junction geometry. In that approximation by integrating Eq. 1c it follows:

$$\phi(x,y) = Kx + \phi_0 \; ; \; (K = \frac{1}{\lambda_J} \frac{H_e}{H_0})$$

From the latter equation and Eq. 1b, by integration over the junction area and taking the maximum versus ϕ_0 it follows:

$$I_J = I_1 \left| \frac{\sin KL/2}{KL/2} \right|$$

that is the well known and very well experimentally verified Fraunhofer-like dependence for small junctions.

A second very interesting case is that of one-dimensional junctions used in many applications.

A junctions can be considered "one-dimensional" if one (or both) of the following conditions is verified:

(A) $\gamma=\delta=0$ - current flowing in the x direction only;

(B) $\dfrac{W}{\lambda_J} \ll 1$ - junction "small" in the y direction.

In fact it is clear from inspection of Eq.1c that in both cases the phase ϕ is constant over the y direction, and since it is:

$$\frac{1}{W} \int_0^W \frac{\partial^2 \phi}{\partial y^2} \, dy = \frac{1}{W} \left[\frac{\partial \phi}{\partial y} \right]_0^W = \frac{1}{\lambda_J^2} \; \frac{I}{I_1} \, (\gamma + \delta)$$

we get:

$$\frac{\partial^2 \phi}{\partial x^2} = \frac{1}{\lambda_J^2} \left\{ \sin\phi - \frac{I}{I_1} \, (\gamma + \delta) \right\} \tag{4}$$

This equation has to be solved with the boundary conditions 3a. Relation 3c is, of course, still valid.

Let us consider now the simple geometrical configurations depicted in fig.2. The configurations a and b are called "in-line" geometries (symmetric and asymmetric respectively) and clearly satisfy the condition A previously discussed. Since $\gamma = \delta = 0$ from Eq.4 the equation to be solved is:

$$\frac{\partial^2 \phi}{\theta x^2} = \frac{1}{\lambda_J^2} \sin\phi$$

with the boundary conditions 3a. This has been done, for the symmetric case, by Owen and Scalapino[2], with the assumption $\alpha = \beta = \text{const.} = \frac{1}{2}$. The assumption of α, β constant in this case, corresponds to a uniform current flow in the superconducting strips forming the junction (so that the field produced at the boundaries is uniform as well). This is not strictly valid if the electrode thickness is larger than the London penetration λ, since in that case the superconducting current density is somewhat higher at the film edges. The effect can be significantly reduced by placing a superconducting ground plane in close proximity to the junction, which tends to uniform the current flow

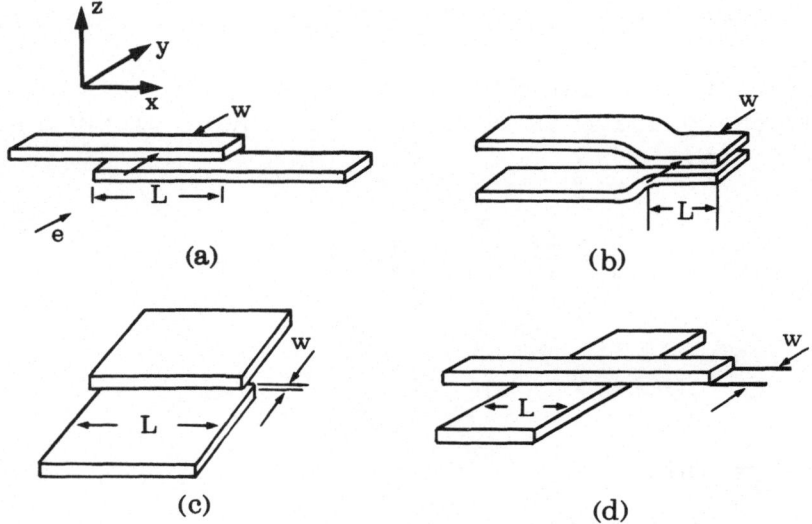

Fig.2. Sketch of typical junction geometries.

in the electrodes. Owen and Scalapino were able to solve
analytically the problem and to deduce both the dependences of the maximum Josephson current I_J on the junction
lenght L and on the applied field H_e. Their results are
in excellent agreement with the experiments. They show
that the currents in the junction are confined in a region of lenght $2\lambda_J$ at the junction edges. For $H_e = 0$
and L $>> \lambda_J$ it is: $I_J = 4W \lambda_J J_1$ (the I_J v.s. L curve "saturates" at large L). The magnetic field dependance is sym-
metric and shows a linear decrease of the critical current
going to 0 at $H_e = H_0$ (Meissener region). At $H_e = (2/\pi)H_0$
a second stable solution is possible corresponding to a
magnetic flux quantum penetrating the junction.

The in-line asymmetric case can be treated in a simi-
lar way assuming $\alpha = 1$ and $\beta = 0$ ($\beta = 0$ since no current
flows in the corresponding edge). Even in this case a good
agreement with the experiments has been found.

For overlap geometry, fig. 2c, the problem can be
considered one-dimensional only if $W/\lambda_J << 1$ (condition B).
In that case, if we assume $\alpha = \beta = 0$ and $\gamma = \delta = 1/2$
const. the equation to be solved is:

$$\frac{\partial^2 \phi}{\partial x^2} = \frac{1}{\lambda_J^2} \left(\sin\phi - \frac{I}{I_1} \right)$$

with boundary conditions

$$\lambda_J \frac{\partial \phi}{\partial x}\bigg|_{x=0} = \lambda_J \frac{\partial \phi}{\partial x}\bigg|_{x=L} = \frac{H_e}{H_0}$$

This corresponds again to assume an uniform current flow in the films. The latter equation (adding time-dependent terms) is generally used in the context of fluxon propagation studies in overlap geometries.

In zero-applied field the only possible solution is $\phi = \phi_0$ = const. The current density is uniform and Josephson current is given by $I_J = I_1$. This accounts for the linear increase of I_J v.s. L in this geometry first observed by Barone and Johnson[3]. In spite of the approximation made ($\gamma = \delta$ = const.) this last result has been carefully experimentally verified in most cases (fig. 3).

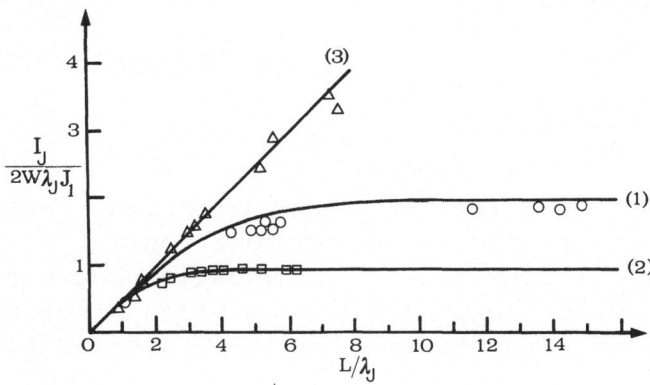

Fig.3. Theoretical maximum Josephson current vs. junction length (full lines). Curves 1-3 correspond to the geometries of fig. 2(a)-2(c) respectively. Circles are experimental data from ref.6, squares from ref. 7, triangles from ref.8.

The situation depicted in fig.2d, "cross geometry",is more difficult to solve. In fact in that case it is not really reasonable to assume constant values for the parameters $\alpha,\beta,\gamma,\delta$ since the direction of the feeding current changes very rapidly close to the boundaries. A first insight in the problem will be only given at the end of the next section. In the final section the problem will be discussed in detail.

APPROXIMATE ANALYSIS BY A LINER CURRENT-PHASE RELATION

To obtain the current density distribution in the junction in relation to the various possible geometrical configurations ofthe electrodes, the use of an approximate linear expression for the current-phase relation has been proposed[4],[5]. This method provides a quick and practical wa to calculate the I_J v.s. L,W and I_J v.s. H_e curves for the various geometries.

The current density is given by:

$$J(x,y) = \frac{J_1}{4} \phi(x,y) \; ; \; \max_{(xy)} J \leq J_1$$

and Eq. 1a can be written as:

$$\frac{\partial^2 \phi}{\partial x^2} + \frac{\partial^2 \phi}{\partial y^2} = \frac{1}{4\lambda_J^2} \phi \tag{5}$$

It can be shown that this approximation conserves the magnetic energy stored in the junction.

Eq. 5 is a linear diffential equation and can be easily solved with the boundary conditions given by Eqs. 3 if the various parameters $\alpha,\beta,\gamma,\delta$ are assumed to be constant.

The resulting current density is:

$$\frac{J(x,y)}{J_1} = \frac{I}{I_1} \frac{\alpha\cosh (L-x)/2\lambda_J + \beta\cosh x/2\lambda_J}{\sinh L/2\lambda_J} \frac{L}{2\lambda_J} +$$

$$+ \; \frac{\gamma \cosh\,(W-y)/2\lambda_J + \delta \cosh\, y/2\lambda_J}{\sinh\, L/2\lambda_J} \cdot \frac{W}{2\lambda_J} \; +$$

$$- \; \frac{H_e}{H_0} \; \frac{\cosh\,(L-x)/2\lambda_J - \cosh\, x/2\lambda_J}{\sinh\, L/2\lambda_J}$$

The first term of this expression represents the polarization current whereas the second one represents a circulating current which screens the applied field H_e (so that $\vec{H} \equiv 0$ inside the junction). The fields are only confined to a region of lenght $2\lambda_J$ close to the edges of the junction.

For $H_e = 0$, imposing the condition $\max\limits_{(xy)} \dfrac{J(x,y)}{J_1} \leqq 1$, it follows:

$$\frac{I_J}{I_1} = \left[\frac{\alpha \cosh L/2\lambda_J + \beta}{\sinh L/2\lambda_J}\;\frac{L}{2\lambda_J} + \frac{\gamma \cosh W/2\lambda_J + \delta}{\sinh W/2\lambda_J}\;\frac{W}{2\lambda_J}\right]^{-1} \quad (6)$$

This expression ($\alpha \geq \beta$ and $\gamma \geq \delta$) is plotted in figure 3 for three one-dimensional geometries:

(1) In-line symmetric geometry $\alpha=\beta=\frac{1}{2}$; $\gamma=\delta=0$

(2) In-line asymmetric geometry $\alpha=1$; $\beta=\gamma=\delta=0$

(3) Overlap geometry $\dfrac{W}{\lambda_J} \ll 1$, $\alpha=\beta=0$; $\gamma=\delta=\frac{1}{2}$

The experimental data are from the literature[6-8]. The agreement is fairly good. It is worth noting that curve (1) is very close to Owen and Scalapino's theoretical results.

With the same tecnique the magnetic field dependences can be easily computed for the geometrical configurations considered above. The results are in very good agreement with the experiments in the Meissner region.

The same sort of analysis can be used to give a first insight in the behavior of two-dimensional junctions, even though in these cases the assumption of constant values for the parameters $(\alpha,\beta,\gamma,\delta)$ is generally rather crude as discussed before. Two-dimensional, cross-type, square junctions have been considered extensively in ref.5 and the calculations showed a rather good agreement with experimental results. To give a practical example of the use of the methods described above, let us consider the geometry indicated in the inset of fig.4 (L=W).

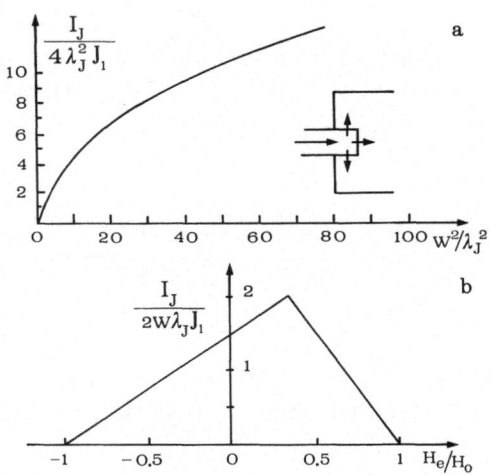

Fig.4. Maximum Josephson current vs. junction dimensions (a) and magnetic field (b) for the geometrical configuration sketched in the inset.

The problem is essentially reduced to determine the right values for $\alpha,\beta,\gamma,\delta$. The current enters from the α-side and it is reasonable to assume that it goes out evenly distributed among the other three edges (see fig.4)

This leads naturally to $\alpha=\frac{1}{2}$ and $\beta=\gamma=\delta=1/6$ (so that condition 3c is satisfied).

By replacing these numerical values in Eq.6 we get:

$$\frac{I_J}{4\lambda_J^2 J_1} = \frac{3 \frac{W}{2\lambda_J} \sinh \frac{W}{2\lambda_J}}{1 + 2\cosh \frac{W}{2\lambda_J}}$$

This expression is plotted in fig.4a. The magnetic field dependence for this geometry can also be easily computed. For $W/\lambda_J \to \infty$ we get:

$$\frac{I_J}{2W\lambda_J J_1} = \frac{3}{2} (1 + \frac{H_e}{H_0})$$

$$\frac{I_J}{2W\lambda_J J_1} = 3 (1 - \frac{H_e}{H_0})$$

(see ref.5). The expression is plotted in fig.4b.

In any case, it is worth to remember that a more correct approach to this matter requires to remove the restrictions made on the parameters $\alpha,\beta,\gamma,\delta$ and to account for the non-linear effects.

EXACT BOUNDARY CONDITIONS FOR TWO-DIMENSIONAL JUNCTIONS

Let us now consider the problem of the current flow in two-dimensional junctions in a more rigorous way. This corresponds to taking into account the non-uniform current density distribution in the superconducting films. This problem has been recently the subject of both theoretical and experimental investigations performed under a cooperative research program between the Physics Department of the Moscow State University and our Institutes[9]. Here we shall follow a somewhat simplified version of the prob-

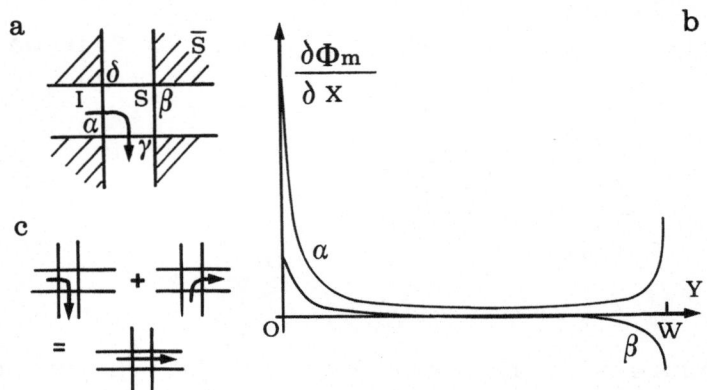

Fig.5. a) Sketch of a cross square junction
       b) Gradient of Φ_m along the junction contour
       c) Current superposition

lem addressing the reader for a complete description to
the aforementioned paper.

As observed at the beginning, we need to solve Eq.1a
with the boundary condition 2. We can make the assumption:

$$\max\ (t,\lambda) \ll \min\ (L,W,\lambda_J)$$

where t is the barrier thickness and λ is the Londono penetration depth in the superconducting films. This condition is always satisfied in the experimental situations
we are interested in.

As a consequence, it can be assumed that the magnetic
field at the boundaries is not affected by what occurs
between the electrodes. Thus, in order to compute $\vec{H}_s$ the
two superconducting layers forming the junctions can be
shorted out leading to a single superconducting film,
without modifying the current distribution ("shorting
principle")[10].

Let us denote by S the portion of the plane occupied

by the superconducting films and by $\overline{S}$ the remaining region (fig.5a). To calculate $\vec{H}_s$ in the half space $z > 0$ we can assume that $\vec{H}_s$ derives from a scalar magnetic potential Φ_m which satisfies the Laplace equation:

$$\nabla^2 \Phi_m (x,y,z) = 0 \tag{7}$$

with the boundary conditions on the $z = 0$ plane:

$$\frac{\partial \Phi_m}{\partial z}\bigg|_{z=0} = 0 \quad , \quad \{x,y\} \; \varepsilon \; S$$

$$\tag{8}$$

$$\frac{\partial \Phi_m}{\partial x}\bigg|_{z=0} = \frac{\partial \Phi_m}{\partial y}\bigg|_{z=0} = 0 \quad , \quad \{x,y\} \; \varepsilon \; \overline{S}$$

These conditions indicate that the regions labeled by $\overline{S}$ are equipotential surfaces. To define uniquely Φ_m we need to know its values at the junction barrier contour. Let us assume a cross (fig.2d) square junction geometry $(L=W)$.

By application of the Ampere's law we have:

$$\Phi_m (0,0) - \Phi_m (0,W) = \frac{2\pi}{c} I$$

$$\Phi_m (0,W) - \Phi_m (W,W) = 0$$

$$\tag{9}$$

$$\Phi_m (W,W) - \Phi_m (W,0) = 0$$

$$\Phi_m (W,0) - \Phi_m (0,0) = \frac{2\pi}{c} I$$

The feeding current I has been assumed to flow through the α and γ sides only (fig.5a). Eq. 7 with conditions 8,9 allows the determination of Φ_m.

The exact boundary conditions for the phase ϕ can then be obtained by the following relations:

$$\lambda_J \left.\frac{\partial \phi}{\partial x}\right|_{x=0} = -\left.\frac{H_y}{H_0}\right|_{x=0} = -\frac{1}{H_0}\left.\frac{\partial \Phi_m}{\partial y}\right|_{x=0}$$

$$\lambda_J \left.\frac{\partial \phi}{\partial x}\right|_{x=W} = \left.\frac{H_y}{H_0}\right|_{x=W} = \frac{1}{H_0}\left.\frac{\partial \Phi_m}{\partial y}\right|_{x=W}$$

$$\lambda_J \left.\frac{\partial \phi}{\partial y}\right|_{y=0} = \left.\frac{H_x}{H_0}\right|_{y=0} = -\frac{1}{H_0}\left.\frac{\partial \Phi_m}{\partial x}\right|_{y=0}$$

$$\lambda_J \left.\frac{\partial \phi}{\partial y}\right|_{y=W} = -\left.\frac{H_x}{H_0}\right|_{y=W} = \frac{1}{H_0}\left.\frac{\partial \Phi_m}{\partial x}\right|_{y=W}$$

(10)

that is from Eq.2 with $\vec{H}\equiv\vec{H}_s = \pm\vec{\nabla}\Phi_m$ the plus or minus sign depending on whether we consider paths lying over or under the superconducting films. Relations 10 are equivalent to Eqs. 3a,b. The solution of Eqs. 7-9 is in any case a non trivial problem and requires very long computing times.

A simple way to obtain Φ_m (i.e. the functions $\alpha(y)$, $\beta(y),\gamma(y),\delta(y)$) has been followed in ref. 9. Let us outline here the main ideas: Eq.7 is equivalent to $\vec{\nabla} \cdot \vec{H}_s = 0$. We can model the field distribution as determined by this latter equation by the distribution of the electric field $\vec{E}$ in the $z \geq 0$ half space, using an electrolytic bath with four flat electrodes lying in the $z = 0$ plane as indicated by the dashed region in fig.5a. The equation for the electric field is again $\vec{\nabla} \cdot \vec{E} = 0$ and it is easy to be convinced that the boundary conditions 8 still hold if we replace Φ_m by the scalar electric potential V. Condition 9 is satisfied by properly fixing the relative constant potentials of the four electrodes in the bath.

The electric field distribution in this "electrolytic analog" is thus the same as the magnetic field distribution in the original problem, and can be easily measured giving straight-forwardly the desired solution. The $\frac{\partial \Phi_m}{\partial x}$ vs. dependence along the junction sides is reported in fig.5b.

As far as the applied magnetic field $\vec{H}_e$ is concerned let us observe that in the experiments it can be either generated by an external solenoid or by an additional current I_H injected into the junction electrode. In the former and more usual situation the field H_e is uniform and can be easily introduced in Eqs. 10 (see also Eq.3a,b). In the latter case the field will be no longer uniform but it can be accounted for by using the superposition principle. In fact the current I_H can be given by the sum of two asymmetric distributions such as that of fig.5c. The problem of the determination of $\vec{H}_e$ in this case is equivalent to that of $\vec{H}_s$, and can be solved using the results of the electrolytic analog. To obtain the I_J vs. H_e (I_H) dependences one needs to solve Eq.1a for any applied $\vec{H}_e(I_H)$.

The following algorithm can be employed: let

$$\phi = \phi_0 + \phi_1$$

where ϕ_0 is solution of

$$\lambda_J^2 \nabla^2 \phi_0 = p_0 \quad ; \quad p_0 = \int_\Gamma \frac{\partial \phi_0}{\partial n} \, dl$$

with the boundary condition on Γ

$$\lambda_J \frac{\partial \phi_0}{\partial n} \bigg|_\Gamma = \frac{1}{H_0} (\vec{z} \times \vec{H})_n \bigg|_\Gamma$$

ϕ_1 is solution of

$$\lambda_J^2 \nabla^2 \phi_1 = \sin(\phi_0 + \phi_1) - p_0 \qquad \qquad (11a)$$

with the boundary condition:

$$\frac{\partial \phi_1}{\partial n} \bigg|_\Gamma = 0 \qquad \qquad (11b)$$

The linear problem for ϕ_0 always has a solution. Then a solution for the non-linear problem 11a with the homogeneous boundary condition 11b is selected numerically.

The I_J v.s. H_e (I_H) dependence is obtained considering for each value of H_e (I_H) the maximum value of I for which a solution there exists.

Typical examples of I_J v.s. H_e (I_H) patterns are reported in fig.6a where a comparison between the two cases of applied magnetic field previously discussed is presented. Experimental results which confirm the theoretical analysis are shown in fig. 6b.

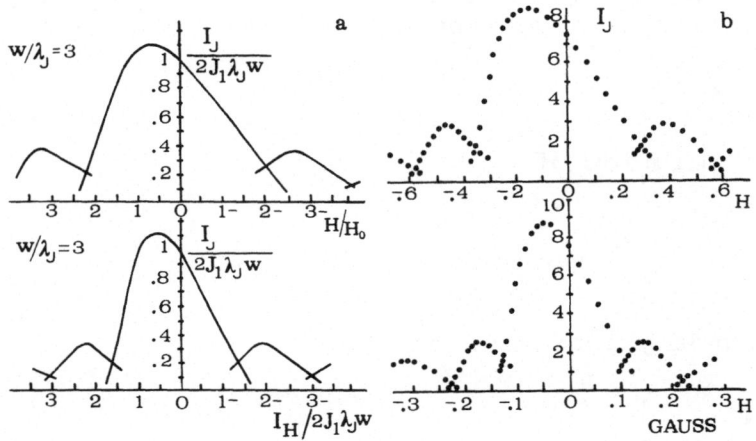

Fig.6. Maximum Josephson current vs. applied magnetic field. a) theoretical, b) experimental. Patterns above correspond to a uniform applied field. Patterns below correspond to a field applied by current injection in the counter-electrode.

CONCLUSIONS

Thus it has been shown the important role of geometrical factors on the junction behaviour. A simplified analysis based on the assumption of a linearized model has been outlined. Although very simple, this approach allows a reasonable understanding of several physical situations otherwise very difficult to handle. In fact it has been used more recently by various authors in various contexts.

A more complete analysis has been discussed in section 4 leading to interesting results fairly closely confirmed by the experiments. As a possible outgrowth of this study we could infer that the case in which the magnetic field is applied by a current flowing in an extra film overlaying the junction (as usually realized in Josephson cryotron) should represent an intermediate situation between the two cases taken in consideration. Furthermore this kind of investigation can give a significant insight into the new proposed computer elements configurations[11] in which the magnetic field is indeed applied by a current injection into the film electrode in order to avoid control lines reaching thereby a further increase of circuits packing density.

ACKNOWLEDGMENTS

Useful suggestions and comments by F.Esposito and S.Pace are gratefully acknowledged. Thanks are due to C. Salvia for technical assistance.

REFERENCES

1. A. Barone, and G. Paternò "Physics and applications of the Josephson effect", Wiley, New York (1982).
2. C. S. Owen, and D. J. Scalapino, Vortex structure and critical currents in Josephson junctions, Phys. Rev. 164:538 (1967).
3. W. J. Johnson, and A. Barone, Effect of junction geometry on maximum zero-voltage Josephson current, J. Appl. Phys. 41:2958 (1970).
4. A. Barone, W. J. Johnson, and R. Vaglio, Current flow in large Josephson junctions, J. Appl. Phys.46:3628 (1975).
5. R. Vaglio, Approximate analysis for stationary current flow in two-dimensional Josephson tunnel junctions, J. Low temp. Phys. 25:299 (1976).

6. C. K. Mahutte, J. D. Leslie, and H. J. T. Smith,
    "Self-field" limiting of Josephson tunnel current,
    Can. J. Phys. 47:627 (1969).
7. S. Basavajah and R. F. Broom, Characteristics of in-
    line Josephson tunneling gates, IEEE Trans. Magn.
    Mag 11:759 (1974).
8. R. Monaco, "Dipendenza dalla temperatura delle singola-
    rità anomale di corrente d.c. su giunzioni Josephson
    estese" Thesis, Univ. of Salerno (1980).
9. A. Barone, F. Esposito, K. Likharev, V. K. Semenov,
    B. N. Todorov, and R. Vaglio, Effect of boundary
    conditions upon the phase distribution in two-
    dimensional Josephson junctions, J. Appl. Phys.
    (1982, in press).
10. K. K. Likharev, and B.T. Ulrich, "Systems with Joseph-
    son junctions: basic theory", Moscow Univ., Moscow
    (1978).
11. S. Masuo, H. Suzuki, and T. Yamaoka, Characteristics
    of a counter-electrode coupled logic gate using
    Josephson junctions, IEEE Trans. Magn. Mag 17:583
    (1981).

WORKSHOP ON ORGANIC MATERIALS AND

SUPERCONDUCTING FLUCTUATIONS

R. L. Greene   and   H. Gutfreund and M. Weger

IBM Research Laboratory   Hebrew University
San Jose, CA 95193   Jerusalem, Israel

I. INTRODUCTION

In his lectures D. Jerome has reviewed the properties of the $(TMTSF)_2X$ class of organic superconductors and has presented experimental and theoretical evidence for the existence of superconducting fluctuation effects to rather high temperature (~40K) in these materials. Since these results are potentially quite significant and are are the same time quite controversial, it was decided to discuss the issue of superconducting fluctuations in a one-hour workshop. The participants in this workshop were D. Jerome, R. L. Greene, H. .Gutfreund and M. Weger. The views expressed by D. Jerome are presented in his paper for this proceedings. In this paper we present the views of the other three participants. No attempt will be made to include rebuttal remarks or otherwise capture the tenor of the discussion. We leave this to the reader's imagination.

II. REMARKS OF R. L. GREENE

1. Introduction

Over the past two years D. Jerome and coworkers[1] have measured properties of the $(TMTSF)_2X$ salts which they have claimed give evidence for the existence of a wide temperature range of superconducting (SC) fluctuation effects above a three dimensional (3D) transition temperature of order $T_c^{3D} \sim 1K$. From his results Jerome proposes that there exists a one-dimensional (1D) mean field T_c at least ten times higher than T_c^{3D}, i.e., $T_c^{1D} \sim 10 T_c^{3D}$. In prior work with P. M. Chaikin and others, we have shown[2] that a wide range of fluctuation effects do <u>not</u> <u>exist</u> in the $(TMTSF)_2X$ superconductors and therefore $T_c^{1D} \sim T_c^{3D} \sim 1K$. My remarks will be organized into three parts:

1. A review of critical magnetic field data which can only be understood if $T_c^{1D} \sim T_c^{3D} \sim 1K$.

2. A demonstration that all the experiments interpreted by Jerome et al.[1,3] in terms of SC fluctuations can be equally well interpreted without fluctuations. Therefore, in my opinion, there is no unambiguous experimental evidence for the existence of superconducting fluctuations in the $(TMTSF)_2X$ materials.

3. Show that the theory[3] which relates 1D SC fluctuation ideas to the $(TMTSF)_2X$ superconductors is not applicable because the interchain interactions are too large.

2. Pauli Limited H_{c2} in $(TMTSF)_2 ClO_4$

This work has previously been reported[2] and will be summarized here. H_{c2} was measured with a 4-probe resistance technique for magnetic field accurately aligned (better than $1/2°$) along the a, b* and c* directions for a single crystal of $(TMTSF)_2ClO_4$ with a $T_c^{3D} = 0.86K$. The results are shown in Figure 1. For orbital pair breaking the critical field is given by

$$H_{c2}^i = \frac{\phi_0}{2\pi \xi_j \xi_k} \qquad (1)$$

where for an open Fermi surface such as calculated[4] for $(TMTSF)_2X$ salts

$$\frac{\xi_i}{\xi_j} = \left(\frac{\sigma_i}{\sigma_j}\right)^{1/2} = \frac{t_i}{t_j} . \qquad (2)$$

Here i,j,k are cyclic permutation of the a, b* and c* directions, σ is the electrical conductivity, t the transfer integral and ξ the coherence length. From the experimental conductivity anisotropy or the calculated transfer integrals[4] and Eqs. (1) and (2) we predict the ratio $H_{c2}^a:H_{c2}^{b*}:H_{c2}^{c*}$ to be 200:20:1 whereas we measure the ratio 28:15:1. Clearly then some mechanism is limiting H_{c2}^a. The logical origin of this is spin pair-breaking, given by the Pauli critical field H_{po}

$$H_{po} = \Delta_o / \mu_B . \qquad (3)$$

Here μ_B is the Bohr magneton and Δ_o is the superconducting order parameter at T=0 [for BCS theory $\Delta_o = 1.76kT_c$]. The measured magnitude of $H_{c2}(T)$ is very close to that predicted by the temperature dependent theory for H_p based on a T_c of order 1K. The implications of this result are clear. The order parameter Δ_o is related to the mean field T_c. If T_c^{1D} is of order ten times T_c^{3D} then Δ_o should be ten times larger and the Pauli limiting field [Eq. (3)] would be ten times greater. In this case we should not have measured any limitation of H_{c2}. Therefore, the critical field data implies that $T_c^{1D} \sim T_c^{3D} \sim 1K$ and thus there cannot be any SC fluctuations very far above 1K.

3. Critique of "Fluctuation" Experiments

Many experiments have been interpreted by Jerome et al.[1,3] in terms of superconducting fluctuations. I discuss a few of the most important ones here.

a) Tunneling with GaSb Schottky Barriers.[5] These results have shown the existence of a large gap (~3.6 meV) in $(TMTSF)_2PF_6$ at P~12 kbar and T~50 mk. With a BCS interpretation ($2\Delta = 3.5 \ kT_c$) this would imply a high mean field superconducting transition temperature, i.e., $T_c^{1D} \sim 12K$. If this interpretation proves to be correct then there can be no doubt that a wide range of superconducting fluctuations exist in the $(TMTSF)_2X$ materials. However, in my opinion, the interpretation in terms of a superconducting gap in the $(TMTSF)_2PF_6$ salt is questionable. Why? Firstly, the geometry of the experiment was such that the tunneling electrons probed the c* direction. In the fluctuation picture the gap is anisotropic and the 1D pseudo-gap should only be observed in the a direction. Secondly, it has not been proven that the

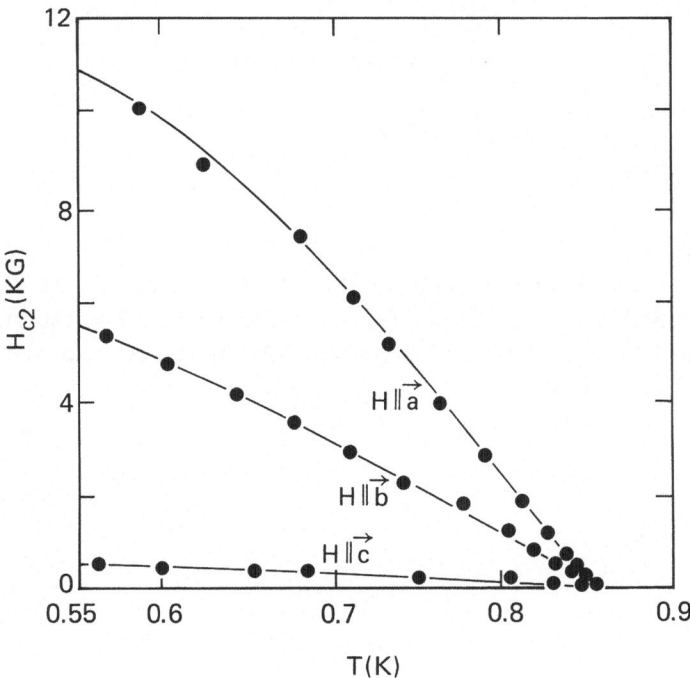

Fig. 1.   The anisotropic critical field for $(TMTSF)_2ClO_4$. Fields were varied in three perpendicular directions to attain the maximum critical field along the a axis (recent data of M. Y. Choi and P. M. Chaikin).

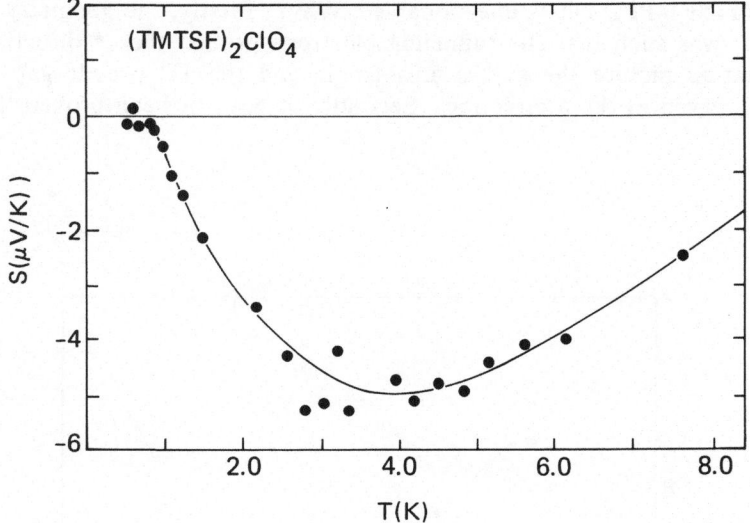

Fig. 2.   Thermopower data of M. Y. Choi and P. M. Chaikin (unpublished) for $(TMTSF)_2 ClO_4$ along the a axis. The thermopower goes positive near 10K and monotonically increases to a value near +25 $\mu V/K$ at 300K.

measured gap is a superconducting gap in the organic salt. It could arise from a band structure gap or a SDW gap or from superconducting amorphous Ga or Sb at the interface. Thirdly, the magnetic field dependence of the gap[5] is not that expected for a superconductor. No change was seen in the magnitude of the gap for perpendicular fields up to 25 kOe, yet large changes in resistivity, thermal conductivity and far infrared absorption are seen for perpendicular magnetic fields of order 2 kOe. Clearly, an unambiguous interpretation of the tunneling data will require more reproducible experiments (hopefully by independent groups) on better characterized junctions and over a wider range of temperature, magnetic field and pressure conditions.

b) Far Infrared Absorption and Reflectivity. Recent experiments by Ng et al.[6] have shown structure near 3.6 meV in the far infrared transmission spectrum of $(TMTSF)_2ClO_4$ at 2K. However, their interpretation in terms of a large superconducting pseudo-gap is contradicted by experiments of Challener et al.[7] who have seen a similar structure in the far infrared reflectivity at 2K. Challener et al. find that this structure does not change in energy up to 80K (where it broadens and disappears into the background) and is unaffected by a magnetic field of 40 kOe in the c* direction. Also the same structure is found in $(TMTSF)_2ClO_4$ single crystals which were irradiated with X-rays to produce ~100 ppm of spin defects. As shown by prior work,[8] at this level of spin defects all traces of superconductivity are destroyed.

c) Specific Heat. Garoche et al.[9] have measured the specific heat of $(TMTSF)_2ClO_4$ at ambient pressure between 0.3K and 2K. A specific heat jump $\Delta C = 1.67 \ \gamma T_c$ is observed at 1.2K which fits the BCS value very well for a T_c of 1.2K. The gap determined from the exponential dependence below T_c is of order $2\Delta \sim 4K$. Thus, there is no evidence of a large gap or SC fluctuations in this data. The specific heat was also measured with a 63 kOe magnetic field along the c* direction. Here, a broad anomaly peaked at 1.4K is observed and it is claimed that the Fermi level density of states N(o) at 2K increases by about a factor of two. The increase in N(o) is interpreted as arising from the suppression of the pseudo-gap (and hence the superconducting fluctuations). I have three objections to the interpretation. First, no large pseudo-gap is measured in the H=0 data. Second, there is no evidence from proton NMR T_1^{-1} data[10] or from magnetic susceptibility data of any change in N(o) as a function of H field or temperature. Finally, the specific heat data in a magnetic field is complicated by the presence of the anomaly at 1.4K and does not extend to high enough temperature to allow an accurate determination of N(o). In summary, the specific heat data seems perfectly consistent with a BCS superconductor with mean field $T_c \sim 1.2K$.

d) Transport Properties Below 40K. Below T~40K the behavior of the magnetoresistance,[11] the temperature dependence[3] of σ and the thermal conductivity[12] have all been interpreted by Jerome et al. as arising from superconducting fluctuations. We believe that a single particle, nonfluctuation, picture gives an equally consistent explanation for all of the transport data. The observation of a large magnetoresistance and Schubnikov oscillations[11] shows that the $(TMTSF)_2X$ materials have a high low temperature mobility (consistent with their high purity) and that electron and hole pockets exist. Several models for the origin of these pockets have been presented, with the common feature that they require a large enough interchain interaction to suppress any fluctuation effects.

The decrease of the thermal conductivity[12] below ~50K is just the opposite of the predictions of 1D fluctuation theory[13] and prior experiments on superconducting filaments.[14] The magnetic field dependence of the thermal conductivity[12] is anomalous and not yet explained. However, it is unlikely to be caused by superconducting fluctuations since the thermopower does not go to zero (and remain there) until the true superconductivity transition occurs at 1K. This is shown in Figure 2. A more complete and convincing explanation of the thermal conductivity (and thermopower) awaits future experiments.

4. Application of Fluctuation Theory to the $(TMTSF)_2X$ Salts

Schultz[3] has proposed a theory of 1D superconducting fluctuations which he and Jerome have used to interpret most of the experiments on the $(TMTSF)_2X$ salts. Our previously published opinion[2] has been that this theory is not applicable to the $(TMTSF)_2X$ superconductors because the interchain interactions are too large. Gutfreund in his remarks below has shown how an anisotropy $(t_\parallel/t_\perp)$ as large as ten takes one out of the 1D fluctuation regime. It now seems to be universally agreed that the ratio t_a/t_{b*} is of order ten in the $(TMTSF)_2X$ compounds. Moreover, with this anisotropy ratio both Grant[4] and Kwak[15] have shown that the parameters used in the Schultz theory are not meaningful.

5. Summary

I have briefly reviewed the arguments which show that superconducting fluctuations do not exist in the $(TMTSF)_2X$ class of organic superconductors. This does not, however, preclude the possibility of such fluctuation effects in some other, more one-dimensional, class of materials. In spite of their rather low T_c's the organic superconductors have generated a lot of interest, which I believe is justly deserved. They exhibit a fascinating variety of phase transitions and many unusual transport properties. A better understanding of these materials may lead us to organic superconductors of more technological interest.

Most of the above discussion has resulted from close collaboration with P. M. Chaikin of UCLA. Useful discussion and experimental inputs have also come from M. Y. Choi, W. Challenger, E. M. Engler, J. F. Kwak, L. Azevedo, J. Schirber, P. M. Grant, P. Haen and S. Z. Huang.

III. REMARKS OF M. WEGER

1. The dc Resistivity

The dc resistivity of organic metals can be described by a single-particle theory, based on electron-phonon scattering. In addition to the ordinary linear Bloch-Frohlich electron phonon coupling, there is a strong coupling quadratic in the phonon operators.[16] This quadratic coupling gives rise to a resistivity $\rho \alpha T^2 / \omega_{ph}^4$, in contrast to the linear coupling, which gives rise to a resistivity $\rho \alpha T / \omega_{ph}^2$. This theory accounts in an excellent way both for the temperature and pressure dependence of the resistivity, down to temperatures below 20K (Figure 3). The pressure dependence is accounted for by this theory without having to use even a single adjustable parameter (Figure 4). This follows from the known strong pressure dependence of the phonon frequencies[17,18] $d \ln \omega_{ph} / d \ln b \approx 10$ to 20. At high pressures, of order 20 to 30 kbar, the quadratic term is reduced considerably more than the linear term, because of the stronger dependence on the phonon frequency. Thus, at these pressures the linear term dominates, giving rise to the well-known Bloch-Gruneisen curve. This curve is shown in Figure 3 as the "Bloch" term.

2. The Frequency Dependence of the Conductivity in the Far Infrared

The strong frequency dependence of the electrical conductivity (real and imaginary components) was observed and studied intensively by Tanner et al.[19] The rapid fall of the conductivity with frequency around 5 to 10 cm^{-1} was attributed to sliding charge density waves. This theory was not consistent with NMR and thermal conductivity data, so a variant was suggested in which there is phonon-drag of the $2k_F$ phonons, which couple very strongly with the electrons. However, in place of a rigid CDW, the electrons and phonons can be described by the Boltzmann transport equation.[20-22] In this theory there is a peak in the conductivity $\sigma(\omega)$ at $\omega=0$, and G(ω) drops at a critical frequency $\omega_c \approx \tau_{ph \rightarrow el}^{-1}$, where $\tau_{ph \rightarrow el}$ is the lifetime of the $2k_F$ phonons with regard to absorption by the electrons. This inverse lifetime $\tau_{ph \rightarrow el}^{-1} \approx \omega_{ph}$ is of order 5 cm^{-1}, in excellent agreement with experiment. The details of the frequency-dependence of the conductivity are accounted for quantitatively by this theory.[22]

3. Magnetoresistance

Jerome mentioned TMTSF-DMTCNQ as the first fluctuating superconductor discovered due to its magnetoresistance. The magnetoresistance of TMTSF-DMTCNQ[23] is almost identical to that of the similar compound HMTSF-TCNQ[24] as shown in Figure 5. The striking

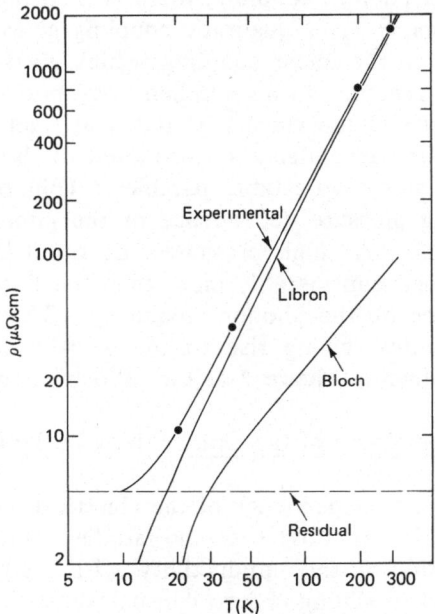

Fig. 3. The resistivity of $TMTSF_2PF_6$ at ambient pressure (from K. Bechgaard, C. S. Jacobsen, K. Mortensen, H. J. Pedersen and N. Thorup, *Solid State Commun.* **33**, 1119 (1980). It consists of a two-phonon ("libron") and a "one-phonon ("Bloch") contributions. The former predominates at ambient pressure and the latter predominates under high pressure.

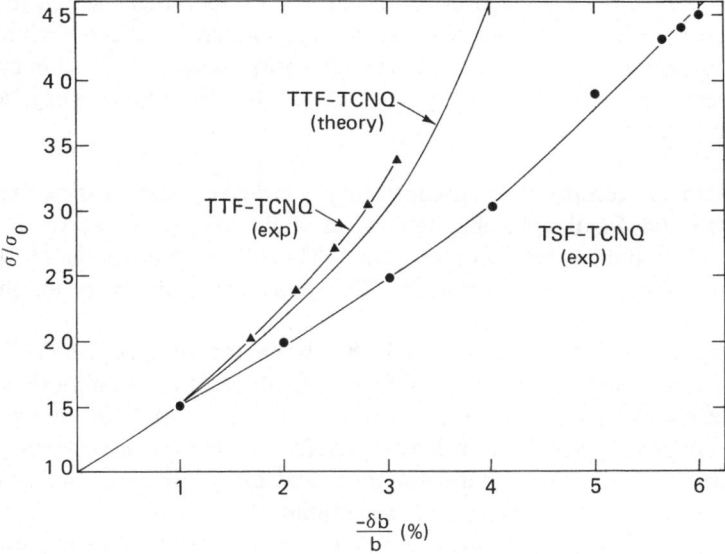

Fig. 4. The pressure dependence of the resistivity of TTF-TCNQ and TSF-TCNQ at ambient temperature. The theoretical curve is calculated for a Lennard-Jones 6-12 potential (from H. Gutfreund, M. Weger and M. Kaveh, *Solid State Comm.* **27**, 53 (1978)).

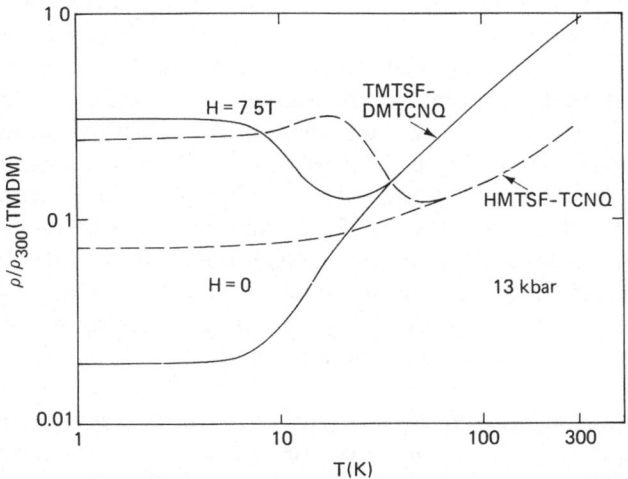

Fig. 5. The resistivity of HMTSF-TCNQ and of TMTSF-DMTCNQ at H=0 and in a magnetic field. The data of HMTSF-TCNQ are extrapolated from 2.8 to 7.5T. The residual resistivity of TMTSF-DMTCNQ is considerably lower because of the much higher crystal quality.

feature is a very large magnetoresistance at low temperature, which disappears suddenly close to 30K to 40K, without a large change in the resistivity. This behavior contrasts sharply with the magnetoresistance of "ordinary" clean metals, where the magnetoresistance is related to the conductivity, and falls with it as the temperature increases.

In both materials the susceptibility changes from a positive value, attributed to the Pauli spin magnetism, at high temperatures, to a negative (diamagnetic) value at low temperature. This diamagnetism is of the same order as the diamagnetism of bismuth. This behavior is also very unique.

Jerome and Weger[25] attributed this behavior in HMTSF-TCNQ to a transition from a metallic state at high temperatures to a semimetallic one at low temperatures, as a covalency gap opens up. In this interpretation, interchain coupling opens a covalency gap $2t_\perp$ at low temperatures. As the temperature rises, and the electron-phonon scattering time $\tau_\parallel$ shortens so that $t_\perp \tau_\parallel \approx \hbar$, this gap is destroyed and a metallic state is observed. While this interpretation does not preclude other, more exotic ones, the phenomenon is the same in TMTSF-DMTCNQ as in HMTSF-TCNQ.

4. Existence of "High" Temperature Gaps

Existence of many physically different states, with nearly degenerate energy, is characteristic of one-dimensional metals. This was already pointed out by Bychkov et al.[26] Actually, singlet superconductivity, triplet superconductivity, charge density wave, spin density wave states, as well as ferromagnetic, "martensitic", "bipolaron" states and many others are nearly degenerate.[27] Thus, in many one-dimensional materials several different states occur at somewhat different temperatures. For example, the tetragonal ("martensitic") state at 42K and the superconducting state at 18K in Nb_3Sn;[28,29] superconducting and CDW states in $NbSe_3$ and similar materials; a fluctuating SDW state giving rise to $4k_F$ reflections around 200K in TTF-TCNQ[30] and a CDW state at 52K. Similarly, a covalency gap opens up in HMTSF-TCNQ (and TMTSF-DMTCNQ) around 40K-50K. In semimetals, a narrow minimum in the density of states around the Fermi surface occurs by definition (zero density of states is by definition a semiconductor, and a high density of states is by definition a metal).

The various states give rise to gaps that can be seen by tunneling (and other techniques). Thus, the existence of a "high temperature" gap in $(TMTSF)_2PF_6$ may be due to various states (CDW, SDW, covalency gap, lattice distortion, semimetals, etc.) and is not necessarily an indication of a superconducting state.

5. Fluctuating One-Dimensional versus BCS Anisotropic Superconductivity

There is no doubt that in $TMTSF_2X$ a very anisotropic BCS

superconducting state is observed at low temperatures (below about 1.4K). The transition temperature of this state is related to the phonon frequency,[31] $T_c \approx (1/50$ to $1/20)\omega_{ph}$. This state is one-dimensional in the sense that $\xi_\parallel >> \xi_\perp$ and this is due to a Fermi surface which is open, but by no means perfectly planar.[32,4] The question is whether (perhaps in some other material) a fluctuating, non-BCS, one-dimensional superconducting state is possible.

The question is, what do we expect from a truly one-dimensional fluctuating superconductor? In principle, there is no reason why a fluctuating one-dimensional superconductor should not be possible. Such a state builds up at the mean-field transition temperature, while the thermodynamic transition temperature is considerably suppressed. This situation was studied[33] in relation with the A-15's. An enhanced NMR relaxation rate is observed[34] considerably above T_c in Nb_3Al, and this may indicate an approach to such a state. In $TMTSF_2PF_6$, an enhanced NMR relaxation rate is due probably to spin fluctuations.

In a truly 1-d system, there are no closed loops, and as a result, no Meissner effect and there is complete magnetic field penetration. As a result, there is no $H^2/8\pi$ term in the expression $F_n - F_s = H^2/8\pi$, and a magnetic field does not suppress superconductivity in the "normal" way. Thus, one should expect a resistivity that falls significantly below the single-particle resistivity as the temperature is lowered. The single particle resistivity has values as indicated in Figure 3. This reduction should not be restored by a moderate magnetic field (since there is no Meissner effect). In a high magnetic field, destruction of superconductivity at the Clogston limit is possible, however, for the nearly-degenerate triplet superconductor, a very high magnetic field may enhance the superconducting state.

The gap (order parameter) must be positively identified as due to superconductivity, rather than other states (such as CDW, SDW, etc.), by methods such as the proximity effect.

IV. REMARKS OF H. GUTFREUND

I would like to make a few general theoretical comments about organic superconductivity. However, I should first remark that organic superconductivity is now a well established subject with a rich phenomenology and will be represented at all future superconductivity conferences.

All the organic superconductors, and for that matter also all the organic conductors, are molecular chain compounds, namely, systems characterized by a one-dimensional (strictly speaking-quasi-one-dimensional) structure. Thus, organic superconductivity is intimately connected with one-dimensionality. The latter has two main consequences. First, the ground state of a one-dimensional electron system can have one of several possible types of order and superconductivity is found only in a certain region of the interaction-parameter space. This was the subject of Weger's seminar at this school. Second, in one-dimensional systems there is a possibility of large fluctuation effects. This was discussed in Jerome's lectures and is the subject of the present discussion.

There are two characteristic temperatures in a system of coupled linear chains. The temperature T_1 at which a certain type of order begins to build up on a single chain. This temperature is determined by the elementary interactions on the chain (forward scattering, backscattering and umklapp) and can be estimated from a mean-field calculation. It is therefore associated with the mean-field temperature. The actual phase transition occurs at a lower temperature T_3, when the phases of the order parameters on the different chains are locked to each other and a three-dimensional order is established. This temperature is determined by the nature and magnitude of the interchain interaction.

The question now is what is the relation between T_1 and T_3. Based on a large number of transport, thermodynamic and critical field experiments, Jerome concludes that in the $(TMTSF)_2X$ compounds $(T_1/T_3) \simeq 30$. If this is indeed the case it is a spectacular result. It means that we have a system at which the basic interactions give rise to superconductivity at $\sim 30°K$ and this raises immediately the question of alternate mechanisms of superconductivity because such a high T_c is hard to reconcile with the phonon mechanism and reasonable estimates of the electron-phonon coupling and the density of states in these materials. Greene and Weger interpret the experimental evidence without involving large fluctuation effects and they conclude the $T_1 \gtrsim T_3$. Their arguments are summarized in their contributions to this panel discussion.

In my contribution I would like to discuss briefly the relation between the ratio T_1/T_3 and the interchain coupling strength and also to consider the question: What do we mean by a one-dimensional system? To this end let us adopt a model which we have discussed previously in various context,[35] in which the interchain tunneling is represented by the electron dispersion

$$\varepsilon(\underline{k}) \;=\; \varepsilon(k_z) - 2k_\perp (\cos ak_x + \cos ak_y) \qquad (1)$$

where $t_\perp$ is the interchain transfer integral, $\underline{a}$ the interchain distance, and $\varepsilon(k_z)$ is a quasi free electron dispersion along the chains, which is characterized either by an effective mass or by the interchain transfer integral $t_{\parallel}$. This electron dispersion gives rise to an open and nested Fermi surface (Figure 6) with a nesting vector $q_0 = (\pi/a, \, \pi/a, \, 2k_F)$. The deviation from strict one-dimensionality, namely, from a flat Fermi surface is proportional to the

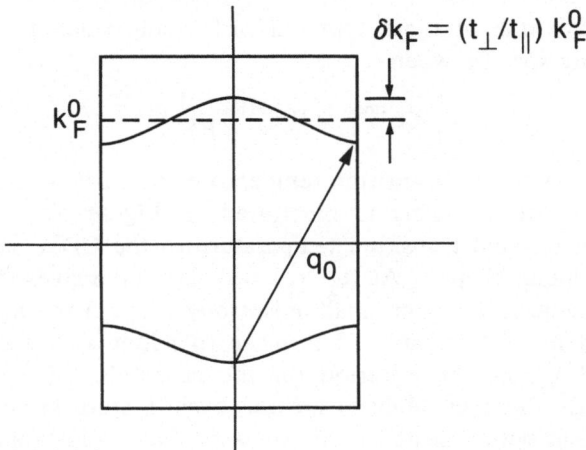

Fig. 6. Cross section of the Fermi surface corresponding to Eq. (1).

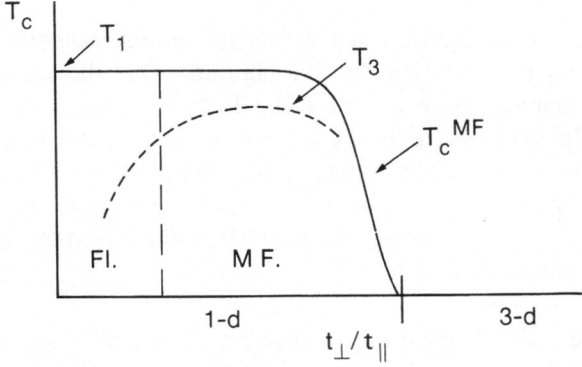

Fig. 7. Mean field (solid line) and actual (broken line) Peierls transition temperature. The distinction between 1-d and 3-d, and between mean-field behavior and fluctuation region is indicated.

ratio $t_\perp/t_\parallel$. The nesting property, i.e., $\varepsilon(\underline{k})=-\varepsilon(\underline{k}-q_0)$ for all $\underline{k}$ on one side of the Fermi surface, is destroyed when the curvature of the Fermi surface is sufficiently large. Since the nesting property is a necessary condition for an electron-hole instability, which can be either CDW or SDW, we conclude that a system can have one of these types of order only when $t_\perp$ is sufficiently small - to be more specific when

$$t_\perp < t_\perp^{max} \simeq 3(T_c^{MF}T_F)^{1/2} \tag{2}$$

where T_c^{MF} is the mean-field transition temperature to CDW or SDW and T_F is the Fermi temperature. This is illustrated in Figure 7. The solid line represents the mean-field transition temperature to the CDW state (the curve for SDW has a similar shape). At $(t_\perp/t_\parallel) = 0$ this curve gives the single chain mean-field T_c, namely, T_1. For small anisotropy $(t_\perp/t_\parallel)$ the mean-field T_c is almost independent of this ratio. I propose to adopt as the definition of a one-dimensional system the criterion for the possibility of an electron-hole instability because these instabilities are the basic feature of one-dimensional systems, or systems with a nested Fermi surface. Pairing instabilities can occur in a system with any value of $(t_\perp/t_\parallel)$. In this sense $(TMTSF)_2PF_6$ is certainly one-dimensional because a SDW is observed at ambient pressure at $T_c \simeq 12K$. We can see that this is consistent with Figure 7 and Eq. (2). Using $T_F \simeq 2000°K$ and taking for a quarter filled band $t_\parallel = T_F$, we find that the system would be one-dimensional in our sense for $(t_\perp/t_\parallel) \lesssim 0.25$. A recent ab-initio calculation by Grant[4] puts this ratio at about 0.1. This value is consistent with many experimental results[2] and has also been quoted by Jerome in his lectures.

The broken line in Figure 7 represents the actual transition temperature, T_3, as calculated by the Ornstein-Zernike method. One distinguishes here two regions. For sufficiently large $t_\perp$ one finds $T_1 \gtrsim T_3$. This is the "mean-field" region in which fluctuation effects extend over a temperature range of 20-50% of T_3 itself. The transition between this region and the large fluctuation regime occurs roughly at $t_\perp \simeq 4T^{MF}$. Quoting again Grant's result[4] $t_\perp \simeq 12$ meV, we find that at least for the SDW transition one does not expect large fluctuation effects in this case.

As for the superconducting transition, I would like to quote the calculation of T_c in a coupled chain system by Maniv and Weger.[33] They have estimated the effect of fluctuations on the mean field T_c and they found that T_c is reduced significantly only when $(t_\perp/t_\parallel)<0.1$.

My remarks can be summarized as follows:

a) One can have interesting "one-dimensional physics" in the sense described above without the extreme one-dimensional fluctuation effects.

b) Theoretically it seems that the value of $(t_\perp/t_\parallel) \simeq 0.1$ is much to large to get such extreme fluctuation effects.

c) The question of existence of large fluctuations has to be decided by experiment but given these theoretical considerations and the alternative interpretations of the experiments discussed by Greene and Weger, I would say at this point that Jerome's chain of $(T_1/T_3) \simeq 30$ in the $(TMTSF)_2X$ superconductors is not established.

V. REFERENCES

1. D. Jerome, Molec. Cryst. Liq. Cryst. <u>79</u>, 155 (1982).
2. R. L. Greene, P. Haen, S. Z. Huang, E. M. Engler, M. Y. Choi and P. M. Chaikin, Molec. Cryst. Liq. Cryst. <u>79</u>, 183 (1982).
3. H. J. Schulz, D. Jerome, A. Mazaud, M. Ribault and K. Bechgaard, J. Physique <u>4</u>, 991 (1981).
4. P. M. Grant, Phys. Rev. B, in press.
5. C. More, G. Roger, J. P. Sorbier, D. Jerome, M. Ribault and K. Bechgaard, J. Physique Lett. <u>42</u>, L313 (1981).
6. H. K. Ng, T. Timusk, J. M. Delrieu, D. Jerome, K. Bechgaard and J. M. Fabre, J. Physique Lett. <u>43</u>, L513 (1982).
7. W. Challener, P. Richards and R. L. Greene, to be published.
8. M. Y. Choi, P. M. Chaikin, S. Z. Huang, P. Haen, E. M. Engler and R. L. Greene, Phys. Rev. <u>B25</u>, 6208 (1982).
9. P. Garoche, R. Brusetti, D. Jerome and K. Bechgaard, J. Physique Lett. <u>43</u>, L147 (1982).
10. L. J. Azevedo, J. E. Schirber and J. C. Scott, Phys. Rev. Lett. <u>49</u>, 826 (1982).
11. J. F. Kwak, J. E. Schirber, R. L. Greene and E. M. Engler, Phys. Rev. Lett. <u>46</u>, 1296 (1981).
12. D. Djurek, M. Prester, D. Jerome and K. Bechgaard, J. Phys. C <u>15</u>, L669 (1982).
13. E. Abrahams, M. Redi and J. F. Woo, Phys. Rev. <u>B1</u>, 208 (1970).
14. S. Wolf and B. S. Chandrasekhar, Phys. Rev. <u>B4</u>, 3014 (1971).
15. J. F. Kwak, Phys. Rev. B, in press.
16. H. Gutfreund and M. Weger, Phys. Rev. <u>B16</u>, 1753 (1977).
17. M. Nicol, M. Vernon and J. T. Woo, J. Chem. Phys. <u>63</u>, 1992 (1975).
18. M. Weger, M. Kaveh and H. Gutfreund, J. Chem. Phys. <u>71</u>, 3916 (1979).
19. D. B. Tanner, K. D. Cummins and C. S. Jacobsen, Phys. Rev. Lett. <u>47</u>, 597 (1981).
20. M. Weger and H. Gutfreund, Comments on Solid State Physics <u>8</u>, 135 (1978).
21. A. J. Heeger, M. Weger and M. Kaveh, in Lecture Notes in Physics, S. Barisic, A. Bjelis, J. R. Cooper and B. Leontic, eds. (Springer, New York, 1979), Vol. 95, p. 223.

22.   S. Marianer, M. Kaveh and M. Weger, Phys. Rev. B25, 5197 (1982).
23.   A. Andrieux, C. Duroure, D. Jerome and K. Bechgaard, J. Physique Lett., L381 (1979).
24.   J. R. Cooper, M. Weger, D. Jerome, D. Lefur, K. Bechgaard, D. O. Cowan and A. N. Bloch, Solid State Commun. 19, 749 (1976).
25.   D. Jerome and M. Weger, in "Physics and Chemistry of One Dimensional Metals," NATO ASI B25, H. J. Keller, ed. (Plenum Press, 1977).
26.   A. Bychkov, L. P. Gorkov and I. E. Dzyaloshinski, Solv. Phys. JET 23, 489 (1966).
27.   B. Horovitz, Solid State Commun. 18, 455 (1976).
28.   M. Weger and I. B. Goldberg, "Solid State Physics," Seitz, Turnbull, Ehrenreich Edt. Vol. 28, p. 1, 1973.
29.   G. Bilbro and W. L. McMillan, Phys. Rev. B14, 1887 (1976).
30.   S. Kagoshima, T. Ishiguro and H. Anzai, Phys. Soc. Japan 41, 2061 (1976); J. P. Pouget, S. K. Khanna, F. Denoyer, R. Comes, A. F. Garito and A. J. Heeger, Phys. Rev. Lett. 37, 437 (1976).
31.   B. Horovitz, H. Gutfreund and M. Weger, Mol. Cryst. Liq. Cryst. 79, 235 (1982).
32.   R. L. Greene and E. M. Engler, Phys. Rev. Lett. 45, 1587 (1980).
33.   T. Maniv and M. Weger, J. Phys. Chem. Solids 36, 367 (1975); 37, 255 (1976).
34.   F. Y. Fradin and G. Cinader, Phys. Rev. B16, 73 (1977).
35.   B. Horovitz, H. Gutfreund and M. Weger, Phys. Rev. B12, 3174 (1975).

SUPERCONDUCTING TUNNEL JUNCTIONS IN

HIGH FREQUENCY RADIATION DETECTORS

Tord Claeson

Department of Physics
Chalmers University of Technology
S-412 96 Gothenburg, Sweden

I   INTRODUCTION

Radio Astronomy - A Demanding Customer

Sensitive detectors are needed in the microwave, millimeter, and sub-millimeter wavelength bands. As the availability of a radio telescope scales as the square of the inverse noise temperature of the detection system, it is obvious that even a moderate improvement of the detector is of great importance. It should be emphasized, though, that it is the total system noise temperature that is the important entity. Hence, in our view, the important element, whether it be a square law video detector, an amplifier, or a mixer transforming the signal to a more convenient frequency, has to fit into the rest of the system. Other properties have also to be considered. An advantage of the superconducting tunnel devices, we will discuss, is the low pump power that is needed. The maintenance of high power, high frequency local oscillators is most costly.

To give an idea of the competition in the field, examples of presently developed detector elements are given in Table I.

Superconducting Tunnel Junctions

Superconducting tunnel junctions promise to be very competitive detector elements - in fact, the ultimate detector in a large wavelength region. They can be operated in varying modes and used in different types of applications. Several good reviews of the field exist[1-6]. One was given in this series of Advances of Superconductivity Institutes by Adde and Vernet[1]. Hence the incompleteness of

241

Table I. Examples of high frequency detector elements.
Approximate values given.

	Freq range (GHz)	Gain (dB)	Conversion loss (dB)	T_N (K)	T_{Rec} (K)	
Maser	2-50	30		1-7	15-25	Narrow bandwidth
Schottky diode	50-700		5-15	80-3000	$\sim$130-4000	
Super-Schottky	30-90		9-18	10-150		Not operative
InSb	100-600				150	Very narrow band

this presentation is justified. Instead of giving a full review, we will discuss two examples of superconducting tunnel junction detectors in more detail. Considerable progress has been reported for these lately. Other high frequency applications of superconducting tunneling but detectors are not covered here. The treatment can also be seen as an illustration to the basic chapter by Barone in this book. All concepts are thus not defined here; the reader has to consult basic texts.

Three main types of superconducting high frequency detectors have been developed: video detectors, parametric amplifiers, and mixers. If we concentrate on superconductor-insulator-superconductor (SIS) junctions, these can be operated in different modes as illustrated in Fig. 1 - unbiased or (voltage or current) biased, bare or shunted junction. Pairs of electrons tunnel in the Josephson mode and single electrons in the quasi-particle mode - or both modes interfere.

Our two examples illustrate disparate aspects. The internally pumped parametric amplifier utilizes a biased, Josephson tunnel junction that is shunted by a resistor and an inductance. It is less well developed at present than the other example that concerns a biased, quasi-particle mixer. Both single and series coupled arrays of tunnel junctions have been extensively studied during the last couple of years. After introducing concepts, models and results for the two specific examples, results for the most successful tunnel devices will be tabulated at the end of this presentation.

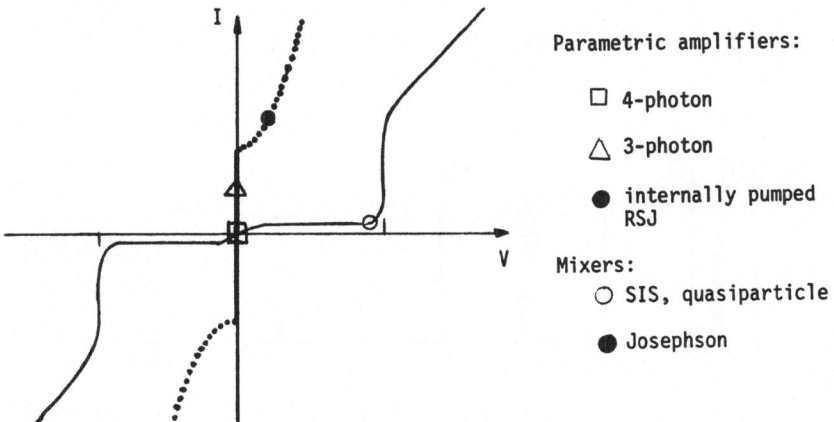

Fig. 1. Different parts of the I-V curve for an S-I-S junction are
         utilized in different types of applications. The elements
         are either externally pumped (quasiparticle and zero voltage
         Josephson) or internally pumped by an average, dc voltage
         over a shunted Josephson junction (RSJ). Video detectors
         are biased like the mixers shown.

II   INTERNALLY PUMPED PARAMETRIC AMPLIFICATION

In a parametric amplifier there is a transfer of power
(photons) from a strong pump (local oscillator) to a weak signal.
Photons are also transferred to idler frequencies. A non-linear
reactance is needed for the amplification to occur, e.g. in a
varactor diode or a Josephson junction.

For the latter:

$$dI/dt = (d/dt)\ (I_c \sin\phi)\ = I_c \cos\phi (d\phi/dt) =$$
$$= I_c \cos\phi (2e/\hbar) V = V/L_J(\phi),$$

i.e. a phase dependent inductance.

In an externally pumped Josephson amplifier, there is a
transfer of power from an externally applied frequency f_{LO} to the
signal at f_0 and to idlers at $f_m = |f_0 + m f_{LO}|$ (the main idlers
being m = -2 for the 4-photon, unbiased, and m = -1 for the
3-photon, current biased, amplifiers). Considerable gain has been
achieved in narrow ranges of pump power[7-10]. However, these amplifiers are generally unstable and very noisy. A so called "noise
rise" often occurs when amplification sets in [11]. The noise level

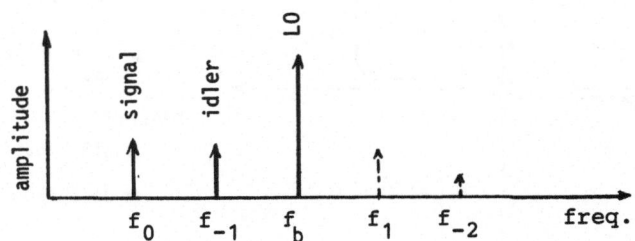

Fig. 2.   Power is transferred from a local oscillator to signal
          and idler frequencies ($f_m = |f_0 + mf_b|$, $m = 0, \pm1,...$).
          In an internally pumped Josephson amplifier, the in-
          ductance is modulated at frequency $f_b = 2\ eV_b/h$ deter-
          mined by the bias voltage V_b.

increases with the amplifier gain, a most undesirable property.
Less well studied than the externally pumped Josephson amplifier
is the internally pumped one.

      A self-pumped amplifier utilizes the Josephson oscillation,
f_b, induced by a constant voltage bias. Power is transferred to
the signal and idler frequencies. Here, we will first restrict the
consideration to a two port model, i.e. neglect small oscillations
other than the signal and one idler at $f_b - f_0$ as shown in Fig. 2.

      We desire a well defined pump, i.e. one having a small har-
monic content and a narrow frequency spread. This is realized with
a constant voltage bias, i.e. for a small source impedance. Hence
we need a small resistance, R_S, shunting the Josephson junction.
However, a small R_S means that it becomes difficult to couple the
high frequency radiation to and from the junction. A solution to
the problem is to include an inductance in the shunt (actually it
is, to a certain extent, included automatically in a microstrip
configuration). At low frequency, and particularly at d.c., the
shunt impedance is small, at high frequency it is large. The equivalent circuit of the inductively shunted junction is given in
Fig. 3.

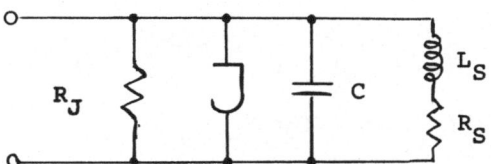

Fig. 3. A Josephson junction, with quasiparticle resistance R_J and capacitance C is shunted by a resistance R_S in series with an inductance L_S. The shunting admittance at frequency f becomes

$$Y = G + jB = 1/R_J + R_S/\left[R_S^2 + (2\pi f L_S)^2\right] +$$

$$j\{2\pi fC - 2\pi f L_S/\left[R_S^2 + (2\pi f L_S)^2\right]\}$$

Amplification Due to Cross-Coupling

With only one idler, the voltage across the junction is[12,13]

$$v(t) = V_b + V_0\cos(2\pi f_0 t + \beta_0) + V_{-1}\cos(2\pi f_{-1}t + \beta_{-1}),$$

where the phases β_m are counted relatively the Josephson oscillation induced by V_b.

The Josephson current becomes

$$i(t) = I_c\sin\{(2e/\hbar)\int v(t)dt =$$

$$= I_c\sin\{2\pi f_b t + \beta_b + (2eV_0/hf_0)\sin(2\pi f_0 t+\beta_0) +$$

$$+ (2eV_{-1}/hf_{-1})\sin(2\pi f_{-1}t + \beta_{-1})\}.$$

Using $e^{j\alpha\sin\phi} = \sum\limits_{n=-\infty}^{\infty} J_n(\alpha)e^{jn\phi}$, we get

$$i(t) = I_c \sum\limits_{m=-\infty}^{\infty}\sum\limits_{n=-\infty}^{\infty} J_m(\alpha_0)J_n(\alpha_{-1})\sin\{2\pi(f_b+mf_0+nf_{-1})t +$$

$$+ m\beta_0+n\beta_{-1}+\beta_b\}$$

where $\alpha_0 = 2eV_0/hf_0$; $\alpha_{-1} = 2eV_{-1}/hf_{-1}$.

Both α's are small entities.

As $f_0 + f_{-1} = f_b$, we obtain for the current component a frequency f_0,

$$i_0(t) = I_c \cdot J_0(\alpha_0) J_{-1}(\alpha_{-1}) \sin\{2\pi(f_b - f_{-1})t - \beta_{-1} + \beta_b\} \approx$$

$$\approx - I_c(eV_{-1}/hf_{-1}) \sin(2\pi f_0 t - \beta_{-1} + \beta_b)$$

In the last step, we retained only the leading terms in the expansion of the Bessel functions as the α's are small. In complex notation

$$I_0 = jI_c(eV_{-1}^*/hf_{-1})e^{j\beta_b} \text{ and, likewise,}$$

$$I_{-1} = jI_c(eV_0^*/hf_0)e^{j\beta_b},$$

where $i_m = \mathrm{Re}(I_m e^{j2\pi f_m t})$, $V_m = |V_m|e^{j\beta_m}$, and the star denotes the complex conjugate.

No power is incident upon the junction at frequency f_{-1}. Terminating the Josephson junction (J in Fig. 3) with a conductance Y_{-1} at that frequency, we have a simple relation between current and voltage:

$$I_{-1} = -Y_{-1}V_{-1} \text{ and thus}$$

$$I_0 = -jI_c(eI_{-1}^*/hf_{-1}Y_{-1}^*)e^{j\beta_b} = (-e^2 I_c^2/hf_0 \cdot hf_{-1})(1/Y_{-1}^*)V_0$$

This is the response of the Josephson junction itself at f_0. However, we must also take into account the shunting elements and the external circuitry. The total current in the shunted junction

$$I_0^{tot} = Y_0^{in} V_0, \text{ where}$$

$$Y_0^{in} = G_0 + jB_0 - e^2 I_c^2/h^2 f_0 f_{-1} Y_{-1}^* =$$

$$= jB_0 + G_0 \{1 - f_c^2 G_0/4f_0 f_{-1} Y_{-1}^*\} ,$$

$G_0 + jB_0$ is the circuit shunting admittance at f_0 as given in Fig. 3,

$$f_c = 2eI_c/hG_0$$

$$Y_{-1} = Y_{-1}^{ext} + G_{-1} + jB_{-1}, \text{ and}$$

Y_{-1}^{ext} is the characteristic impedance of the external world that the circuit sees at f_{-1}.

We see that it is possible to obtain a negative Y_0^{in} and hence gain for certain parameter ranges as the power amplification at f_0 is the usual expression:

$$\Gamma_o = \left| \frac{(Y_0^{ext})^* - Y_0^{in}}{Y_0^{ext} + Y_0^{in}} \right|^2$$

The optimum gain can be achieved by tuning I_c, and thus the characteristic frequency f_c, via a magnetic field.

Note that it is the off-diagonal elements in the Y-matrix for the Josephson junction

$$\begin{pmatrix} I_0 \\ I_{-1}^* \end{pmatrix} = \begin{pmatrix} 0 & Y_{0-1} \\ Y_{-10} & 0 \end{pmatrix} \begin{pmatrix} V_0 \\ V_{-1}^* \end{pmatrix}$$

that are important for the negative effective impedance and the gain.

Circuit and Measurements

We will now discuss a particular experiment[14] in order to get a feeling of how the circuit looks like in real space, how measurements are made and what the results look like.

The physical appearance of the circuit is shown in Fig. 4. The 7 μm x 7 μm Pb tunnel junction was connected via quarter wavelength transformers to the external world and to ground. It was shunted by an Ag strip, which in this experiment was placed close to the junction. (With a longer shunt loop relaxation oscillations occurred[15], an interesting phenomenon which, however, falls outside the scope of this seminar!) The transformer, which was designed for about 10 GHz, transformed the 50-Ω input impedance to about 1-Ω at the junction.

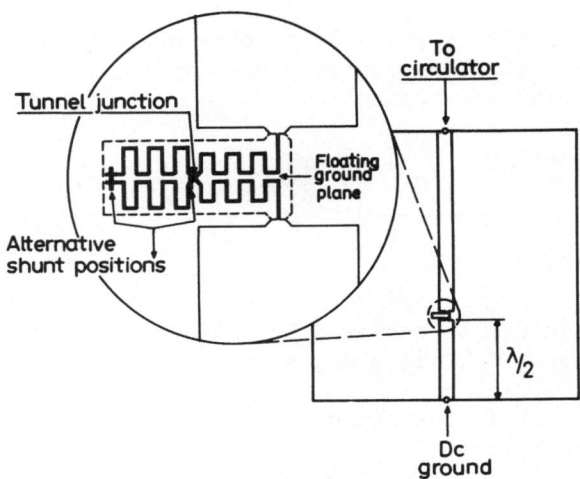

Fig. 4.   A shunted tunnel junction is connected to the external circuitry via microstrip transformers. The lengths of the
strips and their separation to the floating ground plane
determine the input transformer impedance and the inductance of the shunt loop, L_S. The latter could be varied
over a wide range by changing the location of the shunt
– in the example described here, it was placed close to
the junction.

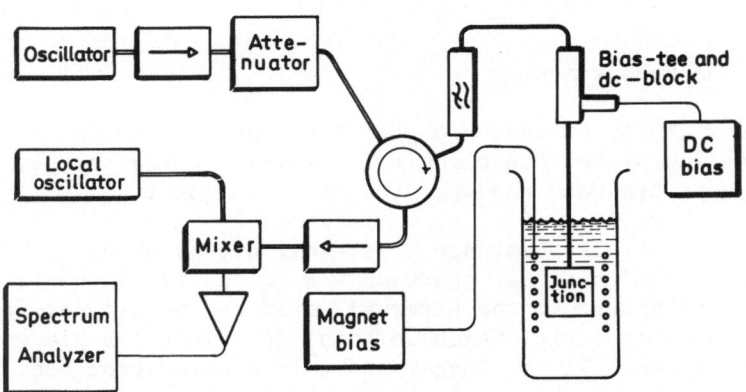

Fig. 5.   Block diagram of the microwave (8–12 GHz) system
used for a self-pumped amplification study.

The amplification of a signal reflected from the device was measured by the circuitry shown in Fig. 5.

An example of the signal amplification is given in Fig. 6. It is most evident from the idler that power has been transferred.

From the measurements we can deduce values of the gain Γ_0 and the noise temperature T_N of the amplifier as functions of controllable parameters, e.g. of I_c as done in Fig. 7. (I_c was varied by a magnetic field from the zero field value of I_c^0).

Several features can be noted from Fig. 7:

(1) T_N is low. The uncertainty in the measurement was rather large, so T_N might be considerably smaller than the upper bounds given in the figure. No "noise rise" was noticed in these amplification experiments. The amplifier became noisier when running in the relaxation mode that appeared for large L_S.

(2)   There is a good agreement between experimental and theoretical gains. For the calculation we have used a somewhat more realistic model with two idlers[13,14] at $f_{\pm 1} = f_b \pm f_0$ while the other, higher harmonic idlers have been accounted for in a changed $f_c = \kappa \cdot 2eI_c/hG_0$, where κ is a measure of how much of the pump voltage is evolved at the fundamental frequency as compared to at harmonics. With two idlers, the expression for the effective admittance at the signal frequency can be written:

$$Y_0^{in} = jB_0 + G_0 \left\{ 1 - \frac{f_c^2 G_0}{4f_0 f_{-1}} \left(\frac{1}{(Y_{-1}^{ext})^* + G_{-1} - jB_{-1}} - \frac{f_{-1}/f_1}{Y_1^{ext} + G_1 + jB_1} \right) \right\}$$

and the gain is calculated the usual way.

(3)   The maximum gain is moderate. However, the fact that the measured gain agrees with the value expected from a simple model indicates that the limited gain was due to a bad impedance match to the external world and not to the amplifier itself.

(4)   The gain is stable over a large range of critical currents. The stability is also evident from Fig. 8 where we plot the gain versus variations in signal frequency, bias voltage (i.e. f_{-1}) and critical current relatively their optimal values. The gain-bandwidth, $\Gamma_0^{\frac{1}{2}} f/f_0$ is large. From the equation above, we expect the gain to vary as a function of the variables signal frequency, bias voltage and the

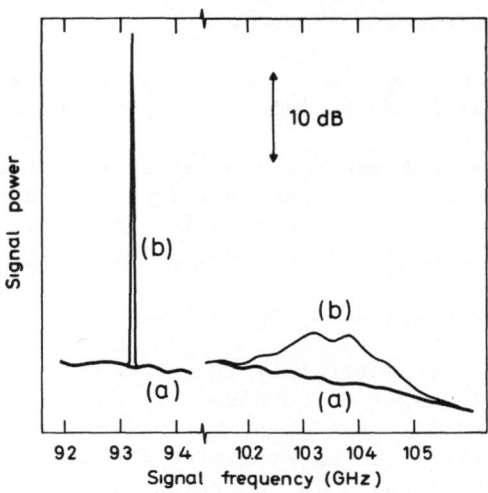

Fig. 6. The output from an irradiated shunted tunnel junction (a)
zero biased and (b) biased such that the Josephson oscillation f_b = 19.68 GHz. The signal is at f_0 = 9.32 GHz. The
idler at $f_{-1} \approx 10.36$ GHz is broad, mirroring the width of
f_b. R_S = 0.03Ω. L_S = 3 pH, I_c^0 = 210 µA.

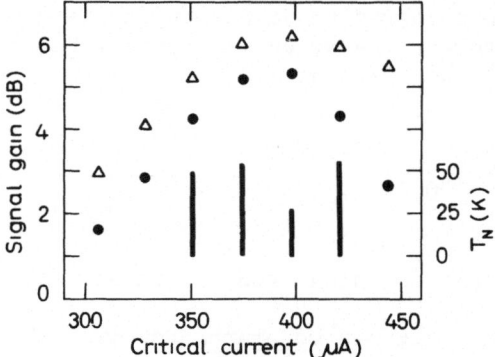

Fig. 7   The critical current dependence of the gain at f_0 = 10.4 GHz
of a parametric amplifier selfpumped at 49 GHz. Experimental values are given by the circles, while the triangles give the theoretical gain taking into account two
idlers and higher harmonics of the local oscillator as
described in the text. The bars give the noise temperatures with their large uncertainties.
R_S = 0.09 Ω, L_S = 4.5 pH, I_c^0 = 930 µA, κ = 0.45,
R_J = 2.3 R_N, the normal state resistance.

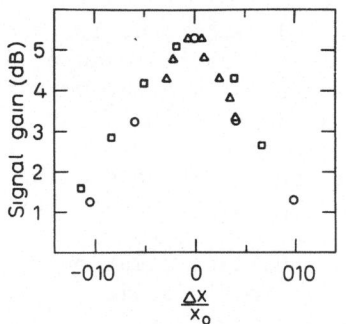

Fig. 8.   Signal gain vs. $\Delta X/X_{opt}$, where X is f_0 (triangles),
bias voltage (circles), or $I_c{}^2$ (squares). At optimum
gain $(f_0)_{opt}$ = 10.4 GHz; $(f_b)_{opt}$= 48.8 GHz, and
$(I_c)_{opt}$ = 395 µA.

square of the critical current. The fact that the variation
is more rapid with deviations in f_0 and bias voltage from
their optimal values than with I_c^2 (more as $I_c{}^2$ as shown in
the figure) must mean that the bandwidth is limited by the
input transformer, not by the amplifier itself. The transformer is probably important in preventing external noise
at idler frequencies from being converted to the signal
frequency.

The saturation levels, i.e. the input signal powers to which
the amplification is linear, of these one-junction amplifiers,
where $f_b \gg f_0$, have been fairly low, of the order of a few pW.

Remaining Problems

What is causing the "noise rise" that often occurs in superconducting parametric amplifiers? If we understand it, can we avoid it?
A review of the problem, amphasizing the importance of pump instabilities, is given by Feldman and Levinsen[11]. One possible cause
could be chaos, a solid state turbulence in the electron pair gas[16,17].
Can this phenomenon be studied by the Josephson effect?

Are the stable and low noise properties of the self-pumped
amplifier at 10 GHz retained at 100 GHz? What happens as the gain
increases? Do we enter a noise rise region?

III. QUASI-PARTICLE MIXERS

We are used to power detectors at low frequencies, while at high frequencies, e.g., in the optical range, single photons can be detected. We will now discuss a mixer working at the borderline between the classical and quantum limits. At high frequency, when the photon energy is larger than the width of the singular region of the I-V characteristic of a superconducting tunnel junction, the quantum nature will give a good conversion, even gain[18].

A non-linear current-voltage relation of an element is the corner-stone of a mixer. In such a device, power is transfered to sum and difference frequencies of the two signals mixed as indicated in Fig. 9. In particular, we are interested in the difference frequency, the intermediate frequency f_{IF}, falling in a range where conventional, low-noise amplifiers can amplify the signal before it is studied in detail.

The conversion loss, or the inverse conversion efficiency, of the mixer is defined as

$$L_C = \text{(Input signal power)}/\text{(Signal power at IF)} = P_S/P_{IF}.$$

The receiver noise temperature $T_{REC} = T_{MXR} + L_C T_{IF}$, where T_{MXR} and and T_{IF} are the noise temperatures of the mixer and of the following intermediate frequency amplifiers. Naturally, we want to keep these quantities, L_C, T_{MXR} and T_{REC}, as small as possible.

Classical Mixer Operation

A local oscillator voltage applied to a non-linear element produces a periodic variation of the small signal conductance dI/dV, which can be analysed as a Fourier series of the form

$$G = G_0 + G_1 \cos 2\pi \, f_{LO} t + G_2 \cos 2 \cdot 2\pi f_{LO} t + \ldots.$$

Adding a signal voltage $v_s \cos(2\pi f_s t)$ vill cause the element current to shift by an amount

$$Gv_s \cos 2\pi f_s t = G_0 v_s \cos 2\pi f_s t + \tfrac{1}{2} G_1 v_s \{\cos 2\pi (f_s + f_{LO}) t$$

$$+ \cos 2\pi (f_s - f_{LO}) t\} + \ldots.$$

There are Fourier components at $f_s + m f_{LO} (m=0, \pm 1, \ldots)$. The IF amplifier may be tuned to accept any of these.

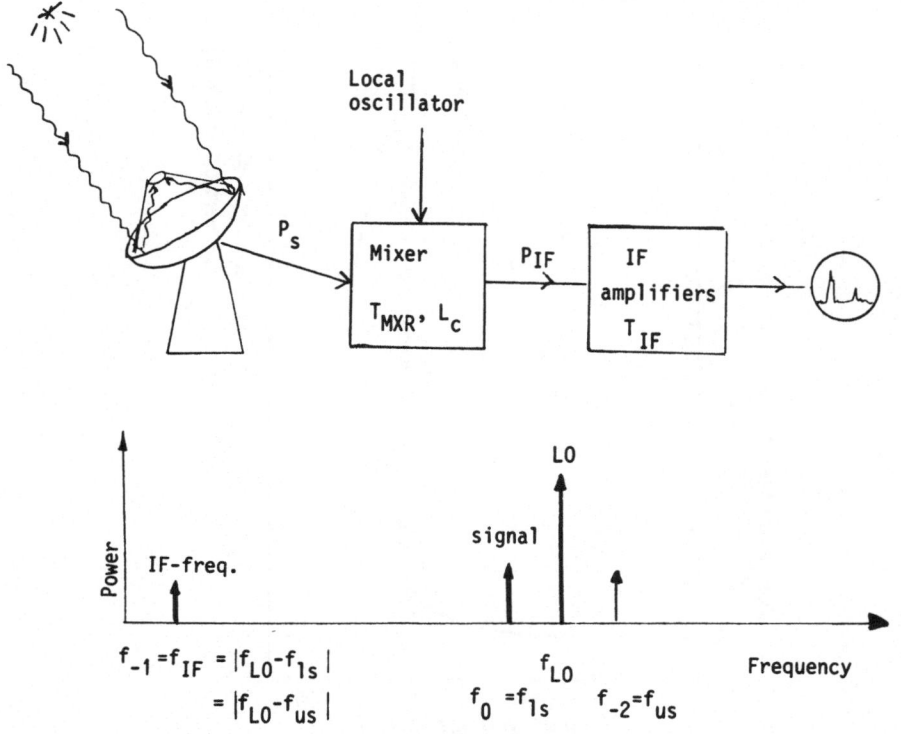

Fig. 9. In a mixer, the current-voltage relation is modulated by a
relatively strong local oscillator signal. A weak signal
admitted into the element will further modulate the perio-
dic response giving rise to current components at $|f_0 + m f_{LO}|$
($m = 0, \pm 1 ..$), where we have the output frequency
$f_{-1} = |f_{LO} - f_s|$. In an SSB mixer, power from only one sideband (the lower sideband f_{1s} in this case) is converted to
the output; in the DSB mode, power from both the lower and
upper sidebands (signal and image frequencies) is converted.

The stronger the non-linearity of the element, the better
the conversion. The curvature parameter, $S = (d^2I/dV^2)/(dI/dV)$, can
be used to compare the efficiencies of different devices. In a
Schottky-barrier diode, $S \sim 50\ V^{-1}$ at room temperature. By making
the metal point a superconductor, i.e. realizing a Super-Schottky
diode, even larger values, about $11.600\ V^{-1}$ at 1 K, have been
obtained[19]. A superconductor-insulator-superconductor tunnel
junction can display a comparable curvature.

An example of an I-V curve of an array of SIS tunnel junctions is
given in Fig. 10. Despite the fact that an array has a less sharp
I-V characteristic than a single junction, the curve looks similar
to one of an ideal switch. Below a threshold voltage there is no

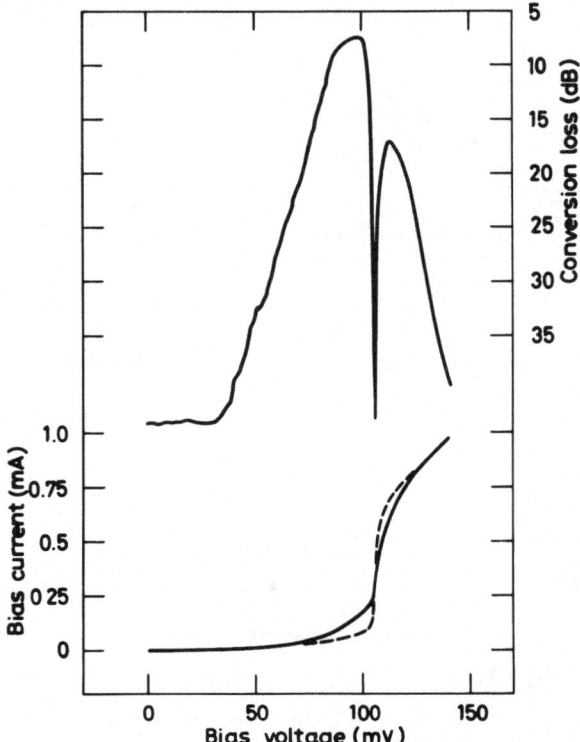

Fig. 10. Quasiparticle current and signal conversion loss as a
function of bias voltage in an array consisting of 40
series coupled Pb-I-Pb tunnel junctions. The signal
frequency = 9 GHz, the intermediate frequency 150 MHz.
The dashed curve gives the bias current without local
oscillator power, the full one the current at an LO
power of about 2 µW. T = 1.5K. From Ref. 20.

(or in our case a small) current, which rises sharply to a relatively
high value above the threshold. The efficiency of an array element
used as a mixer at 10 GHz is also given in Fig. 10. The conversion
loss displays minima in regions around the superconducting energy
gap while it is large at the gap itself.

In the classical limit, the conversion of a non-linear mixer element
must be less than one. However, the small signal analysis[21], upon
which that conclusion is based, relies upon two assumptions:

(1) reciprocity, i.e. the conductances the signal and the intermediate frequency see are the same, and

(2) there is no negative conductance region in the I-V curve.

Before discussing the quantum limit, where a conversion larger than one can be obtained, we shortly present a two-port model.

Mixer Modelling

For levels small compared to the local oscillator drive, currents and voltages at $f_m = f_s + mf_{LO}$, where $m = 0,\pm1,\pm2,..$) are linearly related to each other via the admittance matrix Y defined by

$$i_m = \sum_{m'} Y_{mm'} v_{m'}$$

If we further assume each sideband to be terminated by an admittance Y_m, the small signal voltage response can be related to a set of current generators at each frequency port.

$$v_m = \sum_{m'} Z_{mm'} I_{m'} \qquad , \text{ where}$$

$$||Z_{mm'}|| = ||Y_{mm'} + Y_m \delta_{mm'}||^{-1}$$

Our problem is to calculate the conversion of power from the signal frequency f_0 to the output frequency $f_{IF} = |f_0 - f_{LO}| = f_{-1}$:

$$(L_c)^{-1} = (\text{P out at } f_{IF})/(\text{P in at } f_s) = \tfrac{1}{2} G_{-1}|v_{-1}|^2/(I_0^2/8G_0)$$

where $G_{-1}(=G_L)$ and $G_0(=G_s)$ are the load and input conductances at f_{-1} and f_0 resp.

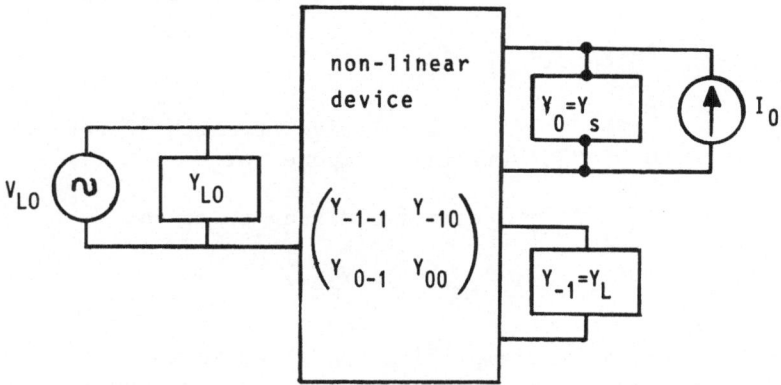

Fig. 11. Two-port model of a mixer. Besides the strong LO frequency, only two weak components at f_0 and f_{-1} are considered.

To simplify the presentation here, we will restrict ourselves to a two-port model, i.e. we will only consider small signals at the signal and output frequencies as pictured in Fig. 11. (It would be more realistic to also include the image frequency, $f_{-2} = |2f_{LO}-f_0| = f_{LO} + f_{IF}$. It may, however, be tuned out in a true single sideband operation.)

In this model we have

$$
\begin{Bmatrix} 0 \\ I_0 \end{Bmatrix} = \begin{Bmatrix} Y_{-1-1}+Y_{-1} & Y_{-10} \\ Y_{0-1} & Y_{00}+Y_0 \end{Bmatrix} \begin{Bmatrix} v_{-1} \\ v_0 \end{Bmatrix}
$$

or

$$
\begin{Bmatrix} v_{-1} \\ v_0 \end{Bmatrix} = \begin{Bmatrix} Z_{-1-1} & Z_{-10} \\ Z_{0-1} & Z_{00} \end{Bmatrix} \begin{Bmatrix} 0 \\ I_0 \end{Bmatrix}
$$

Then the conversion

$$
(L_c)^{-1} = 4G_{-1}G_0|Z_{-10}|^2 = \frac{1}{L_0} \frac{4\eta g_0 g_{-1}}{|(1+y_0)(1+y_{-1})-\eta(1+j\beta)|^2}
$$

$L_0 \quad = G_{0-1}/G_{-10}$ measures the degree of reciprocity,

$y_{-1} \quad = Y_{-1}/G_{-1-1}; \; y_0 = Y_0/G_{00}$ normalized admittances,

$g_{0,-1} \quad = Re\{y_{0,-1}\}$,

$\eta \quad = \dfrac{\dot{G}_{-10}G_{0-1}}{G_{-1-1}G_{00}}$ is the relative strength of off-diagonal (mixing) components,

$\beta \quad = B_{0-1}/G_{0-1}$ a reactive mixing term, and

$G_{mm'} \quad = Re(Y_{mm'}) \; ; \; B_{mm'} = Im(Y_{mm'})$.

β can produce parametric amplification (like in the case we have already discussed) and negative output impedance. However, we want

to show that even in the absence of non-linear reactive elements we
can obtain conversion gain. Hence we tune out β together with the
source and load reactances. This results in an even further simplified expression for the conversion efficiency[18, 22, 23]:

$$(L_c)^{-1} = \frac{1}{L_0} \frac{4\eta \, g_0 g_{-1}}{|(1+g_0)(1+g_{-1}) - \eta|^2}$$

which has a maximum of $\dfrac{1}{L_0} \cdot \dfrac{1-\sqrt{1-\eta}}{1+\sqrt{1-\eta}}$ when $g_0 = g_{-1} = \sqrt{1-\eta}$.

In the classical picture,

$$G_{00} = G_{-1-1} = G_0$$

$$G_{0-1} = G_{-10} = G_1$$

$$\eta = (G_1/G_0)^2 < 1$$

if the I/V curve has no negative region and the reciprocity factor
$G_{0-1}/G_{-10} = 1$. Hence for a classical mixer $(L_c)^{-1} < 1$.

Quantum Theory of Mixing

The so called quantum regime is entered when the width of the
singularity at the energy gap becomes smaller than the energy of a
photon ($\Delta V < hf/e$). (An alternative way is to say that at high enough
frequency the current response to a change in voltage is no longer
instantaneous on the time scale of a photon period. Instead retardation effects have to be considered and the classical theory is no
longer valid[24].) The conversion vs bias does not retain the smooth
appearance as in Fig. 10. Instead fine structure develops. More
important, a conversion from high to low frequency, that is larger
than unity, i.e. gain, appears.

The breaking of reciprocity and the enhancement of the off-
diagonal mixing components that cause the gain are due to photon
assisted tunneling.

Photons can be absorbed or emitted by electrons during the
tunneling process. This photon induced tunneling[25,26] gives
rise to structure within (and above) the gap voltage as illustrated
in Fig. 12. The average current at a bias voltage, V_b, is not only
a function of V_b, but also of $V_b + nhf_{LO}/e(n = \pm1, \pm2, \ldots.)$. In
general, we have:

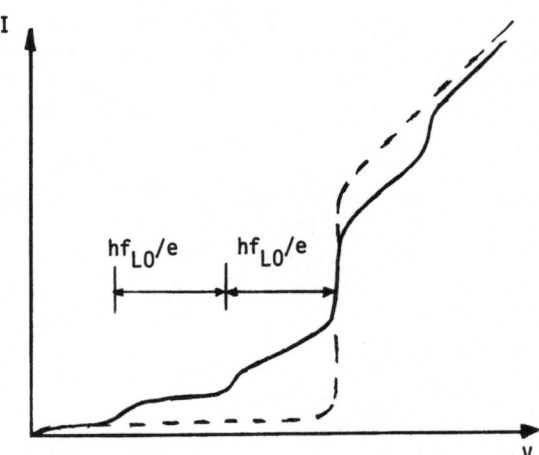

Fig. 12.   The appearance of microwave photon-induced quasi-particle
           tunneling steps. The dashed curve is the I-V curve with no
           microwaves, the full curve the irradiated one. The current
           at a voltage V depends on the I-V curve at V and at points
           separated from it by a multiple of photon energies.

$$I(t) = \sum_{m=-\infty}^{\infty} \sum_{n=-\infty}^{\infty} J_n(\alpha) J_{n+m}(\alpha) I(V_b + nhf_{LO}/e) e^{im \cdot 2\pi f_{LO} t}$$

where $\alpha = eV_{LO}/hf_{LO}$. The dc contribution is given by m=0.

Similar expressions as for the current are also obtained for
the conductance components[18, 24]. Actually, there is also a reactive
current component, being Kramers-Kronig related to the usual quasi-
particle current[27, 28]. It, and hence both components of the complex
admittance of the junction, has photon-assisted structure too[18].

Contributions from voltages separated by photon points have
also to be summed in the calculation of the different admittance
terms. For the conductances we obtain[18]:

$$G_{mm'} = \frac{e}{2hf_{m'}} \sum_{n,n'=-\infty}^{\infty} J_n(\alpha)\, J_{n'}(\alpha)\, \delta_{m-m',n-n'} \ \times$$

$$\{I_{dc}(V_b + n'hf_{LO}/e + hf_{m'}/e) - I_{dc}(V_b + n'hf_{LO}/e)$$

$$+ I_{dc}(V_b + nhf_{LO}/e) - I_{dc}(V_b + nhf_{LO}/e - hf_{m'}/e)\}$$

When f_{LO} is small, the differences in the parenthesis can be approximated by local derivatives on the I-V curve. Then $G_{-10} = G_{0-1}$ and the conversion is less than unity. However, for large f_{LO}, it is no longer valid to use local derivatives; the differences can be large, $G_{-10} \neq G_{0-1}$, and it is possible to obtain conversion gain.

In a two-port mixer we have $\eta \leq 1$ and the available conversion is limited[22]. If we take further idler frequencies into account, it is possible to get an η larger than unity. In that case, the denominator in the expression for $(L_c)^{-1}$ can become very small, i.e. gain appears. The denominator can even be zero, which means that infinite gain is available. Infinite gain is accompanied by an infinite or negative, effective resistance at the output frequency, i.e. a negative resistance region also in the dc I-V-curve. (But note that finite and large gain can be accomplished with a positive IF resistance.)

We have hitherto neglected the reactive part of the quasi-particle tunneling current. But this is not zero. The Kramers-Kronig relation tells us that a discontinuity in the in-phase part of the current must give a sizeable contribution also to the out-of-phase part.

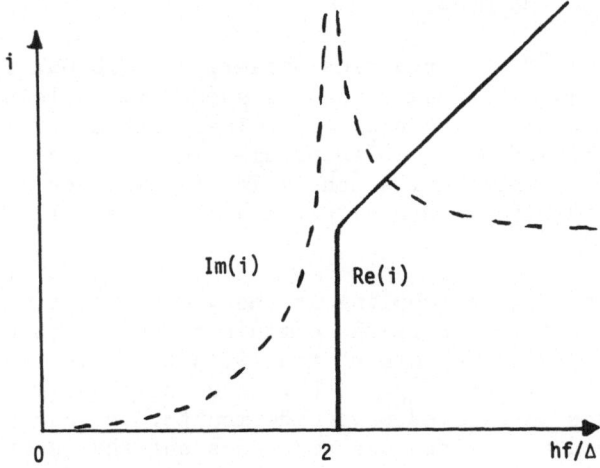

Fig. 13.   Quasi-particle current response function for an ideal
           SIS junction between identical superconductors at T = 0.

The real and imaginary quasi-particle contributions to the
tunnel current (there are similar contributions for the pair current are shown in Fig. 13.

An energy (or phase) dependent reactive part will give
parametric effects and parametric amplification is possible. Calculations show, however, that a high conversion gain can be accomplished without taking into account the reactive part. It must be
present and it may give parametric amplification, but it can usually
be neglected[24]. The crosscoupling strength η is the dominant factor
determining the conversion.

The conversion is a complicated function of frequency, bias
voltage, quality of the I-V curve (sharpness of the gap singularity),
pump power, source and load impedances. It can be calculated, though,
once these properties are known. As already emphasized, the reactive
terms can usually be neglected.

We have concentrated this discussion on the conversion loss.
Of course, the mixer noise temperature is an important entity too.
However, theoretical (as well as experimental) estimates indicate
that the noise temperature is well behaved. It is generally low in
regions of stable gain. In practice, it is usually sufficiently low
that the noise contribution from the IF amplifier, multiplied by the
conversion loss, dominates.

After this lengthy discussion of the conversion gain, we will
discuss a typical experimental set-up and compare with experimental
results.

Experimental Set-Up, Mixers

An example[29] of a practical mixer, a 35-50 GHz receiver, is
shown in Fig. 14. The weak signal is guided via a lens system and
a horn to a mixer block where it is mixed with a local oscillator.
The IF is amplified by a cooled parametric amplifier and a room
temperature FET amplifier. A schematic of the mixer block arrangement and of the circuit deposited on a thin glass is shown in
Fig. 15.

A good microwave coupling to the junctions is necessary.
This may be accomplished by the circuit shown, which uses a broadband antenna. A good feature of the circuit of Fig. 15 is that the
substrate is tractably large. Other successful designs employ miniature substrates in tapered waveguide mounts. The back-short maxi-
mizes the field at the antenna and tunes out the junction capacitance.

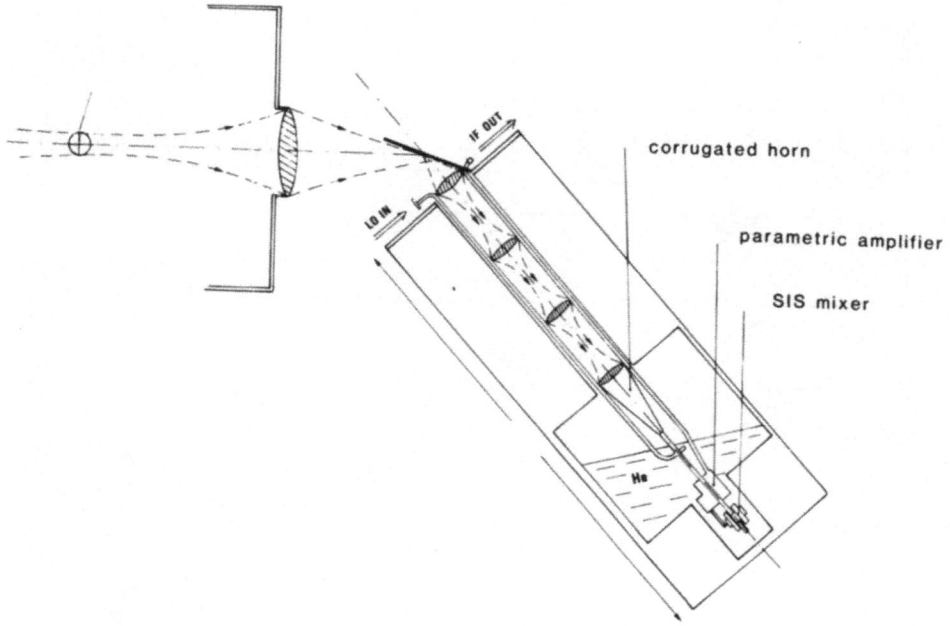

Fig. 14. A schematic of a mixer-receiver tested at the Onsala
          observatory at 35-50 GHz.

        A good matching of a junction to the external world means
that its capacitance should not be too large. Its resistance should
be relatively high at the same time as the current density also
should be high. This means that the junction area must be small.
However, the same requirements may be met in arrays of junctions.

        We will return to the advantages of arrays later. It is
important to keep the array inductance low. This may be done by
placing the junctions in a groove in the waveguide[30] making the
distance between the interconnecting leads and ground very small
(cf. Fig. 15).

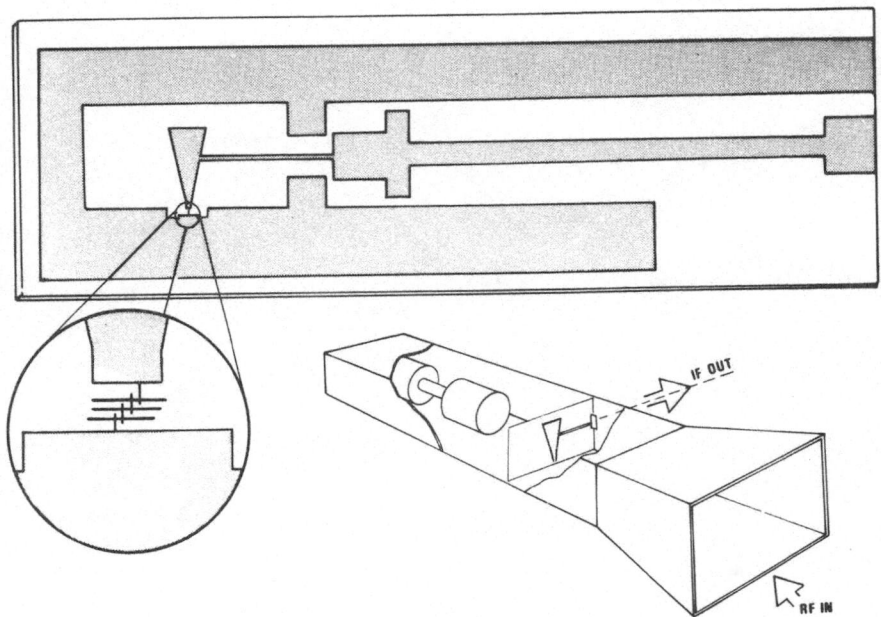

Fig. 15. Añ array of 6 tunnel junctions is coupled to the electro-
          magnetic field via an antenna as shown in the inset. The
          back-short optimizes the coupling. The IF is coupled out
          of the array, and the dc bias into it, via a microstrip
          low-pass filter. The tunnel junctions are about 5 μm.
          Their Josephson critical current densities are typically
          of the order of 10^3 A/cm^2.

Mixer Results

An example[31] of the IF output voltage from a single Pb(In;Au)
electrode junction when a 36 GHz signal was applied is shown in
Fig.16. Note the appearance of conversion maxima spaced by about
hf_{LO}/e in bias voltage. A typical example[32] of the conversion as a
function of bias voltage at 73.5 GHz is shown for a Pb(In) array
in Fig. 17. The optimum conversion at 73.5 GHz (at any bias voltage)
vs array normal tunnel resistance is shown for a large number of
Pb(In) and Pb 6-junction arrays in Fig. 18.

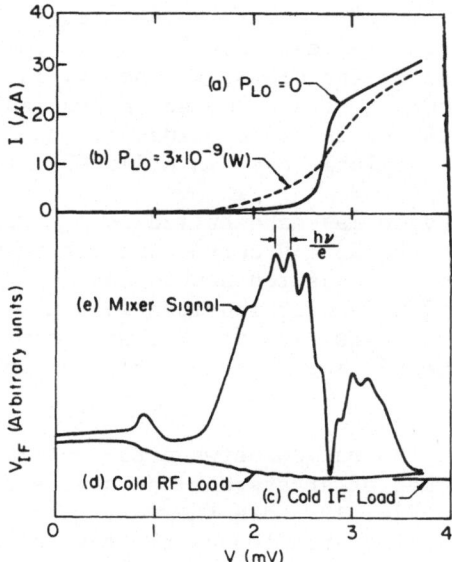

Fig. 16. The upper part shows the dc I-V curve with and without
          35 GHz radiation of a single Pb-alloy junction. The IF
          amplifier output is shown in the lower half. Curves (c)
          and (d) are obtained with cold matched loads in front of
          the IF amplifier and the mixer respectively. Curve (e)
          was taken for an output frequency of 30-80 MHz. From
          ref. 31.

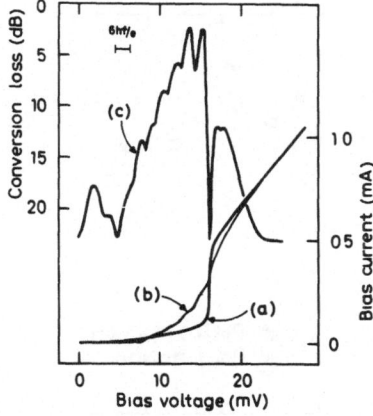

Fig. 17. (a) and (b) dc I-V without and with 73.5 GHz LO power
          (P_{LO} = 0.9 µW) applied to a 6-junction, Pb-alloy array
          with a normal state resistance of 21.5Ω. (c) displays SSB
          conversion at an IF of 30-300 MHz. The optimum conversion
          loss was estimated to 2.2 dB. From ref. 32.

Note from Figs. 17 and 18 that the best measured single sideband (SSB) conversion losses are less than 3dB, i.e. smaller than the minimum loss predicted by the classical theory. In fact, true gain has been reported. McGrath et al[33] noted an available SSB gain larger than 4 dB at 36 GHz in a tin-oxide-tin junction. Kerr et al[34] reported an infinite available gain in a 115 GHz array mixer.

We are now ready to make a detailed comparison between experiment and theory. From the dc I-V curve, and its response to the applied LO power, and the measured source and load impedances we can calculate the optimum conversion loss with no adjustable parameters. For our arrays (Figs. 17 and 18), we assume that the array acts as a lumped element[35] and, furthermore, we neglect reactive terms.

A comparison of the ratios between the optimum experimental conversions for a number of arrays and the corresponding theoretical estimates is given in Fig. 19. Note the very good correspondence between measured and calculated losses for the second peak (from the energy gap) conversion for small array resistances, R_A. The discrepancy is small and remarkably close to the estimated antenna mismatch loss of 0.5 dB. The fall-off of the experimental conversion at large R_A is probably due to too large an $R_A C$ product. The relatively low conversion at the first peak may be explained as due to significant harmonic conversion close to the gap voltage.

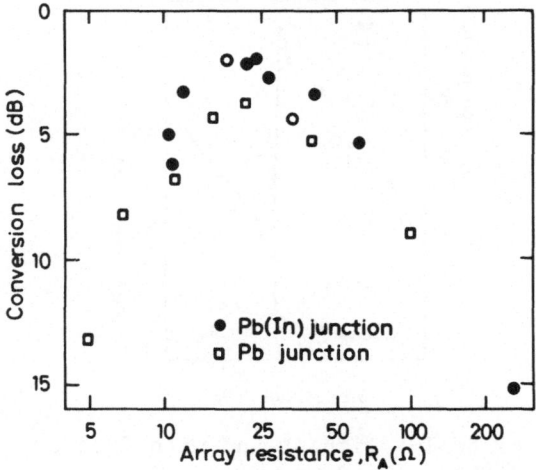

Fig. 18.   Optimum conversion losses for a large number of Pb-alloy and Pb 6-junction arrays. The signal and LO frequencies were about 73.5 GHz. T = 1.5 K.

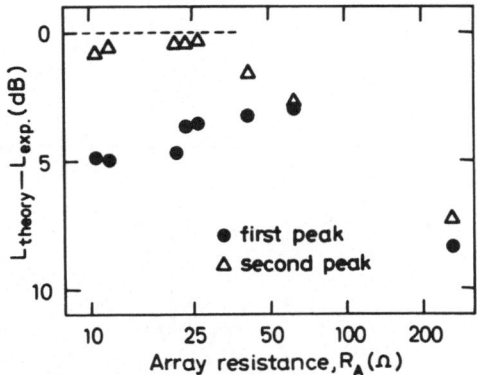

Fig. 19. Discrepancy between the optimum theoretical and experimen-
         tal conversion losses for several Pb-alloy, 6-junction
         arrays at 73.5 GHz. The difference at the second peak is
         remarkably small. From ref. 32.

The good quantitative agreement between theoretical and
experimental conversions assures us that the Tucker model[18] gives
a relevant description of quantum mixing.

The output of a receiver[29] is shown in Fig. 20 for two input
sources - namely noise from an 80 K and a 297 K source respectively.
These data will enable an estimate of the noise temperature of the
mixer and of the receiver. The system is the one shown in Figs. 14
and 15; it operates at 35-50 GHz with an output frequency of about
4 GHz. The large separation between signal and LO frequencies means
that the image frequency could be terminated reactively, i.e. the
receiver could be operated as a true single sideband receiver.

The noise power at the input of the IF amplifier, P_{IF} =
kB $\{(T_A + T_{MXR})/L_c + T_{IF}\}$, B being the bandwidth. In that expression, T_A is the source temperature and L_c includes the losses
due to the input lens guide and the IF mismatch. For simplicity,
T_{MXR} also contains noise contributions from the input lens (or
waveguide) system. The receiver noise temperature can be written as
$T_{REC} = T_{MXR} + L_c T_{IF}$. The different contributions to T_{REC} are given
in Table II. By decreasing the IF mismatch and the input conversion
losses somewhat, by using a better IF amplifier and better lenses,
it should be possible to half T_{REC}.

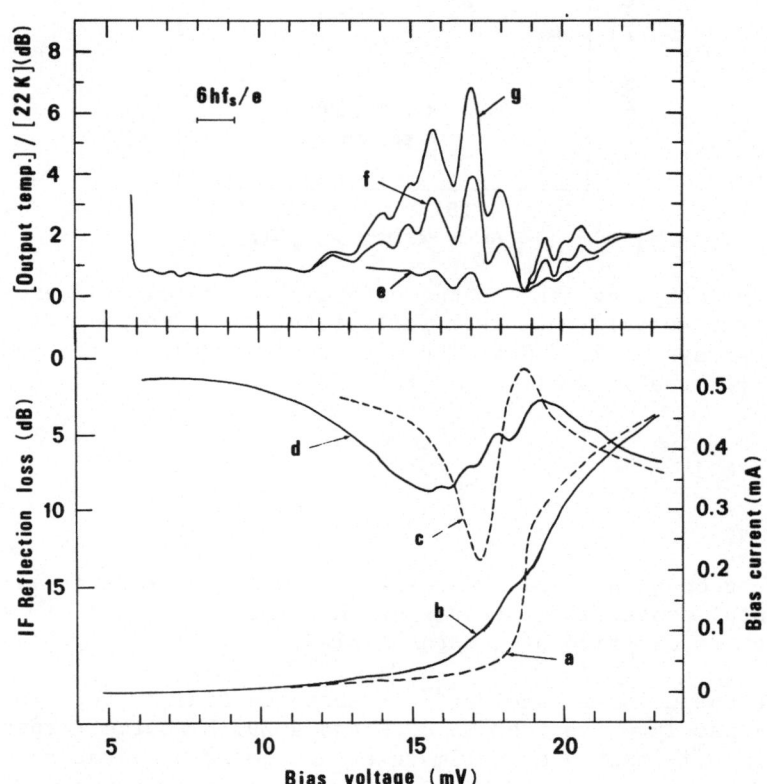

Fig. 20. Dc I-V curves for a lead alloy, 6-junction array are
          given (a) without and (b) with a LO (43 GHz) applied.
          (c) and (d) show the IF reflection loss without and with
          LO. (e) shows the response when a cold load at 2 K replaced
          the input horn. (f) and (g) give the receiver response to
          input sources of 80 and 297 K. The IF was 4 GHz. From
          ref. 29.

Table II.   Properties of an SIS based receiver at 47 GHz. (From
            Ref. 29).

Insertion loss lens guide	1.3 dB
Mixer mount conversion loss	4.0 dB
IF mismatch	1.0 dB
Overall conversion loss	6.3 dB
IF amplifier noise temp.	22 K
Noise from IF ampl. $(=L_c T_{IF})$	94 K
Mixer noise	14 K
Lens guide noise	40 K
T_{REC}	148 K

Fig. 21 shows that the reciever noise temperature decreases
somewhat with increasing frequency, as it should. The higher the
frequency, the more quantum behavior and the better the conversion.

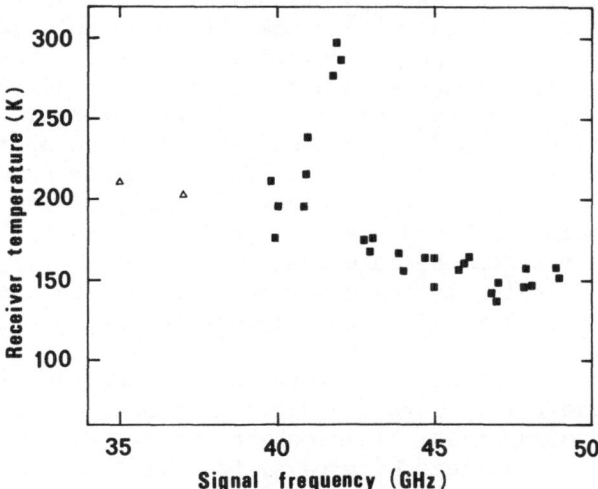

Fig. 21. The frequency dependence of the SSB receiver temperature
         for a device based upon 6 Pb alloy tunnel junctions. The
         intermediate frequency is as high as 4 GHz enabling the image
         frequency to be rejected. The inferior performance at about
         42 GHz is probably due to an unwanted resonance.

Although higher than for a Josephson device, the saturation level of an SIS mixer is limited, of the order of a nW in the device we have just discussed. It should not be a limiting factor in astronomical application. The low saturation level is connected with the low LO power needed. In an array, the pump power, and the saturation level, increases as the square of the number of junctions.

The amplification is stable with small changes in gain for variations in the pump power and the temperature. Examples of results obtained for quasiparticle mixers operating in different frequency regions are shown in Table III. That table contains data reported by several groups active in the field. It covers frequencies up to 230 GHz, conversion efficiencies in a range that includes a value higher than one, and mixer noise temperatures down to a few K, experiments with junctions of different materials, results with single junction and array mixers.

Josephson Noise

Before discussing the frequency limit of quasiparticle mixers, we should shortly treat the subject of Josephson noise. The SIS tunnel junctions are Josephson junctions. If such a junction is biased that closely to zero voltage that the bias voltage plus the time-varying induced voltages at any time hit $V = 0$, the current will follow the zero voltage step until I_c is exceeded. Hence below a so called drop-back voltage we will be in a hysteretic, noisy Josephson regime. Examples[32] of the noise power when operating inside that region are given in Fig. 22. The Josephson effect can be suppressed by a magnetic field. However, if the junction size is small, a large magnetic field must be applied before the flux in the junction has reached the quantized flux value. A large field will cause pair breaking in the superconductor and thus a smearing of the gap edge singularity, an undesirable feature.

An array of junctions is of advantage as the individual elements can be made fairly large. Hence a much smaller magnetic field is needed to suppress the Josephson noise. (We have already discussed several advantages of arrays. Some of these are listed in Table IV.) Other ways to reach the same result would be to use long and narrow junctions or to use SIN junctions. Initial results[32] with the latter type, however, gave relatively large conversion losses at 73.5 GHz (about 9 dB, which can be compared to the optimum value of 2.2 dB for similar SIS junctions). Better performance is expected at higher frequency when the photon energy exceeds the width of the gap edge singularity.

Table III.   Quasiparticle (SIS) Mixer Results.

GHz	Conversion $(L_{SSB})^{-1}$	T_{MIX} (K)	T_{REC} (K)	Ref.
9	0.26	$\sim$10		20
36	2.7	10		33
35–50	0.32	15	220–140	29
55	0.23	27		36
70	0.52	5±94		37
74	0.63	20±90		32
115	0.07	70±40		34
115	0.2*	60*	$\sim$200 $\}$	38
230		<300*	$\sim$500	

* at receiver front end

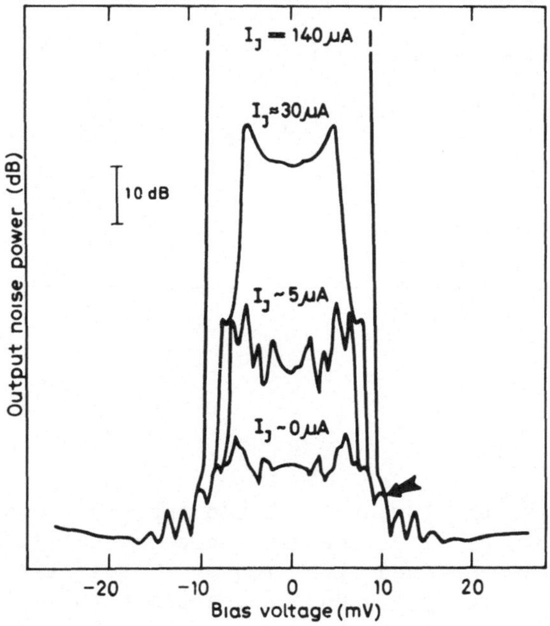

Fig. 22.   Output noise power in the IF (300 MHz) band for an
applied P_{LO} = 1.1 μW to an array of 6 tunnel junctions.
The unpumped I_c's (depressed by a magnetic field) are given.

Table IV.   Arrays of junctions – important properties.

Advantages:

1)   Higher saturation level ($\propto N^2$) and dynamic range
      than for a single junction.

2)   More easily quenched Josephson noise extends high
      frequency limit.

3)   Lower effective shunt capacitance facilitates the
      coupling to the external world, should mean more
      bandwidth and tunability.

4)   Larger junctions more easily fabricated; higher
      quality elements possible.

5)   Survive better. E.g.,discharges are distributed
      over many elements.

Disadvantages:

1)   High junction uniformity is needed in high gain
      mixers with high dynamic resistances.

2)   Experiments less easily compared with theory.

3)   Possibly more noise. Results by Smith et al.[39]
      indicate that $T_{MXR} \propto N^{\frac{1}{2}}$, at least for some types
      of mixers.

4)   Series inductance.

High Frequency Limit

At high f_{LO}, the low order photon induced steps fall close to
the drop-back voltage, i.e. the mixer becomes noisy. Unless the
Josephson noise is quenched, the frequency range is severely
limited. We have already argued the Josephson noise is more easily
suppressed in arrays of junctions than in single junctions. Pb
based SIS array mixers ought to function well up to several hundred
GHz. But how close to the gap frequency (about 700 GHz for Pb) can
an SIS mixer operate? Initial experiments indicate that an appreci-

able conversion is available in an array of Al junctions even at
three quarters of the gap frequency[40].

Remaining Problems

        It still remains an open question how high up in frequency
an SIS mixer can operate. What effect have the neglected imaginary
quasiparticle and pair current components at frequencies close to
the gap frequency? Is it necessary to employ SIN junctions at the
highest frequencies? Can a large gap tunnel junction (giving a high
frequency pair breaking cut-off) be fabricated with a quality high
enough for SIS mixer use?

        What happens at very high gain? Does a noise rise phenomenon
appear?

        Can harmonic pumping[41] also be applied successfully at very
high frequencies? High frequency oscillators are scarce (and expensive); it would be profitable to use a harmonic of a lower frequency pump.

        Fuller quantum noise calculations should be performed. What
is the noise limit? Are the noise properties of an array degraded
relatively a single junction?

IV.   SUMMARY OF SUPERCONDUCTING TUNNEL JUNCTION DETECTORS

        Having discussed the two examples in some detail, we are
ready to summarize the results obtained for several different types
of high frequency detector elements based on tunnel junctions. This
is done in Table V.

        Josephson point-contact junctions have been operated at
higher frequencies than any successful tunnel junction detector[45].
Tunnel junctions, however, are in many respects superior to point
contacts, not the least in stability and ease of operation. SIS
mixers and video detectors show better data than Josephson ones.
The intrinsic noise level is extremely low, close to the quantum
limit as shown in Fig. 23, and the possibility of obtaining gain
is of great importance.

        Quasiparticle parametric amplifiers[46] have not been tested
yet. Their Josephson counterparts sometimes show good data, but
are often plagued by a noise rise. The internally pumped Josephson
amplifier might be an exception to the excessive noise problem -
it should be studied in much greater detail.

Table V.   Superconducting Tunnel Junction Detectors. Examples of
            quasiparticle (SIS) and Josephson junction results.

VIDEO DETECTORS:	f (GHz)	NEP (W/Hz$^{\frac{1}{2}}$)		Ref.
SIS	36	$2.6 \cdot 10^{-16}$		42
	70	$1.7 \cdot 10^{-15}$		43
MIXERS:		L_c (dB)	T_{MIX}^{SSB} (K)	
SIS	35–230	–4.3–+7	10–60	See Table III
Josephson	230		380	44
PARAMETRIC AMPLIFIERS:		Gain (dB)	T_N (K)	
Internally pumped, Josephson	10	5	<30	14
Externally pumped, Josephson, 3 photon	35	4–12	∿50	9
Externally pumped, Josephson, 4 photon	10	20	30	7

    The limited saturation level and dynamic range may be a
handicap for superconducting detectors. This is connected with the
low LO power needed, which in itself can be an advantage. A saturation level of a few pW is definitely of disadvantage while the
level of a few nW obtained with arrays of tunnel junctions should
be quite acceptable.

Future Prospects

    The quasiparticle SIS mixer seems to be today's favorite. It
has already been used routinely at radioastronomical observatories.
It should be possible, in the near future, to decrease the reciver
system noise temperature by a factor of two to three as compared
with the best competitors. This would increase the efficiency of

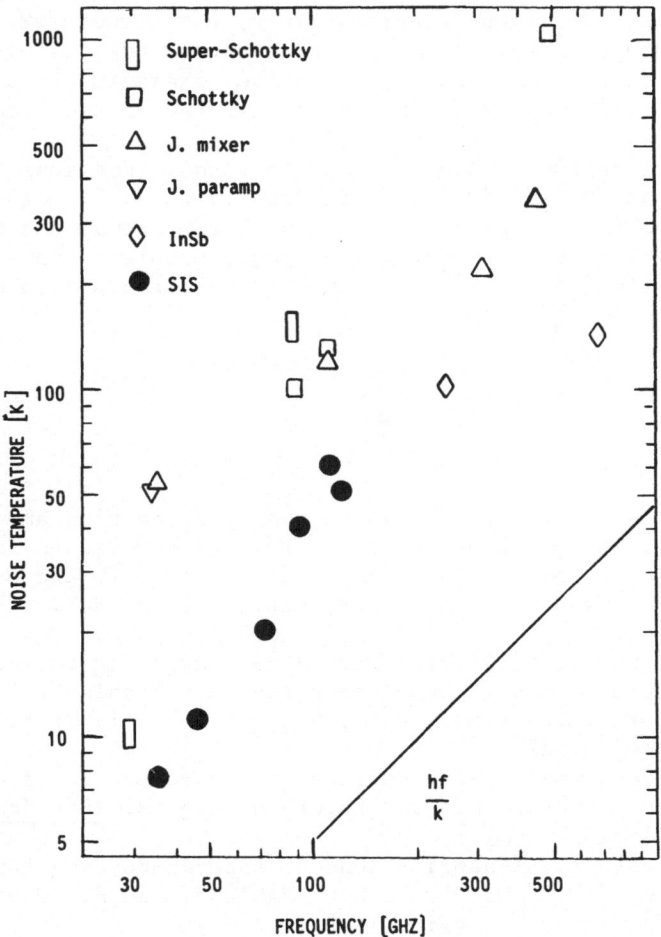

Fig. 23. The (SSB) noise temperature of several high frequency detector elements used or of potential use in radio astronomy. Some of the estimates are very uncertain, error bars
would extend over a large part of the figure, while other
are within a few K. The quantum limit is also indicated.

an observatory by a factor of 5 to 10. However, one can envision even
more improvement. The pump power of a superconducting detector is
low. Using clever antennas it should be possible to pack several
detectors side by side within the focal point of a telescope util-
izing the same local oscillator. Hence several directions could be
mapped at the same time. A 3 x 3 matrix would mean another factor
of ten increase in the telescope capability.

Other applications besides radio astronomy may be coming up.
Plasma physics is such a field[47].

The tunnel junction detectors are planar structures. This is another advantage. Other elements like antennas and filters can be incorporated in the chip design. Low noise mixers and IF superconducting parametric amplifiers may be integrated on the same chip.

The application of superconducting tunnel junctions has extended the quantum limited detection down to the millimeter- and microwave wavelength regions. Although a breakthrough has occurred during the last couple of years, many basic problems and effects remain to be studied and a lot of applied research has to be done before the field is closed.

REFERENCES

1. R. Adde and G. Vernet, High frequency properties and applications of Josephson junctions from microwaves to far-infrared, in "Superconductor Applications: SQUID and Machines", B.B. Schwartz and S. Foner, eds., Plenum, N.Y. (1976), p. 248.

2. P.L. Richards, Millimeter wave superconducting devices, in "Future trends in superconductive electronics", B.S. Deaver, Jr., et al., eds., N.Y. (1978), AIP Conf. Proc. No. 44, p. 223.

3. R.Y. Chiao, Analog applications of the Josephson effect: Recent developments and future prospects, IEEE Trans. Magn. MAG-15:446 (1979).

4. N.F. Pedersen, Rf applications of superconducting tunneling devices, in "SQUID 80", H.D. Hahlbohm and H. Lübbig, eds., W. de Gruyter, Berlin (1980), p. 739.

5. P.L. Richards and T.M. Shen, Superconductive devices for millimeter wave detection, mixing, and amplification, IEEE Trans. El. Devices ED-27:1909 (1980).

6. The SIS (quasiparticle) mixer will be reviewed extensively by a number of authors in "Reviews of infrared and millimeter waves,"Vol. II, K.J. Button, ed., Plenum Press, N.Y. (1982).

7. S. Wahlsten, S. Rudner and T. Claeson, Parametric amplification in arrays of Josephson tunnel junctions, Appl. Phys. Lett. 30:298 (1977); Arrays of Josephson tunnel junctions as parametric amplifiers, J. Appl. Phys. 49:4248 (1978).

8. M.T. Levinsen, N.F. Pedersen, O.H. Soerensen, B. Dueholm and J. Mygind, Externally pumped millimeter-wave Josephson junction parametric amplifier, IEEE Trans. El. Devices ED-27:1928 (1980).

9. J. Mygind, N.F. Pedersen, O.H. Soerensen, B. Dueholm and M.T. Levinsen, Low noise parametric amplification at 35 GHz in a single Josephson tunnel junction, Appl. Phys. Lett. 35:91 (1979).

10. F. Goodall, F. Bale, S. Rudner, T. Claeson and T.F. Finnegan, Parametric amplification in Josephson tunnel junction arrays at 33 GHz, IEEE Trans. Magn. Mag-15:458 (1979).

11. A review is given by M.J. Feldman and M.T. Levinsen, Theories of the noise rise in Josephson paramps, IEEE Trans. Magn. MAG-17:834 (1981).

12. P. Russer, Parametric amplification with Josephson junctions, Archiv der Elektrischen Übertragung 23:417 (1969).

13. H. Kanter, Two-idler parametric amplification with Josephson junctions, J. Appl. Phys. 46:4018 (1975).

14. N. Calander, T. Claeson and S. Rudner, Low-noise self-pumped Josephson tunnel junction amplifier, Appl. Phys. Lett. 39:650 (1981); Shunted Josephson tunnel junctions: High-frequency, self-pumped low noise amplifiers, J. Appl. Phys. 53:5093 (1982).

15. N. Calander, T. Claeson and S. Rudner, A subharmonic Josephson relaxation oscillator - amplification and locking, Appl. Phys. Lett. 39:504 (1981); Relaxation oscillations in inductivity shunted Josephson tunnel junctions, Phys. Scripta 25:837 (1982).

16. B.A. Huberman, J.P. Crutchfield and N.M. Packard, Noise phenomena in Josephson junctions, Appl. Phys. Lett. 37:750 (1980).

17. N.F. Pedersen and A. Davidson, Chaos and noise rise in Josephson junctions, Appl. Phys. Lett. 39:830 (1981).

18. J.R. Tucker, Quantum limited detection in tunnel junction mixers, IEEE J. Quantum Electron, QE-15:1234 (1979); Predicted conversion gain in superconductor insulator-superconductor quasiparticle mixers, Appl. Phys. Lett. 36:477 (1980).

19. M. McColl, M.F. Millea, A.H. Silver, M.F. Bottjer, R.J. Pedersen, and F.L. Vernon, Jr., The Super-Schottky microwave mixer, IEEE Trans. Magn. MAG-13:221 (1977); A. Silver, R.J. Pedersen, M. McColl, R.L. Dickman, and W.J. Wilson, The millimeter wave super-Schottky diode detector, IEEE Trans. Magn. MAG-17:698 (1981).

20. S. Rudner and T. Claeson, Arrays of superconducting tunnel junctions as low-noise 10-GHz mixers, Appl. Phys. Lett. 34:711 (1979).

21. H.C. Torrey and C.A. Whitmer,"Crystal Rectifiers," MIT Radiation Lab. Series, vol. 15, McGraw-Hill, N.Y. (1948).

22. T.C.L.G. Sollner, Large conversion efficiency and dynamic range for SIS mixers, Physica 108B:1365 (1981).

23. T.M. Shen, Conversion gain in mm-wave quasiparticle heterodyne mixers, IEEE J. Quantum Electron. QE-17:1151 (1981).

24. M.J. Feldman, Some analytical and intuitive results in the quantum theory of mixing, J. Appl. Phys. 53:584 (1982).

25. A.H. Dayem and R.J. Martin, Quantum interaction of microwave radiation with tunneling between superconductors, Phys. Rev. Lett. 8:246 (1962).

26.   P.K. Tien and J.P. Gordon, Multiphoton process observed in
      the interaction of microwave fields with the tunneling
      between superconductor films, Phys. Rev. 129:647 (1963).
27.   N.R. Werthamer, Nonlinear self-coupling of Josephson radia-
      tion in superconducting tunnel junctions, Phys. Rev.
      147:255 (1966).
28.   R.E. Harris, Cosine and other terms in the Josephson
      tunneling current, Phys. Rev. B10:84 (1974); Josephson
      tunneling current in the presence of a time-dependent
      voltage, Phys. Rev. B11:3329 (1975).
29.   L. Olsson, S. Rudner, E. Kollberg, and C.O. Lindström, A
      very low noise quasiparticle (SIS) mixer receiver for
      radio astronomical applications, in "12th European Micro-
      wave Conference Proceedings", Microwave Exhibitions and
      Publishers Ltd. (1982) p. 270.
30.   S. Rudner, M.J. Feldman, E. Kollberg, and T. Claeson, SIS
      quasiparticle mixing with long antenna-coupled arrays,
      in "SQUID '80", H.D. Hahlbohm and H. Lübbig, eds., W de
      Gruyter, Berlin (1980), p. 901.
31.   P.L. Richards, T.M. Shen, R.E. Harris, and F.L. Lloyd, Quasi-
      particle heterodyne mixing in SIS tunnel junctions, Appl.
      Phys. Lett. 34:345 (1979).
32.   S. Rudner, M.J. Feldman, E. Kollberg, and T. Claeson, The
      antenna-coupled SIS quasiparticle array mixer, IEEE Trans.
      Magn. MAG-17:690 (1981); Superconductor-insulator-super-
      conductor mixing with arrays at millimeter-wave fre-
      quencies, J. Appl. Phys. 52:6366 (1981).
33.   W.R. McGrath, P.L. Richards, A.D. Smith, H. van Kempen, R.A.
      Batchelor, D.E. Prober, and P. Santhanam, Large gain,
      negative resistance, and oscillations in superconducting
      quasiparticle heterodyne mixers, Appl. Phys. Lett. 39:655
      (1981).
34.   A.R. Kerr, S.-K. Pan, M.J. Feldman, and A. Davidson, In-
      finite available gain in a 115 GHz SIS mixer, Physica
      108B:1369 (1981).
35.   M.J. Feldman and S. Rudner, SIS mixing with arrays, in Ref.
      6.
36.   A.C. Callegari and R.A. Buhrman, Millimeter wave mixing
      with submicron area Nb tunnel junctions, J. Appl. Phys.
      53:823 (1982).
37.   H.J. Hartfuss and K.H. Gundlach, Heterodyne mixing at 70 GHz
      with superconductor-insulator-superconductor tunnel junc-
      tions, Int. J. of Infrared and Millimeter Waves 2:925
      (1981).
38.   T.G. Phillips, D.P. Woody, G.J. Dolan, R.E. Miller, and
      R.A. Linke, Dayem-Martin (SIS tunnel junction) mixers
      for low noise heterodyne receivers, IEEE Trans. Magn.
      MAG-17:684 (1981).

39. A.D. Smith, R.A. Batchelor, W.R. McGrath, P.L. Richards, H. van Kempen, D.E. Prober, and P. Santhanam, Conversion efficiency and noise measurements of SIS array mixers, in "Sixth International Conference on Infrared and milli-meter waves", K.J. Button, ed., Technical Digest (1981) M-4-1.

40. D. Winkler, private communication.

41. H. van Kempen, W.R. McGrath, P.L. Richards, A.D. Smith, R.E. Harris, and F.L. Lloyd, Quasiparticle harmonic mixing with SIS junctions, "Sixth International Conference on Infrared and Millimeter Waves", K.J. Button, ed., Technical Digest (1981) M-4-3.

42. P.L. Richards, T-M Shen, R.E. Harris, and F.L. Lloyd, SIS quasiparticle junctions as microwave photon detectors, Appl. Phys. Lett. 36:777 (1980).

43. H.J. Hartfuss and K.M. Gundlach, Video detection of mm-waves via photon-assisted tunneling between two superconductors, Int. J. Infrared and Millimeter Waves 2:809 (1981).

44. G.M. Daalmans, Th. de Graauw, S. Lidholm, and Fr. v. Vliet, Detection of mm-radiation with high current density sub-micron niobium-niobium Josephson junctions, in "SQUID '80", H.D. Hahlbohm and H. Lübbig, eds., W. de Gruyter, Berlin (1980), p. 863.

45. T.G. Blaney, A theoretical and experimental study of Josephson frequency mixers for heterodyne reception in the submillimetre wavelength range, NPL Report SI, No. 89/0382, National Physical Laboratory, Teddington (1978).

46. G.S. Lee, Superconductor-insulator-superconductor reflection parametric amplifier, Appl. Phys. Lett. 41:291 (1982).

47. B.T. Ulrich and M. Tutter, Josephson junction applications in plasma physics, in "SQUID'80", H.D. Hahlbom and M. Lübbig, eds., W. de Gruyter, Berlin (1980), p. 831.

COEXISTENCE OF SUPERCONDUCTIVITY AND MAGNETISM

M. Brian Maple[*]

Department of Physics and
Institute for Pure and Applied Physical Sciences
University of California, San Diego
La Jolla, California 92093 USA

1. INTRODUCTION

The subject of this series of lectures is the interaction between superconductivity and magnetism. This includes a question that has intrigued both experimentalists and theoreticians alike, namely, can superconductivity and magnetic order occur within the same volume element in a substance and, if so, under what conditions. In recent years, it has been found that these two phenomena do co-exist in certain ternary rare earth compounds, as will be discussed later in some detail.

The earliest investigations into the interrelation between superconductivity and magnetism were made about two and a half decades ago. Ginzburg[1] was the first to address this problem theoretically in 1957, and Matthias, Suhl and Corenzwit[2,3] carried out the first experimental studies in 1958.

The evolution of this subject can be divided into two periods. The first period, which spanned the years 1957 to about 1975, was characterized by experiments on binary and pseudobinary systems formed by dissolving small amounts of rare earth (RE) impurities with partially-filled 4f electron shells into superconducting elements such as La or binary compounds like YOs_2; e.g., $La_{1-x}RE_x$,[2]

---

[*]Research supported by the US Department of Energy under Contract No. AT03-76-ER-70227.

$(Y_{1-x}RE_x)Os_2$.[4] The results were certainly provocative, but
largely inconclusive due to complications associated with chemical
clustering and short range or "glassy" types of magnetic order.
Nonetheless, the experiments stimulated the development of theories
which yielded some very striking predictions, although they were
largely inapplicable to the systems then being investigated. A
"spin off" of this early research (pun definitely intended) was the
achievement of a rather good understanding of the effects of paramagnetic impurities on superconductivity, including crystalline
electric fields (CEF), Kondo scattering, localized spin fluctuations, etc.[5,6]

The second period, around 1976 to the present, originated with
experiments on ternary and pseudoternary compounds of the type
$REMo_6S_8$, $REMo_6Se_8$ and $RERh_4B_4$. These materials have an
ordered RE sublattice plus weak exchange coupling between the
superconducting electron spins and the RE magnetic moments. A
number of antiferromagnetic superconductors in which antiferro-
magnetism and superconductivity coexist in zero applied magnetic
field have been discovered. Several ferromagnetic superconductors
have been found wherein the onset of ferromagnetism causes re-
entrant superconductive behavior with the material losing its superconductivity at a temperature in the neighborhood of the Curie temperature T_M. Moreover, the superconducting-ferromagnetic interactions produce an oscillatory magnetic state with a wavelength
$\lambda \sim 10^2$ Å which coexists with superconductivity. These experiments have prompted a second surge of theoretical activity that can
be characterized as currently being very active and exciting. It is
interesting to note that some of the theories developed in the first
period, 1957-1975, anticipated several of the experimental findings
that emerged during the second period, 1976-present.

In these lectures, the evolution of this subject will be traced
through both periods. Other reviews to which the reader is referred
can be found in references 5-10 (first period) and in references 10-17
(second period). It should also be pointed out that in the superconducting and magnetic systems discussed herein, the localized magnetic moments are associated with RE ions with partially-filled 4f
electron shells.

2. PERIOD 1957-1975

The general subject of the interaction between superconduc-

tivity and magnetism as it developed between 1957 to about 1975 will
be reviewed in this section.

As mentioned in the introduction, the first experiments were
carried out by Matthias et al. in 1958 on $La_{1-x}RE_x$ alloys.[2]   In
its fcc phase, the La matrix has a superconducting transition temperature T_c of 6 K.   A rapid and nearly linear depression of T_c
with concentration n of the RE solute was observed.   The depression of T_c was largest for RE = Gd, with T_c approaching zero for
$n \sim$ 1 at. % Gd.   The dependences of T_c and the Curie temperature
T_M on Gd concentration for the $La_{1-x}Gd_x$ system[2] are shown in
Fig. 1.

Displayed in Fig. 2 are data for the depression of T_c of La
for 1 at.% RE dissolved into La vs RE.[2]   Here, $-\Delta T_c = T_{c_0} - T_c$
(T_{c_0} refers to the La matrix, and T_c to La containing 1 at.% RE)
correlates with the spin S, rather than the effective magnetic
moment μ_{eff}, of the RE solutes.   In addition, $-\Delta T_c$ is anomalously
large for RE = Ce, which, as we will see later, is associated with
the Kondo effect.

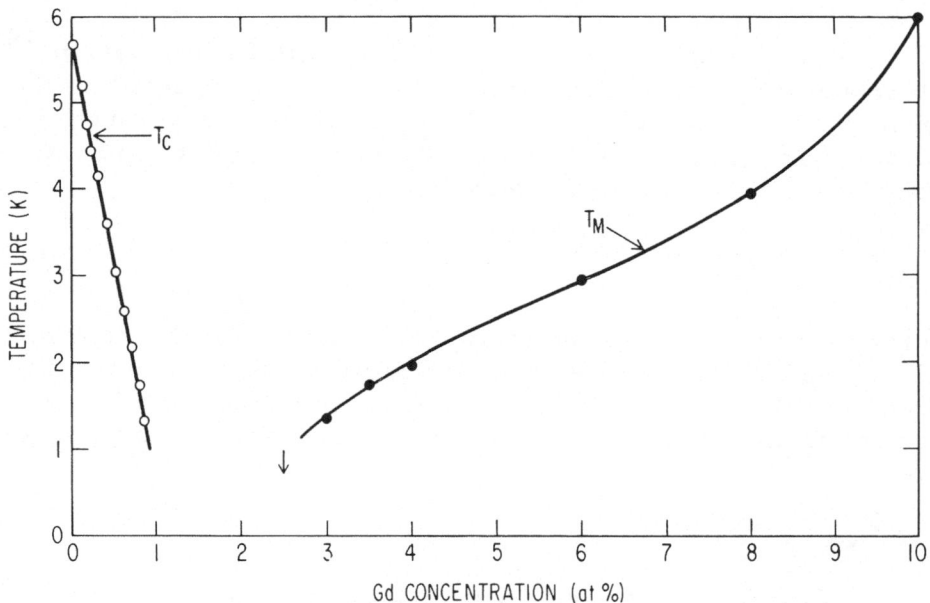

Fig. 1.   Superconducting transition temperature T_c and magnetic
         ordering temperature T_M vs Gd concentration for the
         $La_{1-x}Gd_x$ system (from Ref. 2).

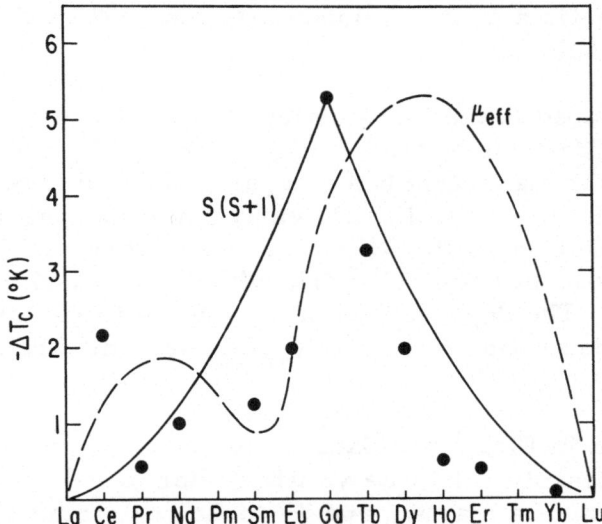

Fig. 2. Depression of the superconducting transition temperature
of La, $-\Delta T_c$, for 1 at.% RE addition.[2] The solid curve
is the variation of the spin factor S(S+1), and the dashed
curve is the variation of the effective magnetic moment
μ_{eff} of the RE additions.

This characteristic variation of $-\Delta T_c$ with RE led Herring[18]
in 1958 and later, Suhl and Matthias[19] in 1959, to suggest that the
exchange interaction between the RE spins $\underset{\sim}{S}$ and the conduction
electron spins $\underset{\sim}{s}$ is responsible for the observed effects on superconductivity. The exchange interaction is given by

$$\mathcal{H}_{ex} = -2\mathcal{J}\underset{\sim}{S} \cdot \underset{\sim}{s} \tag{1}$$

where $\mathcal{J}$ is the exchange interaction parameter that characterizes
both the sign and magnitude of the interaction. For RE ions, the
spin of the RE ion is replaced by its projection onto the total angular momentum operator $\underset{\sim}{J}$ of the Hund's rule ground state which
gives

$$\mathcal{H}_{ex} = -2(g_J-1)\mathcal{J}\underset{\sim}{J} \cdot \underset{\sim}{s} \tag{2}$$

where g_J is the Landé g-factor for the Hund's rule ground state
of the RE ion under consideration.

Matthias and co-workers[3] also studied systems where RE

ions replaced Ce in the matrix compound $CeRu_2$. They found that the depression of T_c by the RE additions to $CeRu_2$ was relatively weak which resulted in a substantial region in which the curves of T_c and the magnetic ordering temperature T_M vs RE concentration overlapped one another. A plot of T_c vs Gd concentration for the $(Ce_{1-x}Gd_x)Ru_2$ system is shown in Fig. 3. This and other $(Ce_{1-x}RE_x)Ru_2$ systems have subsequently been studied in greater detail by Wilhelm and Hillenbrand,[20] Peter et al.,[21] Taylor et al.,[22] Kumagai et al.,[23] Lynn et al.,[24] and others. It has been found that T_c vanishes abruptly in single phase samples above the concentration where the superconducting and ferromagnetic curves intersect one another,[20] and the range of the ferromagnetic correlations is only ~ 80 Å.[24]

As mentioned earlier, these initial experiments that were designed to investigate the interplay between superconductivity and long-range magnetic order stimulated a considerable amount of experimental and theoretical activity that led to a rather good understanding of the effect of paramagnetic impurities on superconductivity.[5,6] The effect of paramagnetic ions on superconductivity, which is of interest in its own right, shall be covered briefly because some of the concepts involved will be useful later in discussions of the interaction between superconductivity and long-range magnetic order.

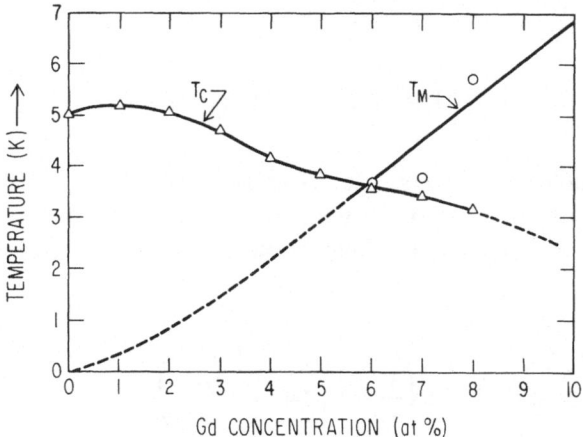

Fig. 3. Low temperature phase diagram for the $Ce_{1-x}Gd_xRu_2$ system (from Ref. 3).

2.1 Effect of Rare Earth Ions on Superconductivity in Zero Magnetic Field

The effect of RE ions on superconductivity in zero magnetic field, both <u>applied</u> and <u>internal</u>, will now be addressed. The relevant interaction is the exchange interaction [Eq. (2)] which, for a RE ion with a fixed orientation of $\underset{\sim}{J}$, has the effect of raising the energy of one member of a Cooper pair $(\underset{\sim}{k}\uparrow, -\underset{\sim}{k}\downarrow)$ and lowering the energy of the other. This results in superconducting electron pair breaking which is manifested in a rapid suppression of T_c with paramagnetic impurity concentration as exhibited, for example, by the $La_{1-x}Gd_x$ system (Fig. 1). There are two cases, one in which $\mathcal{J}$ is positive (ferromagnetic), and the other in which $\mathcal{J}$ is negative (antiferromagnetic) that will be considered in the following.

2.1.1 Positive (Ferromagnetic) Exchange Interaction

In the regime for which $\mathcal{J} > 0$, the T_c of an alloy produced by dissolving RE ions in a superconducting matrix has been calculated by Abrikosov and Gor'kov (AG)[25] and is given by

$$T_c/T_{c_0} = Un(\alpha/\alpha_{cr}) \tag{3}$$

where Un represents a universal function, T_{c_0} is the T_c of the matrix, α is the pair breaking parameter, and α_{cr} is the value of α at which $T_c \to 0$. The expression for α is

$$\alpha = \tau_{ex}^{-1} = \hbar^{-1} nN(0) \mathcal{J}^2 (g_J-1)^2 J(J+1) \tag{4}$$

where τ_{ex} is the exchange scattering lifetime and $N(0)$ is the density of states at the Fermi level, while $\alpha_{cr} = k_B T_{c_0}/4\hbar\gamma$ ($\ln \gamma = 0.57721$ is Euler's constant). Explicitly, the universal relation between T_c/T_{c_0} and α/α_{cr} is given by

$$\ell n (T_c/T_{c_0}) = \psi\left(\frac{1}{2}\right) - \psi\left(\frac{1}{2} + 0.14\frac{\alpha T_{c_0}}{\alpha_{cr} T_c}\right) \tag{5}$$

where ψ is the digamma function, and is displayed in the lower part of Fig. 4.

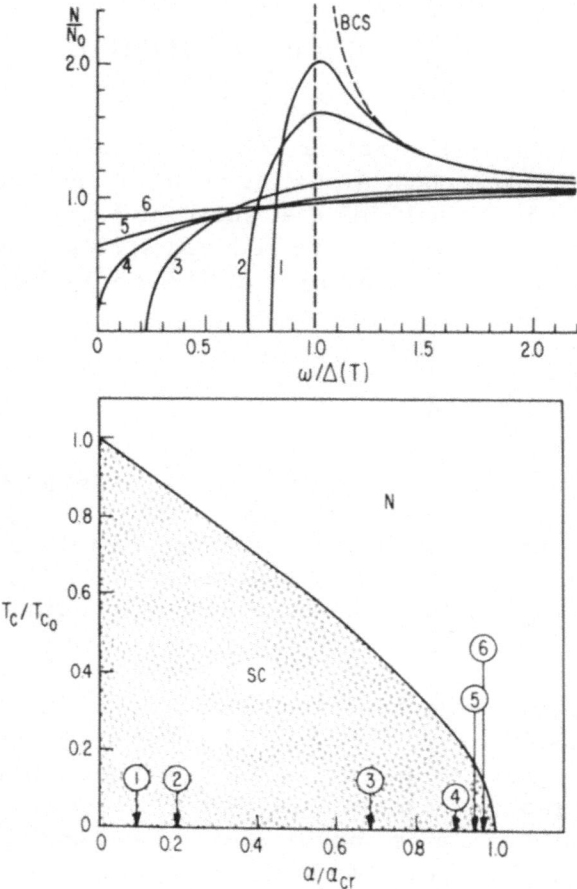

Fig. 4.   (Top) Normalized density of states N/N_0 vs normalized
          energy ω/Δ curves (T = 0 K) corresponding to different
          values of the reduced pair breaking parameter α/α_{cr}
          (from below).   The zero of the energy axis is the Fermi
          energy.   (Bottom) Universal relation between the reduced
          superconducting transition temperature T_c/T_{c_0} and re-
          duced pair breaking parameter α/α_{cr}.   (After Ref. 9.)

In the limit that $\alpha \to 0$, Eq. (5) reduces to the following linear
form

$$T_c/T_{c_0} = 1 - 0.691(\alpha/\alpha_{cr}) = 1 - 0.691(n/n_{cr}) \tag{6}$$

where $n_{cr} = k_B T_{c_0}/4\gamma N(0) \mathcal{J}^2 (g_J-1)^2 J(J+1)$.   In this linear region,
the rate of depression of T_c with RE impurity concentration is
given by

$$\frac{dT_c}{dn}\bigg|_{n=0} = -(\pi^2/2)\, k_B^{-1}\, N(0)\, \mathcal{J}^2 (g_J-1)^2\, J(J+1) \qquad (7)$$

The quantity $(g_J-1)^2 J(J+1)$ is referred to as the deGennes factor which exhibits a characteristic variation with RE and, in particular, attains a maximum at the half-filled 4f shell for RE = Gd. Other predictions of the theory are that the reduced specific heat jump $\Delta C/\Delta C_o$ (ΔC_o refers to the matrix) is a different universal function of T_c/T_{c_o};[26] i.e.,

$$\Delta C/\Delta C_o = Vn(T_c/T_{c_o}) \qquad (8)$$

which deviates markedly from the BCS law of corresponding states $\Delta C/\Delta C_o = T_c/T_{c_o}$. Gapless superconductivity is also predicted where the energy gap Ω_G goes to zero faster than the order parameter Δ with increasing pair breaking parameter.[25] This was verified by Woolf and Rief in 1961 through electron tunneling measurements.[27]

Density of states vs energy curves are shown in the upper part of Fig. 4.[9] Values of α/α_{cr} indicated by numbers in the lower part of the figure correspond to density of states curves labeled by the same numbers in the upper part of the figure. As α/α_{cr} is increased, the density of states curves broaden and states are introduced into the gap. The gap just closes for curve number 4 which corresponds to $\alpha/\alpha_{cr} = 0.91$. "Gapless superconductivity" occurs at all temperatures for $0.91 \leq \alpha/\alpha_{cr} \leq 1$.

Shown in Fig. 5 are T_c/T_{c_o} vs n/n_{cr} data for the (LaGd)Al$_2$ system[28] which are described well by the universal AG function to $T_c/T_{c_o} \sim 0.1$ (solid line in Fig. 5). The reduced Curie-Weiss temperature θ_c/T_{c_o} of a sample with $n/n_{cr} = 1.41$ is indicated in the figure. The low value of θ_c/T_{c_o} reveals that interaction effects are reduced in this system, implying that the RE ions remain paramagnetic to low temperatures.

Displayed in Fig. 6 are data for the reduced specific heat jump $\Delta C/\Delta C_o$ versus T_c/T_{c_o} for both the (LaGd)Al$_2$[29] and ThGd[30] systems. The data are well represented by the solid curve which was computed by Skalski et al.[26] within the context of the AG theory, but deviate markedly from the BCS law of corresponding states which is indicated by the dashed line in the figure.

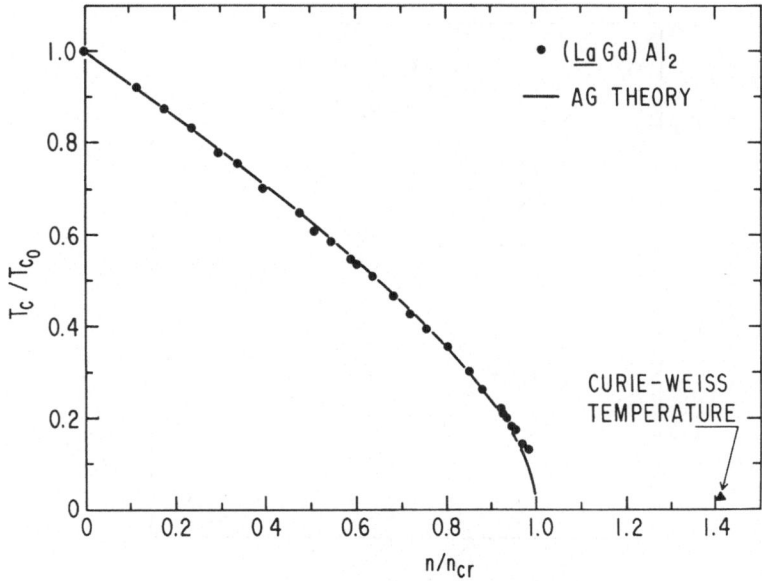

Fig. 5. Reduced superconducting transition temperature T_c/T_{c_0} vs reduced concentration n/n_{cr} for the (LaGd)Al$_2$ system[28] compared to the AG theory (solid curve). T_{c_0} = 3.24 K and n_{cr} = 0.590 at.% Gd substitution for La. The reduced Curie-Weiss temperature θ/T_{c_0} measured at n/n_{cr} = 1.41 is denoted by the solid triangle (after Ref. 6).

Figure 7 contains a plot of the initial rate of depression of T_c, $-(dT_c/dn)_{n=0}$ vs RE, for the LaRE$_2$ and (LaRE)Al$_2$[31] systems. The data are described reasonably well by the simple deGennes scaling except for Ce for which $-(dT_c/dn)_{n=0}$ is anomalously large. The anomalous depression of T_c for Ce in both the La and LaAl$_2$ matrices is due to the Kondo effect which will be discussed next.

2.1.2 Negative (Antiferromagnetic) Exchange Interaction

A negative exchange interaction ($\mathcal{J} < 0$) gives rise to the Kondo effect which, in the normal state, involves the formation of a "quasi-bound" state wherein the spin of the paramagnetic impurity ion tends to be compensated by the conduction electron spins.[32-34] The many-body singlet ground state that develops at temperatures much lower than the so-called Kondo temperature T_K has a binding energy $E_B^K \sim k_B T_K$. Here, T_K is approximately given by

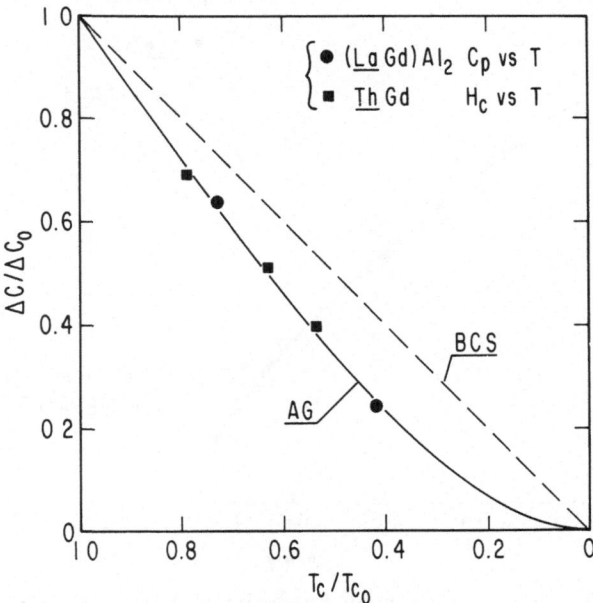

Fig. 6. Reduced specific heat jump $\Delta C/\Delta C_0$ vs reduced transition
temperature T_c/T_{c_0} for the matrix-impurity systems
$(\underline{La}Gd)Al_2$[29] and $\underline{Th}Gd$.[30] The dashed line represents the
BCS law of corresponding states, whereas the solid line is
the result of the AG theory as calculated by Skalski et al.[26]
(after Ref. 6).

$$T_K \sim T_F \exp[-1/N(0)|\mathcal{J}|] \tag{9}$$

where T_F is the Fermi temperature.

The Kondo effect manifests itself in characteristic temperature
dependent anomalies in the magnetic susceptibility, electrical resistivity, specific heat, thermoelectric power, etc., in the vicinity
of T_K. Of these anomalies, the one first noted and addressed theoretically by Kondo[32] in 1964 is the electrical resistivity minimum
phenomenon. The contribution to the electrical resistivity due to
the Kondo effect varies as log T for $T \gg T_K$ and then saturates to
a constant value at low temperatures $T \ll T_K$. When added to the
ordinary nonmagnetic impurity and phonon terms, the Kondo contribution produces a minimum in the electrical resistivity.

The superconducting properties are profoundly affected by the
Kondo effect. For example, the depression of T_c is anomalously
large and attains a maximum value for $T_K \sim T_{c_0}$. Physically, one

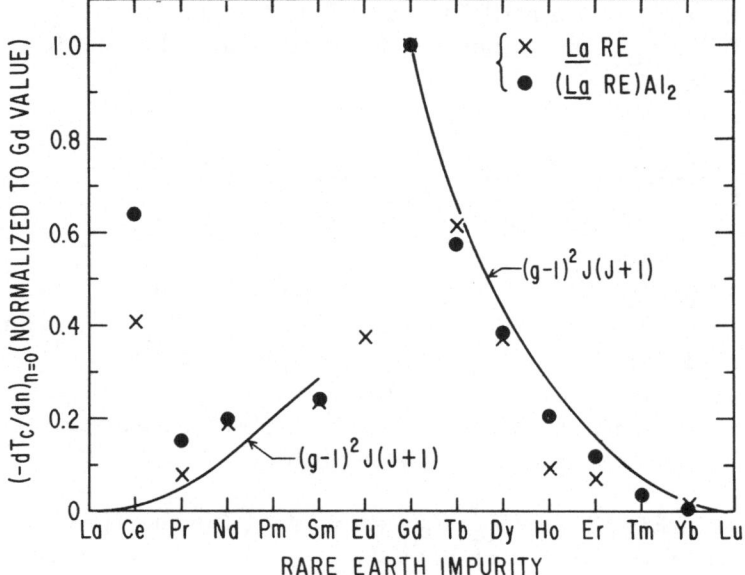

Fig. 7.   Initial rate of depression of the superconducting transition temperature of LaRE alloys[2] and (LaRE)Al$_2$
alloys[31] vs RE impurity (normalized to the Gd value).
The solid line is the deGennes factor $(g_J - 1)^2 J(J+1)$,
normalized to the Gd value. The values of $-(dT_c/dn)_{n=0}$
for Gd impurities are 5.3 and 3.79 K/at.% Gd substitution for La for LaRE and (LaRE)Al$_2$, respectively
(after Ref. 5).

expects the largest effects to occur when $T_K \sim T_{c_0}$ since one is
dealing with the competition between singlet spin pairing of conduction electrons to form Cooper pairs with a binding energy $E_B^{SC} \sim$
$k_B T_c$ and singlet spin pairing between conduction electrons and an
impurity spin with a binding energy $E_B^K \sim k_B T_K$. This explains the
anomalous depression of T_c by Ce impurities in the LaCe and
(LaCe)Al$_2$ systems which each exhibit a Kondo effect in the normal
state with $T_K \sim 0.1$ K.[5,6]

Perhaps the most interesting manifestation of the Kondo effect
in the superconducting state is the phenomenon of reentrant superconductivity that occurs in the regime $T_K \ll T_{c_0}$. An alloy within
a certain range of impurity concentrations will, upon cooling, first
become superconducting at an upper critical temperature T_{c1} and
then reenter the normal state at a lower critical temperature T_{c2}.
The reentrant superconductive behavior is caused by temperature-
dependent pair breaking,[35,36] where, in a certain approximation,

T_c/T_{c_0} and α/α_{cr} can still be related by the universal AG function [Eq. (5)], and α/α_{cr} is a function of temperature that scales with T_K and is given by

$$\alpha/\alpha_{cr} = nB\left\{\frac{\pi^2 S(S+1)}{\ell n^2(T/T_K) + \pi^2 S(S+1)}\right\} \tag{10}$$

where $B = 1/0.14(2\pi)^2 k_B N(0)T_{c_0}$. Other predictions of the theory are strong deviations of the curve of $\Delta C/\Delta C_0$ vs T_c/T_{c_0} from both the BCS law of corresponding states and the AG theory, [37] as well as the existence of a bound state in the energy gap. [38] More detailed discussions of the Kondo effect in superconductors appear elsewhere in this volume in the lectures by J. Ruvalds.

Shown in Fig. 8 are T_c/T_{c_0} vs n data for the (LaCe)Al$_2$ Kondo system[39] for which $T_K \sim 0.1$ K $\ll T_{c_0} \sim 3.3$ K. The dashed line

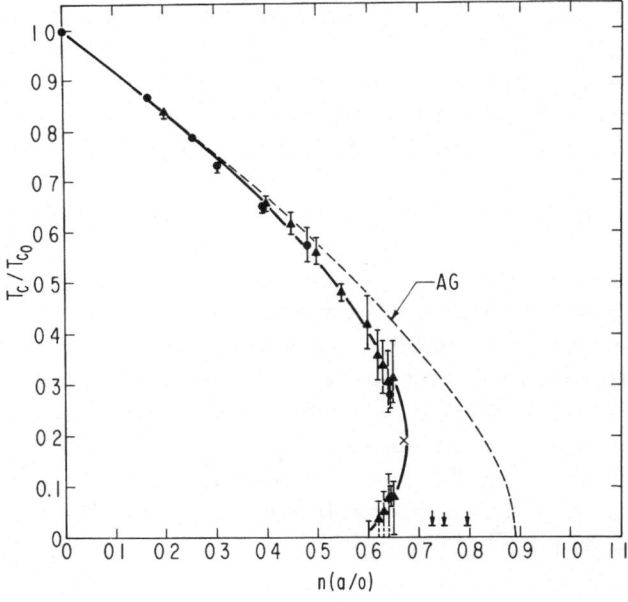

Fig. 8. Reduced superconducting transition temperature T_c/T_{c_0} vs Ce impurity concentration n for the (LaCe)Al$_2$ system. The symbol (x) denotes the estimated turning point of the T_c/T_{c_0} vs n curve, while the solid circles and triangles distinguish two separately prepared sets of alloys. The AG curve (dashed) is shown for comparison (after Ref. 6).

represents the AG behavior for comparison. Reentrant superconductive behavior occurs in a narrow range of Ce concentrations $0.60 \leqslant n \leq 0.68$ at.% Ce. It should be noted that the mechanisms that produce reentrant superconductive behavior of Kondo superconductors, considered here, and ferromagnetic superconductors, discussed later, are distinctly different.

Shown in Fig. 9 are $\Delta C/\Delta C_o$ vs T_c/T_{c_o} data for the (LaCe)Al$_2$ system[40-42] which deviate strongly from both the BCS law of corresponding states and the AG theory.

2.2 Effect of Rare Earth Ions on Superconductivity in Finite Magnetic Field (External and Internal)

The response of superconductivity to a finite magnetic field, which can be one that is externally applied or associated with magnetic order, will now be considered. However, it is first useful to examine magnetic ordering that develops via the RKKY interaction.[43] The RKKY interaction is believed to be the dominant coupling

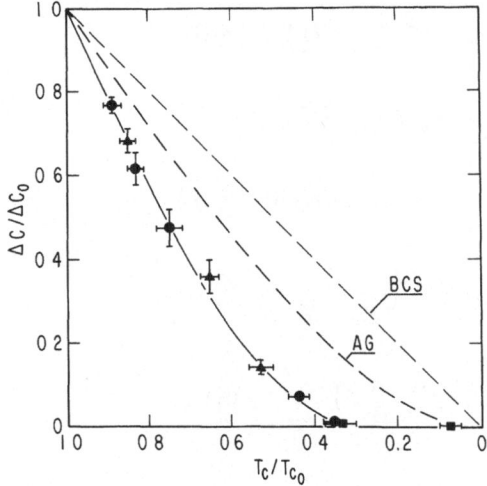

Fig. 9. Reduced specific heat jump $\Delta C/\Delta C_o$ vs reduced transition temperature T_c/T_{c_o} for the (LaCe)Al$_2$ system. The solid triangles, circles and squares represent data from references 40, 41 and 42, respectively. The dashed line describes the BCS law of corresponding states, the dot-dashed line indicates the AG result, and the solid line is a smooth curve drawn through the data (after Ref. 6).

mechanism between the RE magnetic moments in the RE systems of interest, although direct dipole-dipole interactions may also be important in some cases.[44]

2.2.1 Magnetic Ordering via the RKKY Interaction

The interaction between a RE ion with total angular momentum $\underset{\sim}{J}$ located at $R = 0$ and a conduction electron with spin $\underset{\sim}{s}$ is given by the exchange interaction Hamiltonian

$$\mathcal{H}_{int} = -2\mathcal{J}(g_J-1)\underset{\sim}{J} \cdot \underset{\sim}{s}\,\delta(\underset{\sim}{r}) = -[2(g_J-1)\mathcal{J}/g\mu_B]\underset{\sim}{J}\,\delta(\underset{\sim}{r}) \cdot (g\mu_B \underset{\sim}{s})$$

$$= -\underset{\sim}{H}_{eff}(\underset{\sim}{r}) \cdot \underset{\sim}{m} \tag{11}$$

Here, the exchange interaction Hamiltonian has been expressed as an effective field $\underset{\sim}{H}_{eff}$ times the magnetic moment $\underset{\sim}{m}$ of the conduction electron where $\underset{\sim}{H}_{eff}$ is given by

$$\underset{\sim}{H}_{eff}(\underset{\sim}{r}) = [2(g_J-1)\mathcal{J}/g\mu_B]\underset{\sim}{J}\,\delta(\underset{\sim}{r}) \tag{12}$$

The Fourier transform of $\underset{\sim}{H}_{eff}(\underset{\sim}{r})$ yields

$$\underset{\sim}{H}_{eff}(\underset{\sim}{q}) = [2(g_J-1)\mathcal{J}/g\mu_B]\underset{\sim}{J} \tag{13}$$

The conduction electron spin density $\underset{\sim}{s}(\underset{\sim}{r})$ can then be expressed as:

$$\underset{\sim}{s}(\underset{\sim}{r}) = \underset{\sim}{m}(\underset{\sim}{r})/g\mu_B = \frac{1}{g\mu_B V} \sum_{\underset{\sim}{q}} \chi(\underset{\sim}{q})\,\underset{\sim}{H}_{eff}(\underset{\sim}{q})\,e^{i\underset{\sim}{q}\cdot\underset{\sim}{r}}$$

$$= [2(g_J-1)\mathcal{J}/g^2\mu_B^2 V] \sum_{\underset{\sim}{q}} \chi(\underset{\sim}{q})\,e^{i\underset{\sim}{q}\cdot\underset{\sim}{r}}\,\underset{\sim}{J} \tag{14}$$

where $\chi(\underset{\sim}{q})$ is the q-dependent susceptibility which, for a free electron gas, is given by

$$\chi(\underset{\sim}{q}) = \chi_P\,F(q/2k_F)$$

$$= \chi_P\left[\frac{1}{2} + \frac{k_F}{2q}\left(1 - \frac{q^2}{4k_F^2}\right)\log\left|\frac{2k_F+q}{2k_F-q}\right|\right] \tag{15}$$

where $\chi_P = 2N(0)\,\mu_B^2$. From this the form of the spatial dependence of the conduction electron spin density can be deduced

$$\underset{\sim}{s}(\underset{\sim}{r}) \sim \left\{ \frac{\sin(2k_F r) - 2k_F r \cos(2k_F r)}{(2k_F r)^4} \right\} \sim \frac{\cos(2k_F r)}{(2k_F r)^3} \; ; \; k_F r \gg 1 \quad (16)$$

which is the usual RKKY result for a free electron gas. Here, the
spin density is a damped oscillatory function of r which produces
ferromagnetic, antiferromagnetic or complicated magnetic structures depending upon the details of the crystal structure of the
material under consideration.

The interaction between two RE ions can be expressed in
terms of $\chi(\underset{\sim}{q})$ as

$$\mathcal{H}_{RKKY} = - \frac{4(g_J - 1)^2 \, \mathscr{J}^2}{g^2 \mu_B^2 \, V} \sum_{\underset{\sim}{q}} \chi(\underset{\sim}{q}) \, e^{i\underset{\sim}{q} \cdot \underset{\sim}{r}} \, \underset{\sim}{J}_i \cdot \underset{\sim}{J}_j \quad (17)$$

and therefore depends sensitively on the electron band structure,
as well as the crystal structure. Modifications of the electron band
structure by the application of pressure or by alloying can be expected to change magnetic ordering temperatures as well as the
nature of the magnetic ordering, and some examples of this will be
encountered later on in these lectures. Also, the alteration of
$\chi(q)$ in the superconducting state will lead to some interesting magnetic structures in superconductors which will also be discussed.

2.2.2 Origin of the Upper Critical Magnetic Field of Type II Superconductors

The discussion of the effect of RE ions on superconductivity
in finite externally applied or internal magnetic field will begin with
an examination of the origin of the upper critical field H_{c2} of a type
II superconductor.[12] Basically, H_{c2} is determined by the interaction of the magnetic field with the orbits of the conduction electrons

$$\frac{e}{mc} (\underset{\sim}{p} \cdot \underset{\smile}{A}) = \frac{e\hbar}{mc} (\underset{\sim}{k} \cdot \underset{\sim}{A}) \quad (18)$$

which gives rise to the orbital critical field H_{c2}^*, and through the
Zeeman interaction of the magnetic field and the magnetic moments

(or spins) of the conduction electrons

$$-\underset{\sim}{\mu} \cdot \underset{\sim}{H} = -g\mu_B (\underset{\sim}{s} \cdot \underset{\sim}{H}) \tag{19}$$

that produces the Pauli paramagnetic limiting field H_p.

The upper critical magnetic field is determined by the orbital critical field H_{c2}^* or the paramagnetic limiting field H_p, depending upon which one is lower. The orbital critical field H_{c2}^* is given by

$$H_{c2}^* = \Phi_0 / 2\pi\xi^2 \tag{20}$$

where Φ_0 is the flux quantum and ξ is the superconducting coherence length. The paramagnetic limiting field H_p can be discussed by first examining the superconducting F_s and normal state F_n free energies in a magnetic field:

$$F_s(H) = F_s(0) - \frac{1}{2}\chi_s H^2 \tag{21}$$

and

$$F_n(H) = F_n(0) - \frac{1}{2}\chi_n H^2 \tag{22}$$

where χ_s and χ_n are the conduction electron spin susceptibilities in the superconducting and normal states, respectively. A first order transition from the superconducting to normal state in a certain magnetic field H_p will occur when

$$F_n(H_p) - F_s(H_p) = 0 = [F_n(0) - F_s(0)] - \frac{1}{2}(\chi_n - \chi_s)H_p^2$$

$$= \frac{1}{2}N(0)\Delta^2 - \frac{1}{2}(\chi_n - \chi_s)H_p^2 \tag{23}$$

using the result $F_n(0) - F_s(0) = \frac{1}{2}N(0)\Delta^2$ from the microscopic theory of superconductivity. Equation (23) can then be solved for the paramagnetic limiting field H_p which gives

$$H_p = \left[\frac{N(0)}{\chi_n - \chi_s}\right]^{1/2}\Delta \tag{24}$$

For a BCS superconductor at $T = 0$, $\chi_s = 0$ and $\chi_n = N(0) g^2 \mu_B^2 / 2$ which yields the following expression for the Pauli paramagnetic limiting field H_{po}, the well known Clogston-Chandrasekhar limit:[45, 46]

$$H_{po} = \frac{\sqrt{2}\, \Delta(0)}{g\mu_B} = 18.4\, T_c\, (kOe) \qquad (25)$$

Equation (24) shows that H_p can be made arbitrarily large if $\chi_n - \chi_s$ can be made arbitrarily small. The behavior of H_{c2} vs T is depicted schematically in Fig. 10 where the dashed curve represents the behavior of H_{c2} if it were determined solely by the interaction of the magnetic field with the orbits of the conduction electrons, while the solid curve corresponds to the case where H_{c2} is governed by the paramagnetic limiting field for $H_p < H_{c2}^*$. In principle, there are several ways in which one can overcome the limitation imposed by the paramagnetic limiting field, all of which are relevant to the question of whether or not superconductivity and ferromagnetism can coexist and, therefore, are considered below.

The first way is to increase χ_s which can be accomplished by increasing the amount of spin-orbit or exchange scattering.[47, 48]

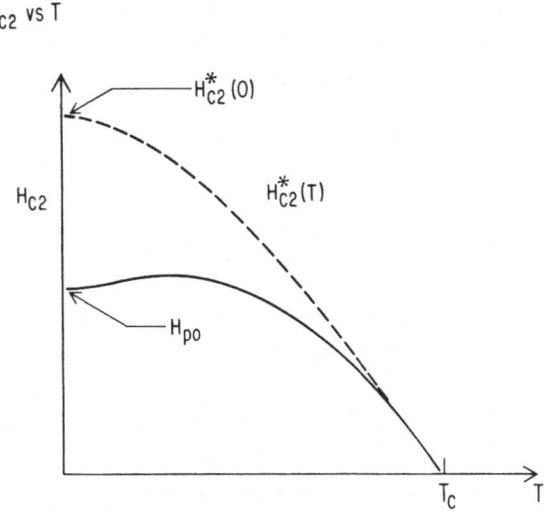

Fig. 10. Schematic representation of the upper critical field H_{c2} vs temperature T for a system in which the Pauli paramagnetic limiting field H_{po} is smaller than the orbital critical field $H_{c2}^*(0)$.

This produces the behavior of χ_s vs T schematically indicated in Fig. 11. The solid curve corresponds to a BCS superconductor ($\tau_{so} = \infty$), while the dot-dashed curve for which $\chi_s = \chi_n$ is for infinite spin-orbit scattering ($\tau_{so} = 0$). An intermediate case (dashed curve) for some finite value of τ_{so} is also shown where χ_s is reduced, but not to zero at $T = 0$. The resultant increase in χ_s decreases the difference $\chi_n - \chi_s$ in Eq. (24), and hence increases H_p so that it eventually exceeds H_{c2}^*.

A second method is for the applied magnetic field $\underset{\sim}{H}$ to be compensated by a negative exchange field (Jaccarino-Peter effect),[49] and a third way is for an oscillatory magnetic state to develop at finite q where $\chi_s(q)$ attains a maximum (Anderson-Suhl cryptoferromagnetism).[50] These two situations will be discussed next in some detail.

2. 2. 3 Exchange Field Compensation Effect

The exchange field compensation effect can occur when a transition metal or RE ion with spin $\underset{\sim}{S}$ interacts antiferromagnetically with the conduction electron spins ($\mathcal{J} < 0$). The total magnetic field $\underset{\sim}{H}_T$ in the material is the sum of the applied magnetic field $\underset{\sim}{H}$ and the exchange field $\underset{\sim}{H}_{eff} = (2\mathcal{J}/g\mu_B)S$ which opposes $\underset{\sim}{H}$ when $\mathcal{J}$ is negative. This produces a compensation of $\underset{\sim}{H}$ that increases with decreasing temperature since the spin polarization of the magnetic ions is proportional to the Brillouin function $B_s (\mu H/k_B T)$; i.e.,

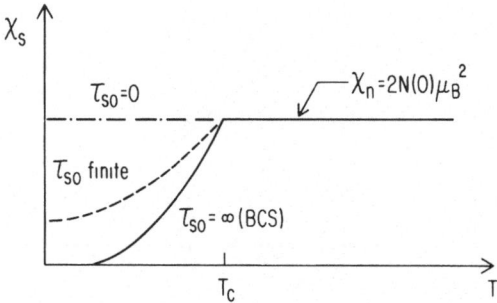

Fig. 11.   Schematic representation of χ_s, the conduction electron magnetic susceptibility in the superconducting state, vs temperature T for spin-orbit scattering lifetimes $\tau_{so} = 0$, τ_{so} finite and $\tau_{so} = \infty$ (BCS).

$$\langle S_z \rangle = S B_S (\mu H / k_B T) \tag{26}$$

In Fig. 12, the behavior of H_{c2} when it is enhanced above the Clogston-Chandrasekhar limit by (1) spin-orbit scattering and (2) exchange field compensation is shown schematically. In the first case, the H_{c2} vs T curve is monotonic and intermediate between the behavior for complete and no paramagnetic limiting. This is to be contrasted to the behavior in the second case where, in zero magnetic field, T_c is decreased by means of paramagnetic spin flip scattering, but H_{c2} increases with decreasing temperature with distinct positive curvature because of the increasing exchange field compensation of the applied field. It should be noted that $\mathcal{J} < 0$ is the condition for the Kondo effect so that those systems that exhibit a Kondo effect would be expected to show an exchange field compensation effect if the matrix is a paramagnetically limited type II high field superconductor. In addition, it should also be possible to induce a ferromagnet with $\mathcal{J} < 0$ to become superconducting in an external field if it were predisposed to become superconducting in the absence of ferromagnetism. This was, in fact, the situation that Jaccarino and Peter originally considered in 1962.

The exchange field compensation effect has been observed in several superconducting-matrix impurity systems. For example, Figs. 13 and 14 display T_c vs n and H_{c2} vs T data for the $Sn_{1.2(1-x)}Eu_x Mo_{6.35}S_8$ system that was first investigated by Fischer et al.[51] The Eu ions reduce T_c in zero field, but enhance

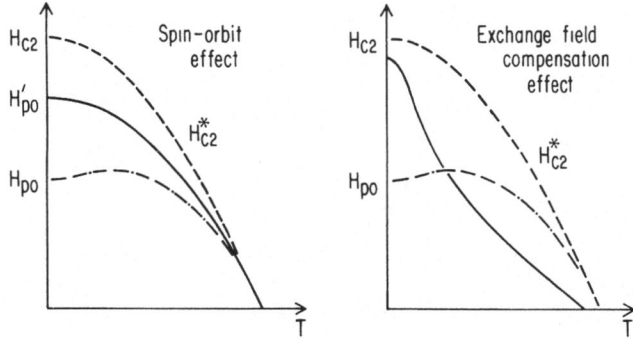

Fig. 12. Schematic representation of the enhancement of H_{c2} by means of the spin-orbit effect and the exchange field compensation effect.

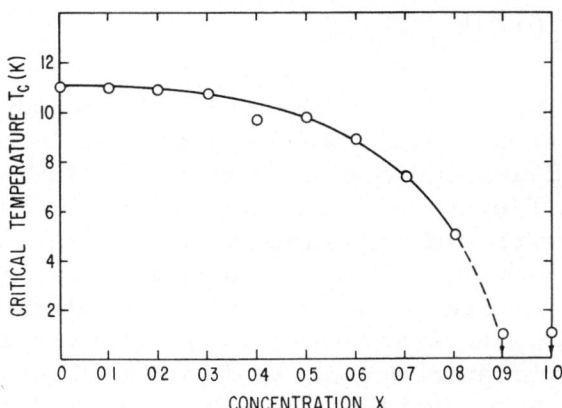

Fig. 13. Superconducting transition temperature T_c vs Eu concentration x for $Sn_{1.2(1-x)}Eu_xMo_{6.35}S_8$ (from Ref. 51).

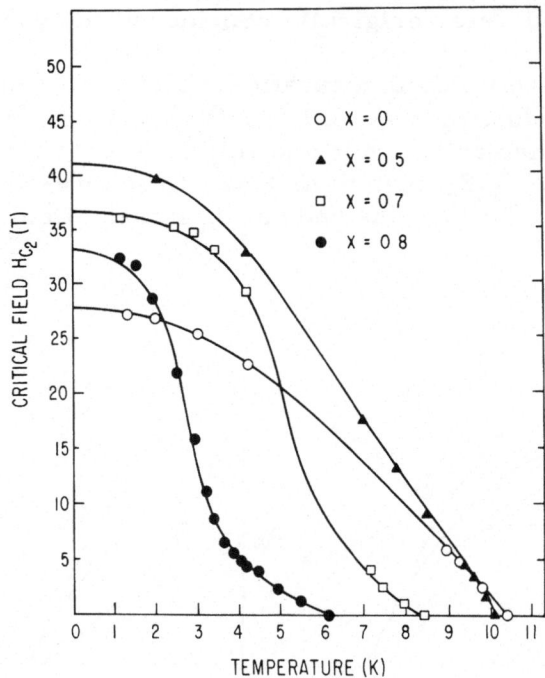

Fig. 14. Upper critical magnetic field H_{c2} vs temperature for $Sn_{1.2(1-x)}Eu_xMo_{6.35}S_8$ for different Eu concentrations x (from Ref. 51).

H_{c2} at low temperatures over that of the $Sn_{1.2}Mo_{6.35}S_8$ matrix and, moreover, induce pronounced positive curvature in the H_{c2} vs T curves that is characteristic of this effect. This interpretation of the data in terms of the Jaccarino-Peter effect is supported by NMR and Mössbauer effect measurements by Fradin et al.[52] Recently, evidence for the exchange field compensation effect was found for the related system $La_{1.2-x}Eu_xMo_6S_8$ by Torikachvili and Maple.[53] The T_c vs n and H_{c2} vs T data for this system are shown in Figs. 15 and 16.

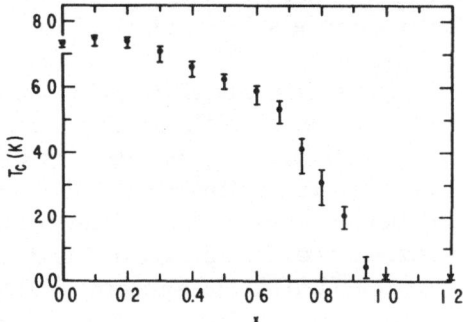

Fig. 15.   Superconducting transition temperature T_c vs Eu concentration x for $La_{1.2-x}Eu_xMo_6S_8$ (from Ref. 53).

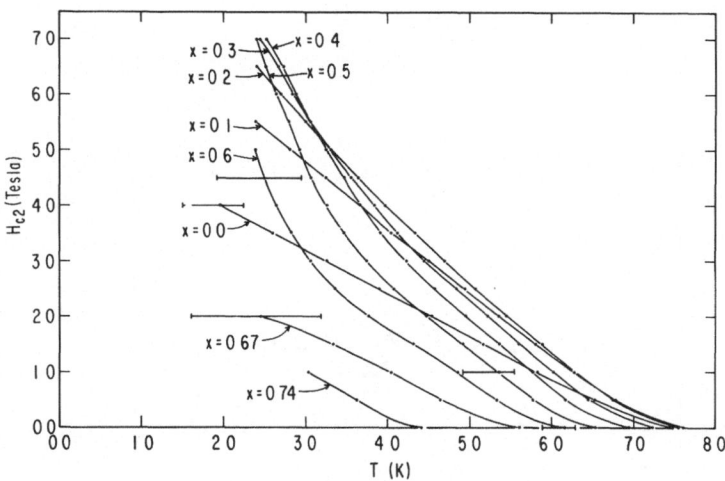

Fig. 16.   Upper critical magnetic field H_{c2} vs temperature T of $La_{1.2-x}Eu_xMo_6S_8$ for different Eu concentrations x (from Ref. 53).

Analysis of H_{c2} vs T data have also been used to provide evidence for the coexistence of superconductivity and some type of magnetic order in the $(\underline{La}Gd)_3In$ system by Crow et al.[54] The $(\underline{La}Gd)_3In$[55] and $\underline{La}Gd$ systems[2,56] are very similar since the T_c vs n curves of both exhibit anomalous behavior which has been attributed to magnetic ordering. The T_c/T_{c_0} vs n/n_{cr} data for these two systems, shown in Fig. 17, conform to AG theory to $n/n_{cr} \sim 0.6$, but deviate in a striking manner from the AG theory above $n/n_{cr} \sim 0.6$. These features have been attributed to magnetic order as evidenced from the T_M vs n/n_{cr} curves which intersect both T_c/T_{c_0} vs n/n_{cr} curves at $T_c/T_{c_0} \gtrsim 0.1$. Benneman has suggested that these features might be explained in terms of magnetic ordering in the following way.[58] First of all, the increase of T_c could be caused by the freezing out of elastic exchange scattering due to the exchange field associated with the magnetic ordering. As the exchange field increases further, a precipitous decrease of T_c occurs which was attributed to the influence of the exchange field on the spins of the conduction electrons. The work provided further evidence for magnetic ordering from the measurements of the H_{c2} vs T curves of the $(\underline{La}Gd)_3In$ system. Displayed in Fig. 18 are the H_{c2} vs T data for the $(\underline{La}Gd)_3In$ system which was investigated by Crow et al.[54] which reveal reentrant behavior for the pseudobinary compounds with n = 0.98 and 1.24 at.% Gd. The H_{c2} vs T data for the n = 0.98 and 1.24 at.% Gd are shown along with H_{c2} vs T data for higher Gd concentration samples in more detail in

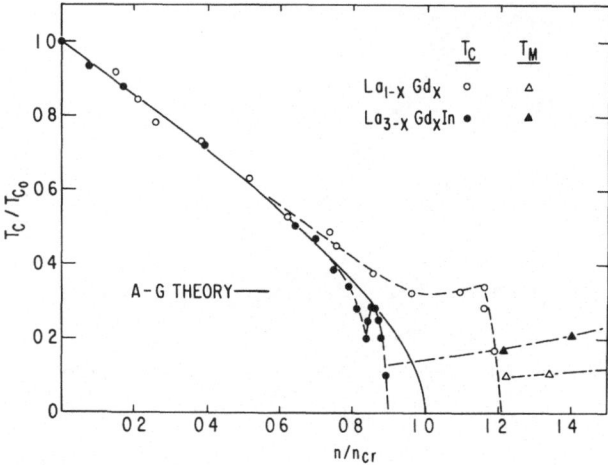

Fig. 17. Superconducting-normal and magnetic-normal phase boundaries for the $La_{1-x}Gd_x$[2,56] and $La_{3-x}Gd_xIn$[55] systems.

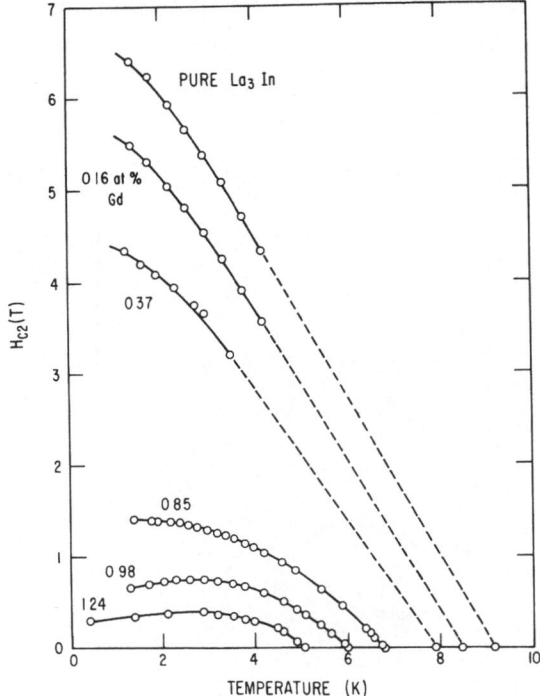

Fig. 18.   Upper critical field H_{c2} vs temperature curves for
$La_{3-x}Gd_xIn$ alloys in the concentration range 0 - 1.24 at.%
Gd (from Ref. 54).

Fig. 19.   Here it can be seen that the sample with n = 1.49 at.% Gd
also exhibits reentrant behavior, while those with n = 1.60 and
1.83 at.% Gd again display a monotonic temperature dependence.
These data were explained[54] in terms of a positive exchange interaction so that the exchange field adds to the applied magnetic field.
The reentrant behavior results from the increase of the exchange
field with decreasing temperature in the paramagnetic state since it
is proportional to the Brillouin function. However, when magnetic
ordering sets in, the magnetic moments saturate so that the exchange field is constant and the critical field curves again become
monotonic functions of temperature.

2.2.4 Cryptoferromagnetic State

The cryptoferromagnetic state proposed by Anderson and
Suhl[50] can be explained by again considering the difference in free
energies between the normal and superconducting states in a

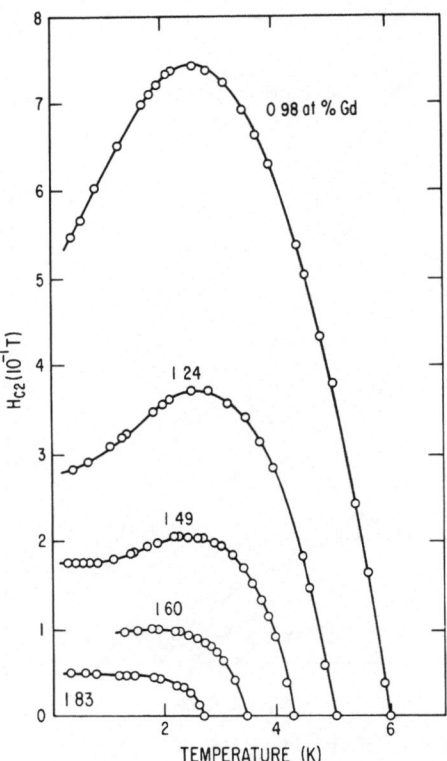

Fig. 19.   Upper critical field H_{c2} vs temperature T curves for
La$_{3-x}$Gd$_x$In alloys in the concentration range 0.98 -
1.83 at.% Gd (from Ref. 54).

magnetic field $\underset{\sim}{H}$ (in this case, the exchange field) and taking into
account the q-dependence of χ_n and χ_s , i.e.,

$$F_n(H) - F_s(H) = \frac{1}{2} N(0)\Delta^2 - \frac{1}{2}[\chi_n(q) - \chi_s(q)]H^2 \qquad (27)$$

It is assumed that for sufficiently large values of $\underset{\sim}{H}$, there exists a
wave number Q such that $F_n(H) - F_s(H) > 0$. This implies that

$$\frac{N(0)\Delta^2}{H^2} > \chi_n(Q) - \chi_s(Q) \qquad (28)$$

In the long wavelength limit,

$$\chi_n(q) \simeq \chi_P \left(1 - \frac{q^2}{12 k_F^2} + \cdots \right) \tag{29}$$

and

$$\chi_s(q) \simeq \chi_P \left(1 - \frac{q^2}{12 k_F^2} - \frac{\pi}{2 \xi q} + \cdots \right) \tag{30}$$

The quantities $\chi_n(q)$ and $\chi_s(q)$ are schematically represented in in Fig. 20. The quantity $\chi_s(q)$ displays a maximum at $Q = (3\pi k_F^2/\xi)^{1/3}$ which, for typical values of k_F and ξ, corresponds to a wavelength $\lambda \simeq 50$ Å. Thus, Anderson and Suhl suggested that the superconductor would break up into domains with characteristic dimensions of the order of 50 Å. Actually, this "cryptoferromagnetic" state anticipated the sinusoidally modulated state that coexists with superconductivity in ferromagnetic superconductors that was discovered by neutron scattering experiments and will be discussed later. However, the origin of the sinusoidally-modulated state may actually be based on the electromagnetic interaction, rather than the exchange interaction that was considered in the Anderson-Suhl theory.

It should also be remarked that if ordering at finite q does not occur, the ground state will surely be ferromagnetic, rather than superconducting, if the mediating influence of spin-orbit and exchange scattering is disregarded. This follows because the lowering of the free energy with respect to the normal paramagnetic state

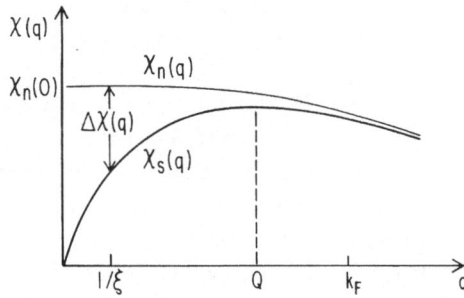

Fig. 20. Schematic representation of the conduction electron spin susceptibility in the normal state $\chi_n(q)$ and in the superconducting state $\chi_s(q)$ (from Ref. 50).

of the ferromagnetic state δF_M and the superconducting state δF_s are $\delta F_M \sim nN(k_B T_M)$ and $\delta F_s \sim (k_B T_c / E_F) N(k_B T_c)$, respectively, where N is the number of atoms. Since $n \gg k_B T_c / E_F$, the ferromagnetic ground state will be favored.

2.2.5 Gor'kov-Rusinov Theory

Another important early theory concerning the coexistence of superconductivity and long range ferromagnetic order was formulated by Gor'kov and Rusinov[59] in 1964. A qualitative discussion of the results of this theory can be made by considering a system in which RE impurity ions are imbedded in a superconducting metallic matrix and by recalling that the magnetic ordering temperature can be expressed approximately as

$$T_M \sim (k_B \mu_B^2)^{-1} n \chi(0) \mathcal{J}^2 (g_J - 1)^2 J(J+1) \tag{31}$$

while the paramagnetic limiting field is given by

$$H_p = \left[\frac{N(0)}{\chi_n - \chi_s} \right]^{1/2} \Delta \tag{32}$$

Figure 21 contains three schematic low temperature phase diagrams corresponding to three different values of $\chi_s(0)$. Figure 21(a) refers to the situation where $\chi_s(0) = 0$ at zero temperature which would be appropriate for a BCS superconductor. Here, the $T_M(n)$ line in the normal state terminates at the $T_c(n)$ curve that is described by the AG theory since magnetic ordering cannot occur via the RKKY interaction in the superconducting state when $\chi_s(0) = 0$. Thus, the low temperature diagram shown in Fig. 21(a) has a normal paramagnetic region, a normal ferromagnetic region, and a superconducting region.

Figure 21(b) is for $\chi_s(0)$ finite at zero temperature which, as mentioned earlier, can be accomplished by either spin-orbit or exchange scattering. Again, the $T_M(n)$ line in the normal state intersects the line of T_c vs n described by the AG theory. However, since $\chi_s(0)$ is finite, this allows magnetic ordering to occur below T_c which gives rise to the $T_M(n)$ curve indicated in Fig. 21(b). In addition, the finite value of $\chi_s(0)$ increases H_p so that superconductivity can persist in the exchange field associated with the magnetic order. However, when the exchange field becomes large enough, superconductivity is destroyed at a second critical

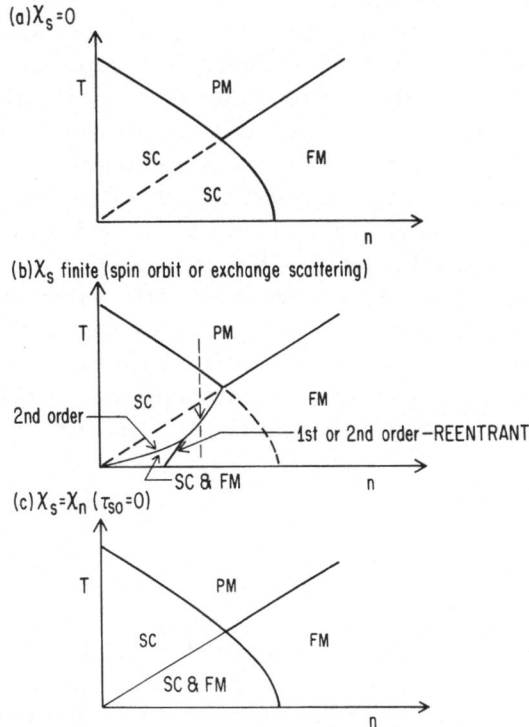

Fig. 21. Schematic representation of low temperature phase diagrams of superconducting-ferromagnetic systems, according to the Gor'kov-Rusinov theory.[59]

temperature T_{c2}, and the line of $T_{c2}(n)$ merges with the line of $T_M(n)$. This produces the interesting low temperature phase diagram shown in Fig. 21(b) which contains a normal paramagnetic region, a normal ferromagnetic region, a superconducting region, and a region in which superconductivity and ferromagnetism actually coexist. Second order transitions occur along the line separating the superconducting region and the region in which superconductivity and ferromagnetism coexist, while either first or second order transitions, depending on the strength of the exchange scattering, occur along the $T_{c2}(n)$ curve.

Finally, in Fig. 21(c), $\chi_s(0)$ equals $\chi_n(0)$, corresponding to the situation where $\tau_{so} = 0$. Since $\chi_s(0) = \chi_n(0)$, the $T_M(n)$ line will pass smoothly through the $T_c(n)$ curve into the superconducting region as shown in the figure. Since $\chi_s(0) = \chi_n(0)$, H_p will become infinite, and, neglecting any considerations involving the orbital critical field H_{c2}^*, superconductivity will persist in any value of the exchange field so that no reentrant superconductive behavior will occur. Therefore, the $T_c(n)$ curve will simply conform to the AG-type behavior. Thus, the low temperature phase diagram in Fig. 21(c) will have a normal paramagnetic region, a normal ferromagnetic region, a superconducting region, and a region in which superconductivity and ferromagnetism coexist. An obvious shortcoming of the theory is that it neglects the q-dependence of χ_s and χ_n as well as electromagnetic interactions of the magnetization with the persistent current. However, the theory has certain plausible features and it was the first theory to predict the phenomenon of reentrant superconductivity caused by the onset of long-range ferromagnetic order, or, for that matter, any mechanism whatsoever.

2.2.6 Other Theories

Two other theoretical developments during this first period should be noted, the multiple pair breaking theory of Fulde and Maki[60] and the theory of antiferromagnetic superconductors by Baltensperger and Strässler.[61] In the multiple pair breaking theory, Fulde and Maki found that under certain circumstances T_c/T_{c_0} can still be related to α/α_{cr} by the universal function of Abrikosov and Gor'kov, i.e. $T/T_{c_0} = Un(\alpha/\alpha_{cr})$. However, in this case, α/α_{cr} could be written as the sum of pair breaking parameters due to (1) spin-depairing caused by exchange scattering of conduction electrons by paramagnetic impurities, (2) momentum depairing resulting from the penetration of an external field into the superconducting matrix in the vortex state, and (3) Pauli or exchange field depairing due to spin polarization of the conduction band by the exchange field of the impurity spins that polarize according to the Brillouin function in the penetrating external field. The reduced pair breaking parameter is then given by

$$\frac{\alpha}{\alpha_{cr}} = \frac{n}{n_{cr}} + \frac{H_{c2}(n,T)}{H_{c2}(0,0)} + \frac{P}{P_{cr}} ; \qquad (\ell_{tr}, \ell_{so} \ll \xi) \qquad (33)$$

where

$$P = \tau_{so} \left[H_{c2}(n,T) + n \, \mathcal{J} \langle S_z \rangle \right]^2$$

where $H_{c2}(n, T)$ is the upper critical field for concentration n and temperature T, τ_{so} is, again, the spin-orbit scattering time, and $\langle S_z \rangle$ is the average z-component of the impurity spin. This expression holds in the limit where the transport mean free path ℓ_{tr} and the spin-orbit mean free path ℓ_{so} are very much smaller than the coherence length ξ. This theory has been used by Crow et al.[54] to analyze their $H_{c2}(n, T)$ data for $(\underline{La}Gd)_3In$ discussed earlier, and by Ishikawa and Fischer[62] to analyze their $H_{c2}(n, T)$ data on $REMo_6S_8$ antiferromagnetic superconductors which will be considered later on.

The first theory for antiferromagnetic superconductors was put forth by Baltensperger and Strässler[61] in 1963. They concluded that superconductivity and antiferromagnetic order were not incompatible with one another in view of the fact that the exchange field averages to zero over the scale of the coherence length. Moreover, they invoked a new pairing state with finite momentum $(k\uparrow, -k+Q\downarrow)$ where Q corresponds to translation by a reciprocal lattice vector. They also calculated the electron-electron interaction via spin waves for antiferromagnetic magnons and found this to be repulsive but weaker than the attractive electron-phonon interaction.

3. PERIOD 1976-PRESENT

The problem of magnetically ordered superconductors experienced a revival in about 1976 when certain ternary and pseudoternary RE compounds were found to exhibit long-range magnetic order below their superconducting transition temperatures. Superconductivity was observed to coexist with antiferromagnetic order but to be destroyed by the onset of ferromagnetic order at a second transition temperature of the order of the Curie temperature. The remarkable physical properties of ternary and pseudoternary RE systems that result from the interaction between superconductivity and long-range magnetic order will be briefly reviewed in the remainder of these lectures.

In the ternary RE compounds considered here, the RE ions are distributed periodically throughout the crystal lattice, and each RE ion has a partially-filled 4f electron shell and a corresponding magnetic moment. Since the compounds are metallic, the RE magnetic moments are coupled via the RKKY interaction. This leads to long-range ordering of the RE magnetic moments, characterized by a sharp magnetic ordering temperature T_M and well-defined features in the physical properties at T_M. For example, there can be a pronounced lambda-type anomaly in the heat capacity at T_M and, in the case of an antiferromagnet, a cusp in the magnetic susceptibility at T_M (in this case, T_M is the Néel temperature T_N).

Another significant feature is that the strength of the exchange interaction parameter $\mathcal{J}$ in these ternary RE compounds is only about 0.01 eV, nearly an order of magnitude smaller than in most binary RE systems. The weak exchange interaction has two important consequences. First, it enables the compounds to retain their superconductivity at a critical temperature T_c, if they are predisposed to become superconducting. This follows from the fact that depression of T_c by elastic exchange scattering of conduction electrons by paramagnetic RE ions scales as $\mathcal{J}^2$ [see Eq. (7)]. Second, the values of T_M are relatively small and comparable to the values of T_c. This is a consequence of the proportionality of T_M to $\mathcal{J}^2$ [see Eq. (31)].

3.1 Ternary Rare Earth Systems

The series of isostructural ternary RE compounds that have been investigated most extensively in connection with the interaction between superconductivity and long-range magnetic order include the rhombohedral RE molybdenum chalcogenides, $REMo_6S_8$[63] and $REMo_6Se_8$,[64] and the tetragonal RE rhodium borides, $RERh_4B_4$.[65, 66] The RE molybdenum chalcogenides belong to a large class of compounds known as Chevrel phases,[67] some of which have very striking superconducting properties that have potential technological applications.[12] The compound $PbMo_6S_8$ has been reported to have a relatively high superconducting transition temperature of ~ 15 K.[68] Many of the Chevrel phase compounds also have exceptionally high upper critical magnetic fields that in some cases approach ~ 60 Tesla, the highest values ever observed.[69-71]

The crystal structures of the rhombohedral RE molybdenum chalcogenides and the tetragonal RE rhodium borides are shown in Figs. 22 and 23, respectively. Within these two crystal structure classes of ternary RE compounds, many examples have been found in which long-range magnetic order occurs at a temperature T_M below the temperature T_c at which the compound becomes superconducting. In most of the compounds studied so far, the magnetic ordering is antiferromagnetic in nature and coexists with superconductivity, although it does modify superconducting properties such as the curve of H_{c2} vs temperature in the neighborhood of the Néel temperature. However, in two of the compounds, the onset of long-range ferromagnetic order at a temperature T_M destroys superconductivity at a second critical temperature $T_{c2} \sim T_M$ below

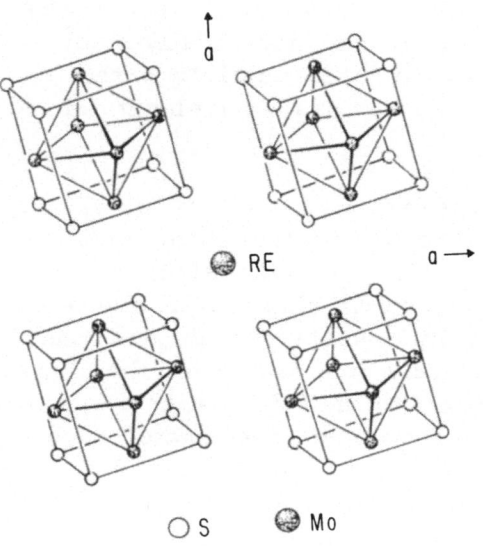

RE

S   Mo

Fig. 22. Crystal structure of
the RE molybdenum chalco-
genides (Chevrel phases).
Two rhombohedral axes (a)
are indicated in the figure;
the third axis projects out-
ward. The RE ion is sur-
rounded by eight Mo_6S_8
units. (After Ref. 72).

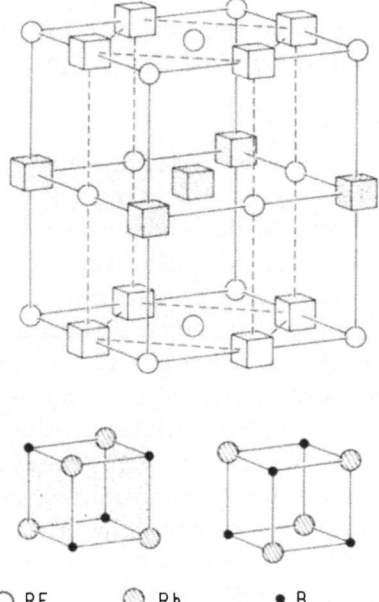

RE   Rh   B

Fig. 23. Primitive tetragonal
crystal structure of RE
rhodium borides $RERh_4B_4$
with the unit cell shown in
dashed outline. For clarity,
the Rh_4B_4 clusters are not
drawn to scale. (After
Ref. 73).

the temperature T_{c1} which the compound originally became superconducting. In the ferromagnetic superconductors, the superconducting-ferromagnetic interactions produce a sinusoidally modulated magnetic state that coexists with superconductivity within a
narrow temperature interval above T_{c2} and has a wavelength of the
order of 100 Å.

The small value of the exchange interaction ($\mathcal{J} \sim 0.01$ eV) in
these compounds appears to be associated with transition metal
molecular units or "clusters" which, along with the RE ions, are
the basic building blocks of these ternary RE phases. The superconductivity is believed to be primarily associated with the transition metal d-electrons that are relatively confined within the clusters and thereby interact only weakly with the RE ions. [15-17]
Recently, other ternary RE systems have been investigated such as
$RERh_{1.1}Sn_{3.6}$,[74] $RE_2Fe_3Si_5$,[75] $RERuB_2$,[76] etc.

3.2 Exchange Interaction

The weak exchange interaction between the conduction electron
spins and the RE magnetic moments in the RE molybdenum chalcogenides and RE rhodium borides enables these compounds to retain
their superconductivity, in spite of relatively large concentrations
of RE ions of ~ 7 and ~ 11 at.%, respectively. For both classes of
compounds, the AG theory provides a qualitative description of the
systematic variation of T_c with RE and an estimate of the magnitude
of $\mathcal{J}$.

With the exception of Ce and Eu, all of the $RE_xMo_6S_8$ and
$RE_xMo_6Se_8$ ($x = 1.0$ or 1.2) compounds are superconducting with
T_c's within each series that display a systematic dependence on RE
as shown in Fig. 24 (the T_c of $Yb_xMo_6S_8$ is anomalously high in
comparison to the other superconducting $RE_xMo_6S_8$ compounds,
since the Yb ions are divalent).[77] The dashed lines in the figure
are linear interpolations between the T_c's of the corresponding non-
magnetic La and Lu compounds, and represent estimates of the T_c's
of the RE compounds with partially-filled 4f electron shells in the
absence of pair breaking interactions due to the magnetic moments
of the RE ions. The depressions of T_c relative to these linear interpolations are also shown in Fig. 24. Here it can be seen that the
variation of ΔT_c with RE can be described qualitatively by the
deGennes scaling factor $(g_J-1)^2 J(J+1)$ which is indicated by the solid
lines. For both series of $RE_xMo_6S_8$ and $RE_xMo_6Se_8$ compounds,
the value of $|\mathcal{J}|$ inferred from the ΔT_c vs RE data is of the order

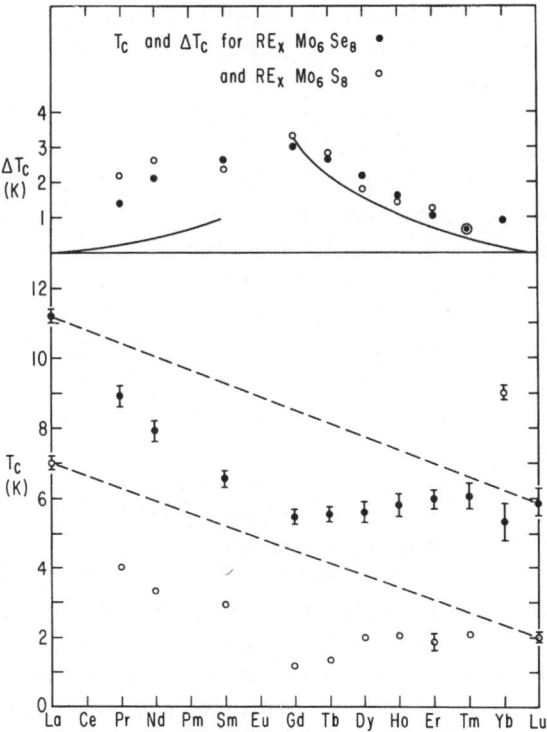

Fig. 24. Superconducting transition temperature T_c and depression of T_c (ΔT_c) vs RE for $RE_xMo_6S_8$ and $RE_xMo_6Se_8$ compounds. The dashed lines represent the interpolated T_c's of the RE molybdenum chalcogenide compounds in the absence of magnetic pair breaking interactions. The solid line represents the deGennes factor, $(g_J-1)^2J(J+1)$, normalized to the data for $Gd_xMo_6S_8$. (After Ref. 77).

of 0.01 eV. Electron paramagnetic resonance experiments on the compound $Gd_xMo_6Se_8$ yielded the value $\mathcal{J} \sim +0.01$ eV, in support of the analysis of the T_c vs RE data in terms of paramagnetic pair breaking effects.[78] The failure of the Ce and Eu compounds to become superconducting down to 50 mK (at zero pressure) may be due to anomalously strong exchange scattering associated with the Kondo effect as evidenced by Kondo-like anomalies in their normal state properties,[77] and the enhancement of H_{c2} by means of the exchange field compensation effect when Eu is doped into the Chevrel phase superconductors $Sn_{1.2}Mo_{6.35}S_8$,[51] $PbMo_{6.35}S_8$,[51] $La_{1.2}Mo_6S_8$,[53] and $Yb_{1.2}Mo_6S_8$.[79] However, the situation for Eu appears to be rather complex and remains to be completely resolved.[80]

The superconducting critical temperatures T_c and magnetic ordering temperatures T_M (in this case, Curie temperatures) of the $RERh_4B_4$ compounds are plotted vs RE in Fig. 25 as originally reported by Matthias et al.[65] No data are shown for $RERh_4B_4$ compounds with RE = La, Ce, Eu and Yb since they could not be formed. With increasing RE atomic number, the low temperature behavior switches from superconductivity for Nd and Sm to ferromagnetism for Gd, Tb, Dy and Ho and back to superconductivity for Er, Tm and Lu. Moreover, all the superconducting $RERh_4B_4$ compounds in which the RE 4f electron shell is partially-filled undergo some type of magnetic ordering below their T_c's at temperatures T_M in the vicinity of 1 K. Whereas $ErRh_4B_4$ becomes ferromagnetic,[81,82] $NdRh_4B_4$,[83,84] $SmRh_4B_4$,[85] and $TmRh_4B_4$[86,87] exhibit antiferromagnetic transitions.

If one disregards the ferromagnetic $RERh_4B_4$ compounds near the middle of the RE series, T_c appears to display a minimum near Gd at the half-filled 4f electron shell. This general behavior is again in compliance with scaling of the depression of T_c by paramagnetic impurities with the deGennes factor that attains a maximum value at Gd. Besides the depression of T_c, other properties (e.g., the magnetic ordering temperature and the spin disorder resistivity) of an isostructural series of metallic RE compounds in which the RE

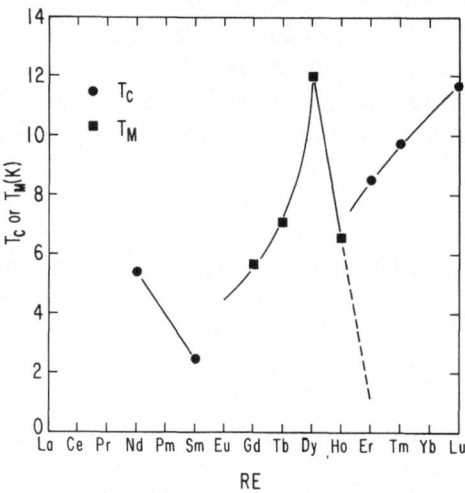

Fig. 25. Superconducting (T_c) and magnetic (T_M) transition temperatures of $RERh_4B_4$ compounds as originally reported by Matthias et al. (After Ref. 65).

magnetic moments interact with one another via the indirect RKKY
mechanism should scale with the deGennes factor[88] (see Section
2.2.1), as observed, for example, in the $REAl_2$ compounds.[31]
However, the peak in the magnetic ordering temperatures of the
$RERh_4B_4$ compounds occurs at RE = Dy rather than Gd as would be
expected from the variation of the deGennes factor with RE. This
suggests that other factors including CEF, magneto-elastic and,
possibly, even dipolar interactions affect the superconducting and
magnetic properties of the $RERh_4B_4$ compounds. Normal state
properties such as magnetic susceptibility, heat capacity and thermal expansion are, in fact, replete with features that can be attributed to CEF effects where the splittings of the energy levels of the
RE ions are also of the order of T_c and T_M. These features include pronounced Schottky anomalies in the specific heat, reduced
entropies of ordering and magnetic moments (compared to Hund's
rule values) and magnetic anisotropy.

The magnitude of $\mathcal{J}$ of the $RERh_4B_4$ compounds has been estimated by means of the AG theory from the depression of T_c of
$LuRh_4B_4$ by RE impurities with partially-filled 4f electron shells.[89]
Data for the rate of depression of T_c of $LuRh_4B_4$, $[T_{c_0} - T_c(x)]/x$,
are plotted vs RE in Fig. 26 where $T_{c_0} = T_c (x=0) = 11.36$ K and
x is the atomic fraction of Lu ions replaced by the indicated RE
ions.[89] Since the depression rate for the nonmagnetic RE = La^{3+}
ion should be zero, a linear baseline (dashed line in Fig. 26) was
drawn between its measured depression rate and the zero point of
$LuRh_4B_4$. The solid line in the figure is the deGennes factor normalized to the rate of depression of T_c for Gd, for comparison with
the measured values of $[T_{c_0} - T_c(x)]/x$. The rates of depression of
T_c of the heavy RE ions from RE = Gd to Tm scale with the deGennes
factor, indicating that the quantity $N(0)\mathcal{J}^2$ is approximately constant
for these heavy rare earth compounds and thus cannot be used to explain why the maximum in T_M is found at $DyRh_4B_4$, rather than at
$GdRh_4B_4$, on the basis of RKKY ordering. The depressions of T_c
in Fig. 26 generally lie above the deGennes curve for the lighter RE
ions to the left of Gd and slightly below the deGennes curve for the
heavy RE ions including and to the right of Gd with the exception of
Yb. This may be due to a general decrease in the magnitude of $\mathcal{J}$
with increasing RE atomic number and/or the splitting of the RE ion
energy levels by CEF which is generally larger for the lighter RE
ions. The depressions of T_c for both the Ce and Yb impurities are
anomalously large and may be associated with the Kondo effect (see
Section 2.1.2). The negligibly small depression of T_c for Eu impurities suggests that Eu is trivalent when substituted for Lu in

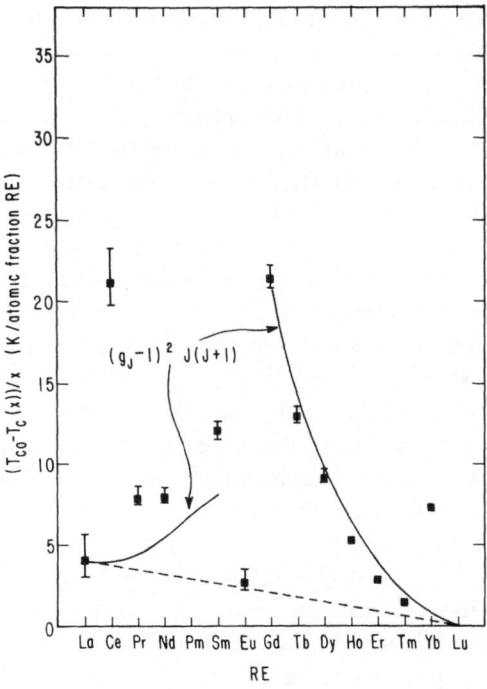

Fig. 26. Rate of depression of T_c vs RE in $(Lu_{1-x}RE_x)Rh_4B_4$. The
dashed line is the assumed baseline upon which the
deGennes curve (solid line), normalized to the depression
rate for RE = Gd, is drawn. The values of the ordinate
are the depressions of T_c in K from the T_c value of
$LuRh_4B_4$ that would be expected if the RE ion totally replaced the Lu ion (from Ref. 89).

$LuRh_4B_4$ with a $4f^6$ configuration and corresponding nonmagnetic
J = 0 Hund's rule ground state.

In principle, the most accurate estimate of the magnitude of
$\mathcal{J}$ can be made from the depression of T_c of $LuRh_4B_4$ by Gd impurities since Gd is an S-state ion and CEF effects can be neglected in
the temperature range of interest. The initial depression of T_c of
$LuRh_4B_4$ by Gd impurities, $(dT_c/dx)_{x=0} = -19$ K per atomic fraction
Gd in Lu, yields the value $|\mathcal{J}| = 2.3 \times 10^{-2}$ eV atom assuming
N(0) = 0.35 states/eV-atom-spin direction.

3.3 Ferromagnetic Superconductors

Two ternary RE compounds, $HoMo_6S_8$[90,91] and $ErRh_4B_4$,[81,82] have been observed to exhibit reentrant superconductive behavior due to the onset of long-range ferromagnetic order. The compounds become superconducting at an upper critical temperature T_{c1} and then lose their superconductivity at a second lower critical temperature T_{c2} that is of the order of the magnetic ordering (Curie) temperature T_M. The superconducting-normal transition at T_{c2} is thermally hysteretic,[81] and there is a spike-shaped feature in the heat capacity at T_{c2} indicating a first order transition.[73,92] In the following, the physical properties of the compound $ErRh_4B_4$, which are similar to those of $HoMo_6S_8$, will be described in some detail.

Typical ac magnetic susceptibility and electrical resistance vs temperature data for $ErRh_4B_4$ are shown in Fig. 27.[93] The thermal hysteresis at T_{c2} is evident in both properties. The failure of the resistance below T_{c2} to attain its full normal state value may be due to the presence of filaments of another phase, although other explanations such as superconducting fluctuations have been suggested.[81,94] Shown in Fig. 28 are electrical resistance vs temperature data for $ErRh_4B_4$ in various magnetic fields between 0 and 12 kOe.[95] The data show that the applied magnetic field has the effect of lowering T_{c1} and raising T_{c2} and changing the transition

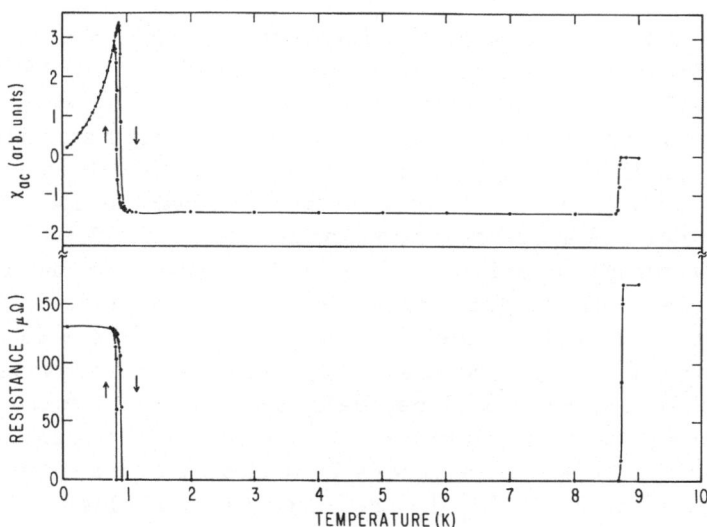

Fig. 27. Typical ac magnetic susceptibility χ_{ac} and electrical resistance vs temperature data for $ErRh_4B_4$ (after Ref. 93).

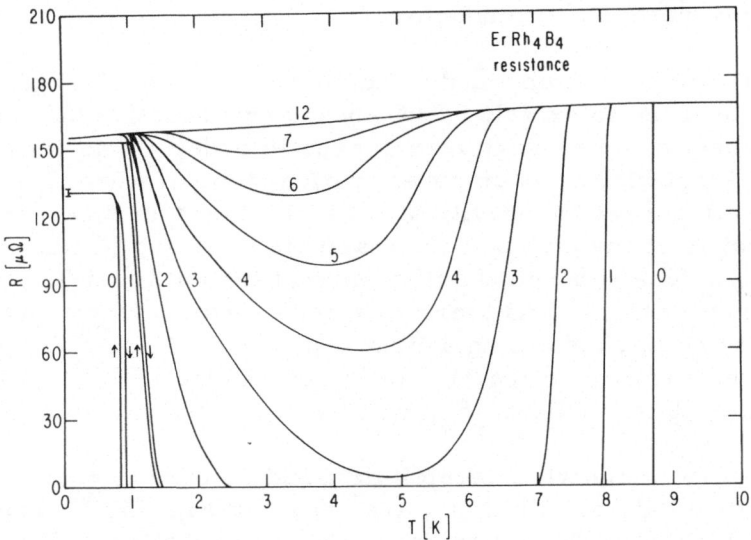

Fig. 28. Electrical resistance R vs temperature T data for
ErRh$_4$B$_4$ in applied magnetic fields between 0 and 12 kOe
(from Ref. 95).

at T_{c2} from first order to second order. The curve of H_{c2} vs temperature deduced from these data is shown in Fig. 41 (see Section
3.5).

Shown in Fig. 29 is a plot of the heat capacity of ErRh$_4$B$_4$ and
the isostructural nonmagnetic compound LuRh$_4$B$_4$ below 18 K.[73] The
data reveal a jump in the heat capacity at T_{c1} = 8.7 K on a broad
background with negative curvature that is a Schottky anomaly arising from the partial lifting by the CEF of the 16-fold degeneracy of
the Er^{3+} J = 15/2 Hund's rule multiplet. At lower temperatures
(see inset), there is a spike-shaped feature at T_{c2} = 0.93 K (measured upon warming) superimposed on another anomaly that is apparently associated with the long-range ferromagnetic ordering of the
Er^{3+} magnetic moments in the vicinity of T_{c2}.[82] There is also a
shoulder in the heat capacity above T_{c2} whose origin may be attrib-
utable to the formation of a sinusoidally modulated magnetic state
that coexists with superconductivity that will be discussed in the
following Section (3.4). The excess heat capacity per mole, $\Delta C/R$,
vs temperature for ErRh$_4$B$_4$ that results after the electronic and
lattice contributions, estimated from the normal state heat capacity
of LuRh$_4$B$_4$, have been subtracted is shown in Fig. 30.[73] The curve
represents a Schottky anomaly that was fitted to the data and

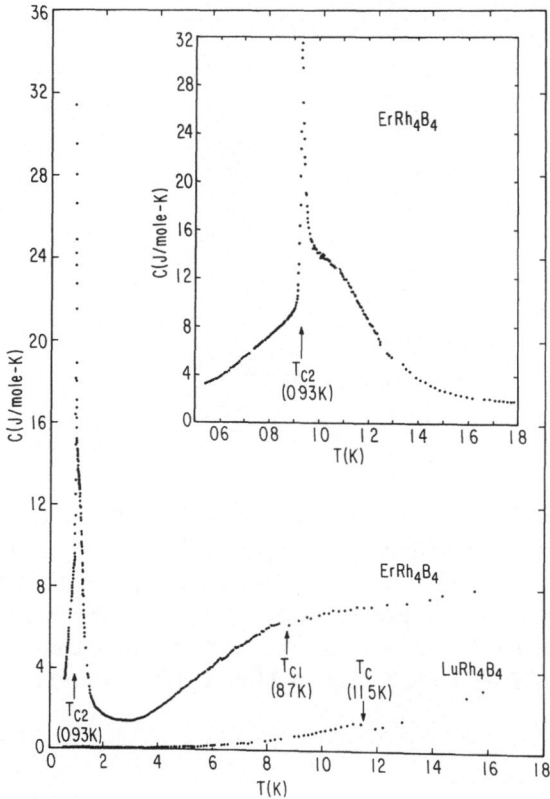

Fig. 29. Specific heat C vs temperature T for $ErRh_4B_4$ and $LuRh_4B_4$. The inset shows a detailed plot of C vs T for $ErRh_4B_4$ in the vicinity of the reentrant superconducting transition at T_{c2} (from Ref. 73).

corresponds to the energy level scheme that is indicated in the figure. The ground state has been taken to be a quartet (or a ground state doublet separated from a nearby excited state doublet by a few K) since the entropy associated with the magnetic anomaly is $\sim R\ln 4$ per mole Er. Schottky anomalies are found in the heat capacity of other $RERh_4B_4$ compounds and demonstrate that CEF effects must be taken into account in order to achieve a detailed understanding of the remarkable physical properties of the $RERh_4B_4$ compounds.[93, 96]

Neutron scattering experiments have been particularly informative in studies of magnetic superconductors. Shown in Fig. 31

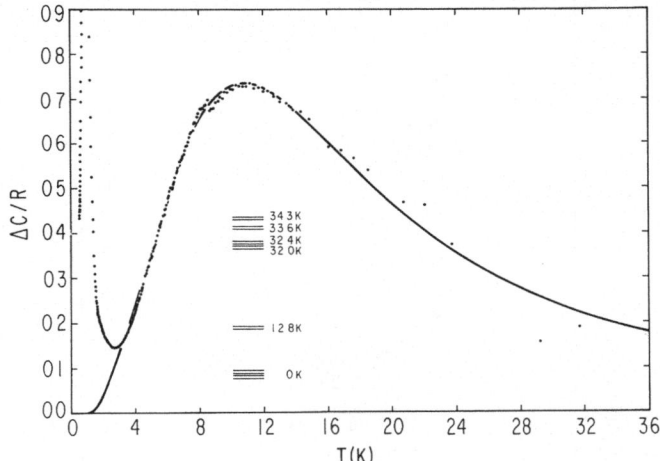

Fig. 30.   Excess heat capacity ΔC in units of the molar gas constant
R vs temperature T for $ErRh_4B_4$. The solid line represents a calculated fit of the Schottky anomaly with a CEF
energy level scheme consisting of a ground state quartet
and excited state doublets at temperatures of 12.8, 32.0,
32.4, 33.6, 34.3 and > 100 K (from Ref. 73).

are the normalized neutron scattering intensity for the (101) reflection and the heat capacity vs temperature for the reentrant superconducting compound $ErRh_4B_4$. [97]   The neutron diffraction data
show that the ferromagnetic transition is broad, extending up to of
the order of 1.4 K, well above the temperature of the reentrant
superconducting transition at $T_{c2} \sim 0.9$ K. Also, there is definite
thermal hysteresis between ~ 0.8 and ~ 1.4 K. The width of the
ferromagnetic transition has been attributed to a distribution of
effective Curie temperatures within the material, while the hysteresis may be caused by the nucleation of normal ferromagnetic domains within the paramagnetic superconducting regions between T_{c2}
and ~ 1.4 K. It should be noted that there are no features in the
neutron scattering intensity vs temperature curve that reflect the
superconducting normal transition at $T_{c2} \sim 0.9$ K other than the disappearance of the hysteretic behavior below T_{c2} (measured upon
cooling). The heat capacity data also reveal thermal hysteresis,
part of which appears to be associated with the formation of ferromagnetic domains discussed above in connection with the neutron
scattering experiments (between ~ 0.9 and ~ 1.3 K) and part of
which is due to the reentrant transition at T_{c2}.

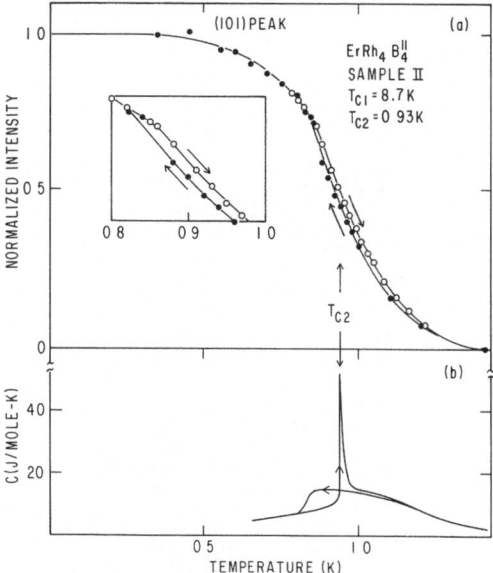

Fig. 31. Ferromagnetic Bragg intensities for $ErRh_4B_4$ compared with the specific heat data of the identical sample (from Ref. 97).

Displayed in Fig. 32 are magnetization M vs magnetic field H data for $ErRh_4B_4$[95] which reveal an evolution of the M vs H curves from Type II towards Type I superconductive behavior with decreasing temperature. Tachiki et al.[98] have explained this with the following argument. For one vortex, the condition for flux quantization of the magnetic induction $b(\underset{\sim}{x})$ is

$$\int b(\underset{\sim}{x})\, d^2x = \int h(\underset{\sim}{x})d^2x + 4\pi \int m(\underset{\sim}{x})\, d^2x = \Phi_0 \tag{34}$$

where $h(\underset{\sim}{x})$ and $m(\underset{\sim}{x})$ are the magnetic field and magnetization, respectively, and $\Phi_0 = hc/2e$ is the flux quantum. As the temperature approaches the Curie temperature, $m(x)$ increases and, therefore, $\int h(\underset{\sim}{x})d^2x$ must decrease, in order to satisfy Eq. (34). This implies that $h(\underset{\sim}{x})$ develops an oscillation with a negative region as shown schematically in Fig. 33. Since the mutual interaction between two vortices is proportional to $h(\underset{\sim}{x})$, the positive and negative regions of $h(\underset{\sim}{x})$ correspond, respectively, to repulsive and attractive forces between the vortices. Thus, as the interaction between the vortices changes from repulsive to attractive with decreasing

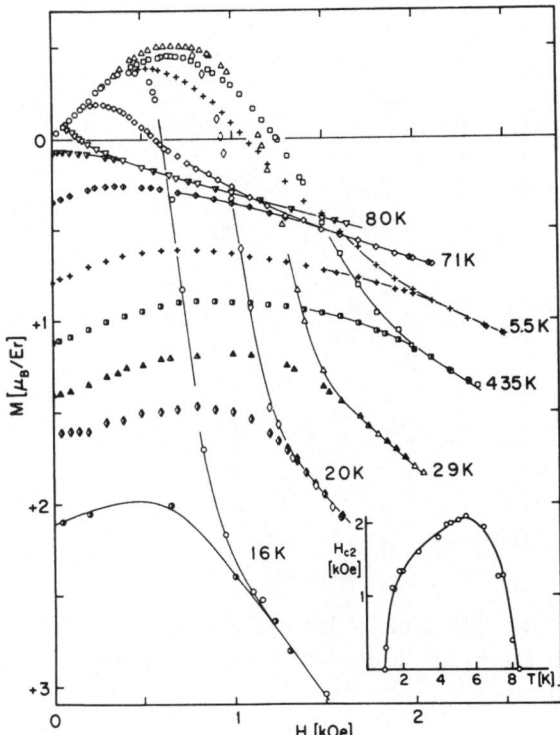

Fig. 32. Low field magnetization M vs magnetic field H isotherms
for $ErRh_4B_4$. The diamagnetic magnetization is plotted
on the positive y-axis. The inset shows the $H_{c2}(T)$ curve
as derived from magnetization and magnetostriction (not
shown) measurements (from Ref. 95).

temperature, the magnetization curves evolve from Type II/2 to
Type II/1 (or even Type I) behavior, as indicated schematically in
Fig. 33. Crabtree et al. [99] have recently made H_{c2} vs T measurements on a single crystal specimen of $ErRh_4B_4$ with the applied
magnetic field oriented both parallel and perpendicular to the tetragonal c-axis. The H_{c2} vs T data with H ⊥ c differ considerably
for those with H ∥ c and reflect the change from Type II towards
Type I behavior as the temperature is lowered.

Many other experimental investigations on $ErRh_4B_4$ have been
carried out in recent years. [15-17] Among these, electron tunnelling
studies have been made by Rowell et al. [100] and Umbach et al. [101] on
polycrystalline thin film specimens through oxide barriers and by
Poppe [102] on single crystals through vacuum. The studies by

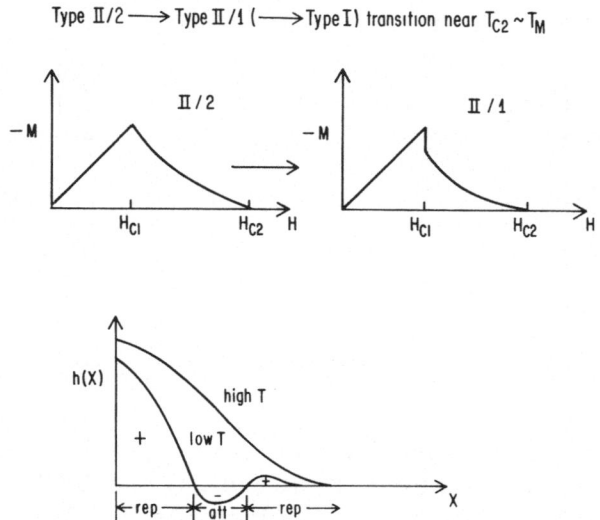

Fig. 33. Schematic diagram of M vs H and h vs x for Type II/2 and Type II/1 behavior.[98]

Umbach et al.[101] and Poppe[102] indicate that $ErRh_4B_4$ is a strong-coupled superconductor, and yield values for $2\Delta/k_BT_c$, of at least 4.2 and 3.8, respectively. Josephson pair tunneling has also been observed by Umbach and Goldman.[103] Their data for the maximum dc Josephson current I_c vs temperature T, along with electrical resistance R vs T data on the same $ErRh_4B_4$ specimen, are shown in Fig. 34.

3.4 Sinusoidally Modulated Magnetic State

Perhaps the most provocative of the neutron scattering results emerged from the small angle scattering studies of Moncton et al.[97] Shown in Fig. 35 are their neutron intensity vs scattering angle data at various temperatures on a polycrystalline $ErRh_4B_4$ sample whose reentrant superconducting normal transition occurs at ~ 0.7 K. A peak near 1° is seen to develop which grows in intensity as the temperature is decreased and then disappears abruptly when the sample becomes normal below T_{c2}. The data have been interpreted in terms of fluctuations into a state proposed by Blount and Varma[104] in which the magnetization is sinusoidally modulated with a wavelength $\lambda \sim 100$ Å that was assumed to take the form of a spiral. These spatially modulated magnetic fluctuations, which coexist with

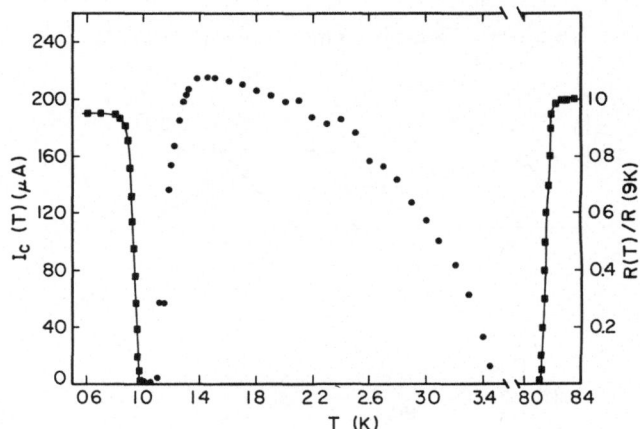

Fig. 34.   Maximum dc Josephson current $I_c(T)$ of an $ErRh_4B_4$-
Lu_xO_y-In junction (filled circles) and $R(T)/R(9\ K)$ (filled
squares) of the $ErRh_4B_4$ electrode (from Ref. 103).

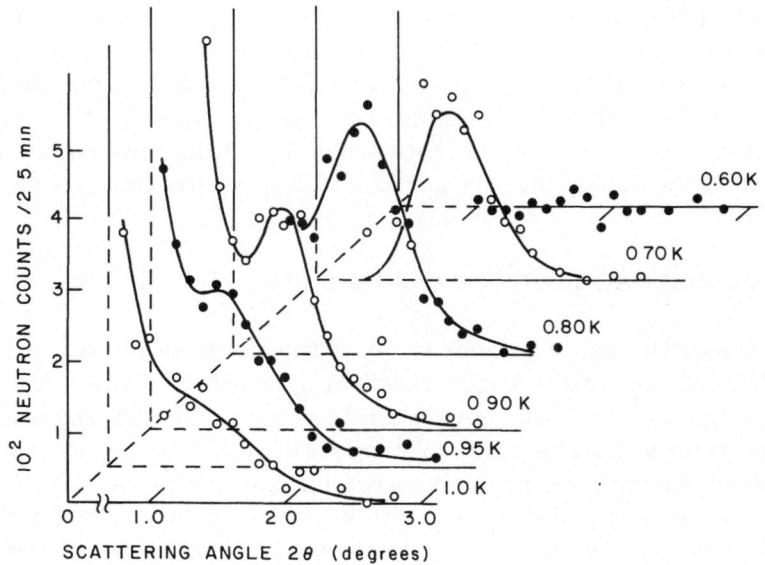

Fig. 35.   Small angle neutron scattering results on $ErRh_4B_4$ obtained at various temperatures.   The peak at $2\theta = 1.4°$
indicates an oscillatory magnetization with a wavelength
of ~ 100 Å (from Ref. 97).

and apparently owe their existence to superconductivity, are reminiscent of the cryptoferromagnetic state suggested by Anderson and
Suhl.[50] Several other theoretical investigations of spiral magnetic
states with wavelengths $\sim 10^2$ Å that coexist with superconductivity
in ferromagnetic superconductors above T_{c2} have recently appeared
in the literature.[105-111] On the other hand, it has also been suggested that the periodic magnetic structure above T_{c2} may actually
be a spontaneous vortex lattice.[112, 113] The formation of a self-
induced vortex lattice in a magnetic superconductor was first considered about a decade ago by Krey.[114] Recently, other possibilities have been examined such as a laminar structure, stabilized by
the RE magnetization in a self-consistent manner,[14] and combined
spiral-magnetic and spontaneous vortex states.[115] Similar small
angle scattering results have been obtained on polycrystalline specimens of the reentrant ferromagnetic superconductor $HoMo_6S_8$.[116, 117]

Quite recently, neutron diffraction and electrical resistivity
measurements were carried out on a single crystal of $ErRh_4B_4$ by
Sinha et al.[118] Their findings are consistent with the neutron scattering experiments on polycrystalline $ErRh_4B_4$ by Moncton et al.,[97]
and revealed some important new features as well. From the data
on the single crystal, Sinha et al.[118] concluded that superconductivity and long-range ferromagnetic order coexist, but in a spatially
inhomogeneous manner between 0.71 K (T_{c2} measured upon cooling)
and 1.2 K. Moreover, they found that the sinusoidally-modulated
magnetic state is a transverse linearly polarized long-range magnetic structure with a wavelength of ~ 100 Å. The linearly polarized
sinusoidal modulation lies along the [010] axis and the propagation
directions are at 45° to the [001] and each of the [100] and [010]
axes. With decreasing temperature, the modulated magnetic moment increases faster than the ferromagnetic moment and then disappears abruptly with the loss of superconductivity and transition
to the normal ferromagnetic state. The higher temperature transition between the purely superconducting and magnetically-ordered
phases appears to be continuous. Greenside et al.[119] have shown
that when the magnetic anisotropy is sufficiently strong, a linearly
polarized sinusoidal state can be favored over both spiral and
vortex states.

Several of the recent theories dealing with oscillatory magnetic states that coexist with superconductivity are based on the
electromagnetic interaction between the RE magnetic moments and
the persistent current.[104, 108-111] The treatment by Tachiki et
al.[109] emphasizes the q-dependence of the exchange interaction and

will be briefly discussed. The molecular field $h_m(q)$ acting on a RE ion can be expressed as the sum of the exchange field and the magnetic field $h(q)$ generated by the persistent current

$$h_m(q) = \gamma(q)\, m(q) + h(q) = \gamma(q)\, m(q) - 4\pi F(q)\, m(q)$$

$$= [\gamma(q) - 4\pi F(q)]\, m(q) = \tilde{\gamma}(q)\, m(q) \qquad (35)$$

where $\gamma(q)$ is the exchange interaction constant which includes all the contributions from magnetic dipolar, RKKY and superexchange interactions, and $m(q)$ is the RE magnetization. The normal state exchange constant $\gamma(q)$ is parameterized by

$$\gamma(q) = (T_M - Dq^2)/C \qquad (36)$$

where T_M is the Curie temperature and C is the Curie constant, while the function $F(q)$ is given by

$$F(q) = \frac{\exp(-\xi^2 q^2/2)}{\lambda_L^2 q^2 + \exp(-\xi^2 q^2/2)} \qquad (37)$$

where λ_L is the London penetration depth. Equation (35) implies that the effective exchange interaction constant in the superconducting state is given by

$$\tilde{\gamma}(q) = \gamma(q) - 4\pi F(q) \qquad (38)$$

where the second term describes the screening of $\gamma(q)$ at long wavelengths by the persistent current. The q-dependence of $\gamma(q)$ and $\tilde{\gamma}(q)$ is depicted schematically in Fig. 36; whereas $\gamma(q)$ attains a maximum at $q = 0$, $\tilde{\gamma}(q)$ has a maximum at $Q \sim (4\pi C/D)^{1/4}\lambda_L^{-1/2}$ which indicates that periodic magnetic order with wave number Q will develop in the superconducting state. The free energies of the oscillatory magnetic, ferromagnetic normal, and superconducting paramagnetic states, normalized so that the energy of the paramagnetic normal state is zero, are schematically indicated in Fig. 36. The superconducting paramagnetic to superconducting oscillatory magnetic state transition is second order, while the superconducting oscillatory magnetic to ferromagnetic normal state transition is first order, in accord with experiment.

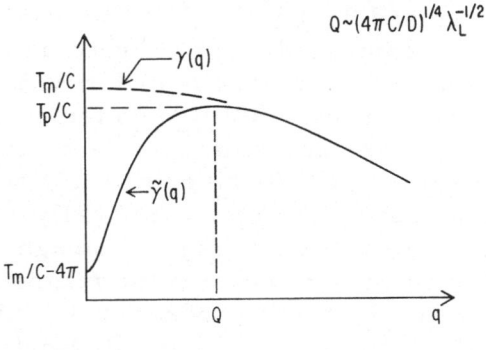

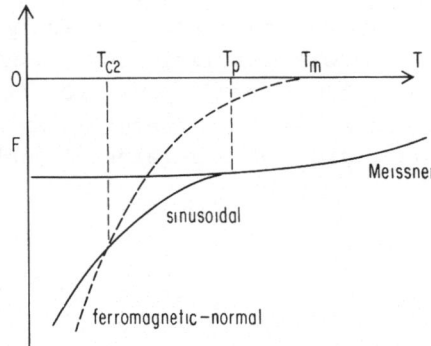

Fig. 36.   Schematic representation of the screened exchange interaction $\tilde{\gamma}(q)$ and the superconducting paramagnetic,
oscillatory magnetic superconducting, and ferromagnetic
normal free energies (after Ref. 120).

Although the exchange interaction is operative in these materials, as discussed earlier, the electromagnetic interaction appears
to be primarily responsible for the sinusoidally-modulated magnetic
state that coexists with superconductivity.   This conclusion follows
from estimates of the wavelength λ of the sinusoidally modulated
magnetic state in $ErRh_4B_4$ which are $\leqslant 10$ Å for the exchange interaction and $\sim 10^2$ Å for the electromagnetic interaction.

3.5 Antiferromagnetic Superconductors

Coexistence of superconductivity and long-range antiferromagnetic order was initially observed in the RE molybdenum chalcogenides $REMo_6S_8$ (RE = Gd, Tb, Dy, and Er)[17,62] and $REMo_6Se_8$
(RE = Gd, Tb and Er)[17,121] and later in the RE rhodium borides

RERh$_4$B$_4$ (RE = Nd, Sm and Tm).[17,83-87] The occurrence of magnetic order in REMo$_6$S$_8$ compounds was inferred by Ishikawa and
Fischer[62] from a feature at the magnetic ordering temperature T$_M$
in the curve of the upper critical field H$_{c2}$ vs temperature. Low
temperature magnetization measurements[122] and neutron diffraction
measurements[123-126] on the REMo$_6$S$_8$ compounds revealed that the
RE magnetic moments order antiferromagnetically. The magnetic
structure is characterized by alternating ferromagnetic (100) planes
in the nearly cubic RE sublattice in which the magnetic moments are
parallel or antiparallel to the rhombohedral [111] axis. Plots of the
sublattice magnetization vs temperature for REMo$_6$S$_8$ compounds
with RE = Gd, Tb and Dy are shown in Fig. 37 where they are compared with the data for H$_{c2}$ vs temperature that are displayed in the
inset of the figure. Ishikawa and Fischer[62] analyzed their H$_{c2}$ vs T
data for the REMo$_6$S$_8$ antiferromagnetic superconductors in terms
of the multiple pair breaking theory[60] and extracted an additional
pair breaking parameter whose temperature dependence resembles
that of the antiferromagnetic order parameter (sublattice magnetization).

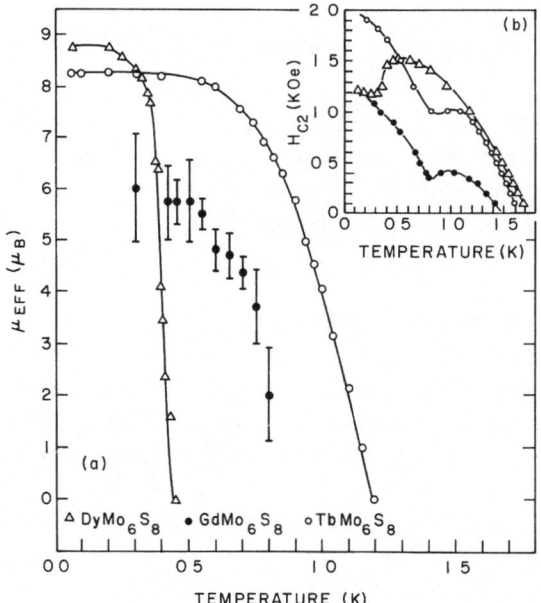

Fig. 37. Temperature dependence of (a) the magnetic moment μ_{eff}
         as determined from neutron diffraction measurements,
         and (b) upper critical field H$_{c2}$ of REMo$_6$S$_8$ compounds
         with RE = Gd, Tb and Dy (after Ref. 126).

In the $REMo_6Se_8$ compounds, the long-range antiferromagnetic ordering in the superconducting state was inferred from pronounced lambda-type specific heat anomalies and cusp-like features in the magnetic susceptibility at T_M.[121,127,128] Specific heat data for the compound $GdMo_6Se_8$[128] are shown in Fig. 38, while magnetic susceptibility data for the same compound[121] are shown in Fig. 39. Magnetic susceptibility measurements in the superconducting state are possible in many of the $REMo_6Se_8$ compounds because they are extreme Type II superconductors that permit nearly complete penetration of the magnetic fields in which the magnetization measurements are made.[71] Neutron scattering measurements have confirmed the development of long-range antiferromagnetic order in $GdMo_6Se_8$,[129] but have not succeeded in establishing the type of magnetic ordering that occurs in $ErMo_6Se_8$ because of complications associated with impurity phases.[130]

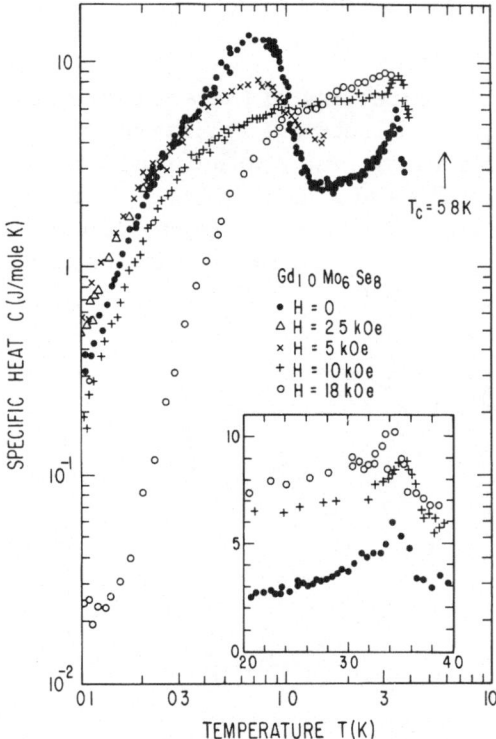

Fig. 38. Specific heat C vs temperature T of $GdMo_6Se_8$ in several applied magnetic fields between 0 and 18 kOe. The arrow denotes the superconducting critical temperature $T_c = 5.8$ K (from Ref. 128).

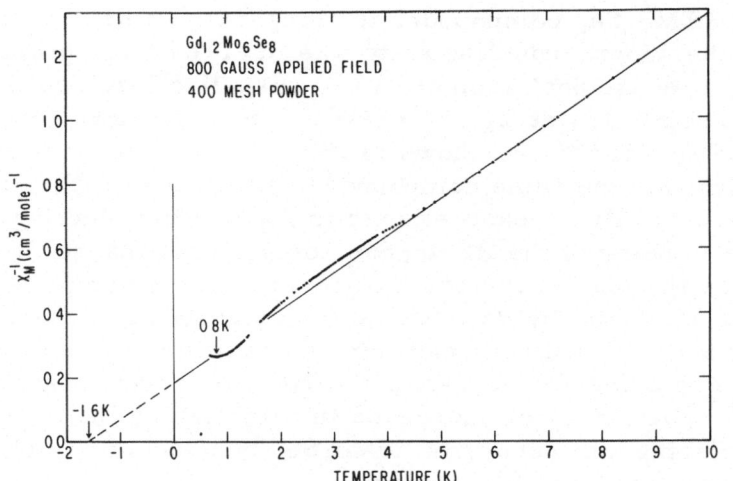

Fig. 39.   Inverse molar magnetic susceptibility χ_M^{-1} vs temperature
for $Gd_{1.2}Mo_6Se_8$ between 0.7 and 10 K (from Ref. 121).

One of the most interesting of the antiferromagnetic $RERh_4B_4$
compounds is $NdRh_4B_4$.   Electrical resistance vs temperature data
for a $NdRh_6B_6$ sample in various applied magnetic fields are shown
in Fig. 40 (the excess Rh and B is required in order to stabilize the
desired $NdRh_4B_4$ phase).[83]   In magnetic fields between 3 and 6 kOe,
the sharp increase in resistance near 1.3 K and the sharp decrease

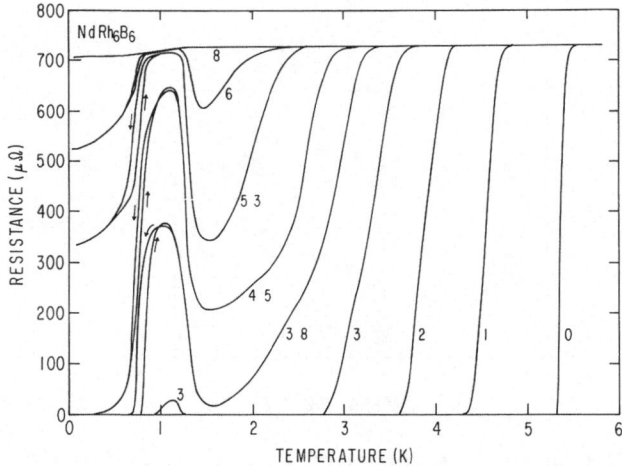

Fig. 40.   ac electrical resistance vs temperature for $NdRh_6B_6$
in various applied magnetic fields between 0 and 8 kOe
(from Ref. 83).

in resistance in the neighborhood of 0.9 K suggest the occurrence of two separate antiferromagnetic transitions. Heat capacity measurements in zero applied magnetic field reveal two lambda-type anomalies that peak at T_{N1} = 1.31 K and T_{N2} = 0.89 K, respectively, indicating that two magnetic phase transitions are associated with the superconducting to normal state transitions below T_c which are manifested in the resistance data. The presence of thermal hysteresis in the R vs T data near T_{N2} suggests that the lower temperature transition is first order, whereas the absence of any such hysteresis near T_{N1} would imply that a second order transition occurs at this temperature.

By defining the superconducting transition temperature as the temperature at which R is 50% of its normal state value, the upper critical field H_{c2} vs T data for $NdRh_4B_4$ shown in Fig. 41 have been obtained.[83] The critical field curve shows an abrupt depression near T_{N1} followed by a rapid recovery below T_{N2}, although $H_{c2}(0)$ = 5.4 kOe remains below the value of 6.5 kOe expected from an extrapolation of the data above 1.6 K. Therefore the onset of magnetic order at T_{N1} causes additional pair breaking, whereas the magnetic structure below T_{N2} is more compatible with superconductivity.

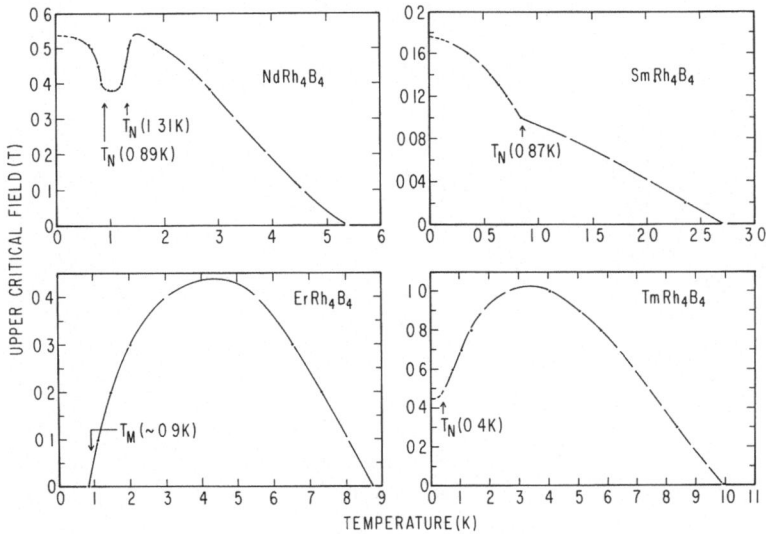

Fig. 41. Upper critical magnetic field vs temperature for $NdRh_4B_4$, $SmRh_4B_4$, $ErRh_4B_4$ and $TmRh_4B_4$ (after Ref. 96).

Neutron diffraction experiments[84] indicate that the magnetic phases of $NdRh_4B_4$ in zero applied magnetic field are body-centered tetragonal antiferromagnetic structures in which the Nd^{3+} moments are alternately parallel and antiparallel to the c-axis, with a sinusoidal modulation along the [100] direction with $\lambda = 46.5$ Å in the high temperature magnetic phase, and along the [110] direction with $\lambda = 45.2$ Å in the low temperature magnetic phase. The temperature dependences of the neutron diffraction intensity of a representative satellite from each of the two magnetic phases of $NdRh_4B_4$ are shown in Fig. 42.[84]

Also shown in Fig. 41 are resistively determined H_{c2} data for the other two antiferromagnetic $RERh_4B_4$ superconductors, $SmRh_4B_4$[85] and $TmRh_4B_4$,[86] as well as the ferromagnetic $RERh_4B_4$ superconductor, $ErRh_4B_4$.[95] Whereas the $NdRh_4B_4$ data show both a suppression and an enhancement of H_{c2} below the two Néel temperatures, H_{c2} of $SmRh_4B_4$ is enhanced below the Néel temperature, while H_{c2} of $TmRh_4B_4$ shows very little change below the Néel temperature. This indicates that there is no universal behavior of H_{c2} vs T for

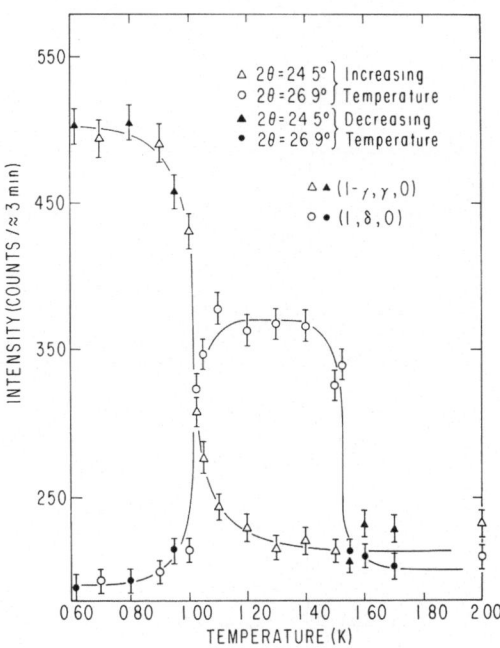

Fig. 42. Temperature dependence of the neutron scattering intensity of a representative satellite from each of the two magnetic phases of $NdRh_4B_4$ (from Ref. 84).

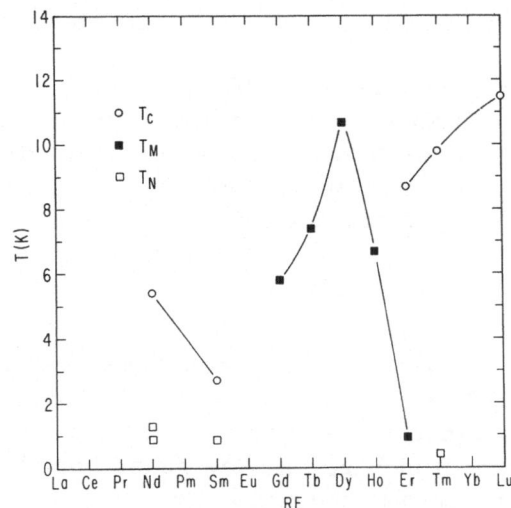

Fig. 43. Superconducting T_c, ferromagnetic T_M, and antiferromagnetic T_N transition temperatures vs RE for $RERh_4B_4$
compounds (after Ref. 96).

antiferromagnetic superconductors. Both enhancements and depressions of H_{c2} are found below T_N, which appear to be determined
by a combination of the mechanisms enumerated below.

The anomalous depression of H_{c2} in the vicinity of T_N and
other properties of antiferromagnetic superconductors have been
addressed recently by numerous theories.[131] Several mechanisms
by means of which superconductivity is modified by antiferromagnetic order have been considered. These include (1) the reduction
in pair breaking due to the decrease in the mean magnetization and,
in turn, the conduction electron spin polarization below T_N; (2) the
increase in pair breaking due to the magnetic moment fluctuations
in the vicinity of T_N; (3) the decrease of the attractive phonon mediated electron-electron pairing interaction by antiferromagnetic
magnons; (4) the reduction of the available phase space for virtual
pair scattering by the change in lattice periodicity associated with
antiferromagnetic order; and (5) the pairing of electrons with finite
momentum. Finite momentum pairing of electrons into states
$(\underset{\sim}{k}\uparrow, -\underset{\sim}{k}+\underset{\sim}{Q}\downarrow)$, where $\underset{\sim}{Q}$ corresponds to translation by a reciprocal
lattice vector, was originally proposed by Baltensberger and
Strässler[61] in 1963, as discussed earlier, and has since been incorporated into several recent theories.[131]

3.6 Superconductivity and Competing Magnetic Interactions

Experiments on pseudoternary RE compounds provide an alternate method for studying the interaction between superconductivity and long-range magnetic order as well as for exploring the effects of competing types of magnetic moment anisotropy and/or magnetic order. Two types of $RERh_4B_4$ pseudoternaries have been formed, one in which a second RE element is substituted at the RE sites, and another in which a different transition element is substituted at the Rh sites.

An example of the first type of pseudoternary $RERh_4B_4$ system is $(Er_{1-x}Ho_x)Rh_4B_4$ whose low temperature phase diagram, delineating the paramagnetic, superconducting, and magnetically ordered phases, is shown in Fig. 44. The phase boundaries have been determined from ac magnetic susceptibility[96,132] and neutron diffraction measurements.[133,134] The phase diagram displays regions in which the Er^{3+} and Ho^{3+} magnetic moments independently order ferromagnetically within the basal plane and along the tetragonal c-axis, respectively, separated by a region of mixed magnetic phases. The temperature interval above T_{c2} in which the sinusoidally modulated magnetic phase in $ErRh_4B_4$ coexists with normal ferromagnetic domains is also indicated in the figure.   This inhomogeneous

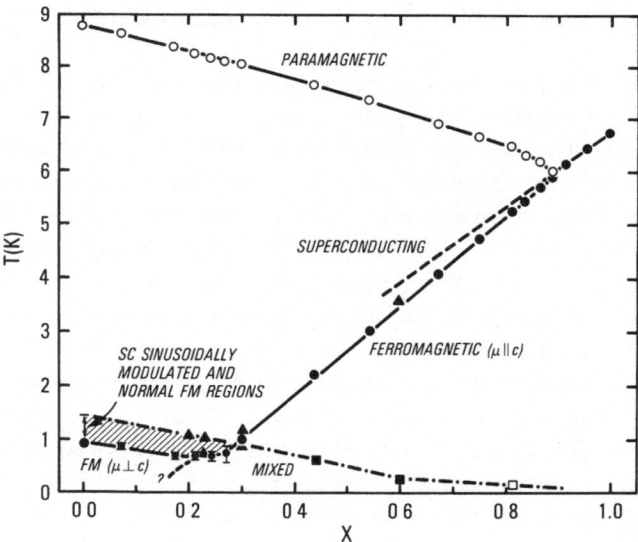

Fig. 44.   Low temperature phase diagram for the system
           $(Er_{1-x}Ho_x)Rh_4B_4$ determined from ac magnetic susceptibility[96,132] and neutron diffraction[133,134] measurements.

phase presumably persists within a certain region in the T-x plane (shaded area in the figure). There is a tricritical point at the concentration x_c = 0.89 at which T_{c1}, T_{c2} and T_M become coincident. The T_{c2} vs x phase boundary for x smaller than x_c is depressed relative to the linear extrapolation of T_M vs x for $x > x_c$ (dashed curve in Fig. 44). Analysis of neutron diffraction data on an $(Er_{0.4}Ho_{0.6})Rh_4B_4$ sample indicates that the actual T_M of 3.67 K is about 0.2 K less than would have occurred in the absence of superconductivity, in accord with the dashed line extrapolation as well as with theoretical predictions.[135] The temperature dependence of the magnetization M(T) of $(Er_{0.4}Ho_{0.6})Rh_4B_4$, derived from the (101) peak intensity, is shown in Fig. 45. The solid line in the figure is calculated from mean-field theory.

Specific heat measurements performed on selected $(Er_{1-x}Ho_x)$-Rh_4B_4 samples by MacKay et al.[92] are shown in Fig. 46. The mean-field like behavior of $HoRh_4B_4$ greatly simplifies the interpretation of these data. For x = 0.912, just above x_c, the anomaly associated with the magnetic transition has a shape similar to that of $HoRh_4B_4$. The specific heat anomaly of $HoRh_4B_4$ due to ferromagnetic ordering is shown in Fig. 47, where it is compared to the

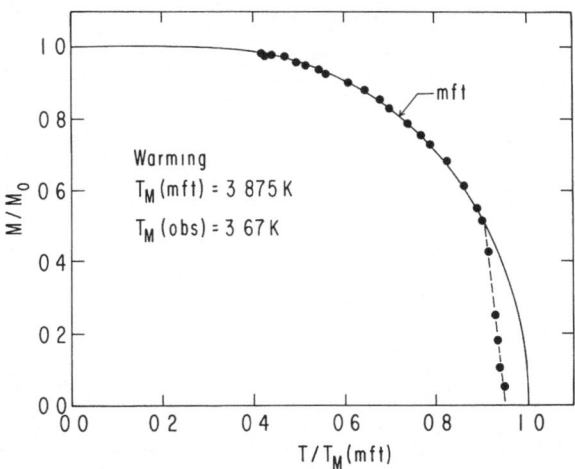

Fig. 45. Temperature dependence of the magnetization M(T) derived from the (101) peak intensity for $(Ho_{0.6}Er_{0.4})Rh_4B_4$. The solid curve is calculated from mean-field theory (from Ref. 135).

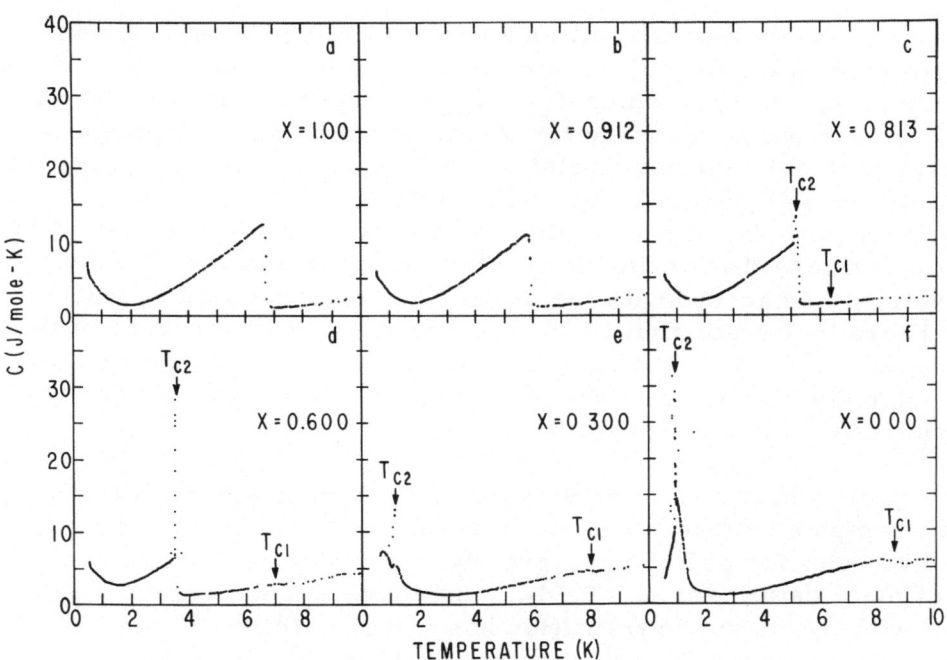

Fig. 46. Heat capacity C vs temperature for certain members of
the $(Er_{1-x}Ho_x)Rh_4B_4$ system. The arrows indicate the
upper (T_{c1}) and lower (T_{c2}) superconducting-to-normal
state transition as measured by ac magnetic susceptibility
(from Ref. 92).

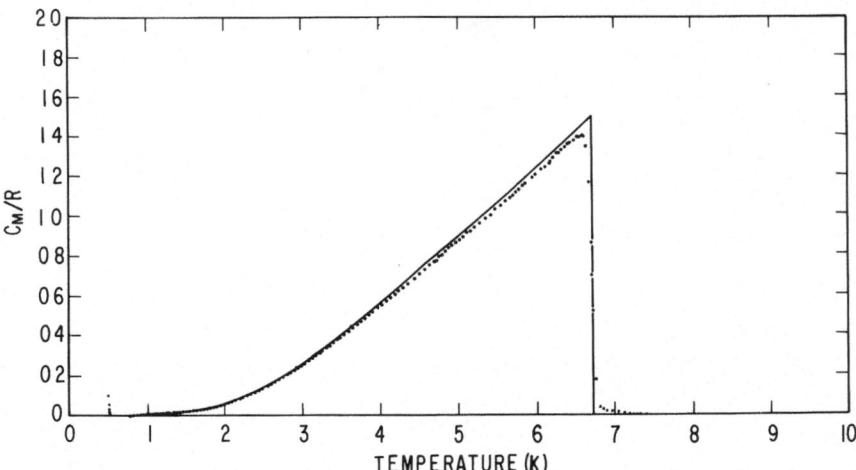

Fig. 47. Temperature dependence of the specific heat anomaly of
$HoRh_4B_4$ due to the ferromagnetic ordering. The solid
line is the mean-field result for an $S = 1/2$ ferromagnet
(from Ref. 136).

predictions of mean-field theory for spin one-half.[136] However, at $x = 0.813$, just below x_c, a spike-shaped feature appears to be superimposed on top of the mean-field like magnetic transition. Since the sample with $x = 0.813$ is reentrant, whereas that with $x = 0.912$ is purely ferromagnetic, the spike-shaped features must be associated with the first-order reentrant superconducting to normal ferromagnetic transition at T_{c2}. For the four superconducting members of the system presented in Fig. 46, the spike-shaped feature was coincident with the destruction of superconductivity as measured by ac magnetic susceptibility. The low temperature heat capacity of $(Er_{0.7}Ho_{0.3})Rh_4B_4$ shows two maxima indicative of two distinct orderings. Also, Schottky anomalies present in each of the pseudoternary compounds could be described as the weighted sum of the respective Schottky anomalies for $ErRh_4B_4$ and $HoRh_4B_4$, indicating the additive nature of the crystal field effects.[137]

A striking example of the second type of pseudoternary $RERh_4B_4$ system is $Ho(Rh_{1-x}Ir_x)_4B_4$. This system was first investigated by Ku et al.[138] and provided evidence for the coexistence of superconductivity and antiferromagnetic order with $T_N > T_c$ for $x \geqslant 0.6$. Subsequently, several detailed investigations of a Ho-$(Rh_{0.3}Ir_{0.7})_4B_4$ compound were carried out; heat capacity measurements[139] confirmed the bulk character of the antiferromagnetic order in this material, while neutron diffraction experiments[140] revealed the magnetic structure. Recently, low temperature specific heat, ac magnetic susceptibility and electrical resistance measurements were used to investigate the nature of magnetic ordering and the dependence of the magnetic ordering temperature on x in the $Ho(Rh_{1-x}Ir_x)_4B_4$ system.[141] The low temperature phase diagram that emerged from this study is shown in Fig. 48. The first study of this system yielded a similar phase diagram, but with no evidence for magnetic ordering at temperatures $T > 1.2$ K for $0.2 < x < 0.6$.

Shown in Fig. 49 are low temperature specific heat data for the $Ho(Rh_{1-x}Ir_x)_4B_4$ system from which the low temperature phase diagram of Fig. 48 has been partially derived.[141] The anomalies due to magnetic ordering for the compounds with $x = 0$ and $x = 0.15$ have a mean-field ferromagnetic shape, whereas the anomalies for the compounds with $x = 0.25$ to 0.70 have a different shape which is indicative of antiferromagnetic order. Particularly interesting are the curves for $x = 0.30$ and 0.35 where there are two peaks, suggesting that there are two types of antiferromagnetic order, similar to what occurs in $NdRh_4B_4$. The antiferromagnetic structure of the

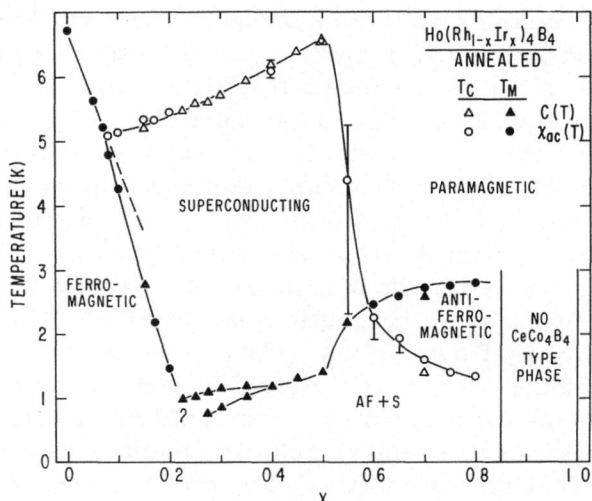

Fig. 48.   Low temperature phase diagram for the system
Ho(Rh$_{1-x}$Ir$_x$)$_4$B$_4$ determined from ac magnetic susceptibility and heat capacity measurements (from Ref. 141).

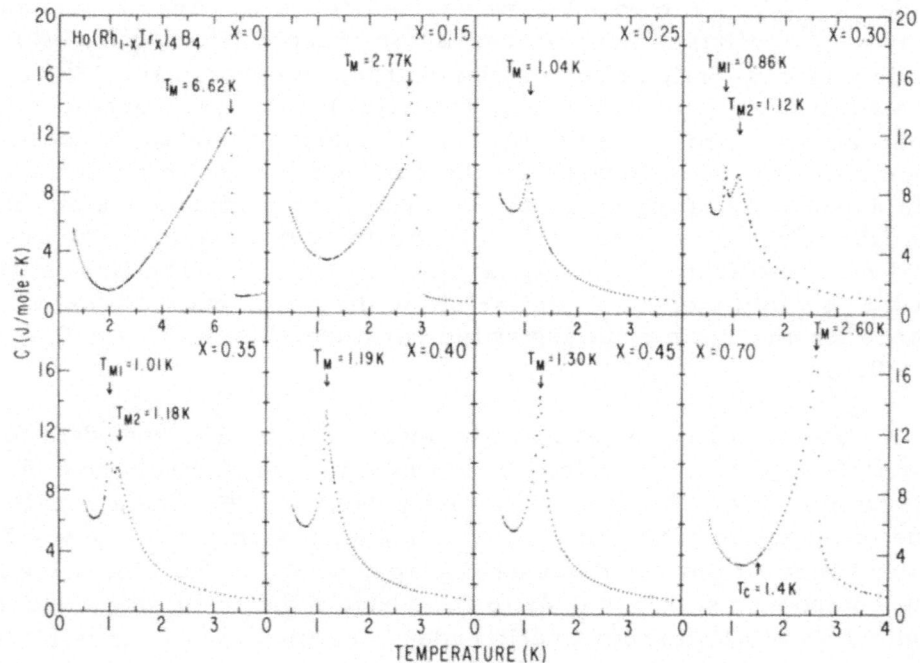

Fig. 49.   Heat capacity C vs temperature for various Ho(Rh$_{1-x}$-
Ir$_x$)$_4$B$_4$ compounds (from Ref. 141).

compound with $x = 0.7$, determined by neutron diffraction measurements, is shown in Fig. 50.[140]

As a final example, the low temperature phase diagram of the $(Sm_{1-x}Er_x)Rh_4B_4$ system, which has been deduced from ac magnetic susceptibility and specific heat data, is displayed in Fig. 51.[142] Resistively determined H_{c2} vs temperature data for this system, measured by Lambert et al., are shown in Fig. 52.[143]

4. CONCLUDING REMARKS

In these lectures, the evolution of the subject of the interplay between superconductivity and magnetism, particularly the coexistence of these two phenomena, has been traced during the past two and a half decades. The major developments and key issues have been emphasized in order to provide the reader with an overview of

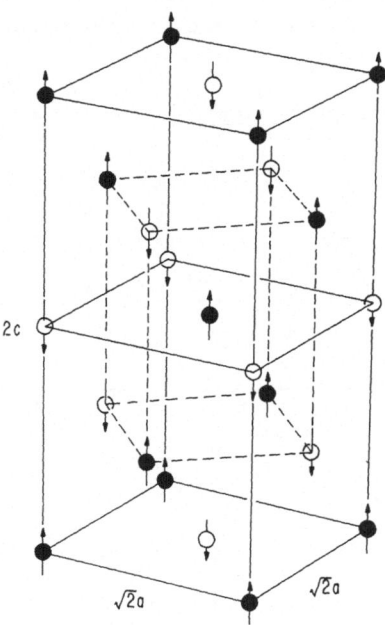

Fig. 50. Proposed magnetic unit cell for $Ho(Rh_{0.3}Ir_{0.7})_4B_4$. The crystallographic unit cell is outlined in dashed lines and the Rh, Ir and B atoms have been removed for clarity (from Ref. 140).

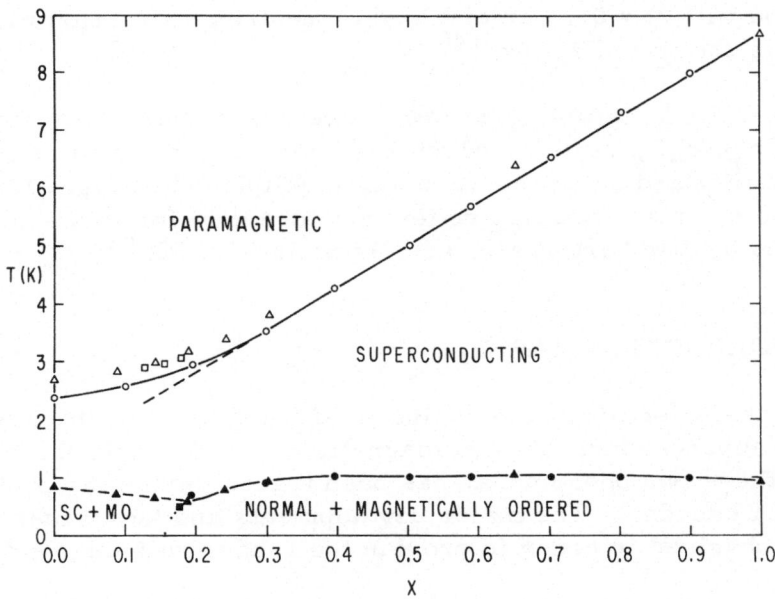

Fig. 51.   Low temperature phase diagram for the system
           $(Sm_{1-x}Er_x)Rh_4B_4$ determined from ac magnetic sus-
           ceptibility and heat capacity measurements (from
           Ref. 142).

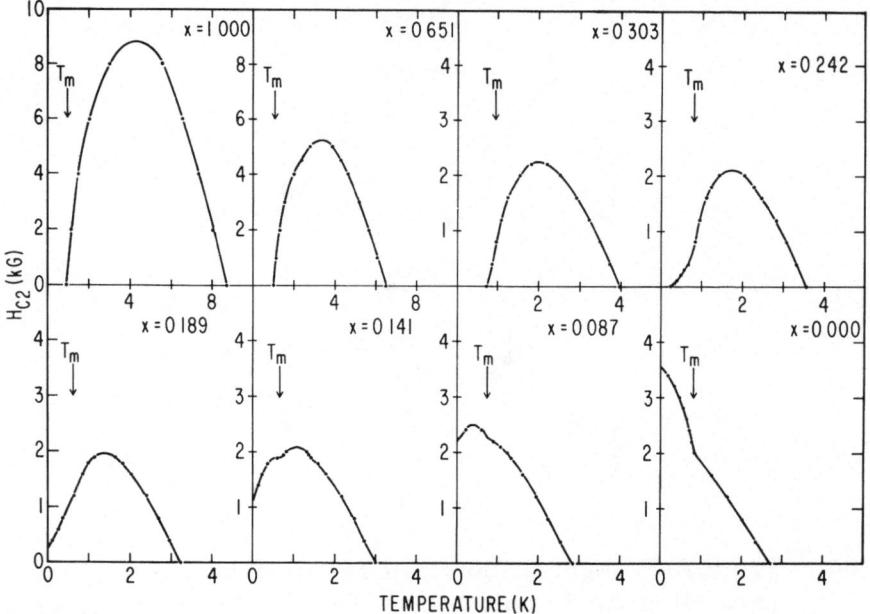

Fig. 52.   Upper critical field H_{c2} vs temperature for various
           $(Sm_{1-x}Er_x)Rh_4B_4$ compounds (from Ref. 143).

this intriguing field and the sense of excitement that has pervaded it since its inception in the late 1950's. Thus, this chapter does not constitute a complete review of the subject; the reader is referred to several review articles and many of the original references that have been cited herein for further information and bibliographies. For example, some very interesting recent developments have been omitted which are somewhat beyond the scope of this article. These include the observation of superconductivity in two rather unusual substances, a "Kondo-lattice" heavy fermion system, $CeCu_2Si_2$,[144] and a weakly magnetic material, Y_9Co_7.[145]

Looking back over the past twenty-five years or so, it is fair to say that some truly remarkable phenomena are associated with interacting superconducting and magnetic order parameters. In view of the rapid progress involving the ternary RE compounds after 1976, this subject should fascinate and challenge experimentalists and theoreticians alike for some time to come.

ACKNOWLEDGMENT

The author would like to express his appreciation to Nancy McLaughlin and Annetta Whiteman for their invaluable assistance in the preparation of this manuscript.

REFERENCES

1. V. L. Ginzburg, Sov. Phys. JETP $\underline{4}$, 153 (1957).
2. B. T. Matthias, H. Suhl and E. Corenzwit, Phys. Rev. Lett. $\underline{1}$, 92 (1958).
3. B. T. Matthias, H. Suhl and E. Corenzwit, Phys. Rev. Lett. $\underline{1}$, 449 (1958).
4. H. Suhl, B. T. Matthias and E. Corenzwit, J. Phys. Chem. Solids $\underline{19}$, 346 (1959).
5. M. B. Maple, in "MAGNETISM: A Treatise on Modern Theory and Materials," H. Suhl. ed., Vol. V, pp. 289-325 (Academic Press, New York, 1973).
6. M. B. Maple, Appl. Phys. $\underline{9}$, 179 (1976).
7. M. A. Jensen and H. Suhl, in "MAGNETISM: A Treatise on Modern Theory and Materials," G. T. Rado and H. Suhl, eds., Vol. IIB, pp. 183-214 (Academic Press, New York, 1966).

8. Ø. Fischer and M. Peter, in "MAGNETISM: A Treatise on
    Modern Theory and Materials," H. Suhl, ed., Vol. V,
    pp. 327-352 (Academic Press, New York, 1966).
9. R. D. Parks, in "Superconductivity," P. R. Wallace, ed.,
    Vol. 2, pp. 625-690 (Gordon and Breach, New York, 1966).
10. S. Roth, Appl. Phys. $\underline{15}$, 1 (1978).
11. M. B. Maple, J. Physique $\underline{39}$, C6-1374 (1978); M. Ishikawa,
    Ø. Fischer and J. Müller, J. Physique $\underline{39}$, C6-1379 (1978).
12. Ø. Fischer, Appl. Phys. $\underline{16}$, 1 (1978).
13. M. B. Maple, in "Science and Technology of Rare Earth
    Materials," E. C. Subbarao and W. E. Wallace, eds.,
    pp. 167-193 (Academic Press, New York, 1980).
14. M. Tachiki, Physica $\underline{109\ \&\ 110B}$, 1699 (1982).
15. "Ternary Superconductors," G. K. Shenoy, B. D. Dunlap and
    F. Y. Fradin, eds. (North Holland, Amsterdam, 1981).
16. "Superconductivity in Ternary Compounds I," Topics in Cur-
    rent Physics, Vol. 32, Ø. Fischer and M. B. Maple, eds.
    (Springer, Berlin, Heidelberg, New York, 1982).
17. "Superconductivity in Ternary Compounds II," Topics in Cur-
    rent Physics, Vol. 34, M. B. Maple and Ø. Fischer, eds.
    (Springer, Berlin, Heidelberg, New York, 1982).
18. C. Herring, Physica $\underline{24}$, S 184 (1958).
19. H. Suhl and B. T. Matthias, Phys. Rev. $\underline{114}$, 977 (1959).
20. M. Wilhelm and B. Hillenbrand, J. Phys. Chem. Solids $\underline{31}$,
    559 (1970); Phys. Lett. A $\underline{31}$, 448 (1970); Z. Naturforsch.
    Teil A $\underline{26}$, 141 (1971).
21. M. Peter, P. Donzé, Ø. Fischer, A Junod, J. Ortelli,
    A. Treyvaud, E. Walker, M. Wilhelm and B. Hillenbrand,
    Helv. Phys. Acta $\underline{44}$, 345 (1971).
22. R. D. Taylor, W. R. Decker, D. J. Erickson, A. L. Giorgi,
    B. T. Matthias, C. E. Olsen and E. G. Szklarz, in
    "International Conference on Low Temperature Physics,
    Boulder, Colorado, 1972, W. J. O'Sullivan, K. D.
    Timmerhaus and E. F. Hammel, eds., Vol. 2, pp. 605-
    609 (Plenum Press, New York, 1974).
23. K. Kumagai, T. Matsuhira and K. Asayama, J. Phys. Soc.
    Japan $\underline{45}$, 422 (1978).
24. J. W. Lynn, D. E. Moncton, L. Passel and W. Thomlinson,
    Phys. Rev. B $\underline{21}$, 70 (1980).
25. A. A. Abrikosov and L. P. Gor'kov, Sov. Phys. JETP $\underline{12}$,
    1243 (1961).
26. S. Skalski, O. Betbeder-Matibet and P. R. Weiss, Phys.
    Rev. A $\underline{136}$, 1500 (1964).

27. M. A. Woolf and F. Reif, Phys. Rev. A 137, 557 (1965).

28. M. B. Maple, Phys. Lett. A 26, 513 (1968).

29. C. A. Luengo and M. B. Maple, Solid State Commun. 12, 757 (1973).

30. W. R. Decker and D. K. Finnemore, Phys. Rev. 172, 430 (1968).

31. M. B. Maple, Solid State Commun. 8, 1915 (1970).

32. J. Kondo, Progr. Theoret. Phys. 32, 37 (1964).

33. A. J. Heeger, Solid State Phys. 23, 283 (1969).

34. See various Chapters in MAGNETISM: A Treatise on Modern Theory and Materials, H. Suhl, ed., Vol. V (Academic Press, New York, 1973).

35. E. Müller-Hartmann and J. Zittartz, Phys. Rev. Lett. 26, 428 (1971).

36. A. Ludwig and M. J. Zuckermann, J. Phys. F 1, 516 (1971).

37. E. Müller-Hartmann and J. Zittartz, Solid State Commun. 11, 401 (1972).

38. E. Müller-Hartmann, in "MAGNETISM: A Treatise on Modern Theory and Materials," H. Suhl, eds., Vol. V, Ch. 12 (Academic Press, New York, 1973).

39. M. B. Maple, W. A. Fertig, A. C. Mota, L. E. DeLong, D. Wohlleben and R. Fitzgerald, Solid State Commun. 11, 829 (1972).

40. C. A. Luengo, M. B. Maple and W. A. Fertig, Solid State Commun. 11, 1445 (1972).

41. H. Armbrüster, H. v. Löhneysen, G. Riblet and F. Steglich, Solid State Commun. 14, 55 (1974).

42. S. D. Bader, N. E. Phillips, M. B. Maple and C. A. Luengo, Solid State Commun. 16, 1263 (1975).

43. R. M. White, Quantum Theory of Magnetism, p. 197 (McGraw-Hill, New York, 1970).

44. M. Redi and P. W. Anderson, Proc. Natl. Acad. Sci. USA 78, 27 (1981).

45. B. S. Chandrasekhar, Appl. Phys. Lett. 1, 7 (1962).

46. A. M. Clogston, Phys. Rev. Lett. 9, 266 (1962).

47. R. A. Ferrell, Phys. Rev. Lett. 3, 262 (1959).

48. P. W. Anderson, Phys. Rev. Lett. 3, 325 (1959).

49. V. Jaccarino and M. Peter, Phys. Rev. Lett. 9, 280 (1962).

50. P. W. Anderson and H. Suhl, Phys. Rev. 116, 898 (1959).

51. Ø. Fischer, M. Decroux, S. Roth, R. Chevrel and M. Sergent, J. Phys. C 8, L474 (1975).

52. F. Y. Fradin, G. K. Shenoy, B. D. Dunlap, A. T. Aldred and C. W. Kimball, Phys. Rev. Lett. 38, 719 (1977).

53. M. S. Torikachvili and M. B. Maple, Solid State Commun. 40, 1 (1981).
54. J. E. Crow, R. P. Guertin and R. D. Parks, Phys. Rev. Lett. 19, 77 (1967).
55. J. E. Crow and R. D. Parks, Phys. Lett. 21, 378 (1966).
56. R. A. Hein, R. L. Falge, Jr., B. T. Matthias and E. Corenzwit, Phys. Rev. Lett. 2, 500 (1959).
57. R. P. Guertin, J. E. Crow and R. D. Parks, Phys. Rev. Lett. 16, 546 (1966).
58. K. H. Bennemann, Phys. Rev. Lett. 17, 438 (1966).
59. L. P. Gor'kov and A. I. Rusinov, Sov. Phys. JETP 19, 922 (1964).
60. P. Fulde and K. Maki, Phys. Rev. 141, 275 (1966).
61. W. Baltensperger and S. Strässler, Phys. Cond. Materie 1, 20 (1963).
62. M. Ishikawa and Ø. Fischer, Solid State Commun. 24, 747 (1977).
63. Ø. Fischer, A. Treyvaud, R. Chevrel and M. Sergent, Solid State Commun. 17, 21 (1975).
64. R. N. Shelton, R. W. McCallum and H. Adrian, Phys. Lett. 56A, 213 (1976).
65. B. T. Matthias, E. Corenzwit, J. M. Vandenberg and H. Barz, Proc. Natl. Acad. Sci. USA 74, 1334 (1977)
66. J. M. Vandenberg and B. T. Matthias, Proc. Natl. Acad. Sci. USA, 74, 1336 (1977).
67. R. Chevrel, M. Sergent and J. Prigent, J. Solid State Chem. 3, 515 (1971).
68. B. T. Matthias, M. Marezio, E. Corenzwit, A. S. Cooper and H. Barz, Science 175, 1465 (1972).
69. Ø. Fischer, H. Jones, G. Bongi, M. Sergent and R. Chevrel, J. Phys. C7, L450 (1974).
70. S. Foner, E. J. McNiff and E. J. Alexander, Phys. Lett. 49A, 269 (1974).
71. S. Foner, E. J. McNiff, R. N. Shelton, R. W. McCallum and M. B. Maple, Phys. Lett. 49A, 269 (1974).
72. M. Marezio, P. D. Dernier, J. P. Remeika, E. Corenzwit and B. T. Matthias, Mat. Res. Bull. 8, 657 (1973).
73. L. D. Woolf, D. C. Johnston, H. B. Mackay, R. W. McCallum and M. B. Maple, J. Low Temp. Phys. 35, 651 (1979).
74. J. P. Remeika, G. P. Espinosa, A. S. Cooper, H. Barz, J. M. Rowell, D. B. McWhan, J. M. Vandenberg, D. E. Moncton, Z. Fisk, L. D. Woolf, H. C. Hamaker, M. B. Maple, G. Shirane and W. Thomlinson, Solid State Commun. 34, 923 (1980).

75. H. F. Braun, Phys. Lett. 75A, 386 (1980).

76. H. C. Ku and R. N. Shelton, Mat. Res. Bull. 15, 1441 (1980).

77. M. B. Maple, L. E. DeLong, W. A. Fertig, D. C. Johnston,
       R. W. McCallum and R. N. Shelton, in "Valence Instabili-
       ties and Related Narrow Band Phenomena," R. D. Parks,
       ed., pp. 17-29 (Plenum Press, New York, 1977).

78. S. Oseroff, R. Calvo, D. C. Johnston, M. B. Maple, R. W.
       McCallum and R. N. Shelton, Solid State Commun. 27,
       20 (1978).

79. M. S. Torikachvili, J. Beille, S. E. Lambert and M. B.
       Maple, Superconductivity in d- and f-Band Metals 1982,
       W. Buckel and W. Weber, eds., KfK 1982, p. 241.

80. For a review, see M. B. Maple, M. S. Torikachvili, R. P.
       Guertin and S. Foner, in "The Rare Earths in Modern
       Science and Technology," G. J. McCarthy, H. B. Silber
       and J. J. Rhyne, eds., Vol. 3, pp. 301-313 (Plenum
       Press, New York, 1982).

81. W. A. Fertig, D. C. Johnston, L. E. DeLong, R. W.
       McCallum, M. B. Maple and B. T. Matthias, Phys. Rev.
       Lett. 38, 387 (1977).

82. D. E. Moncton, D. B. McWhan, J. Eckert, G. Shirane and
       W. Thomlinson, Phys. Rev. Lett. 39, 1164 (1977).

83. H. C. Hamaker, L. D. Woolf, H. B. MacKay, Z. Fisk and
       M. B. Maple, Solid State Commun. 31, 139 (1979).

84. C. F. Majkrzak, D. E. Cox, G. Shirane, H. A. Mook, H. C.
       Hamaker, H. B. MacKay, Z. Fisk and M. B. Maple,
       Phys. Rev. B 26, 245 (1982).

85. H. C. Hamaker, L. D. Woolf, H. B. MacKay, Z. Fisk and
       M. B. Maple, Solid State Commun. 32, 289 (1979).

86. H. C. Hamaker, H. B. MacKay, M. S. Torikachvili, L. D.
       Woolf, M. B. Maple, W. Odoni and H. R. Ott, J. Low
       Temp. Phys. 44, 553 (1981).

87. C. F. Majkrzak, S. K. Satija, G. Shirane, H. C. Hamaker,
       Z. Fisk and M. B. Maple, Phys. Rev. (submitted).

88. P. G. deGennes, J. Phys. Rad. 23, 510 (1962).

89. H. B. MacKay, L. D. Woolf, M. B. Maple and D. C. Johnston,
       J. Low Temp. Phys. 41, 639 (1980).

90. M. Ishikawa and Ø. Fischer, Solid State Commun. 23, 37 (1977).

91. J. W. Lynn, D. E. Moncton, W. Thomlinson, G. Shirane and
       R. N. Shelton, Solid State Commun. 26, 493 (1978).

92. H. B. MacKay, L. D. Woolf, M. B. Maple and D. C. Johnston,
       Phys. Rev. Lett. 42, 918 (1979).

93. M. B. Maple, H. C. Hamaker, L. D. Woolf, H. B. MacKay,
       Z. Fisk, W. Odoni and H. R. Ott, in "Crystalline Electric

Field and Structural Effects in f-Electron Systems, "
J. E. Crow, R. P. Guertin and T. W. Mihalisin, eds.,
pp. 533-543 (Plenum Press, New York, 1980).

94. G. K. Shenoy, B. D. Dunlap, F. Y. Fradin, S. K. Sinha,
C. W. Kimball, W. Potzel, F. Pröbst and G. M. Kalvius,
Phys. Rev. B 21, 3886 (1980).

95. H. R. Ott, W. A. Fertig, D. C. Johnston, M. B. Maple and
B. T. Matthias, J. Low Temp. Phys. 33, 159 (1978).

96. M. B. Maple, H. C. Hamaker and L. D. Woolf, Chapter 4
of reference 17.

97. D. E. Moncton, D. B. McWhan, P. H. Schmidt, G. Shirane,
W. Thomlinson, M. B. Maple, H. B. MacKay, L. D.
Woolf, Z. Fisk and D. C. Johnston, Phys. Rev. Lett.
45, 2060 (1980).

98. M. Tachiki, H. Matsumoto and H. Umezawa, Phys. Rev.
B 20, 1915 (1979).

99. G. W. Crabtree, F. Behroozi, S. A. Campbell and D. G.
Hinks, to be published.

100. J. M. Rowell, R. C. Dynes and P. H. Schmidt, in "Superconductivity in d- and f-Band Metals," H. Suhl and M. B.
Maple, eds., pp. 409-418 (Academic Press, New York,
1980).

101. C. P. Umbach, L. E. Toth, E. D. Dahlberg and A. M.
Goldman, Physica 108B, 803 (1981).

102. U. Poppe, Physica 108B, 805 (1981).

103. C. P. Umbach and A. M. Goldman, Phys. Rev. Lett. 48,
1433 (1982).

104. E. I. Blount and C. M. Varma, Phys. Rev. Lett. 42, 1079 (1979).

105. H. Suhl, J. Less-Common Metals 62, 225 (1978).

106. L. N. Bulaevski, A. I. Rusinov and M. Kulić, Solid State
Commun. 30, 59 (1979).

107. K. Machida and T. Matsubara, Solid State Commun. 31, 791 (1979).

108. H. Matsumoto, H. Umezawa and M. Tachiki, Solid State
Commun. 31, 157 (1979).

109. M. Tachiki, A. Kotani, H. Matsumoto and H. Umezawa,
Solid State Commun. 31, 927 (1979).

110. R. A. Ferrell, J. K. Bhattacharjee and A. Bagchi, Phys. Rev.
Lett. 43, 154 (1979).

111. H. Suhl, in Felix Bloch Festschrift, M. Chodorow,
R. Hofstadter, H. Rorschach, A. L. Schawlow, eds.,
Rice University Press (1980).

112. C. G. Kuper, M. Revzen and A. Ron, Phys. Rev. Lett. 44,
1545 (1980).

113. M. Tachiki, H. Matsumoto, T. Koyama and H. Umezawa,
Solid State Commun. 34, 19 (1980).

114. U. Krey, Int. J. Magnetism 3, 65 (1972); Int. J. Magnetism 4, 153 (1973).
115. C. R. Hu and T. E. Ham, Physica 108B, 1041 (1981).
116. J. W. Lynn, J. L. Raggazoni, R. Pynn and J. Joffrin, J. Physique Lett. 42, L45 (1981).
117. J. W. Lynn, G. Shirane, W. Thomlinson, R. N. Shelton and D. E. Moncton, Phys. Rev. B24, 3817 (1981).
118. S. K. Sinha, G. W. Crabtree, D. G. Hinks and H. A. Mook, Phys. Rev. Lett. 48, 950 (1982).
119. H. S. Greenside, E. I. Blount and C. M. Varma, Phys. Rev. Lett. 46, 49 (1981).
120. H. Matsumoto and H. Umezawa, Cryogenics 23, 37 (1983).
121. R. W. McCallum, D. C. Johnston, R. N. Shelton and M. B. Maple, Solid State Commun. 24, 391 (1977).
122. M. Ishikawa and J. Müller, Solid State Commun. 27, 761 (1978).
123. D. E. Moncton, G. Shirane, W. Thomlinson, M. Ishikawa and Ø. Fischer, Phys. Rev. Lett. 41, 1133 (1977).
124. W. Thomlinson, G. Shirane, D. E. Moncton, M. Ishikawa and Ø. Fischer, J. Appl. Phys. 50, 1981 (1979).
125. W. Thomlinson, G. Shirane, D. E. Moncton, M. Ishikawa and Ø. Fischer, Solid State Commun. 31, 773 (1979).
126. C. F. Majkrzak, G. Shirane, W. Thomlinson, M. Ishikawa Ø. Fischer and D. E. Moncton, Solid State Commun. 31, 773 (1979).
127. R. W. McCallum, D. C. Johnston, R. N. Shelton, W. A. Fertig and M. B. Maple, Solid State Commun. 24, 501 (1977).
128. L. J. Azevedo, W. G. Clark, C. Murayama, R. W. McCallum, D. C. Johnston, M. B. Maple and R. N. Shelton, J. Physique 39, C6-365 (1978).
129. M. B. Maple, L. D. Woolf, C. F. Majkrzak, G. Shirane, W. Thomlinson and D. E. Moncton, Phys. Lett. 77A, 487 (1980).
130. J. W. Lynn, D. E. Moncton, G. Shirane, W. Thomlinson, J. Eckert and R. N. Shelton, J. Appl. Phys. 49, 1389 (1978).
131. For a review, see P. Fulde and J. Keller, Chap. 9 of Ref. 17.
132. D. C. Johnston, W. A. Fertig, M. B. Maple and B. T. Matthias, Solid State Commun. 26, 141 (1978).
133. H. A. Mook, W. C. Koehler, M. B. Maple, Z. Fisk, D. C. Johnston, and L. D. Woolf, Phys. Rev. B25, 372 (1982).
134. H. A. Mook, O. A. Pringle, S. Kawarazaki, S. K. Sinha, G. W. Crabtree, D. G. Hinks, M. B. Maple, Z. Fisk, D. C. Johnston, L. D. Woolf and H. C. Hamaker,

·   Superconductivity in d- and f-Band Metals 1982,
W. Buckel and W. Weber, eds., KfK 1982, p. 201.

135. L. D. Woolf, D. C. Johnston, H. A. Mook, W. C. Koehler,
M. B. Maple and Z. Fisk, Physica 109 & 110B, 2045
(1982).

136. H. R. Ott, L. D. Woolf, M. B. Maple and D. C. Johnston,
J. Low Temp. Phys. 39, 383 (1980).

137. M. B. Maple, H. C. Hamaker, D. C. Johnston, H. B.
MacKay, M. S. Torikachvili and L. D. Woolf, to be
published.

138. H. C. Ku, F. Acker and B. T. Matthias, Phys. Lett. 76A,
39 (1980).

139. L. D. Woolf, S. E. Lambert, M. B. Maple, H. C. Ku,
W. Odoni and H. R. Ott, Physica 108B, 761 (1981).

140. H. C. Hamaker, H. C. Ku, M. B. Maple and H. A. Mook,
Solid State Commun. 43, 455 (1982).

141. K. N. Yang, S. E. Lambert, H. C. Hamaker and M. B.
Maple, Superconductivity in d- and f-Band Metals 1982,
W. Buckel and W. Weber, eds., KfK 1982, p. 217.

142. L. D. Woolf and M. B. Maple, pp. 181-184 of reference 15.

143. S. E. Lambert, L. D. Woolf and M. B. Maple, J. Low
Temp. Phys. (submitted).

144. F. Steglich, Superconductivity in d- and f-Band Metals 1982,
W. Buckel and W. Weber, eds., KfK 1982, p. 145.

145. B. V. B. Sarkissian, Superconductivity in d- and f-Band
Metals 1982, W. Buckel and W. Weber, eds., KfK 1982,
p. 311.

ION IMPLANTATION - A PROMISING TECHNIQUE FOR THE PRODUCTION OF NEW SUPERCONDUCTING MATERIALS

Bernd Stritzker

Institut für Festkörperforschung
Kernforschungsanlage Jülich
5170 Jülich, W. Germany

INTRODUCTION

Since the discovery of superconductivity the research of many experimentalists was motivated by the desire to raise the superconducting transition temperature, T_C, to high enough values that the widespread technological potential of superconductivity could be used in an economically feasable manner. This was the reason why a large variety of newly developed techniques for materials preparation were immediately applied in superconductivity. In this manner methods like rapid quenching from the vapor or the liquid phase, sputtering, chemical vapor deposition and high pressure techniques resulted in a high variety of interesting superconducting phases. During the midsixties the semiconductor industries applied very successfully ion implantation for the doping of semiconducting materials. For this reason the group with Prof. Buckel at the university in Karlsruhe started to introduce this technique in superconductivity. Although they originally intended to use ion implantation only for the well-defined doping of superconductors with magnetic impurities, the applicability of this technique to the production of new materials became quite obvious. Hence up to now mainly three experimental groups situated in the nuclear research centers in Karlsruhe and Jülich as well as in Orsay have joined the work on ion implantation in superconductors. The interesting results achieved by these groups have been reviewed recently by several authors.[1-5] For this reason in the following I do not intend to give a complete overview of this field.

347

Instead I want to restrict myself to a few selected
examples demonstrating the widespread applicability of
the ion implantation technique in superconductivity.
The discussion will not only include interesting superconducting properties but also new metallurgical aspects
as well as new insights in the implantation technique.

In the first part, the ion implantation technique,
an extremely non-equilibrium method, is compared with
other more widely studied non-equilibrium methods like
vapor and liquid quenching. It will be shown that the
electrical and superconducting properties of the resulting alloys can be used for an easy identification and
comparison of the different resulting phases. It is
demonstrated that all these techniques can be looked
upon as more or less rapid quenching processes. However,
for the ion bombardment also a collisional aspect has
to be taken into account as will be shown in the case of
Pd. In the second part the production of interesting
superconducting materials by means of ion implantation
will be described. Besides stoichiometric, ordered alloys,
highly concentrated metal-H systems with T_C values up
to 17 K can be produced. In summary I want to emphasize
that ion and also laser bombardment techniques yield an
interesting potential for the development of interesting
superconducting materials.

Basic Theoretical Considerations:

A superconductor is characterized by its transition
temperature, T_C, below which the electrons condense into
Cooper-pairs forming a macroscopic quantum state. T_C can
be expressed as:

$$T_C \propto <\omega> \exp - \frac{1+\lambda}{\lambda - \mu^* - \mu_p} \qquad (1)$$

Below T_C the repulsive Coulomb interaction μ^* as well as
the perhaps present Cooper-pair-breaking paramagnon interaction is overcome by the attractive electron-electron
interaction λ mediated by the phonon system with average
phonon frequency $<\omega>$. This electron-phonon interaction
can be calculated in two different ways, depending on
which part of the interaction should be emphasized:

a) phonon picture: $\lambda = 2 \int \alpha^2(\omega) F(\omega) \frac{1}{\omega} d\omega$ $\qquad (2)$

with $\alpha^2(\omega) F(\omega)$ = electron-phonon interaction parameter times phonon density of states

b) electron picture: $\lambda = \dfrac{N(0)<J^2>}{M<\omega^2>}$ (3)

with $N(0)$ = electronic density of states at the
              Fermy energy

  J^2 = electron-phonon matrix element

  M = atomic mass

Both descriptions a and b are equivalent and it depends on the special system which description is more
adequate to achieve a better understanding.

It can be easily seen that T_C can be enhanced
either by

1) reduction of μ_p (compare irradiated Pd)

2) increase of λ by:

  a) increase of α^2 or $<J^2>$ (compare Pd-H)

  b) increase of $N(0)$ (compare A15 materials)

  c) decrease of $<\omega>$, i.e. weakening of $F(\omega)$
          (compare amorphous non-transition metals)

  d) coupling to additional phonon modes, i.e. optic
     modes (compare metal-H systems)

The examples in brackets indicate already that
each of these alternatives will be covered in this
review, thus I want to show that ion beams can influence all these important superconducting parameters.

METASTABLE AND AMORPHOUS ALLOYS

As a highly non-equilibrium process, ion implantation can result in metastable alloys which are interesting for superconducting purposes due to metastability
with respect to two different properties:

1) Equilibrium solubility limits can be highly exceeded,
   i.e. a concentration regime where a special, interesting phase exists can be substantially extended.

2) New non-equilibrium crystal structures or even amorphous phases can be achieved in a metastable state.

Both "kinds of metastability", i.e. metastability with respect to composition or with respect to structure, are desirable in superconductivity, since

1) variation of the composition within the same crystal structure allows the variation of the electronic density of states at the Fermi energy, $N(0)$, and

2) a change of the crystal structure influences dominantly the phonon spectrum, $F(\omega)$, of the system under investigation.

Thus it it possible to alter both $N(0)$ and $F(\omega)$, i.e. two important parameters determining the superconducting properties. This potential of the implantation technique resulted for example in a disordered metastable system with a T_C as high as 17 K [6] consisting of Pd, Cu and H, i.e. three non-superconducting elements.

Non-Equilibrium Preparation Methods

Since ion implantation is not at all the only technique to produce metatsable systems we have first to consider the competing techniques like vapor quenching, liquid quenching and laser quenching. The application of these techniques to the production of a superconductor have been reviewed extensively.[7-12]

a) Vapor Quenching: The desired material, in general consisting of two or more elements, is evaporated from different evaporation sources like electron gun or resistivity heated filaments in a vacuum chamber ($\leqslant 10^{-8}$ mbar). The vaporized atoms are condensed as thin films (~ 1000 Å thickness) onto a substrate. This condensation process can be considered as a rapid quenching technique since the hot atoms loose their kinetic energy instantaneously. By cooling the substrate to 4 K it is possible to achieve quenching rates up to $\sim 10^{14}$ K/sec. Thus metastable or even amorphous films can be produced. The low substrate temperature does not only increase the quenching rate but it also hinders the diffusion or clustering of the different components of the frozen-in alloy system. The quenching rate naturally depends not only on the substrate temperature but also on special properties of the alloys, such as melting point, thermal conductivity and so on. The vapor quenching method was first used for the production of superconductors by Buckel and Hilsch.[13]

b) Liquid Quenching: This non-equilibrium method,
which was first developed by P. Duwez,[14] has generated
much interest in recent years because the economical production of interesting amorphous materials became possible. Since there are a large number of various technical realizations of this method only the basic process
will be described. The melt of the desired alloy is
forced into good thermal contact with a water-cooled surface of Cu. There the alloy solidifies with quenching
rates of the order of $\sim 10^8$ K/s into ~ 20 μm thick material. The quenching rate can be influenced by the temperature of the Cu surface, by the velocity of the Cu with
respect to the liquid, i.e. by the special liquid quenching technique, and by the thermal properties of the
alloy.

c) Laser Quenching: A technique which just recently
arose very much interest in semiconductor physics is
laser annealing.[15] There a heavily disordered surface
layer on top of a single crystalline semiconductor is
molten with a very short laser pulse (~ 20 ns) with high
power. Then this molten layer recrystallizes epitaxially
on top of the crystalline substrate, restoring a perfect
single crystalline surface. The velocity of the recrystallization, i.e. the resolidification depends on the
substrate temperature as well as on material properties.
This is the reason that the same technique yields rather
a quenching process than an annealing process if it is
applied to metals with high thermal conductivity. Then
the rapid resolidification leads to quenching rates of
about 10^9 K/s. A major difficulty with this technique is
the coupling of the laser light to the metal. This problem can perhaps be overcome by the use of laser frequencies which penetrate more into a conducting material.
A Russian group was the first to use laser processing
for the production of superconducting alloys several
years ago.[16]

d) Ion Implantation and Ion Irradiation: For this
non-equilibrium method ion beams of the order of 100
keV and fluences in the range of 10^{14} to 10^{17} ions/cm^2
are used for the preparation of metastable alloys. The
material of interest, mostly an evaporated thin film,
is mounted in a vacuum chamber on a substrate holder
the temperature of which can be varied between 2 K and
room temperature or even up to several hundred degrees
C. In the implantation case the accelerated ions come
to rest within the thin film thus altering the composition. In contrast the composition of the film is not
changed during the irradiation because the ions pene-

trate nearly all through the evaporated material and
stick inside the substrate. Both ion implantation and
irradiation may result in a change of the structure.
This can be interpreted within different models; introduction of lattice defects, radiation enhanced diffusion
or rapid quenching of a locally heated collision cascade
region. The quenching rate of the latter "thermal spike"
model can be influenced by the substrate temperature,
local energy due to the incoming ions as well as by parameters of the sample material. In most of the following
cases the ion beams are applied to targets at temperatures below 10 K resulting in a very high quenching rate.
This substrate temperature is very suitable for the production of superconductors, since it is easy to measure
the superconducting transition temperature as a function
of increasing ion fluence. The measured T_C value represents the transition temperature of a bulk superconductor
despite the fact that i.e. ion implantation yields an inhomogeneous, Gaussian shape concentration profile. The
reason is that the maximum of the distribution is generally several hundred Ångström wide. On the other hand
the superconducting coherence length amounts ≤ 100 Å in
such a disordered material. Thus the superconductivity
in the middle of the implanted region is not affected
by proximity effects of the surrounding material.

Internal Energy of Metastable Phases:

In order to simplify the discussion of the metastable phase I want to introduce an extended "Ostwald
step rule", a very old model[17] which was already used
for the description of the vapor quenching experiments.[13]

The Ostwald step rules states the following. After
condensation of a vapor, the condensate has to pass consequently through all possible high temperature phases
until the equilibrium phase is reached. This is shown
schematically as solid line in Fig. 1 where the internal
energy of the system (without any T-dependence) is plotted versus temperature. Starting from the high internal
energy of the vapor phase at high temperatures the internal energy is reduced during a slow cooldown passing
through the liquid phase and then through the different
high temperature phases, i.e. HTI and HTII, until the
equilibrium phase is reached. The important conclusions
from this step rule for the present considerations are
the following, which include rapid quenching processes.
By means of sufficiently rapid cooling it should be
possible to quench in some of the intermediate phases.

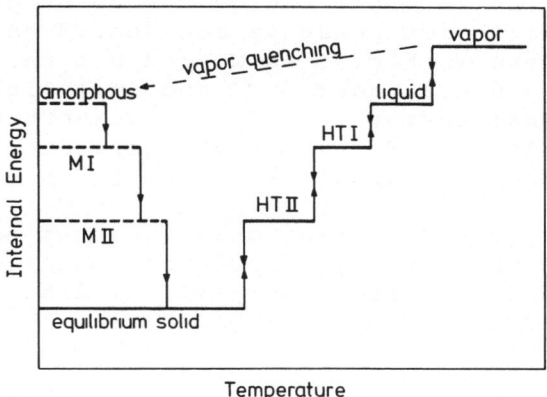

Fig. 1. Schematic representation of the internal energy
        as a function of temperature

That means it should be possible that these intermediate phases as identified by the internal energy can be
stabilized at low temperatures. For instance, the experimental results suggested that amorphous phases as obtained by vapor quenching onto He-cold substrates can
be described as frozen liquids. The liquid phase on the
other hand is the high temperature phase following the
vapor phase.

     In order to demonstrate more clearly the applica-
tion of the Ostwald step rule, a very fast quenching
process is included in the schematic phase diagram of
Fig. 1 as dashed lined. The vapor phase is extremely
rapidly cooled down to very low temperatures as indicated by an arrow. Then an amorphous metastable state
can be stabilized. The properties of this amorphous
phase suggest that it is identical with a frozen liquid
phase. That means in our simple phase diagram in Fig. 1
that the value of the internal energy (without T-dependence) of the amorphous phase is equal to that of the
liquid phase. Similarities in diffraction patterns, resistivities and Hall coefficients in both the amorphous
and the liquid state prove this analogy.

By means of careful anealing of this amorphous,
i.e. frozen liquid phase it is possible in many cases
to pass consecutively through other metastable phases,
having their analogues in intermediate high temperature
phases. Thus the internal energy can be stepwise reduced
until the equilibrium phase is reached. Such metastable
phases are shown in Fig. 1 as MI and MII in analogy to
the high temperature phases HTI and HTII with the corresponding internal energy values. The structure of MI and
MII is equivalent to HTI and HTII temperatures like the
amorphous phase is equivalent to the frozen liquid phase.

Generally the different phases cannot only be distinguished by their x-ray properties, but also from
their different electrical properties, like residual
resistivity or T_C.

Comparison of Ion Bombardment and Vapor Quenching

The most striking comparison between the two extreme non-equilibrium techniques, ion implantation and
vapor quenching, has been performed recently for Ga. Ga
is an especially interesting metal because it is the
only pure metal besides Bi which can be vapor quenched
into a liquidlike amorphous state.[13] In contrast amorphous metals generally need about 10 at.% of a second
element to stabilize the amorphous state[7].

The resistance of vapor quenching Ga versus temperature is shown in Fig. 2.[13] The resulting amorphous
phase has both a high T_C = 8.5 K and a high residual
resistivity ρ_o = 29 $\mu\Omega$cm. Annealing to more than 15 K
results in an irreversible decrease of the resistance.
The amorphous phase has transformed into ß-Ga with T_C
= 6.3 K and a low resistivity of ρ_o = 3 $\mu\Omega$cm. This
crystalline metastable ß-phase resembles very much the
structure and electronic properties of a high pressure
phase of Ga. Above 60 K this metastable phase transforms
into the equilibrium α-phase with a low T_C = 1.07 K and
a medium resistivity of 12 $\mu\Omega$cm. This example demonstrates that the measurements of the electric resistivity readily provide evidence not only for phase transitions but also for the characterization of the different phases by the different values for $d\rho/dT$. Once after
the corresponding structure of the different phases has
been determined by for instance electron diffraction
like in the case of Ga[18] it is only necessary to measure
the resistivity as function of temperature to identify
the various phases.

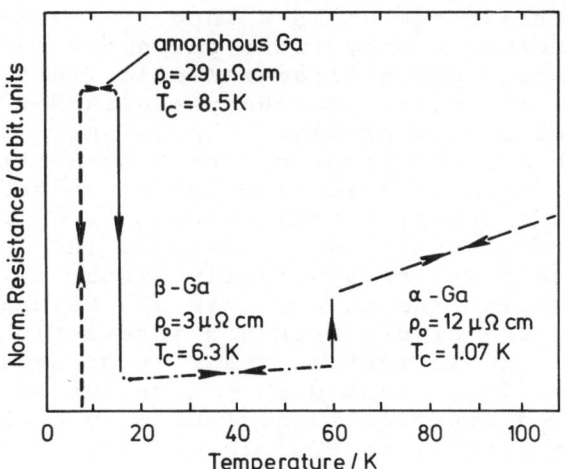

Fig. 2. Normalized resistance of vapor quenched Ga as a
        function of temperature (after ref. 13)

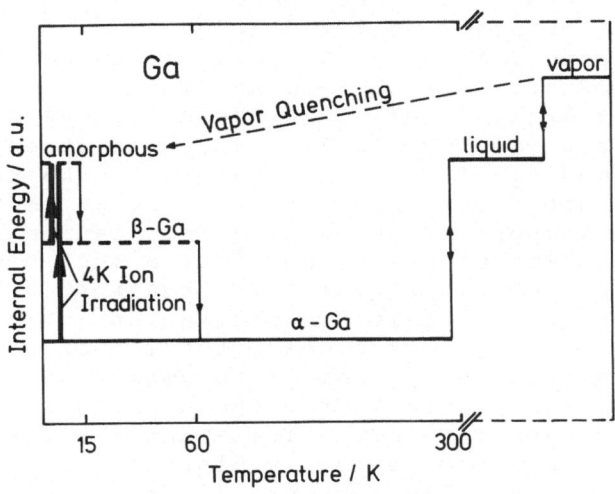

Fig. 3. Schematic representation of the internal energy
        of Ga as a function of temperature

Fig. 3 transfers the experimental results of Fig. 2 into the simple internal energy model. Vapor quenching results in the amorphous, i.e. the frozen liquidlike phase, which transforms at 15 K into the metastable ß-Ga with lower internal energy. This phase has no analogon in the equilibrium phase diagram (solid lines). The reason is that the corresponding phase is not a high temperature but a high pressure phase, which cannot be included in this phase diagram. The ß-phase transforms at 60 K into the equilibrium α-phase which has the lowest T_C value. At this point a common property of the non transition metals should be emphasized: The liquidlike amorphous phase has a substantially higher T_C than the crystalline equilibrium phase. This can be understood by considering the electron-phonon interaction λ in the phonon picture. Amorphization means in general a sub-stantially less dense packed structure due to many in-ternal holes and imperfections. Thus many atoms can oscillate more freely, so $F(\omega)$ is smeared and shifted to lower energies ω in comparison to the crystalline mate-rial. In addition the electron-phonon coupling $\alpha^2(\omega)$ is also enlarged. Since λ is determined by the low energy part of $\alpha^2(\omega)F(\omega)$, this means that the amorphous phase has a higher λ and thus a higher T_C.

In a recent experiment[20] the influence of low tem-perature ion irradiation was examined in Ga in order to compare the result with that obtained by vapor-quenching. All three Ga phases were irradiated with heavy Ar^+-ions of 275 keV penetrating the Ga films at temperatures below 10 K. Starting with α-Ga an immediate transformation (only $\sim 10^{14}$ Ar ions/cm^3) into the amorphous phase was achieved. This is shown in Fig. 4 by the steep increase of both T_C and ρ_0. T_C and ρ_0 saturate at the values typical for amorphous Ga. Starting from an amorphous Ga film low temperature heavy ion irradiation did not change the amorphous phase. However, a much higher dose of $\sim 10^{16}$ ions/cm^2 is required to transform ß-Ga into the amorphous phase. This transformation is more diffi-cult to achieve, since the nearest neighbor coordination in ß-Ga is already very similar to amorphous Ga. Thus there is a strong tendency for the amorphous regions to transform back into ß-Ga. The authors argue that the high ion dose homogeneously distributes 0-impurities throughout the Ga-films thus stabilizing the amorphous phase also in this case.

All these results mean that ion irradiation at low temperatures produces the same amorphous, i.e. frozen liquid phase as vapor quenching. This can be easily

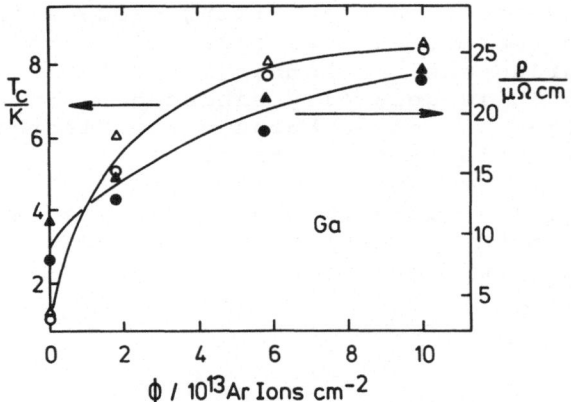

Fig. 4. T_C and ϱ_0 of α-Ga versus dose of irradiation Ar-
        ions (at 4 K) indicated the transformation into
        amorphous Ga (after ref. 20)

understood by the assumption of a thermal spike model:
The high energy density in a cascade as produced by a
heavy ion leads to a locally restricted hot, i.e. molten
or even evaporated region within a cold matrix. Due to
the small molten value ($\emptyset \sim 20$ Å) a high temperature
gradient to the cold matrix leads to an extremely high
cooling rate comparable to that of vapor quenching
($\sim 10^{14}$ K/s). The validity of this model for the explanation of the Ga results is underlined by the fact,
that the amorphous Ga phase could not be achieved by
He-irradiation which does not produce extended cascades.[20] The transformation from α- and ß-Ga into amorphous Ga by low temperature ion irradiation is indicated by the thick solid arrows at 4 K in the simple
phase diagram in Fig. 3. This description does not include the intermediate high temperature state within
the thermal spike region.

Besides Ga there are numerous other non-transition metals which can be prepared in an amorphous or heavily disordered state by vapor quenching together with about 10 at.% of a stabilizing element. Many of these systems have been studied either by ion irradiation of the alloy or by implantation of a stabilizing element into the non-transition metal. In both cases the targets remain below 10 K during the ion bombardment. For non-transition metals T_C is generally enhanced by lattice disorder as in the case of Ga because the disorder enhances $\alpha^2(\omega)F(\omega)$ at lower energies ω.[7]

The following table show some examples which have been examined both by vapor quenching and heavy ion bombardment at low temperatures.

Table I

non-transition element	T_c/K			references
	crystalline	vapor quenched	ion bombarded	
In	3.41	4.35	4.45	21,22
Be	0.026	9	6 - 9	23-25
GeCu	-	3.3	3.7	26

Table I shows that both non equilibrium methods yield similar T_C values, i.e. phases with a similar degree of disorder. From this result we can conclude again that ion bombardment at low temperatures is a rapid quenching process with a similar quenching rate as vapor quenching. Concerning the results in table I one has to distinguish carefully, within the pure element has been investigated. Then in general one will obtain only a disordered material (except Ga, Bi). However, if a second element is present (willingly or unwillingly) one has a good chance to produce an amorphous phase. In this respect the method of vapor quenching is very sensitive to simultaneously condensed impurities which are present in the evaporation system. The method of ion bombardment tends to distribute impurities, which have been already present as clusters in the material or which have been accumulated on the cold surface of the target. In both cases the homogeneously distributed impurities stabilize more effectively, thus resulting in a considerably more disordered or even amorphous phase. These intrinsic difficulties impede a direct comparison, since very little is known about the real impurity concentration or the structure of the investigated material. For this reason most ion implantation groups working at the moment on better characterization and definition of their specimens. The examples in Table I have been chosen because the same experimental groups investigated the same material both with vapor quenching and low temperature ion bombardment. For pure In in both cases most probably no amorphous phase has been obtained. In the Be case the high T_C phase, which seems not be amorphous is stabilized by impurities, i.e. Ge or O. For Ge-Cu the ion implantation technique was able to reveal the reasons for superconductivity. Whereas vapor quenched alloys need about 50 at.% Cu to stabilize an amorphous phase, only $\sim$ 15 at.% Cu is necessary with the implantation technique to achieve an even higher T_C. Thus it could be shown that the superconductivity is due to an amorphous Ge phase and not caused by a speculated electron-exiton interaction.

There are quite a number of different examples which show that there is a strong resemblance between the phases obtained by vapor quenching or by heavy ion beam techniques applied to samples at 4 K. It depends on the special material which of the two techniques is more favorable for superconductivity, i.e. for a higher degree of lattice disorder.

Comparison Ion Bombardment and Liquid Quenching

In the following only few examples will be shown where ion bombardment has been applied to systems which had been produced in an amorphous phase by liquid quenching.

The system ZrCu was the first superconducting metallic glass. Although the x-ray diffraction pattern looks "amorphous" low temperature ion bombardment had a considerable effect on both T_C and resistivity ρ of this metallic glass.

Table II	x(at%)	$\Delta\rho/\rho$	T_c/T_c
Increase of resistivity $\Delta\rho/\rho$ and	26	0.39	0.068
transition temperature T_c/T_c of	30	0.43	0.10
$Zr_{1-x}Cu_x$ metallic glasses after	35	0.51	0.28
low temperature He irradiation	40	0.72	0.29
($\sim 10^{16}$ cm^{-2}) (ref. 27)	50	0.34	0.69

Table II shows the resulting increses in both T_C and ρ for various Cu-concentrations after He-irradiation at 4 K.[27] In all cases low temperature irradiation yields a positive influence, i.e. the system becomes even more amorphous due to the irradiation. Similar T_C values could be obtained by Cu-implanttation into Zr. Fig 5 shows ΔT_C, the increase of T_C due to irradiation, as a function of annealing temperature. In all cases ΔT_C vanishes for annealing to room temperature. This result can be explained by a higher quenching rate involved in the ion beam method compared to liquid quenching resulting in a "more amorphous" phase.

Another example where ion beam techniques could be very successfully applied is the (Ti, Zr, Hf)-Fe system. Whereas it is easy to produce (Ti, Zr, Hf)$_{80}$-Cu$_{20}$ metallic glasses it is difficult to prepare the

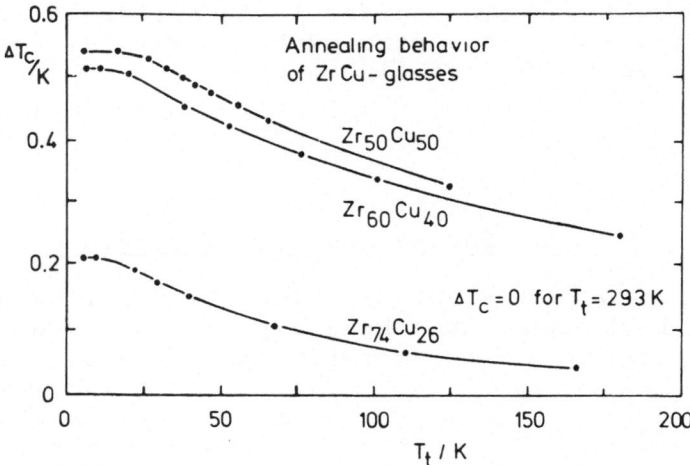

Fig. 5. Reduction of the T_C increase ΔT_C induced by 4 K He$^+$ irradiation as a function of annealing temperature T_t (after ref. 27)

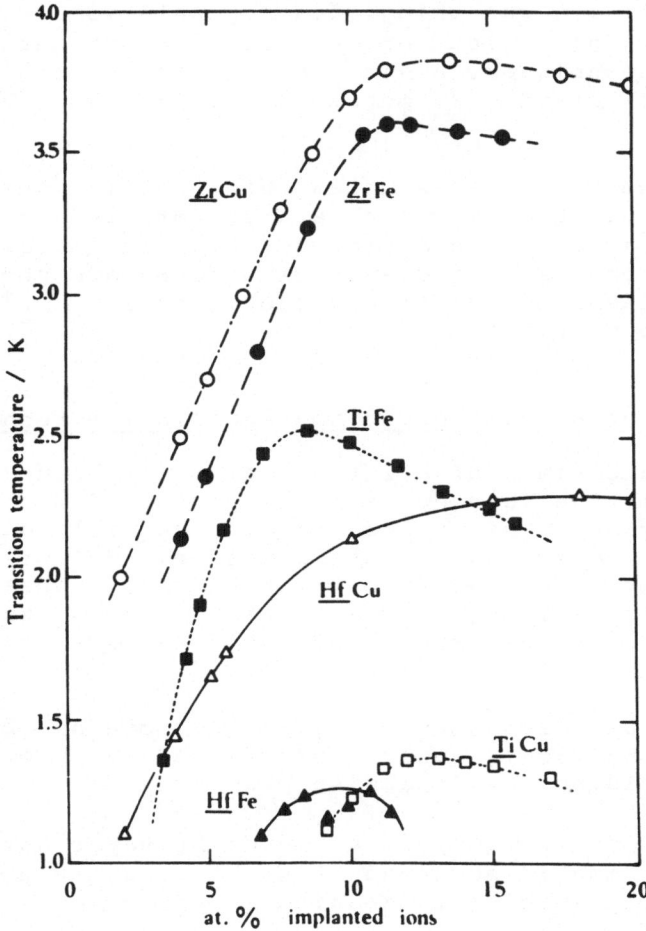

Fig. 6. T_C for Ti, Zr and Hf as a function of low tem-
       perature implanted Cu or Fe concentration
       (after ref. 28)

corresponding Fe-stabilized alloys by liquid quenching.
For TiFe it is even not yet possible. Due to the higher
quenching rate of the heavy ion implantation at 4 K
it was expected that it should be possible to produce
also the Fe alloys in an amorphous state. The results
are shown in Fig. 6 where T_C is plotted as a function
of the concentration of implanted Fe and Cu (for comparison). As can be seen all three Fe alloys can be
produced and have a similar concentration dependence
as the corresponding Cu alloys. TiFe, not producible
by liquid quenching, has a substantially higher T_C value
compared to TiCu. These results[28] show that ion implantation is indeed a technique with a higher quenching

rate than liquid quenching. Ion implantation at 4 K can produce a large variety of metallic glasses than liquid quenching techniques. A detailed systematic study of (Ti, Zr, Hf) based - 3d metal alloys will soon be published.

In summary these examples show that ion beam techniques can produce amorphous states more effectively than the liquid quench technique. This is understandable since also the estimated quenching rates are several orders of magnitude different (for example $\sim 10^{14}$ K/s compared to $\sim 10^9$ K/s).

Comparison of all different non-equilibrium techniques

The comparison of all four non-equilibrium methods described in the introduction was performed for the $Te_{1-x}Au_x$ system. This system has structurally different, metastable, solid phases with completely different values for ρ and its temperature derivative $d\rho/dT$. These properties make $Te_{1-x}Au_x$ very attractive for an examination based on a measurement of the electrical resistivity.

Previously metastable $Te_{1-x}Au_x$ alloys had been produced both by liquid quenching[29,30] and by vapor quenching[31] yielding the following phases:

I)   Amorphous $Te_{1-x}Au_x$ is achieved by vapor quenching. This phase is semiconducting, i.e. ρ increases exponentially with decreasing temperature.

II)   Simple cubic $Te_{1-x}Au_x$ ($0.15 \leqslant x \leqslant 0.35$) is produced by liquid quenching. This phase has a slight negative temperature coefficient of the resistivity ($d\rho/dT < 0$) like a metallic glass and becomes superconducting at about 2.3 K.

III) In contrast the equilibrium phase consists of a phase mixture of Te and Au and Te_2Au. This phase mixture behaves as metal or semiconductor depending on the composition, but is not superconducting in any case.

This summary of the different properties of the three phases makes it obvious that the three phases can be easily distinguished just by measuring ρ and $d\rho/dT$. Thus it is very easy to study thin films of $Te_{1-x}Au_x$, which have been processed by such non-equilibrium techniques as ion bombardment and laser quenching.

First the results of ion irradiation will be described.[32] Homogeneous Te-Au films ($\sim$1000 Å thick) consisting of the equilibrium phase III were irradiated at
4 K with He$^+$-ions penetrating totally through the sample.
Fig. 7 shows the resulting change of the resistivity
versus irradiation dose for Te$_{75}$Au$_{25}$. An initial steep
increase of ϱ due to the introduced lattice disorder is
followed by a smooth further increase of ϱ. After irradiation with $3 \cdot 10^{15}$ He$^+$/cm^2 the metallic character of
phase III vanishes and the sample behaves semiconducting
like phase I. After $6 \cdot 10^{15}$ He$^+$/cm^2 ϱ has increased 3
orders of magnitude. The whole sample has been trans-
formed into the amorphous phase I. Then the sample was
stepwise thermally annealed. The resulting resistance
behavior is shown in Fig. 8 as a function of temperature.
The annealing was interrupted at the temperature indicated by the open points and the sample was cooled down to
1.1 K. The resulting reversible temperature dependences
of the resistance are plotted as dashed curves. The
sample remains in Phase I up to annealing temperatures
of 230 K. Then an irreversible drop of the resistance
indicates a phase change which is completd at about
250 K. Phase II has been formed as demonstrated by both
the slight negative value of dϱ/dT and the superconducting transition. This phase is stable up to 400 K where
a transformation into the equilibrium metallic phase III
occurs (dϱ_{III}/dT > 0 and ϱ_{III} > ϱ_{II}).

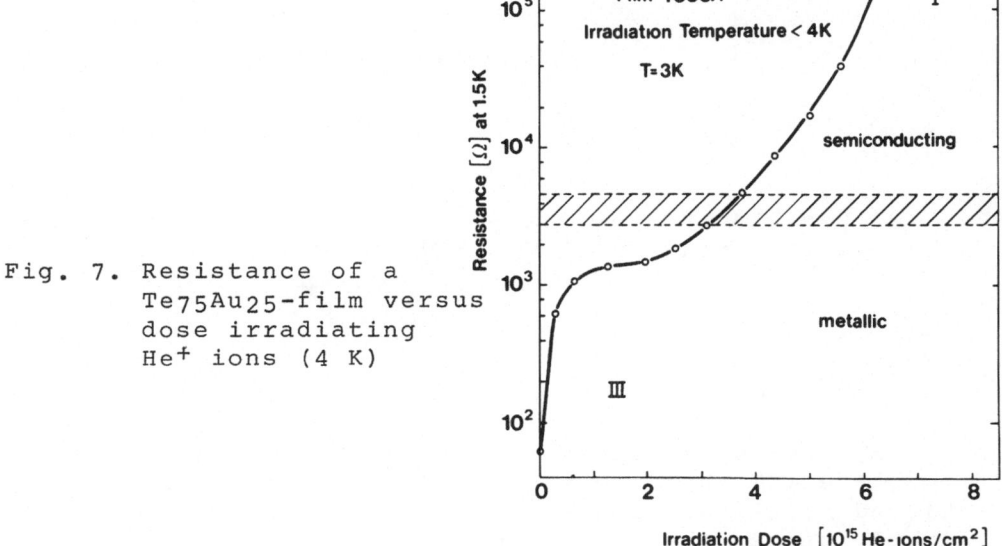

Fig. 7. Resistance of a
        Te$_{75}$Au$_{25}$-film versus
        dose irradiating
        He$^+$ ions (4 K)

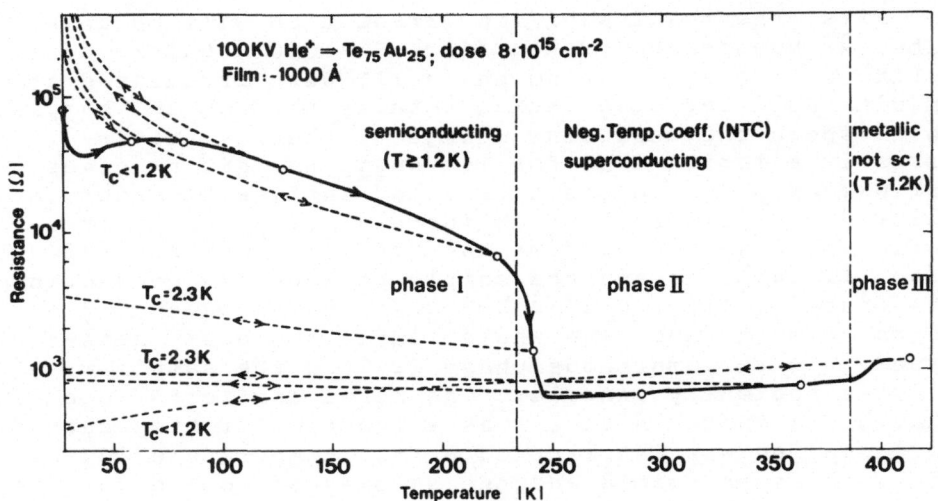

Fig. 8. Annealing behavior of Te₇₅Au₂₅ after low tempe-
        rature irradiation with He⁺ ions. Double (single)
        arrows indicate reversible (irreversible) changes.

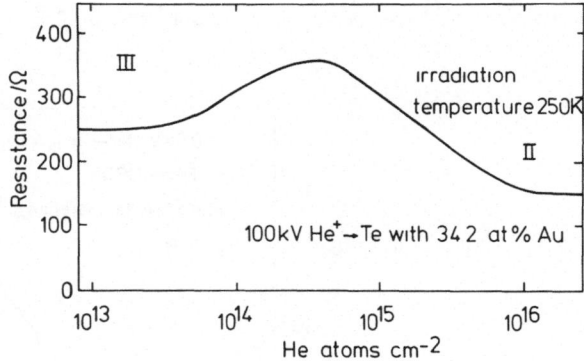

Fig. 9. Resistance of a Te₆₆Au₃₄ film versus dose of
        irradiating He⁺ ions (250 K)

        In a second experiment a direct transformation
from phase III into the superconducting phase II was
achieved by irradiation at 250 K. This temperature is
slightly higher than the transformation of I into II.

Fig. 9 shows the resulting change of the resistance as a function of He dose. The resistance passes through a maximum, dsecreases, and saturates at a value below the initial resistance. This behavior ($\rho_{II} < \rho_{III}$) and the occurrence of superconductivity at 2.3 K assures that phase II has been formed directly out of Phase III.

The results of both ion bombardment experiments suggest again that ion irradiation at 4 K is comparable to vapor quenching. On the other hand ion irradiation at 250 K is similar to liquid quenching.

Now the results obtained by laser quenching will be discussed.[33] Te-Au sandwich films of $\sim$1000 Å (nominal 30 at.% Au content for homogeneous distribution of Te and Au) were mounted into a liquid He-cryostat. The sample was irradiated with a Q-switched Ruby laser. After irradiation the superconducting transition temperature was measured resistively after the shutters in the radiation shields of the cryostat had been closed. In order to ensure a better coupling of the laser energy to the Te-Au sandwich layers the outermost layer consisted of Te. By means of this laser quenching procedure the superconducting phase II could be achieved.

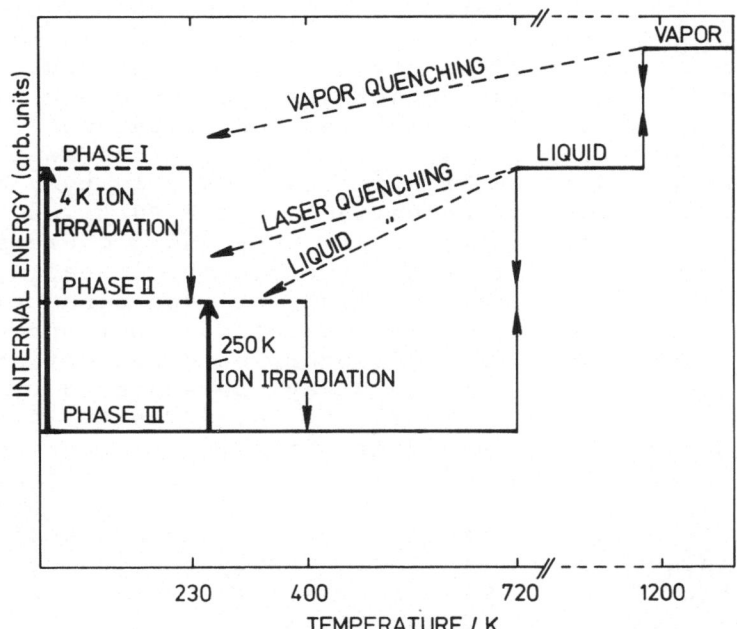

Fig. 10. Schematic representation of the internal energy of Te-Au as a function of temperature

     Fig. 10 summarizes a schematic phase diagram for
$Te_{1-x}Au_x$ based on the simple internal energy model. The
equilibrium phases are indicated by solid lines, the metastable ones by dashed lines. The various non-equilibrium processes are included. It is reasonable to assume
that similar quenching rates are involved in those processes leading to the same phase:

1) The amorphous phase I can be achieved by both vapor
   quenching or 4 K ion irradiation.

2) The superconducting, simple cubic phase II can be
   produced by liquid and laser quenching at 4 K and
   250 K ion irradiation.

The analogy to the other non-equilibrium methods involving rapid quenching of vapor or liquid phases support
again the explanation by a thermal spike model for the
ion irradiation.

Irradiated Palladium - an Exception or a New Class of
Superconductors ?

     The ion bombardment experiments described so far
are in good agreement with a thermal spike model, i.e.
a locally very hot cascade region which is cooled down
extremely fast. This model does not take into account
the collisional aspect of the ion bombardment. However,
there is one example up to now i.e. low temperature ion
irradiation in Pd, which cannot just be explained by rapid quenching. Vapor quenching and ion bombardment give
quite different results as will be seen in the following.
Vapor quenched Pd has a high initial resistivity ϱ and
is not superconducting. Annealing of such Pd films results in a decreasing lattice disorder and ϱ decreases
as shown by the solid line in Fig. 11. The annealing was
often interrupted in order to look for superconductivity
at various values of ϱ_0. In all cases no superconductivity was found above 0.1 K. However, superconducting
transitions up to 3.2 K could always be achieved by
He^+ irradiation of these films at 4 K.[34] The irradiated and
superconducting films show quite different annealing
curves (dashed line in Fig. 11). This different behavior
in connection with other observations led to the assumption that the main difference between both non-equilibrium techniques is the following: Only the ion bombardment can kick Pd atoms into interstitial positions within the Pd lattice. Such interstitial atoms <u>cannot</u> be
produced by vapor or liquid quenching for energetic
reasons. These interstitial Pd atoms are thought to be
the reason for the observed superconductivity.

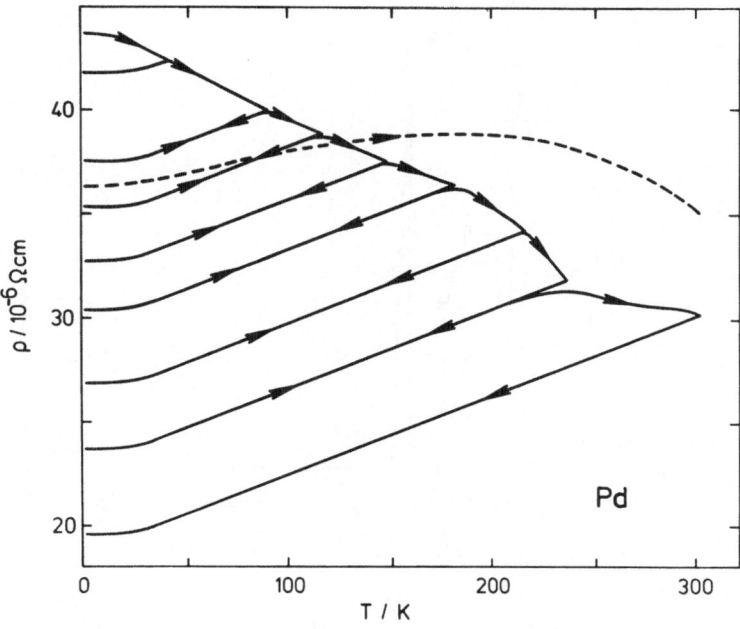

Fig. 11. Resistivity ϱ of vapor quenched 400 Å thick Pd-
films as a function of annealing temperature for
as-condensed Pd (solid curve) and after 4 K He$^+$
irradiation into the superconducting state
(dashed curve)

A very simple model explains the superconductivity
of irradiated Pd in the following way. For crystalline
Pd it is commonly accepted that the absence of superconductivity can be explained by the occurrence of strong
spin fluctuations, i.e. a large μ_p in equation (1). So it
is reasonable to assume that the low temperature irradiation reduces μ_p, thus allowing the observed T_c. This can
be understood by examining Fig. 12 which shows the electronic bandstructure of Pd. The solid line represents
calculations for crystalline Pd.[35] The Fermi level is located in a very narrow peak (<0.1 eV width) of the density
of states. This high value for N(0) is one of the reasons
for the large Stoner enhancement factor $S=1/[1-U\,N(0)]$
with the exchange parameter U. It is reasonable to assume that disorder would lead to a smearing of the bandstructure and thus to a reduction of N(0) as indicated
by the dotted line. This reduction of N(0) seems not yet
sufficient to diminish the spin fluctuations since vapor
quenched Pd is not superconducting. However, the interstitial Pd atoms as produced by the irradiation will

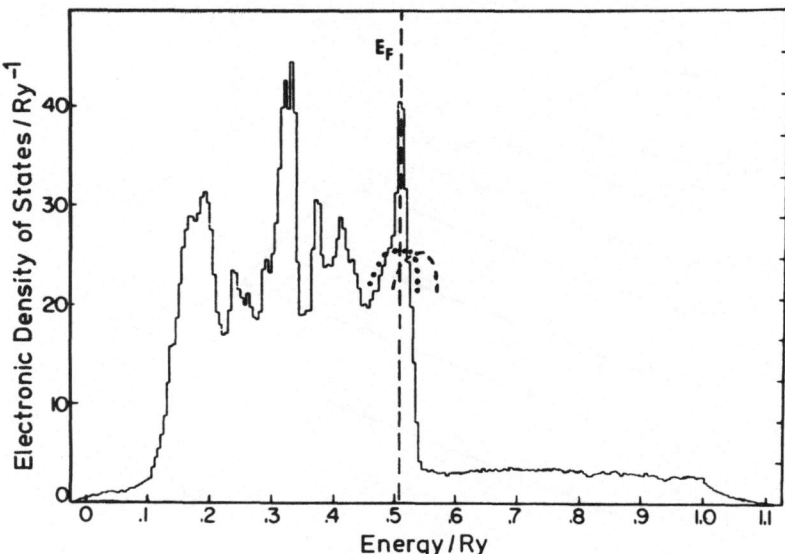

Fig. 12. Electronic density of states versus energy of
         crystalline Pd (solid line: calculation after
         ref. 35), disordered Pd (dotted line: assumed
         shape) and irradiated Pd dashed line: assumed
         slope)

have an additional effect: the change of the local symmetry. As a consequence the Fermi energy will be shifted
with respect to the bandstructure. One possibility is
indicated as a dashed line in Fig. 12. It can be seen
that this change in symmetry will lead to a further reduction of N(0). Then S would be so small, meaning the
spin fluctuations would have vanished, that superconductivity might become possible. Recent experiments[36]
show that the susceptibility decreases at least an order
of magnitude due to low temperature irradiation. This is
in good agreement with the simple model based on the
influence of interstitial Pd atoms.

     The preceeding experiments on disordered metastable
systems have given a strong hint that low temperature irradiation with heavy ions can be well understood by the
assumption of rapid quenching of the thermal spike region. Thus similar results are obtained as by vapor
quenching, where the condensed vapor atoms are rapidly
quenched by the cold substrate. However the Pd experiment has shown that additional collisional effects have
to be taken into account in order to describe the ion

techniques. This technique can perhaps produce a new
type of superconductor like irradiated Pd. Good candidates are other elements with spin-fluctuations, i.e.
Sc, Cr and Rh.

Besides the experiments described above a very de-
detailed knowledge of metastable and sometimes even amorphous phases of transition metals like Mo[37], V[38], Nb[39]
and Re[39] have been achieved by ion implantation. These
very interesting results could not be covered by this
paper since there is little known about the production
of those disordered phases by other non-equilibrium
techniques.

HIGHLY ORDERED SYSTEMS:

In this chapter the production of well-ordered,
stoichiometric superconductors by means of ion beams
will be discussed rather briefly. A more detailed discussion is given in references 2 and 4. Lattice defects,
which are inevitably created during ion implantation, are
the main reason opposing the production of ordered superconductors. In order to get rid of these defects the specimen has to be annealed during or after ion bombardment.
This chapter will not deal with the large number of experiments using ion beams for the study of the influence
of lattice defects on T_C (compare references 2,3,4).

NbC was the first ordered superconductor to be produced by ion implantation.[41] As starting material under-
stoichiometric single crystals of $NbC_{0.69}$ (B1 structure)
were used. This material has a T_C of only 3.7 K since T_C
depends very sensitively on the stoichiometry in these
B1 materials.[42] Then two different experiments were performed in order to increase the C-content further. In
one case C-ions were implanted at different substrate
temperatures whereas in the other case a C layer was evaporated on top of the single crystal. Then the specimens
were annealed in order to remove the irradiation defects
or to diffuse the C overlayer into the crystal. The resulting T_C values are shown in Fig 13 as a function of
annealing temperature. A maximum T_C of 11.5 K is obtained in the implanted specimen after annealing to
1050°C. In contrast the interdiffused sample has to be
heated to 1450°C resulting in a lower T_C of 11.1 K. The
higher T_C in the implanted specimen is due to a C content
closer to stoichiometry which could be achieved by the
lower annealing temperature corresponding to a smaller
number of vacancies in the thermodynamic equilibrium.

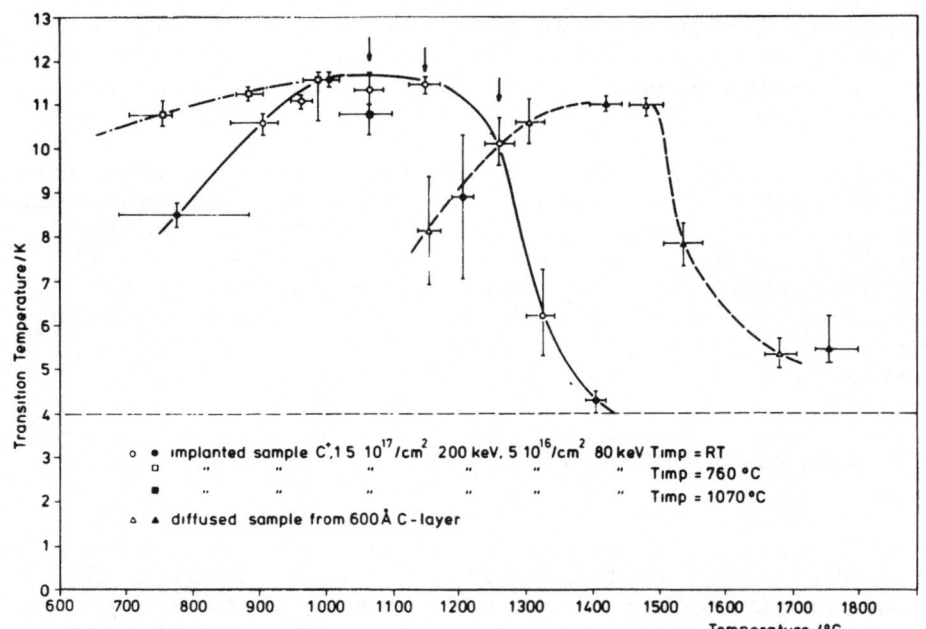

Fig. 13. T_C of a single crystal $NbC_{0.89}$ implanted with
C^+ ions (solid line) and after interdiffusion
of C (dashed line) versus annealing temperature
(after ref. 41)

In a recent experiment the production of Nb_3Sn was
attempted by means of ion implantation.[43] Nb_3Sn was
chosen because it is an easyly producible A15 superconductor. The A15 materials are very interesting superconductors because all high T_C materials crystallize
in this structure. In contrast to the rather simple B1
structure (like NbC), the A15 represents a very complicated crystal structure which is very sensitive to radiation damage. Thus it was interesting to test if such a
material can be produced by ion implantation too. Evaporated thin Nb films served as starting material. Also
in this experiment Sn was introduced not only by implantation but also by interdiffusion of a Sn overlayer.
The result is shown in Fig. 14 where superconducting
transitions are plotted for different preparation steps.
It can be clearly seen, that the implanted specimens
need a higher annealing temperature than the diffused
ones (925°C against 850°C) in order to obtain the same
T_C of 17.8 K. It should be emphasized that the as-im-
planted material already consists locally of A15 material,
as could be shown by Mössbauer effect experiments[44],
but the long-range order and thus a high T_C is missing.
These results demonstrate that it is more difficult to

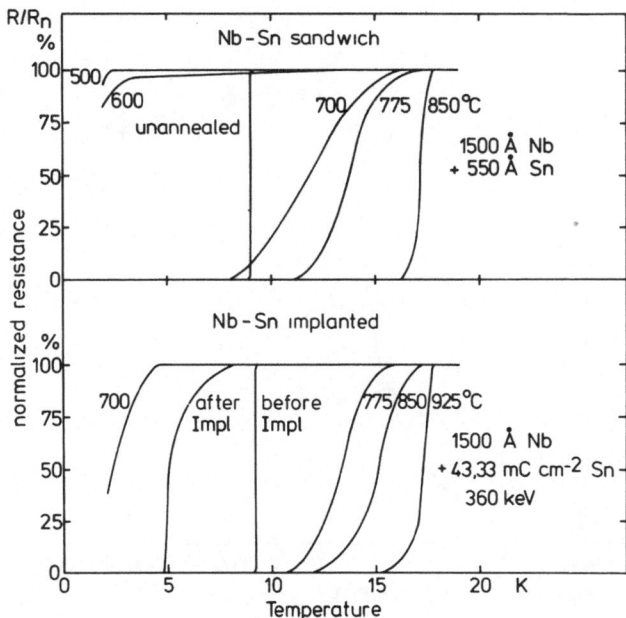

Fig. 14. Normalized resistance versus temperature of a
Nb-Sn sandwich film (upper part) and a Sn-im-
planted Nb film (lower part) for the indicated
preparation stages and annealing temperatures
(after ref. 43)

produce a well ordered A15 phase out of a heavily disordered phase than out of unirradiated Nb. It was not
possible to produce Nb_3Ge by implanting Ge into Nb and
subsequent annealing since Nb_3Ge is not stable at temperatures above $\sim 800°C$.

From these experiments one can argue that the implantation method is able to produce well-ordered stoichiometric material as long as the defects induced by
irradiation anneal out before the desired phase is
formed.

These difficulties were overcome by a different
attempt where Ge ions were implanted into bulk material
of $Nb_3(Ge_{0.8}Nb_{0.2})$ with A15 structure and a T_C of 6.5
K.[45] In this case annealing up to 750°C was successful
because the Ge enriched and highly damaged layer could
recrystallize epitaxially in the A15 phase. Thus T_C
values up to 16.2 K could be achieved. This result is
demonstrated in Fig. 15.

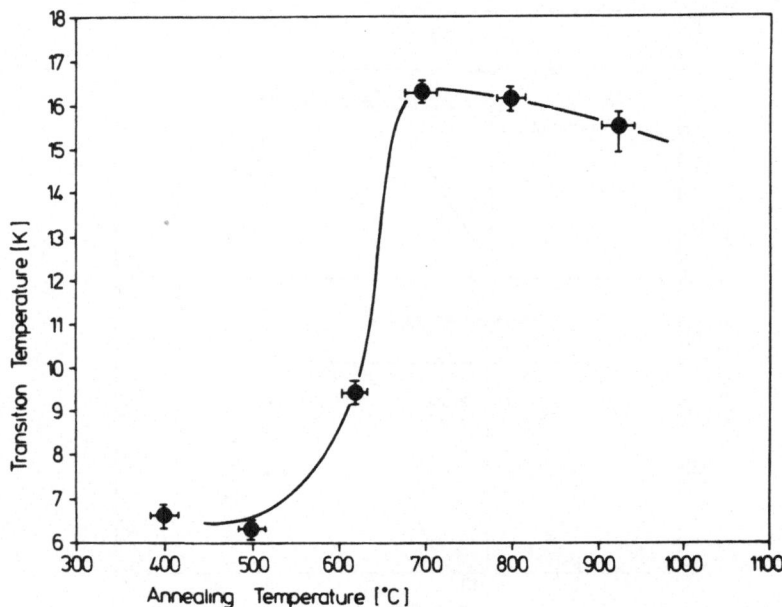

Fig. 15. T_C of a Nb_3 ($Ge_{0.8}Nb_{0.2}$) crystal with A15 struc-
         ture implanted with Ge^+ ions as a function of
         annealing temperature (after ref. 45)

A similar experiment involving epitaxial regrowth
was performed with the aim of producing Nb_3Si with A15
structure. Nb_3Si is speculated to have a higher T_C than
Nb_3Ge (T_C= 23.2 K) and Nb_3Sn (T_C=18.2 K) according to
the sequence Si-Ge-Sn in the periodic table. For this
implantation experiment[46] an A15 material $Nb_3(Al_{0.9}Si_{0.1})$
was implanted with further Si. Annealing resulted in a
surface layer of A15 $Nb_3(Al_{0.8}Si_{0.2})$. However, T_c was
only 5.6 K.

All these different attempts show, that a difficult crystal structure like A15 can in principle be
made. In the case of a stable A15 material (Nb_3Sn)
annealing to high enough temperatures is sufficient
to form the ordered material. This is not possible for
metastable A15 phases (Nb_3Ge). In these cases epitaxial
regrowth of the implanted material onto well ordered
crystals is helpful. However, it is no difficulty to
produce stoichiometric, well ordered material with
simpler crystal structures by ion implantation. There,
ion implantation can be advantageous. It can yield better
stoichiometry since the post annealing requires lower
temperatures compared to diffusion methods. Thus a smaller number of vacancies is built into the resulting
compound.

At the moment the situation is not yet so clear for the production of superconductors by means of pulsed laser and electron beams. However, the difficulties seem to be very similar to ion implantation, because the rapid resolidification hinders a perfect recrystallization. The references 47-51 and 12 are recommended for better information about the present situation.

METAL HYDROGEN SYSTEMS

This chapter will deal with the interesting super-conducting properties of metal-hydrogen systems. The production and investigation of most of these systems has only become possible since the method of hydrogen or deuterium implantation was applied to the metals of interest. This interest in the superconducting properties of metal-hydrogen arose with the discovery of the re-markably high T_C values of Th_4H_{15}[52] and PdH[53,54]. In addition the observation of an inverse isotope effect,[54] i.e. a higher $T_C \simeq 11$ K in PdD compared to $T_C \simeq 9$ K in PdH, and the astonishing increase of T_C to 17 K in hy-drogenated PdCu alloys[55] stimulated further experimental and theoretical work. Comparably high T_C values have only been found in Nb based well ordered materials.[56]

In the following the influence of H on the super-conducting properties will be briefly summarized for the well studied PdH system. The other examples, both transition and non-transition hydrogen systems where hydrogen excerts a positive influence will be described. It will become evident that both electronic and phonon effects, especially coupling to optic H phonons, deter-mines superconductivity in metal hydrogen systems. I will only describe some common features for a personal choice of experimental results. A more complete overview can be obtained in several review articles[57] including all the references, which cannot be cited here. Theo-retical considerations are summarized in reference 58. In addition to the equations (1)-(3) one has to consider that the H interstitials lead to optic phonons in the metal-H system. Since the mass of the H is at least one order of magnitude smaller than the mass of the last atom the electron phonon coupling constant λ can be split into an acoustic and an optic part:

$$\lambda = \lambda_{ac} + \lambda_{op} \tag{4}$$

In general the occurrence of λ_{op} should lead to an enhancement of T_C as long as other properties remain unchanged. However, the influence of optic phonon modes
due to H(D) interstitials decreases with increasing
frequency λ_{op} of the optic modes. In addition the H
interstitials will change the electronic properties
of the host, because in general the electron of the
H will be partially transferred into the conduction
band of the metal host.

In the following I want to consider mainly those
superconducting metal-H systems which have been produced
and investigated by H(D) implantation into metallic
targets at low temperatures (<10 K). This non-equilibrium technique has to compete with the conventional
H(D) charging methods, like electrolytic or pressure
methods which depend on the H(D) solubility of the metal.
However, this solubility is very low in most of the
interesting superconducting systems. Thus only the H(D)
implantation method can be applied without problems.
The disadvantages of this technique, i.e. the introduction of lattice defects and the inhomogeneous, not
well-known H(D) distribution, have to be taken into
account.

Pd-H System[57]:

A large variety of both experimental and theoretical work yielded a good understanding of the microscopic
mechanisms causing superconductivity in PdH(H). These
mechanisms will be briefly summarized in the following.

As already mentioned before pure crystalline Pd is
not a superconductor (at least above the lowest applied
temperature of 1.7 mK) although its high value of N(0)
(compare Fig. 12) and its suitable phonon spectrum should
result in a reasonable T_C. The reason is that Pd is a
strongly exchange enhanced paramagnet. The resulting
pair breaking effect of the spinfluctuations μ_p is sufficient to explain the absence of superconductivity.
The addition of H(D) to the Pd decreases N(0), because
besides the creationof new s-states far below the Fermi
energy the s and d bands of the Pd at the Fermi energy
are filled. Thus μ_p decreases with increasing H content
and the system becomes diamagnetic for concentrations
H/Pd $\gtrsim$ 0.65. However, this decrease of μ_p is only a
precondition for the occurrence of superconductivity,
but it is not sufficient to explain the high observed
T_C values. These result from an especially

high electron-phonon coupling due to the H(D) interstitials. These light interstitials give rise to optic
phonon modes at comparable low energies, i.e. $\omega_{opD} \simeq 45$
meV for H and $\omega_{opD} \simeq 31$ meV for D. By means of superconducting tunneling experiments and measurements of
the temperature dependence of the electrical resistance
it could be demonstrated that the coupling to the optic
modes dominates the attractive electron-phonon interaction. This strong electron-phonon coupling to the
optic modes is caused by the large density of s-elec-
trons, $N(0)_H$, at the site of the H or D interstitials.
Whereas the overall value for $N(0)$ decreases $N(0)_H$ increases with increasing H(D) concentration. Thus the
coupling of s electrons to the optic phonons results
in a large value of $\lambda_{op}=(1-3)\cdot\lambda_{ac}$. The herewith calculated dependence of T_c on the H(D) concentration agrees
qualitatively with the measured values. However, there
is some disagreement with regard to the observed inverse
isotope effect. There are two main models to explain
the higher T_c of PdD compared to PdH:

a) Anharmonic effects lead to a ratio $\omega_{opH}/\omega_{opD} > \sqrt{2}$.
   Thus the optic phonons due to D interstitials couple
   more effectively to the electrons and $\lambda_{opD} > \lambda_{opH}$.
   Ratios $\omega_{opH}/\omega_{opD} > \sqrt{2}$ have been detected by superconducting tunneling, neutron scattering and resis-
   tance measurements.

b) The electronic properties of PdD and PdH are different because more bonding states are created in
   the H case compared to the D case. Consequently the
   electronic density of states at the interstitial
   sites are different for the two isotopes: $N(0)_H <
   N(0)_D$. This reasoning is in agreement with NMR and
   de Haas van Alphen measurements as well as with the
   observed scaling of T_c.

Model b has received a lot of support from recent neutron
scattering experiments[59] which indicate that the H(D)
atoms vibrate in totally harmonic potentials. However,
the potentials are quite different for H and D leading
to the observed ratio $\omega_{opH}/\omega_{opD} > \sqrt{2}$.

     In summary, the reasons for superconductivity in
the PdH system are quite obvious:

   i) $\mu_p = 0$
  ii) substantial coupling to the optic H(D) phonons
 iii) high s-electron density of states at the intersti-
      tial site.

The situation is not as clear in the case of hydrogenated Pd-noble metal alloys where T_C increases up to
17 K for PdCu alloys. Up to now, these rather high superconducting transition temperatures could be only
achieved with ion implantation methods. A systematic
study of the variation of T_C with Cu concentration was
performed by low temperature implantation of H and D.
The result is shown in Fig. 16 where the maximum T_C with
respect to H(D) implantation is shown for different PdCu
alloys. The inverse isotope effect not only vanishes but
also changes sign with increasing Cu content. In a recent
experiment about 25 at.% Cu has been implanted into the
surface layer of rather thick Pd foils. Then the Pd
foils were charged electrolytically with H at 77 K.
Broad superconducting transitions starting at about
16 K have been observed.[60] The increase of T_C with increasing Cu content is not yet fully understood. There
are some hints that the acoustic phonon spectrum is
weakened and thus λ_{ac} increases with the addition of
noble metals. A more important effect might be that
$N(0)_H$ increases with increasing noble metal content.
However, the reasons for a maximum T_C with respect to
noble metal concentration and the change of sign of the

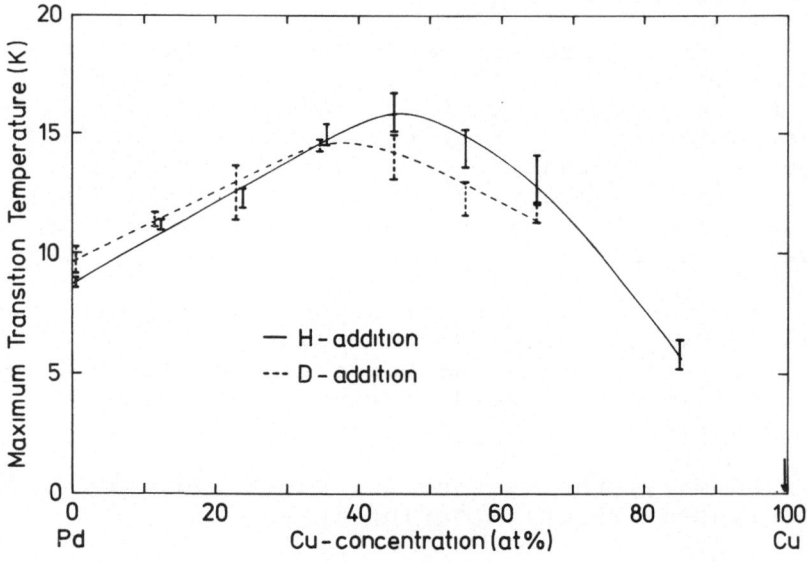

Fig. 16. Maximum T_C with respect to H (solid line) or
          D (dashed line) implantation as a function of
          composition of the PdCu host (after ref. 6)

isotope effect are not fully understood. Recent experiments[61] show that there is some correlation between the
high temperature resistivity and T_C or λ, respectively.
Perhaps further experiments of this kind will yield a
better understanding of the microscopic processes.

Assumptions that the disorder produced by the implantation is respondible for the high T_C seem to be
rather unlikely for the following reasons. In pure
PdH(D) the measured T_C values are independent of the
H(D) charging method. In addition it has been shown that
ordering effects influence T_C only very slightly.[62] On
the other hand, high pressure H charging (up to 20 kbar)
of PdCu alloys yielded no T_C values above 9 K, the value
for PdH.[63]

Other Transition Metal-Hydrogen Systems:

The only new transition metals where an enhancement
of T_C by addition of H or D has been found in recent
years are the elements of groups IV B, i.e. Ti, Zr and
Hf.[64] No improvement of the superconducting properties
have been found for Pt[65], V[66], Cr[67]. The results for Ti,
Zr and Hf are shown in Table 3. The H(D) charging was
performed by low temperature implantation. The effect
of lattice disorder on T_C was simulated by implantation
of inert He atoms. It can be seen that T_C increases

Table 3

Maximum T_C values for group IVB elements implanted with
H(D) and corresponding experimental parameters (ref. 64)

Element (T_c/K)	Implanted Species	Maximum T_c/K	Corresponding H(D)/metal	Isotope effect
Ti	H_2^+	4.95	0.14	Normal
(0.4)	D_2^+	4.89	0.13	sign
	He	1.00	–	
Zr	H_2^+	3.14	0.13	Inverse
(0.6)	D_2^+	4.65	0.13	
	He	1.49	–	
Hf	H_2^+	1.75	0.18	Inverse
(0.13)	D_2^+	2.23	0.16	
	He	1.00	–	

substantially by the addition of H or D whereas disordering effects T_C much less. Although the implanted
H(D)/metal concentrations amount to $\sim$0.15, preliminary
x-ray measurements show only the presence of α-phase.
For both Zr and Hf a remarkable inverse isotope effect
is observed, whereas a slight normal isotope effect occurs in the Ti case. Similar results have been obtained
by co-deposition of H(D) and Zr onto targets at 4 K[68].
Recent neutron scattering experiments on α-phase hydrogenated and deuterated Zr showed no sign of any anharmonic effects in this system[69]. Thus the model developed
for the explanation of superconductivity in PdH does
not seem to be appropriate. It is assumed that mainly
electronic effects dominate the superconducting properties of H(D) implanted Ti, Zr and Hf.

Non-Transition Metal-Hydrogen Systems:

     A very interesting system, which had been studied
in the past, is the Al-H(D) system[70,71]. H or D implantation at low temperatures yielded a remarkable increase
of T_C from 1.17 K to 6.75 K or 6.05 K. Simulation of
the influence of lattice defects by He implantation
resulted only in a $T_C \simeq$ 3.7 K. Recently we started a
systematic study of the H(D) influence on the superconducting non-transition metals in order to get a better
understanding of the effect of H(D) interstitials on
superconductivity[72]. Since most of the non-transition
metals have no solubility for hydrogen we used the low
temperature implantation method. Besides H and D, He
also was implanted to simulate the plain radiation effect on T_C. In all cases we found T_C enhancement due to
H or D implantation far exceeding the increase caused
by lattice defects.

     In this connection, Pb is an especially good ex-
ample since it is the only non transition metal superconductor whose T_C value is not enhanced but slightly
reduced by lattice defects. Fig. 17 shows the result of
the low temperature implantation experiment. T_C of a
Pb-foil is plotted as a function of the implanted dose
of H, D or He[73]. It is obvious that the hydrogen isotope considerably increase T_C in contrast to He. A pronounced inverse isotope effect, i.e. a higher T_C for
the deuterated sample compared to the hydrogenated
sample, can be detected. However, it is not yet clear if
this isotope effect is real or if D can be accumulated
to higher concentrations in the implanted region due to
a lower D mobility at 4 K compared to H. The initial decrease of T_C can only be detected in a thin film where

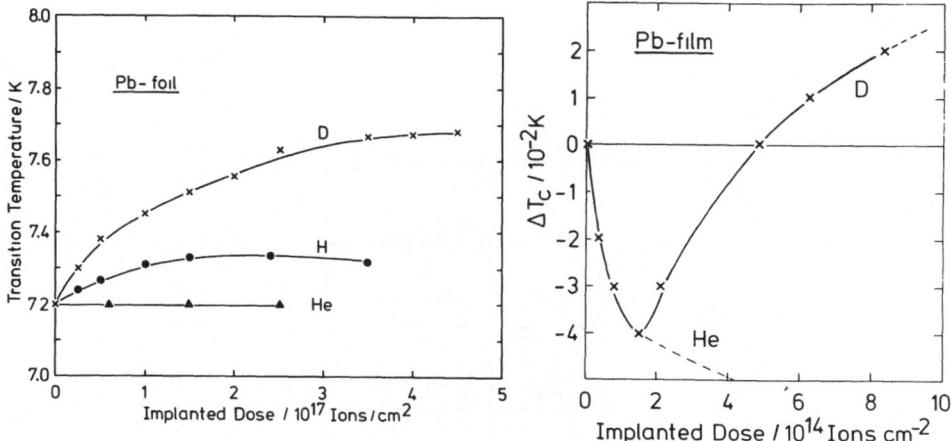

Fig. 17. T_C of a Pb-foil (a) and T_C of a Pb-film (b)
as a function of the implanted dose of H, D
and He (after ref. 73, 72)

the whole Pb area is affected by the He implantation.
Fig. 17b cleary demonstrates the initial T_C depression
due to lattice defects for both He and D implantation.
But soon the T_c enhancement of the D atoms takes over.

Besides Pb various other non transition metals
like Be, Zn Cd, In and Sn, were previously investigated.[72] The results of these implantation experiments
are included in Fig. 18, showing part of the periodic
table. The results for all elements which have been
investigated for the influence of H or D according to
my present knowledge are included. For these elements
the known T_C values for the crystalline,[56] the vapor
quenched[56] and the He irradiation metal as well as for
the H and D charged metal are plotted.[57] In addition
the method of H, D charging is included and the sign
of the isotope effect is indicated by a small arrow.

For the H(D) charged non transition metals the
following common properties can be summarized:

a)   H and D excert a special influence, enhancing T_C.

b)   The sign of the isotope effect seems to be quite
       accidental.

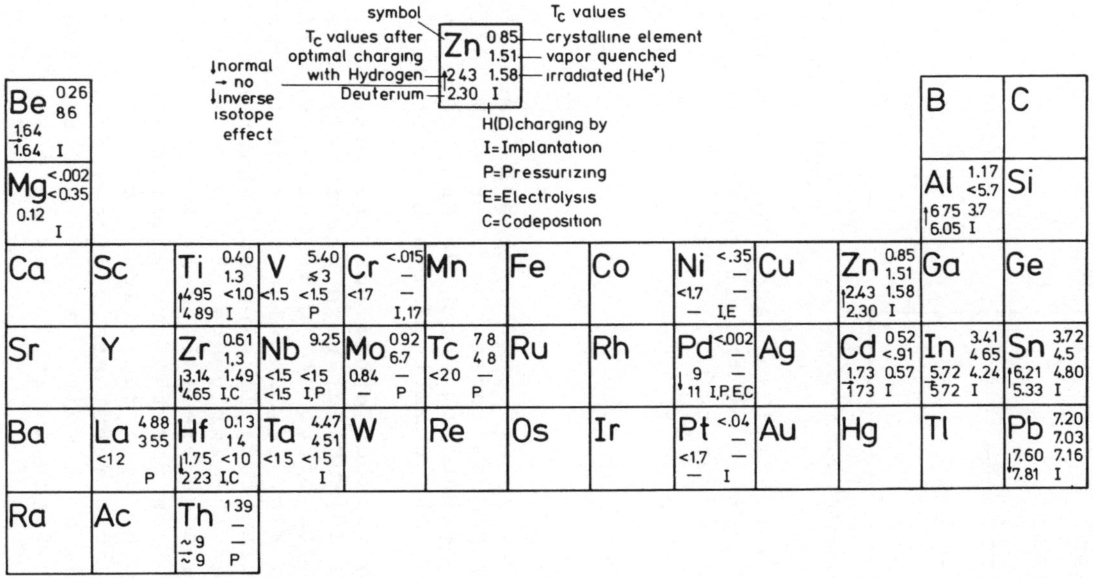

Key (example cell):

- symbol
- T_c values after optimal charging — with Hydrogen / Deuterium
- ↓ normal → no ↓ inverse isotope effect
- T_c values — crystalline element / vapor quenched / irradiated (He⁺)
- H(D) charging by: I = Implantation, P = Pressurizing, E = Electrolysis, C = Codeposition

Example:
Zn | 0.85 | 1.51 | 1.58 I | ↓2.43 | 2.30 I

Periodic chart data:

Element	crystalline	H	D (method)	vapor quenched	irradiated (method)
Be	0.26	8.6 I		1.64 ↓1.64 I	
Mg	<.002			0.12 I	
Ti	0.40	1.3	<1.0 I	↓4.95	↑4.89 I
V	5.40	≤3	<1.5	<1.5 P	
Cr	<.015	—	<17	1,17	
Ni	<.35	—	—	<17	— 1E
Zn	0.85	1.51	1.58 I	↓2.43	2.30 I
Al	1.17	<5.7		↑6.75 3.7	6.05 I
Zr	0.61	1.3	1.49 IC	↓3.14	↑4.65
Nb	9.25	<1.5	<1.5 I.P		
Mo	0.92	6.7	0.84	— P	
Tc	7.8	4.8		<20	— P
Pd	<.002	—		↓.9	11 I,P,EC
Cd	0.52	<.91	—	1.73	1.73 I
In	3.41	4.65	5.33 I	5.72 4.24 ↑6.21	5.72 I
Sn	3.72	4.5	4.80 I	5.72 ↑6.21	5.33 I
La	4.88	3.55		<12	P
Hf	0.13	1.4		↓1.75 ↑2.23 IC	<10 <15 I
Ta	4.47	4.51		<15	<15 I
Th	1.39	—		↓9 ↑9	P
Pb	7.20	7.03		↓7.60 7.16	7.81 I

Other elements shown without data: Ca, Sc, Mn, Fe, Co, Cu, Ga, Si, B, C, Ge, Sr, Y, Ru, Rh, Ag, Ba, Re, Os, Ir, Pt, Au, Hg, Tl, Ra, Ac.

Pt | <.04 | — | — | <17 | — I

Fig. 18. Part of the periodic chart including data for those elements which have been investigated for superconductivity after H(D) charging

In the present situation these results are rather puzz-
ling because there is no obvious explanation for all the
observed effects on superconductivity. However the increase of T_C can be fitted quite nicely using a McMillan
formalism.[72] In order to achieve a detailed understanding
of the microscopic mechanisms determining the superconducting properties it is necessary to obtain much more
detailed information, like tunneling experiments and
detailed band structure calculations. Recent tunneling
experiments on Pb and In films doped with H by the
method of vapor quenching proved the electron-phonon
coupling to the optic phonons.[74] Thus a model as developed for PdH seems to be most applicable. However, the
T_C values of these investigated films are slightly lower
compared with the H free films. Perhaps the H concentration was too low to increase T_C noticeably. In any
case more information about changes in the phonon spectra
and the electronic properties is required.

CONCLUSIONS

     In this preceeding article I have attempted to give
an impression of the enormous potential which is implied
in ion beam techniques for the production and improvement of superconductors.

     It has been shown that a large number of low tempera-
ture ion bombardment experiments can be explained by a
thermal spike consideration, i.e. a locally very hot
cascade region which cools down very fast to the surrounding lattice temperature. The quenching rate involved in this process is comparable to that of vapor quenching. However, in some experiments the collisional aspect of the ion bombardment plays an important role.
Thus pure Pd could be transformed into a new type of
material which becomes superconducting.

     In addition it was shown that ion implantation can
be used also for the production of stoichiometric, well
ordered superconductors. However, post-annelaing is necessary to remove the introduced lattice defects. This
annealing hinders the formation of interesting metastable phases, an obstacle which could be overcome partially by epitaxial regrowth.

     In the last chapter metal-hydrogen systems have
been discussed which could be produced only by H(D)
implantation. Their very interesting properties under-

line the importance of the implantation technique for
superconductivity.

In summary I wanted to demonstrate that there is
a rapidly growing interest in ion beam techniques for
the production and also the investigation of new materials with interesting superconductivity properties.
Since the ion beam process is a highly non-equilibrium
technique it has the potential to produce nearly any
element combination in a well defined manner. Amorphous
or well-ordered superconductors can be produced in a
well-defined, reproducible and clean manner. Concentration dependences can be very easy studied on one
and same sample in one experiment. All these advantages
of the ion beam techniques will be used further for
the production of new superconductors and will deliver
in the future exciting results.

REFERENCES

1. B. Stritzker, J. Nucl. Mat. 72: 256 (1978).

2. O. Meyer, Superconductivity, in: "Treatise on Mate-
    rials Science and Technology," J.K. Hirvonen, ed.,
    p. 415, Academic Press, New York (1980).

3. H. Bernas and P. Nedellec, Nucl. Instr. and Meth.
    183: 845 (1981).

4. O. Meyer, to be published in Proc. EPS-Meeting,
    Istanbul, Sept. 1981.

5. B. Stritzker, Metastable Alloys, in: "Surface Modi-
    fications and Alloying", Nato Institute, to be pub-
    lished.

6. B. Stritzker, Z. Physik 268: 261 (1974).

7. G. Bergmann, Physics Rep. 276: 161 (1976).

8. W. Buckel, Liquid-like Amorphous Thin Films, in:
    "Diffraction Studies on Non-Crystalline Substances",
    p. 713, Publishing House of the Hungarian Academy
    of Sciences, 1980.

9. T.H. Geballe and J.M. Rowell, Thin Solid Films 91:33
    (1982).

10. H.J. Güntherodt, P. Oelhafen, R. Lapka, et al.,
    Inst. Phys. Conf. Ser. 55:619 (1981).

11. W.L. Johnson, Superconductivity of Strongly Disor-
    dered and Glassy Materials, in: "Superconductivity
    in d- and f-Band Metals 1982", W. Buckel and W. Weber,
    eds., Kernforschungszentrum, Karlsruhe (1982), p.341.

12. B. Stritzker, Superconducting Effects in Rapidly Quenched Materials Formed by Ion and laser Bombardment, in: "Laser and Electron-Beam Interactions with Solids," B.R. Appleton and G.K. Celler, eds., Elsevier Science. Publishing Company, New York (1982), p. 363.

13. W. Buckel and R. Hilsch, Z. Physik 138: 109 (1954).

14. P. Duwez, R.H. Willens and W. Klement, J. Appl. Phys. 31:1136 (1960).

15. B.R. Appleton and G.K. Celler, eds.,"Laser and Electron-Beam Interactions with Solids", Elsevier Science Publishing Company, New York (1982).

16. V.N. Gridnew, I.Ya. Dekhtyar, L.I.Ivanov et al., JETP Lett. 18:154 (1973).

17. W. Ostwald, Z. Physik. Chem. 22:289 (1897).

18. W. Buckel, Z. Physik 238: 136 &1954).

19. A. Defrain, J. Chimie Physique 74:851 (1977).

20. U. Görlach, M. Hitzfeld, P. Ziemann and W. Buckel, Z. Physik B 47:227 (1982).

21. G. Heim, W. Bauriedl and W. Buckel, J. Nucl. Mat.72: 263 (1978).

22. A. Hofmann, P. Ziemann and W. Buckel, Nucl. Instr. Meth. 183:943 (1981).

23. A. Comberg, S. Ewert and W. Wühl, Z. Physik B20:165 (1975).

24. B. Stritzker and S. Ewert (1978), to be published.

25. J. Klein, A. Leyer, J. Chaumont and H. Bernas (1976), unpublished.

26. B. Stritzker and H. Wühl, Z. Physik 243:361 (1971) and Z. Physik B24:367 (1976).

27. J.D. Meyer, J. Physique 41:68-762 (1980).

28. J.D. Meyer, F. Ochmann and B. Stritzker, Solid State Comm. 39: 419 (1981).

29. H.L. Luo and W. Klement, J. Chem. Phys. 36: 1870 (1962).

30. C.C. Tsuei and L.R. Newkirk, Phys. Rev. 183: 619 (1969).

31. G. Krauss, W. H.-G. Müller, F. Baumann and W. Buckel, J. Less-Common Metals 43: 13(1975).

32. J.D. Meyer and B. Stritzker, Nucl. Instr. Meth. 183:965 (1981).

33. B. Stritzker, C.W. White, B.R. Appleton, and S.T.
    Sekula (1980), to be published.

34. B. Stritzker, Phys. Rev. Lett. 42: 1769 (1979) and
    Inst. Phys. Conf. Ser. 55:529 (1980).

35. D.A. Papaconstantopoulos, B.M. Klein, E.N. Economou
    and L.L. Boyer, Phys. Rev. B17:141 (1978).

36. J.D. Meyer and B. Stritzker, Phys. Rev. Lett. 48:502
    (1982).

37. G.Linker and O. Meyer, Solid State Comm. 20:695 (1976).

38. G. Linker, J. Nucl. Mat. 72:275(1978).

39. G. Linker, Rad. Effects 47: 225 (1980), and ref. II,
    p. 367.

40. A. ul. Hag and D. Meyer, to be published, J. Low.
    Temp. Phys.

41. J. Geerk and K.-G. Langguth, Solid State Comm. 23:
    83 (1977).

42. J.K. Hulm and R.D. Blaugher, Transition-Metal Super-
    conductors, in: "Superconductivity in d- and f-Band
    Metals", D.H. Douglass ed., AIP Conference Proceed-
    ings, New York (1972).

43. J.M. Soeder and B. Stritzker, J. Nucl. Mat. 72: 270
    (1978).

44. J.M. Soeder, B. Stritzker, J. Bolz, J.C. Glass and
    F. Pobell, J. Low Temp. Phys. 33: 9 (1978).

45. J. Geerk, Solid State Comm. 33: 761 (1980).

46. M.T. Clapp and R.M. Rose, J. App. Phys. 51: 540
    (1980).

47. B.R. Appleton, C.W. White, B. Stritzker, O. Meyer,
    J.R. Gavaler, A.I. Braginski, and M. Ashkin in:
    Laser and Electron Beam Processing of Materials,
    White and Peercy eds. (Academic Press, New York
    1980), p. 714.

48. O. Meyer, J.R. Thompson, B.R. Appleton, C.W. White,
    and S.T. Sekula, to be published.

49. B. Stritzker, B.R. Appleton, C.W. White, and S.S.Lav,
    Solid State Comm. 41: 321(1982).

50. B. Pannetier, T.H. Geballe, R.H. Hammond, and J.F.
    Gibbons, Physica 107B 471 (1981).

51. G. Linker and J. Geerk, to be published.

52. C.B. Satterthwaite and I.L. Toepke, Phys. Rev. Lett.
    25: 741 (1970).

53. T. Skoskiewicz, Phys. Status Solididi (a) 11: K123 (1972).

54. B. Stritzker and W. Buckel, Z. Physik 257: 1 (1972).

55: B. Stritzker, Z. Physik 268: 261 (1974).

56. B.W. Roberts, J. Physical and Chemical Reference Data 5: 581 (1976) and NBS Technical Note 983.

57. Review papers with additional references:
B. Stritzker and H. Wühl in: Topics in applied Physics, Vol. 29, Hydrogen in Metals II, G. Alefeld, and J. Völkl, eds., Springer-Verlag, Berlin (1978) p. 243.
D.G. Westlake, C.B. Satterthwaite and J.H. Weaver, Physics Today 31: 32 (1978).
W. Buckel, Z. Phys. Chem. 116: 135(1979).
B. Stritzker, to be published in Proc. Int. Symp. on the Electronic Structure and Properties of Hydrogen in Metals, Richmond, Virginia 1982.

58. M. Gupta, to be published in Proc. Int. Symp. on the electronic Structure and Properties of Hydrogen Metals, Richmond, Virginia 1982.

59. J.J. Rush, D. Richter and R. Hempelmann, to be published.

60. A. Leiberich, W. Scholz, W.J. Standish and C.G. Homan, Phys. Lett. 87A: 57 (19817.

61. A.F. Rex, J. Ruvalds and B.S. Deaver, to be published in Proc. Int. Symp. on the Electronic Structure of Hydrogen in Metals, Virginia 1982.

62. R.W. Standley, and C.B. Satterthwaite, to be published in Proc. Int. Symp. on the Electronic Structure and Properties of Hydrogen in Metals, Virginia 1982.

63. V.E. Antonov, I.T. Belash, E.G. Ponyatovskii and V.I. Rashupkin, JETP Lett. 31: 422 (1980).

64. J.D. Meyer and B. Stritzker, Nucl. Instr. and Meth. 182/183: 933 (1981).

65. A. Traverse, H. Bernas and J. Chaumont, Solid State Comm. 40: 725 (1981).

66. W. Däumer, K. Lüders, Z. Szücs and H. Weber, J.Less-Common Metals 78: 91 (1981).

67. H. Bernas and P.Nedellec, Nucl. Instr. and Methods 182/183: 845 (1981) and references therein.

68. P. Plein and S. Ewert, to be published.

69. R. Khoda-Bakhsh and D.K. Ross, J. Phys. F. 12: 15 (1982).
    R.Hempelmann, D. Richter and B. Stritzker, J. Phys. F, 12: 79 (1982).

70. A.M. Lamoise, J. Chaumont, F. Meunier and H. Bernas, J. Phys. Lett. (Paris) 36: L 271 and L 305 (1975).

71. S.T. Sekula and J.R. Thompson, Nucl. Instr. Meth. 182/183: 937 (1981).

72. F. Ochmann and B. Stritzker, to be published in Proc. IBMM 82, Grenoble.

73. F. Ochmann, J.D. Meyer and B. Stritzker, Physica 107B: 655 (1981).

74. B.W. Nedrud and D.M. Ginsberg, Physica 108B: 1175 (1981).

EMPIRICAL APPROACH TO SUPERCONDUCTIVITY*

T. H. Geballe

Department of Applied Physics
Stanford University
Stanford, California 94305

Bell Laboratories,
Murray Hill, New Jersey 07974

1. INTRODUCTION

The study of superconductivity has been characterized by distinctly different approaches. The empirical approach which I emphasize in these lectures, has roots which go back to the discovery of superconductivity at Leiden over 70 years ago. Even with all the analytical advances made since then it is easy to enumerate reasons why such an approach continues to be vital:

(i) There is the challenge with the possibly staggering impact upon technology of finding superconductivity at temperatures higher than the present 23 K limit. The usual way of discovering a new superconductor is (1) you prepare a sample which is a derivative of (i.e., by some logical way is derived from) a known superconductor, or is a sample made of an entirely new combination of elements, (2) identify the phase(s) present in so far as possible, and (3) observe the response of the sample to an electrical, magnetic or thermal probe at low temperatures. The observance of a superconducting response must have been an exhilirating experience 70 years ago - it still is! Unanticipated discoveries continue to provide clues for understanding superconductivity.

*Research at Stanford supported by the Air Force Office for Scientific Research Contract No. F49620-82-C-0014.

(ii) There is the possibility of uncovering new physics whenever superconductivity is discovered and the behavior is analyzed in new classes of materials. The pairing interaction which causes the superconducting condensation can, in principle be due to the (virtual) exchange of any Bose excitation - phonon, exciton, plasmon or magnon. The pairing can be between carriers with opposite spins (s-type pairing), or with parallel spins (p-type pairing). So far the original BCS model of phonon-mediated, s-type pairing describes all known superconductors although there have been suggestions that the behavior of some superconductors may involve other mechanisms.

(iii) There are new insights to be gained by the interplay between experiment and theory. Investigations of strong-coupling superconductivity using tunneling spectroscopy have been particularly effective. Properties which are averages of microscopic parameters of the theory can be systematically measured; in fact it is frequently more convenient to measure normal-state properties by making measurements of superconductivity than vice versa. Improvements in vacuum techniques and the control of thin film depositions and characterization have led to a more widespread and reliable use of tunneling spectroscopy.

In the following lectures, I will first review some successes and failures in the past history of the empirical approach, then new methods of synthesis and characterization using thin film technology which bring a wider range of superconducting compounds and composites and finally, recent applications of superconducting tunneling spectroscopy to some of the more interesting high transition temperature (T_c) superconductors will be given in conjunction with the more traditional methods such as specific heat.

2.   EXPERIMENTAL METHODS

The convenience and the extreme sensitivity with which superconductivity can be detected are major reasons why the empirical approach is so viable. I will review well known techniques for observing superconductivity that are of practical value as well as some specialized methods that are particularly useful in thin-film measurements.

A.   The Observation of the Transition

The most powerful tool in the search for superconductivity is the observation of a transition into the superconducting state - a second-order transition at a temperature T_c in zero magnetic field. Discontinuities in electrical properties and heat capacity are well-defined for pure materials, a consequence of there being little short range order above the transition due to the large superconducting coherence length.   In fact T_c's of selected elements are available from the National Bureau of Standards, Washington, D.C. as secondary temperature standards. However, in impure multiphase samples, in which new superconductors are usually discovered, the transition frequently occurs over a range of temperature corresponding to a range of composition, strain, and other variables which affect the transition. The breadth can depend upon the method used for detecting the superconductivity; its magnitude and shape provide useful information.   When reporting the occurrence of high-temperature superconductivity, researchers tend to quote the onset temperature, i.e., the highest temperature at which an unambiguous superconducting signal has been detected - possibly representative of only a few percent or less of the sample. This practice can be misleading but it is justifiable because the intent is to establish the existence of some superconductivity. One must bear in mind that the quoted temperature need not be representative of the bulk of the sample, and perhaps not even of the quoted composition and structure.

In fact, it is quite possible for only a small atom-fraction of the sample to give a full superconducting signal. On the one hand, this makes the detection of superconductivity a powerful method for identifying trace amounts of minority phases if they happen to have higher T_c's than the parent phase, and on the other hand the results are easily misinterpreted if the minority phase is not identified.

Convenient electric, magnetic and thermal methods for detecting T_c are indicated schematically in Fig. 1. Each has advantages which depend upon the purpose of the investigation and the material under study. They are classified as bulk or non-bulk, depending upon whether or not the signal is proportional to the atom fraction of superconductor present. Non-bulk methods are generally the first to be employed in any survey of new materials because of the convenience, sensitivity and speed with which the measurements can be made and interpreted. If superconductivity is detected by one of the non-bulk methods, then further methods should be employed. I have found the powdering method to be a very useful compromise between convenience and the likelihood of being able to make a correct interpretation.

3.   HISTORICAL DEVELOPMENTS

Let me illustrate with some past case histories, and some presently active studies, how the search for superconductivity and studies of its phenomenology can lead to insights. First let us consider some of the more notable discoveries made in the pre-war era (Table I).

In retrospect, there was much to be learned from just noting the circumstances under which the superconductivity was found. Important clues as to the mechanism of superconductivity and evidence for an energy gap were within experimental reach for a long time.

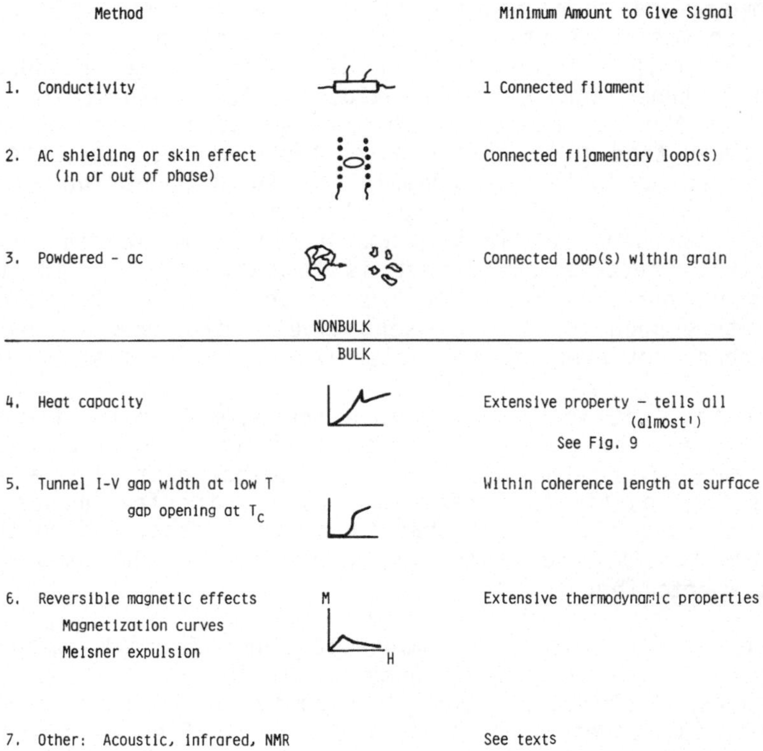

Method		Minimum Amount to Give Signal
1. Conductivity		1 Connected filament
2. AC shielding or skin effect (in or out of phase)		Connected filamentary loop(s)
3. Powdered - ac		Connected loop(s) within grain
	NONBULK	
	BULK	
4. Heat capacity		Extensive property - tells all (almost!) See Fig. 9
5. Tunnel I-V gap width at low T gap opening at T_c		Within coherence length at surface
6. Reversible magnetic effects Magnetization curves Meisner expulsion		Extensive thermodynamic properties
7. Other: Acoustic, infrared, NMR		See texts

FIG. 1 Conventional methods for detecting superconductivity.

Even prior to 1922, it was clear that high electrical conductivity in the normal state was in no way correlated with superconductivity; in fact there was an anticorrelation. At that time Onnes and Tuyn[1] undertook a study to see if the lower - energy vibrational modes of a heavy isotope of Pb (obtained by the decay of uranium) would result in a T_c detectably different from that of ordinary Pb. In their words,

"The object of the investigation was to establish the vanishing point of [the resistivity of] Pb more accurately, as well as to trace a possible difference in the vanishing point of Pb and uranium - Pb. Regarding a difference of vanishing point temperature for isotopes it seemed not impossible that the occurrence of superconductivity might be influenced by the mass of the nucleus."

TABLE   I

Some T_c Discoveries and Their Significance

DATE	MATERIAL	SIGNIFICANCE	REFERENCE
1913	Amalgam of Sn and Hg; and many eutectics	Superconductivity not confined to pure elements;	(a)
1929	CuS $T_c = 1.6$	A compound with poor electrical conductivity becomes superconducting even when its components, which include one of the best electrical conductors known, do not.	(b)
1930	Pb-Bi $T_c = 8$	T_c can be raised above components in solid solutions	(c)
1928	Ta $T_c = 4.4$	T_c occurs in transition metals with unfilled d-shells	(d)
1930	NbC $T_c = 10.5$	Highest T_c's found in intermetallic compounds between Nb and s-p elements	

(a)   K. Onnes, Communications Kammerling Onnes Lab Leiden 13, 133d (1913-14); W. J. De Haas and J. Voogd. ibid 18, 199c (1928-30).

(b)   W. Meissner, Z. Phys. 58, 570 (1929).

(c)   J. C. McLennan, J. F. Allen, J. Q. Wilhelm, Trans. Roy. Soc. Canada, 24, 25 (1930).

(d)   W. Meissner, Phys. Z, 29, 987 (1928).

(e)   W. Meissner and H. Franz, Z. F. Phys. 65, 30 (1930).

It can now be estimated[2] that the mass difference should have caused a $\Delta T_c \sim 0.01$ K - just outside the limits of reproducibility of the He gas thermometer used by Onnes and Tuyn. The isotope effect, not discovered until a generation later, was a major factor in establishing the electron-phonon interaction as the cause of superconductivity.

Even though many investigations of the phenomenological relationships between magnetic and thermal properties were carried out following the discovery of the reversible nature of the transition by Meissner and Ochsenfeld in 1933 (Meissner Effect) there was no immediate discovery of an energy gap. The electronic contributions to the specific heat in the superconducting state were all presumed to go as T^3, a prediction of the two-fluid model of Gorter and Casimir. The fact that that model gave reasonably good agreement kept the true exponential dependence over the temperature range well below T_c, the consequence of a relatively temperature independent energy-gap, from being recognized for twenty years.[3]

Investigations of the occurrence of superconductivity were undertaken by Hulm and Matthias in the early 1950's; some of these they describe in Vol. 68 of this series.[4] From those studies came the realization that superconductivity is a common phenomenon which occurs in most non-magnetic metallic elements and intermetallic compounds. By 1957 Matthias was able to summarize the occurrence of superconductivity throughout the periodic table in terms of the number of valence electrons per atom, the atomic or molecular volume, and the average mass of the constitutents. The average number of valence electrons per atom (e/a) proved to be useful in much the same way that the same parameter was used to predict the magnetic properties of alloys of the 3d transition metals by use of the Slater-Pauling curve.[5] The qualitative behavior of T_c as a function of e/a throughout the periodic system was depicted by Matthias as shown in Fig. 2.

Systematic investigations of superconductivity in transition metal alloys by Hulm and Blaugher[6] showed in fact, two maxima occurring in alloys between Zr-Nb at e/a = 4.7 and between Mo-Ru at e/a = 6.5 in the 4d series. Closely related behavior is found in other transition metal alloys made from elements adjacent or nearly adjacent to each other in the periodic table, and with some privisos, even to high T_c compounds. Those with the highest known T_c's occur in the A15 structure and with e/a between 4.5 and 4.75. Pseudo binary-compounds with the B-1 (NaCl) structure such as Nb(CN), which have the next highest range of T_c's, also have e/a $\sim$ 4.75 - 5.0. The occurrence of maxima in any system as a function of e/a, with values near to those quoted above has become known as Matthias' Rule. Using e/a as a guide for discovering new superconductors has been fruitful as can be seen

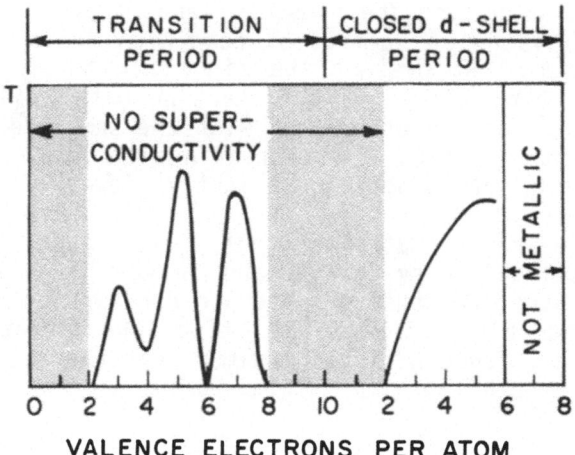

FIG. 2 Superconducting transition temperatures of the
elements throughout the periodic system. After B.
T. Matthias: "Low Temperature Physics" Vol. 2, C.
J. Gorter, Editor (1957).

in review articles[7] and in the references in Robert's
tables [8]. Its virtue is simplicity but its literal application
requires a rigid density of states of energy bands whose filling
depends only upon e/a. There is much evidence that specific
elements such as Nb have a strong influence on superconductivity
which implies that local atomic (chemical) configuration is
significant. The more recent empirical approach of Miedema[9]
retains the idea of universal curves for the behavior of T_c and
other properties which depend upon the density of states at the
Fermi level. Miedema allows for a transfer of charge between
atoms, proportional to the difference of the respective
electronegativities, and thus recognizes the importance of
chemical effects. This requires an additional set of
experimentally accessible parameters.

The e/a approach requires only the Periodic Table. For
example it predicts that transition elements to the right of Ti,
Zr or Hf, which can be dissolved in them, will raise T_c (Fig.
3). The rule works well for near-neighbors in the periodic table
such as Zr-Nb-Mo[6] but breaks down where there is charge-
transfer and other local interactions between the constituents.
Even in a seemingly simple system such as alloys of V and Nb a
minimum in T_c (at constant e/a) is found. The minimum may well be
associated with changing many-body spin fluctuations which compete
with superconducting interactions[10] in Nb,V and some

A15 compounds; tunneling spectroscopy lends support.[11] At best
an optimum e/a is not a sufficient condition for a maximum T_c in
any given system. At present the highest known T_c's occur with
compounds for which e/a ~ 4.75, starting with Nb_3Ge, (T_c onset
at 23.2 K). V_3Ge on the other hand with the same e/a and the same
A15 structure has T_c ~ 6.   Disordering the Nb_3Ge (at constant
e/a), or changing the composition slightly, lowers T_c drastically.

A break-down of Matthias' rule was found in amorphous
transition metal alloys by Collver and Hammond.[12] In amorphous-
systems where band structure effects are non-existent, or minimal,
and where one might expect to find the least amount of
complication they found a single maximum near the half-filled band
rather than the double maxima in Fig. 3.

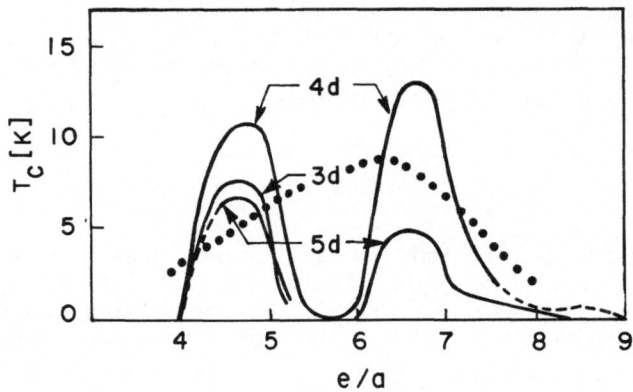

FIG. 3 Experimental T_c curves for elements and near
          neighbor alloys of the 3d, 4d and 5d transition
          metals.   Dotted line for amorphous 4-d.

An initially misleading, but eventually informative study
using the e/a approach was made in observing the effect of
dissolving Fe in Ti.[13]   In dilute Fe-solutions T_c increased
roughly 10 times as fast as expected from the behavior found for
other transition elements. This apparent enhancement of T_c by Fe
gave support to a hypothesis in vogue at the time that transition-
metal superconductivity might be due to, or enhanced by,
mechanisms other than phonon-induced pairing.   That suggestion
also derived support from the observed independence or vastly
reduced dependence of T_c upon mass found for the isotopes of Ru,
Zr, Os and Mo ("0-isotope effect") and other transition metals, in
contrast to the $M^{-1/2}$ dependence found for s-p metals.[14]

A thorough investigation which required very long annealing times finally gave a simple explanation of the seemingly enhanced T_c's in Ti(Fe) and required no unusual mechanism. The high-T_c was due to a distributed second phase of bcc β-Ti connected in grain boundaries and present in too small a concentration in the majority hexagonal Ti phase to be detected by X-ray diffraction techniques. As can be seen in Fig. 4 only 0.045 at % of the Fe remained in the hexagonal phase; the rest of the nominal 1 or 2% concentration of Fe was in the bcc phase which had an appropriate e/a ratio for the observed T_c. Upon powdering connectivity of the grain boundaries of bcc Ti was broken and the signal was vastly reduced.

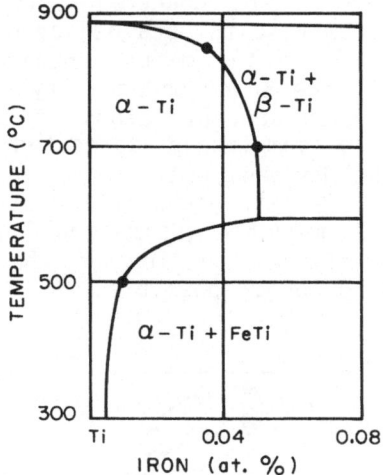

FIG. 4 Solubility of Fe in hexagonal (α) Titanium.
Temperature dependent phase boundaries well below
the melting point are difficult to establish; in
the present case months of careful annealing were
required (Ref. 13).

A study to test the limitations of the rigid band approach, which still has unexplored aspects, was carried out by synthesizing phases formed between the early d- and f-band elements and the Pt group metals.[15] Large differences in electronegativity and consequently substantial charge transfer occurs in such cases. The powdering method was introduced as a standard semiquantitative procedure in order to avoid the difficulties encountered in Ti(Fe). Powdering is only easily done

when the grain-boundaries are made of brittle, intermetallic
compounds, but that is frequently the case. However powdering is
not fool-proof. There are cases where the material is ductile and
subject to strain-induced lowering of T_c, or the topological
distribution of the superconducting phase is not destroyed by the
powdering. Usually at the stoichiometric composition the powered
sample can be expected to give the full signal. Some data for
LaRh are shown in Fig. 5. As little as 1% La in the melt produced
interconnected LaRh$_5$ regions in the grain boundaries and gave a
full diamagnetic signal. Upon powdering, as expected, only a small
fraction of the full signal remained. When the La
concentration was reduced < 0.1% there was considerable
broadening of the T_c even in the bulk. Transmission electron
microscopy[16] showed LaRh$_5$ filaments were still present, but no
longer continuous; shielding currents were sustained by proximity-
induced superconductivity in the connecting Rh phase. The TEM
results were confirmed from results obtained by dissolving a small
amount (~ 1 at. %) of Fe in the melts containing 0.1% La, and
0.5% or greater La. All the superconductivity in the 0.1 sample
was destroyed. With higher concentrations of La (> 0.5%) where
the LaRh$_5$ filaments were completely connected, no significant
change in T_c due to added Fe was found. One can conclude that if
Fe is soluble in LaRh$_5$ it is in a non-magnetic state, while in the
Rh phase it must be in a magnetic (pair-breaking) state. In the
latter case, proximity effect coupling of the LaRh$_5$ filaments
through the Rh phase is no longer possible.

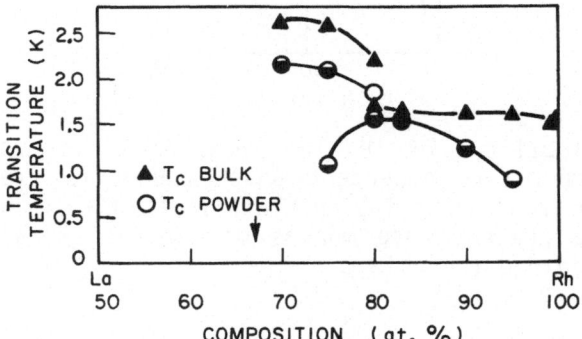

FIG. 5 Transition temperatures for arc melted (bulk)
        samples and for the same samples powdered and
        sieved. The filling of the circles represents the
        volume fraction of the powder which was
        superconducting. At 83% Rh 100% of the volume of
        the powdered sample registered a superconducting
        signal indicating the pure LaRh$_5$ phase was present
        At 75% there are two superconducting phases
        present. (Ref. 15).

4.   STILL OPEN QUESTIONS

A.   Unknown Structures

The dependence of T_c upon composition can signal interesting behavior.   For instance, in the Y-Ir (Fig. 6) system it can be seen that superconductivity exists when very small quantities of Ir are added to the arc-melted Y.   No signal survived during the ensuing powdering, again suggesting the existence of a small amount of an unidentified superconducting phase perhaps in grain boundaries as in the LaRh system.   Further work such as scanning-transmission electron microscopy (STEM) is needed in order to identify the phase.   The powder method also indicates the existence of 1, (or even 2) compounds in the high Ir side of the system.   It has recently been suggested that the superconductivity in that compositional range is unconventional, perhaps due to phonon softening related in some way to a very fine-scaled eutectic.[17]   However a metastable compound has since been isolated which has the composition and superconducting properties needed to explain the observations[18] and unless proven otherwise, is presumed to account for the observations.

There are other compositions where trace superconductivity was detected and which await further work to identify the cause. These can be seen in the Y-Co, and Y-Pt systems and also in the Sc-Co system (Fig. 6b).   While no superconductivity was found in the Y-Pd system, a ternary Heusler-alloy $Y-Pd_2Sn$ has recently been discovered to be superconducting.[19]

It is apparent from the rich behavior shown in Figs. 5, 6a and 6b that such studies are valuable and informative ways of discovering new phases and new superconductors.   Since there are many metastable systems that can be prepared by the techniques of rapid cooling, ion implantation, and variations it is likely that many superconductors remain to be discovered.

B.   Known Structures With Interactions Of Special Interest

$SrTiO_3$ is a perovskite-structure semiconductor which when made degenerate by replacing the Ti (4 valence electrons) with Nb(5 valence electrons), or by reducing the oxygen below stoichiometry, becomes superconducting.[20]   The carrier density is in the 10^{19} - 10^{20} per cc range, the lowest known concentration of any superconductor. T_c increases with roughly the 2/3 power of the carrier concentration in the decades below ~ 1K.   The dependence extrapolates to the T_c of another perovskite $Ba(PbBi)O_3$, where the superconductivity has been discovered at the surprisingly high temperature of 12 K even though it still has an

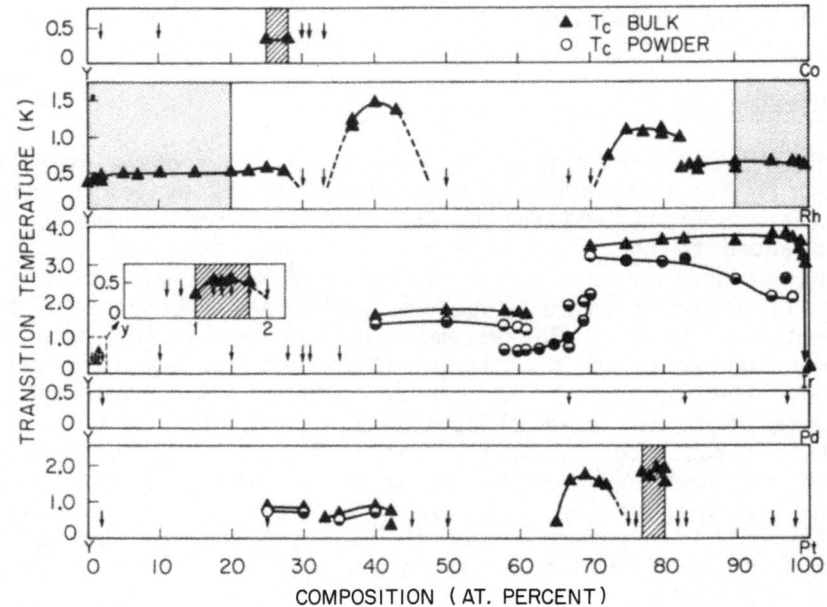

FIG. 6a Transition temperatures of yittrium-platinum metal
phases. The symbols have the same meaning as in
Fig. 5. (Ref. 15).

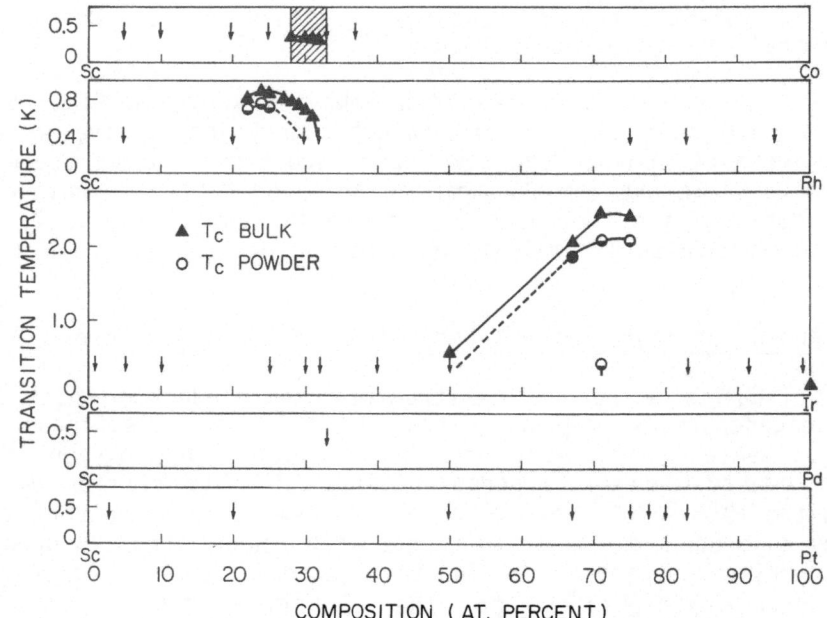

FIG. 6b Superconductivity   in   scandium-platinum   metal
phases. The symbols have the same meaning as in
Fig. 5. (Ref. 15).

order of magnitude fewer carriers than ordinary metals.[21] Tunneling data show coupling to extremely low-energy phonons which probably is important in causing this compound to have a higher T_c than any other s-p-band material (Batlogg et al., Ref. 21).

$CeCu_2Si_2$ is superconducting at T ~ 0.5 K.[22] The role of Ce is of particular interest. The present evidence indicates that the 4f electron is mixed into the conduction band giving a very large density of states at the Fermi level (effective mass ~ 200). The mechanism by which this very heavy Fermi liquid condenses into a superconducting state remains a puzzle. T_c is very sensitive to pressure at low pressures, perhaps related to similar behavior observed in α-U[23] in which charge density waves have recently been observed.[24] While charge density waves can be a common occurrence in anisotropic nearly 2-dimensional metals,[25] their occurrence in 3-dimensional metals is rare.

The above examples involve structures where there are more than two components, with more than two inequivalent sites in the unit cell (commonly referred to as ternary systems although there can be more than 3 sites components in systems such as intercalated crystals of TaS_2.)[26] In such cases it is possible to have one set of sites magnetic while another set supports superconductivity. The methods of investigation given in Fig. 1 do not probe within the unit cell; in such cases neutron scattering is most useful. Professor Maple discusses many interesting ternary systems in his lectures in this volume.

Organic superconductors are the newest arrival on the scene. They are covered in the lectures of Dr. Jerome also in this volume. Their characteristics will undoubtedly be much better understood in a few years; at present there are new compounds being synthesized which show a challenging combination of spin density wave and superconducting phenomena.

5. HIGH TEMPERATURE SUPERCONDUCTIVITY

What can be deduced concerning the probability of finding higher temperature superconductors and where might they be found? These questions have been around for a long time. One certainty is that if T_c's significantly above 23 K are found, there will be no lack of available models to explain the results. Two of the most studied are the one-dimensional model of W. A. Little[27] which invokes virtual excitons in the side chains of long organic molecules, and the two-dimensional Ginzburg model which invokes polarization charges induced across metal-semiconducting surfaces.[28] In these models moderate coupling constants are estimated which put kT_c within one or two orders of magnitude of the energy of the virtual excitations. Such models

above. Even though arguments can be given for believing such
models will never be realized,[29] they are also model
dependent.   There is no equivalent of the second law of
thermodynamics' ruling out of perpetual motion which exists to
rule out the possibility of excitonic superconductivity. In fact,
in my laboratory at Stanford, and by others elsewhere,[30]
transient signals have been observed above liquid nitrogen
temperatures in CuCl samples under pressure which were suggestive
of a superconductivity.   It was postulated they might have been
due to a polarization-induced electron pairing at a dynamic (i.e.,
a moving) metal-semiconductor interface.[30]   Effects which are
suggestive of superconductivity have also been observed in
pressure-quenched CdS.[31] Unfortunately it has not been possible
in CuCl or CdS or any other system to obtain reproducible signals
above ~ 23K that can be unambiguously attributed to superconductivity.   It will always be prudent to be alert for the
possibility of new pairing mechanisms which are not necessarily
restricted to low T_c's.   Innovative methods of detection going
beyond those in Fig.1 may be necessary to make further progress,
particularly when as is likely, the superconducting regions are
dispersed in a non-conducting medium rather than connected in a
nice way which would give a large volume signal. The only pairing
mechanisms established at present are phonon-induced.   Phonon
frequencies are typically $\lesssim$ 400 K; phonon-induced superconductivity in a typical metal is expected to be limited to
temperatures an order of magnitude less - not far from the present
limit.

   There have been many extrapolations to the limits of T_c in
given systems using the various parameters of the strong-coupling
theory, or trends in the periodic table.   None are particularly
distinguished for providing an increased understanding or
guidance. As we shall see, in the more detailed discussion of A15
superconductors to follow, when the coupling in an A15 phase as a
function of concentration reaches a certain magnitude some phase
change intervenes even though the stoichiometric concentration may
not have been reached.

   Today we are unable to define the occurrence of superconductivity much more precisely than we could 25 years ago in spite of
the fact that there are many more superconductors known and there
is a much better theoretical understanding. The empirical facts
may be stated as follows:

   (1) The highest T_c's exist in almost single A15 phase
metastable Nb_3X compounds (X = Ge, Ga, Al, Si) where the X-element
concentration is present in an amount exceeding the stable phase
boundary.

   (2) The 4-d transition metals have higher T_c's than

comparable 3d or 5d metals. The 3d metals have stronger electron-electron interactions which reduce the net pairing interaction. The 5d metals are heavier; their lower-T_c relative to the 4d's can be understood in terms of the mass (normal isotope) effect.

(3) T_c in the s-p metals reaches a maximum ~ 9 K in alloys of the heavy elements Pb and Bi. Tunneling spectroscopy,[32] shows that the electron-phonon coupling becomes stronger as e/a increases until finally in amorphous films phonon softening lowers T_c. The already mentioned $Ba(PbBi)O_3$ system is also based upon heavy metals and very low frequency phonons.

(4) There are however, classes of high-T_c materials in which the superconductivity can be attributed to coupling to much higher frequency phonons, for example quench-condensed amorphous Be films (T_c ~ 10 K). Of particular interest are the hydrides of Pd and its alloys where the optical-mode vibrations are coupled to the conduction electrons, as discussed in the lectures by Dr. Stritzker. T_c ~ 17 K can be reached in PdCu alloy films implanted at low temperatures with deuterium; the frequencies are unusually low for optical modes. There is still not enough known about this class to suggest that the limit of T_c has been reached. The fact that the Fermi-level cuts through the hydrogen s-band is probably essential.

(5) Also to be considered are ternary (and greater) phases with 3 or more intra unit cell sites which can act more or less independently of each other. Among the first discovered were non-cubic tungsten bronze oxides[33], and spinel chalcogenides.[34] The Chevrel-phase chalcogenides ($PbMo_6S_8$) and the ternary rare earth borides[35] have moderately high T_c's and are subject to chemical modifications of many kinds. Chevrel, in his lectures, discusses the crystal chemistry by means of which the Mo-cluster building-blocks are joined to make chains of varying length over a wide range. Unfortunately, T_c decreases as the length of the chains increases.

6.   THIN FILM SUPERCONDUCTIVITY

A.   Synthesis

Thin film deposition is a powerful technique for dealing with instabilities; with proper control it provides the means for producing and retaining metastable and unstable phases. Methods of synthesis rely upon codeposition or sequential deposition from

independently controlled sources.   These methods include chemical
vapor deposition, where the reacting species are introduced by a
flowing gas stream, and a chemical reaction takes place on the
substrate; and physical vapor deposition, where the components are
produced as beams from various sources by either sputtering or by
evaporation.   The most highly refined of these methods is
molecular-beam epitaxy which is employed in the fabrication of
semiconductor heterojunctions and multilayered quantum-well
structures.[36]   Progress has also been made in obtaining metallic
films with new structures, extended compositions and with
multilayered coherent interfaces.   It is likely that there will be
further progress in the near future.

I would like to describe the deposition system at Stanford
developed under the leadership of R.H. Hammond.[37]   The system is
particularly useful for handling high-melting refractory metals
such as Nb, and for fabricating tunnel junctions and microbridges.

The evaporator is shown schematically in Fig. 7.   The three
electron beams (whose trajectories are shown as dashed curves) are
drawn from tungsten filaments, magnetically deflected thorugh 270°
and focused on to the source materials.   The sources are in a
line; the center source has been compactly designed to reduce the
distance between the sources and hence the angular spread
subtended by the sources at the substrate.   It is possible to
steer the electron beam over the surface of the source, and to
sweep at frequencies up to 100 Hz.   The rate monitor signals are
fed back to control both the power in the electron beam and the
amplitude of the sweep.   The latter controls the power density at
the surface of the source with time-constant (~ 10 ms) and gives
control of composition at the monolayer scale.   This control is
particularly important when depositing films with metastable
compositions where a fluctuation could nucleate an unwanted phase.

Further details can be found in articles by Hammond and in
the doctoral theses of A. Hallak, D. Moore, J. Kwo, R. Feldman, K.
Kihlstrom A. de Lozanne and D. Rudman all of whom have made
significant contributions to the evaporation system.

The sample holder holds two or three rows of substrates,
each usually ten single crystal (6 mm x 6 mm x 1/2 mm) sapphires
and sits inside a furnace.   The substrate rows can be aligned
parallel to the sources as shown in Fig. 7, or perpendicular to
them.   In the former case with two adjacent sources active there
is a linear compositional variation along the row of ~ 0.7 at. %
per sapphire, a useful gradient for studying compositional
dependences.   In the perpendicular configuration, there is still
some variation in composition across each sample due to the
spatial geometry, but very little from sample to sample.   This
configuration is useful for studying the effects at constant

THREE COLINEAR SOURCE CONFIGURATION

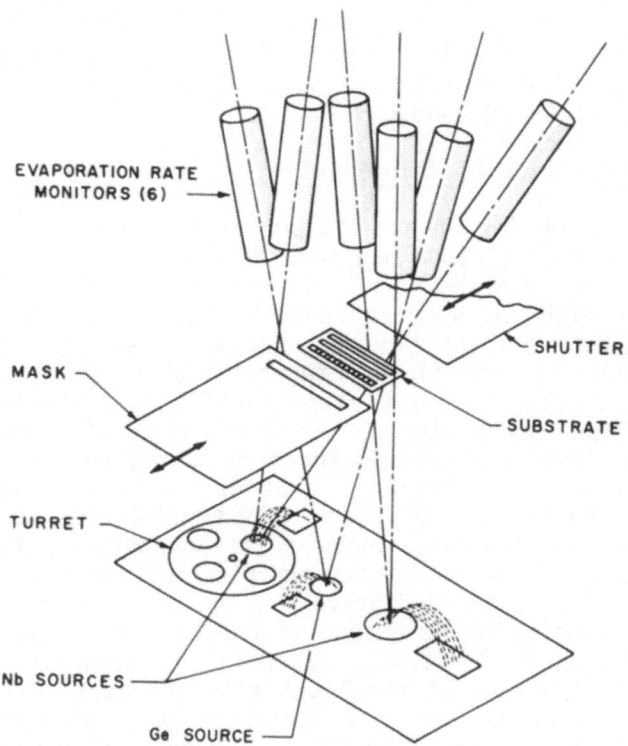

EVAPORATION RATE
MONITORS (6)

SHUTTER

MASK

SUBSTRATE

TURRET

Nb SOURCES

Ge SOURCE

FIG. 7 Schematic representation with symmetrical three
source evaporation.

composition, for instance the effect of sample thickness. The
compositional spread within a given sample due to the geometrical
consequence of the separated sources can be greatly reduced if 3
sources are employed, the outside sources being of the same
material. This reduces the geometrically-induced spread to almost
zero, a particularly nice feature for heat capacity measurements
where compositional variation can mask sharp transitions. The
same arrangement is also of importance in suppressing
compositional spread due to geometrical shadowing by non-
atomically smooth surfaces during growth.

The evaporator is designed for multi-use operation; frequently 4 or more different evaporations, using different sources take place in a week. Each investigation usually has its own substrate holder and furnace. After an overnight bake-out the system pressure is 2×10^{-8} torr. During deposition the typical pressure ranges from 0.5 to 2.5×10^{-7} torr, the principal gases being H_2, CH_4, H_2O, CO, and CO_2.

B. Characterization of Thin Films

It is helpful for diagnostic purposes to be able to vary the composition in a continuous way keeping other deposition variables constant. This can be done in a single evaporation in the configuration with the sources parallel to the line of substrates - the so-called phase-spread orientation shown in Fig. 7. The electrical and structural parameters should then reflect the continuously varying composition in any single phase regime. The behavior of T_c, the width of the transition ΔT_c, which may be different for the different methods of detection (Fig. 1), the resistivity, and the lattice constants all are functions of composition; their behavior gives a good indication of the degree of homogeneity and the crossing of phase boundaries. Collectively they can provide a reliable characterization upon which to base further studies of superconductivity. The composition of the film is usually determined by electron microprobe analysis. While the absolute accuracy is ± 1.5%, the relative accuracy is better. However, concentrations of the light element gases, C, H, O and N are difficult to establish. The variation in lattice constant due to the compositional spread in Nb-Al evaporation is shown in Fig. 8. Samples above 21% had some second phase present; compositions in such cases were then estimated by inter-polation. The slope $\Delta a/\Delta c = 3.2$ $\times 10^{-3}$ Å/at. % Al is the prediction obtained using the effective radii for the A15 structure assigned by Geller,[38] and later revised somewhat,[39,40] which assumes the contact distance between the Nb and Al is $\sqrt{5}/4$ a_0. The calculated slope assumes the maximum order allowed by the stoichiometry - i.e., all the Al are on Al sites. The radius of Al has been slightly adjusted - well within the 1% consistency of Geller radii - in order to obtain the agreement in magnitude.[41]

The temperature of the substrate and the deposition rate are important variables. Not only can the phase boundaries be temperature dependent at relatively low temperatures (note Fig. 4), but more important, particularly in metastable situations, is the extent of thermally-activated surface diffusion during the deposition. In more recent work with metastable Nb_3Ge it has been found that transient temperatures of the substrate, during the initial deposition where the film emissivity is changing, can

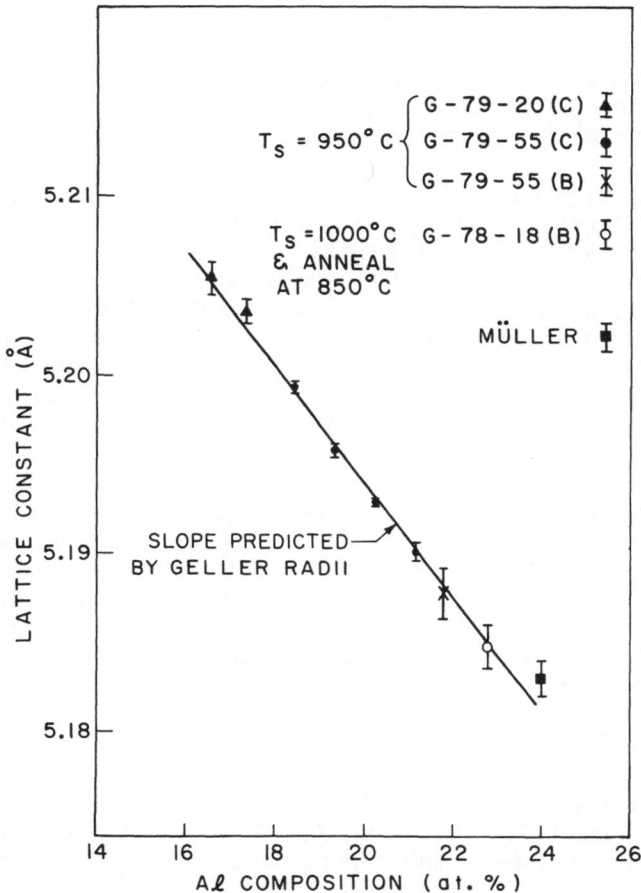

FIG. 8   Lattice constants for Nb-Al samples deposited at
         950 and 1000 C (with in situ annealing at 850 C).
         The plotted straight line is calculated from the
         Geller radii model with a slope of 0.0032 Å/at.%
         Al. (After Kwo).

cause a nucleation of the stable equilibrium phases and a lowering
and broadening of T_c.[42]

   Another cause of the broadening of T_c can be a fine-scaled
inhomogeneity due to shadowing caused by surface roughness.   The
resulting microscopic inhomogeneity can easily nucleate an
unwanted second phase which will then continue to grow during the
deposition.   Microscopic inhomogeneities as well as the
macroscopic compositional gradient mentioned earlier are both
reduced by the employment of 3 sources during deposition (again
with the outside two sources being the same element).

We employ two deeper probes for characterizing films in terms
of superconducting behavior, namely heat capacity and tunnel
junction current-voltage (I-V) curves. It is helpful to measure
heat capacities on the same samples (or adjacent ones made in the
same evaporation) on which the tunneling is done.

Heat capacity has a long and honorable history in the annals
of superconductivity - largely because it is an extensive property
that cannot be disguised by shielding currents. It reflects all
the thermal excitations in the sample which can include spurious
signals originating from unwanted second phases. McMillan
revitalized heat capacity when he showed how the analysis can
yield the strong coupling parameters (see equations (2) and (4)
below). We illustrate some of the parameters which can be
obtained in Fig. 9, which is a traditional plot of C/T vs. T^2.

The tunneling technique is particularly suitable for thin
films. Fabrication of a tunnel junction becomes possible when the
surface can either be oxidized in a controlled fashion or
protected by a barrier which can be deposited over the film. In
either case the barrier must have the proper characteristics for
obtaining tunnel currents at reasonable applied voltages. The
former is usually referred to as a natural and the latter as an
artificial barrier.

The tunnel current probes only a few coherence lengths into
the surface in contrast to the heat capacity. Since the
superconductivity near the surface is more likely to be degraded
rather than enhanced, the tunneling is more apt to underestimate
rather than overestimate the superconductivity. The structure of
the I-V curves gives a wealth of information.[43] The T_c of the
material being probed by the tunneling is measured by the
reduction which an opening of a gap in the density of states has
on the conductance at zero bias. At much lower temperatures the
width of the rise in current at the sum of the gaps of the
electrodes (when both are superconducting), above the small width
expected from thermal smearing, reflects either gap anisotropy or
inhomogeneity; in some cases a second superconducting phase
contributes a distinct second rise. Any conductance below the sum
of the gaps in excess of that expected from thermally excited
carriers is due to tunneling into some second-phase normal metal
or lower gap or gapless superconductor. The conductance at zero
bias is termed leakage current and is due to metallic shorts
through the barrier.[43]

Until recently, with the noteable exception of Ta,[44]
experimental difficulties with barriers and surface degradation
prevented good junctions from being fabricated with transition
metal films in any reliable and systematic way. The application
of tunnelling spectroscopy was limited to non-transition metals.

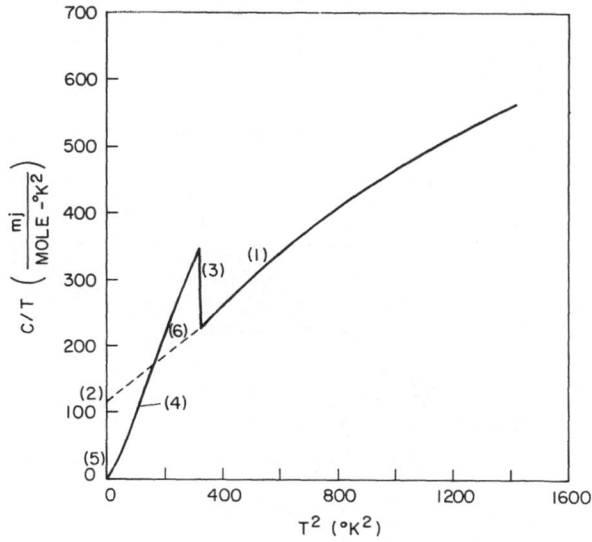

FIG. 9 Heat capacity of a 2 μm film of Nb_3Sn illustrating
limits and constraints. (After F. Hellman).

Diagnostic attributes of $C/T = \gamma T + \beta T^3$ plot
(1) → Slope gives vibrational T^3 term
(2) → Intercept gives electronic linear term
(3) → Sharpness, magnitude, → homogeneity, coupling strength
(4) → Quasiparticle excitations; measure of gap
(5) → Non-zero intercept means normal electrons present
(6) → 3rd Law consistency constrains extrapolation

The latter have natural surface oxides which make excellent barriers or can be evaporated easily over Al/Al_2O_3 which is by far the most reliable junction oxide.[45] There is very little leakage through the barrier; the conductance at 0-bias for a Pb-I-Pb junction is only 10^{-5} of the normal state conductance. Starting with the work of Moore, et. al.[46] and the development of reliable oxidized amorphous silicon barriers by Rudman and Beasley[46] it has been possible to make systematic tunnel junction studies with A15 compounds. Figure 10(b) illustrates the data for junctions made with NbGe in the A15 phase over a wide range of composition, T_c's and energy gaps. They are not as ideal as the barriers on Al or Pb junctions; however the leakage current is small (~ 0.2% of the normal conductance), not enough to obscure the further analysis of the tunnel current. The excess current due to tunneling from the Pb counterelectrode into some normal material at the A15 interface is appreciable but also does not obscure the main features. In Fig. 10(b) the excess is ~ 5% of the normal conductance except for sample 6 which is 10%. Sample 6 undoubtedly has more of the stable nonsuperconducting Ge-rich sigma phase present.

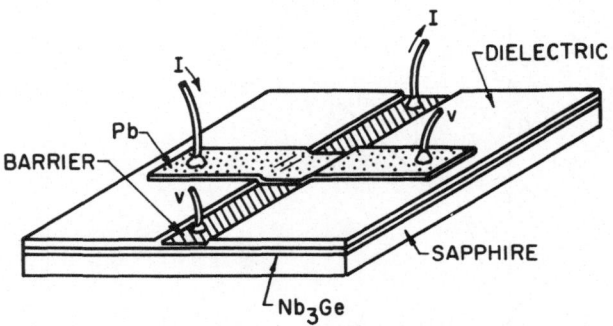

FIG. 10a Schematic of tunnel junction prepared by sequential
         depositions. Frequently 4 Pb cross-electrodes are
         deposited on the 6 mm sapphire.

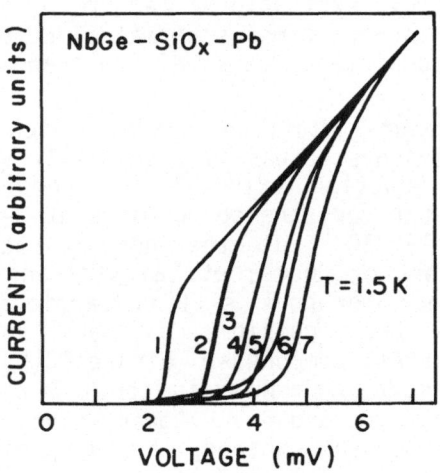

FIG. 10b Current voltage characteristics of NbGe junctions
         of varying composition.

7.   METASTABLE HIGH T_C A15 SUPERCONDUCTORS

Before discussing the metastable A15's we should review briefly the established properties of A15 compounds. They are covered in detail in a number of good review articles.[47] Those of most interest here are Nb_3X and to some extent V_3Si where X = Al, Ga, Si, Ge, and Sn. The most careful studies have been made on Nb_3Sn (T_c = 18.1K) and V_3Si(T_c = 17.1 K) because these are the only two that can be obtained in the A15 phase as stoichiometric single crystals. Both have anomalous temperature dependent Knight shifts and magnetic susceptibilities. Clogston and Jaccarino[48] have suggested the temperature dependence is due to the magnitude of kT becoming appreciable with respect to sharp structure in the density of states near the Fermi-energy; "thermal smearing" occurs as T increases. Calculations of the electronic band structures which have been done using quite different approaches are in good agreement. The structure is difficult to calculate on a fine enough scale to compare with kT; there are ~ 20 bands which must be taken into account. Klein et al.[49] have found very flat bands in certain directions around Γ_{12} which could account for the results. However while their results show almost no difference in the density of states between Nb_3Sn and Nb_3Al, Stewart et al.[50] have found that the best available NbAl and NbGe A15 samples have electronic heat capacity coefficients only 60% that of Nb_3Sn. Since NbAl and NbGe can only be obtained as a non-stoichiometric, Nb-rich phases there is still some possibility that ordered stoichiometric phases might behave more like the prediction of the band calculation. Mattheiss and Weber have found that hybridization in the flat Γ_{12} band is strong and that compensating d-d and p-d overlap near the Fermi-level can account for the anomalous electronic properties.[51]

It is found that the A15's fall naturally into two groups. The first consists of V_3Si, Nb_3Sn and others that exist as stoichiometric compounds. The second consists of NbAl, NbGe, NbGa, and NbSi.* None of the latter can be made as stoichiometric compounds. All except Si exist as the equilibrium ground state over niobium-rich compositional ranges. The Si compound is metastable at all compositional ranges. All have low T_c's in their equilibrium ranges of composition but by various techniques can be obtained in metastable compositions closer to stoichiometry where they have the highest T_c's known. They have other characteristics in common, and each has its own idiosyncracies as well.

---

*We will frequently use NbX rather than "Nb_3X" to avoid implying the stoichiometric composition for the Nb-rich A15 phases.

The root of the metastability can be traced to the too-small size of X = Al, Ge, Ga, Si for the stoichiometric structure. By placing some larger Nb on the X site, the unit cell is expanded. It is reasonable that the dominant effect of the increased volume is in the lowering of the kinetic energy of the conduction electrons.[52] The phase diagram for Nb_3Al (Fig. 11) shows features which are common to all the systems - namely the retrograde boundary on the X-rich side, and the strong deviation from stoichiometry below ~ 1500 K. The approach to stoichiometry at high temperatures is due to entropy stabilization, and an increase in volume. Webb et. al.[53] have shown that the disorder introduced by deviations from stoichiometry (as distinct from Matthias' rule) is the major cause in the concentration dependence of T_c which is between -1 and -2K per atomic % of Nb for all the A15 NbX systems.

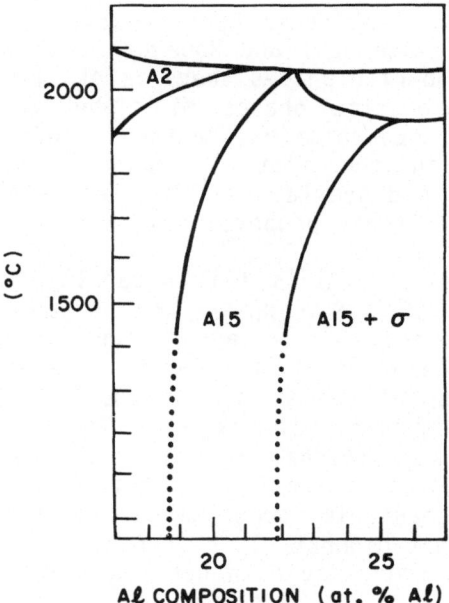

Fig. 11 Phase diagram for NbAl showing the strong retrograde solubility below 2000 K. The lines below 1500 are shown as dotted because equilibria at lower temperatures are difficult to establish. (After Flugiker). NATO Advanced Study Institute Vol. 68, (Plenum Press) N.Y. 1981, pg. 576.

It is possible to approach stoichiometric Nb_3X by different techniques which suppress the unwanted nucleation of the equilibrium phase. In bulk materials rapid quenching from high temperatures where the stoichiometric composition exists (Fig. 11) followed by a lower temperature annealing to remove the anti-site disorder has been successful in producing Nb-Ga with $T_c >$ 20K.[54] Nb-Si has been made at very high pressures where the smaller volume of the A15 phase is favored by explosive compression techniques.[55]

Vapor-deposited films are inherently extremely rapidly quenched and therefore well-suited for obtaining metastable structures. There are further ways of suppressing the formation of the unwanted phases and obtaining an increased level of metastability. I will discuss two - the presence of a third element (gas) during growth which affects the growth kinetics and/or shifts the free energy of the A15 relative to competing phases, and the presence of a prepared surface which by epitaxy favors the A15 phase.

Many workers[56] have shown that oxygen in small concentration (about 10^{-6} Torr for 30 Å/sec deposition rates) during the deposition is helpful in stabilizing more stoichiometric A15 NbGe and NbSi, although its exact role is not clear. Feldman and Hammond have shown that the oxygen affects the NbSi film throughout its growth. A tentative explanation, consistent with known results so far, is that the oxygen blocks grain boundary diffusion paths. The beneficial role of oxygen in the growth of NbGe has been attributed by different authors to a variety of causes, some due to its presence during the initial deposition, and others due to its presence throughout the growth.[57]

Epitaxy has also been helpful in stabilizing the A15 phases. Epitaxial growth is normally on single-crystal substrates; the interest is usually in preparing high-quality semiconductor layers. However, grain boundaries do not necessarily degrade superconductors. Recognizing this, Dayem et al.[58] have used polycrystalline A15 Nb-Ir substrates to stabilize high T_c Nb-Ge. Nb-Ir forms an equilibrium A15 phase over a wide composition range with lattice constants from 5.13 to 5.17 Å, the lower value being appropriate for epitaxial growth of stoichiometric Nb_3Ge.

It has also been possible to obtain polycrystalline epitaxial growth of A15 phase Nb-Si[59] on Nb_3Ir. The epitaxial Nb-Si films adopt the lattice constant of the underlying Nb-Ir film to within about 0.001 Å (Fig. 12) even if the substrate is prepared so as to put the Nb-Si under a tensile stress corresponding to a positive strain of up to 0.01 Å. The surprising result is that the NbSi remains cubic - it is expanded in all three cubic axes rather than

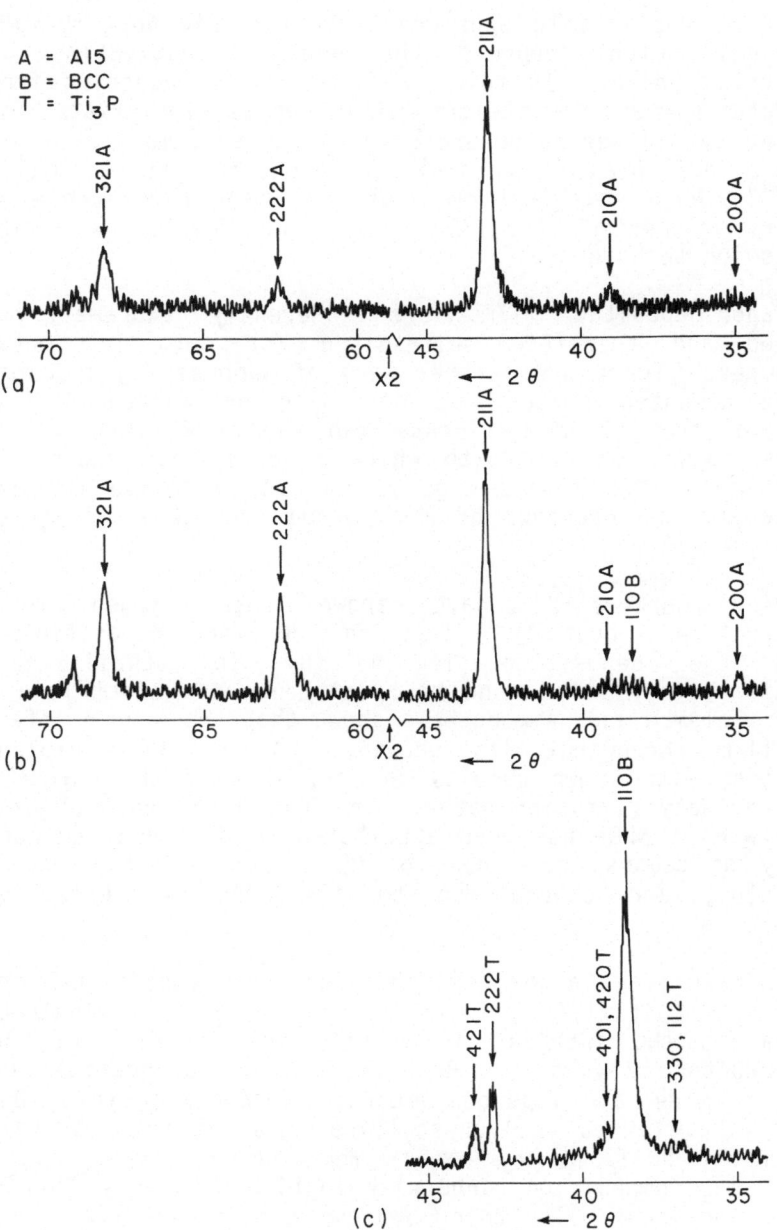

FIG. 12 Diffraction patterns from 3 samples made in a single run: Nb₃Ir was initially deposited on rows (a) and (b) with (c) masked. NbSi was then deposited on rows (b) and (c) and (a) masked. The results show A15 patterns with identical lattice constants on (a) and (b) and the Ti_3P and bcc phases on row (c) (After R. D. Feldman).

being contracted in the direction normal to the epitaxial plane as would be expected for an elastic strain (Poisson's ratio)[60]. A simple interpretation, subject to further testing, is that a small and variable amount of residual gas in the evaporator is incorporated into the lattice in order to adjust the lattice constant of the Nb_3Si to that of the Nb_3Ir. It is well known that A15 compounds can accommodate relatively large quantities of interstitial hydrogen.

It is possible to use compositional grading, or self-epitaxy, to increase the B element concentration further once the A15 phase has been deposited at a more stable composition. This combination of polycrystalline epitaxy plus compositional grading has resulted in T_c's of A15 NbSi films up to about 13 K as compared to explosive compression preparation with T_c's of 18K.

Oxygen does not have a beneficial role in the growth of A15 NbAl, presumeably in its presence Al_2O_3 forms in preference to anything else. However it is possible to improve T_c of the NbAl films by self epitaxy - i.e., compositional grading during growth. The initial tunneling experiments on such material showed no corresponding improvement in the gap. This was then shown to be due to the inability of the metastable phase to maintain itself during growth by tunneling into a series of films with the epitaxial layer thickness as a variable.[61]

The tunneling current probes within a coherence length or so (i.e. about 50 Å) of the surface of the epitaxial layer. The gap (Fig. 13) is seen to increase as expected on the basis of the proximity effect for thicknesses up to about 150 Å. The improved growth is only maintained for thicknesses of about 200 Å or less. The degradation beyond 200 Å is obscured by the higher T_c material when employing the non-bulk methods of detection of T_c (Fig. 1). As the film grows beyond 200 Å it evidently relaxes back to the composition it would have without epitaxy. By about 600 Å, diffusion (presumeably surface diffusion during the growth) results in the nucleation of Nb_2Al (a non-superconducting sigma phase compound), which causes the excess current observed below the gap. The competition between surface diffusion and film growth can be altered by changing the deposition conditions, for example, a faster rate of deposition would presumeably lead to less degradation.

In contrast to the epitaxially-stabilized A15 phases, it appears as if the gas-stabilized phases can maintain themselves over considerable thicknesses. NbGe films of 1 micron make good junctions and it is not until above 10 microns that the tunnel features start being washed out. There have been no tunneling studies carried out on Nb_3Si but X-ray diffraction data indicate

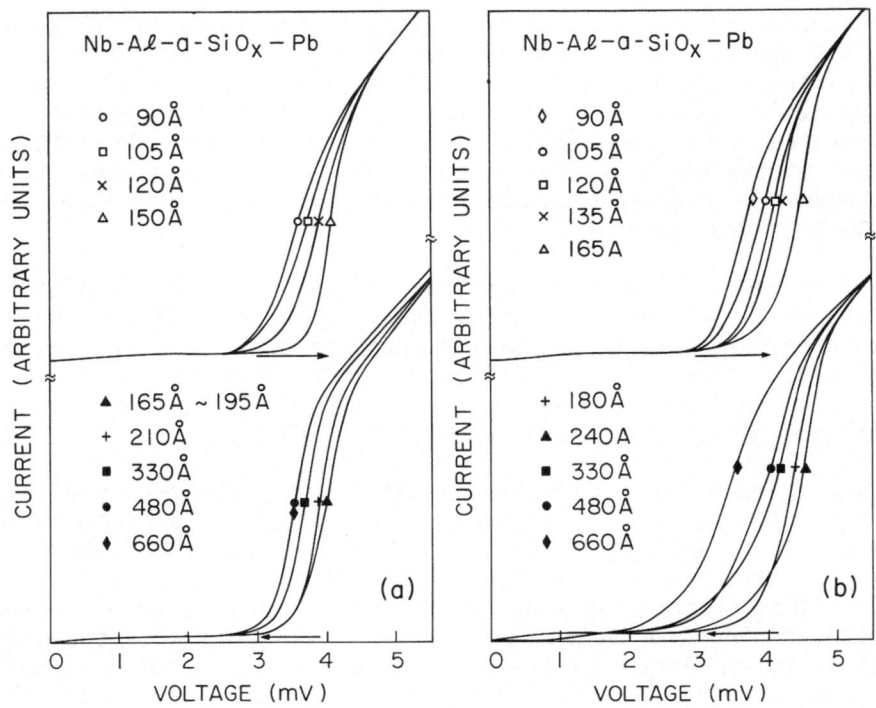

Fig. 13 The current-voltage (I-V) characteristics of NbAl
         junctions as a function of epilayer thickness for
         epilayers of composition $Nb_{0.77}Al_{0.23}$ and
         $Nb_{0.76}Al_{0.24}$ prepared on substrates of $Nb_{.81}Al_{.19}$
         using the self-epitaxy method.

that the O_2-stabilized compositions maintain themselves for at
least 1 micron.

     Further evidence for the improved homogeneity and non-
degradation of the oxygen stabilized NbGe films is given by the
critical current measurements which are significantly higher than
previously reported results (Fig. 14).   These results of course
require that there be a high pinning force (which we are not
considering here) as well as good material throughout the film.
More direct evidence has been obtained by heat capacity
measurements[62] which are as sharp or somewhat sharper than
comparable NbSn films containing the same excess of Nb.

A.   Tunneling Spectroscopy - Strong Coupling and Metastability

     In BCS (weak-coupled) superconductors the quasi particle life
times are long.   When T_c becomes appreciable with respect to the
Debye temperature (~ 5% or more) and the quasi particle energies

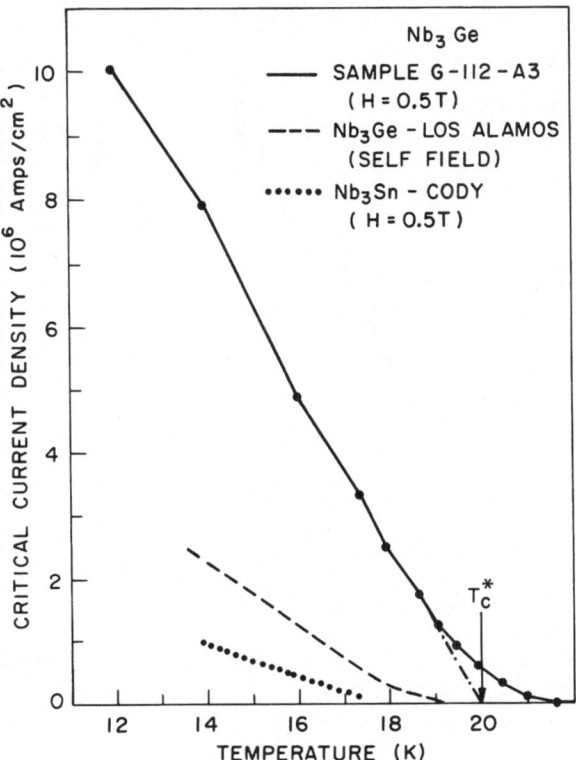

FIG. 14 Critical current density versus temperature showing
T_c^* obtaining by a linear extrapolation to the
temperature axis. Empirically it is found that a
high T_c^* and J_c go hand in hand.

are broadened then the gap equations must be solved in the strong
coupling limit.[63]   This requires a knowledge of the electron-
phonon spectral (Eliashberg) function, $\alpha^2(\omega)F(\omega)$ which can be
found experimentally by tunnelling. Here $\alpha(\omega)$ is the coupling
integral and $F(\omega)$ the phonon density of states.

The ratio $2\Delta/kT_c$ is a measure of the coupling strength. The
values shown in Fig. 15 all reach a value ~ 4.5 (about 30% above
the BCS weak-coupled value) at the maximum concentration of X (=
Al, Ge, Sn) which can be maintained in the A15 structure. The
value ~ 4.5 seems to be the experimental limit achievable in the
A15 structure. In terms of the analytical expression obtained by
Kresin et. al.[64] who used an Einstein frequency ω_0 to model the
phonon spectrum,

$$2\Delta/kT_c = 3.53 \left[(1 + 5.3 (T_c/\omega_0)^2 \ln \omega_0/T_c\right] , \qquad (1)$$

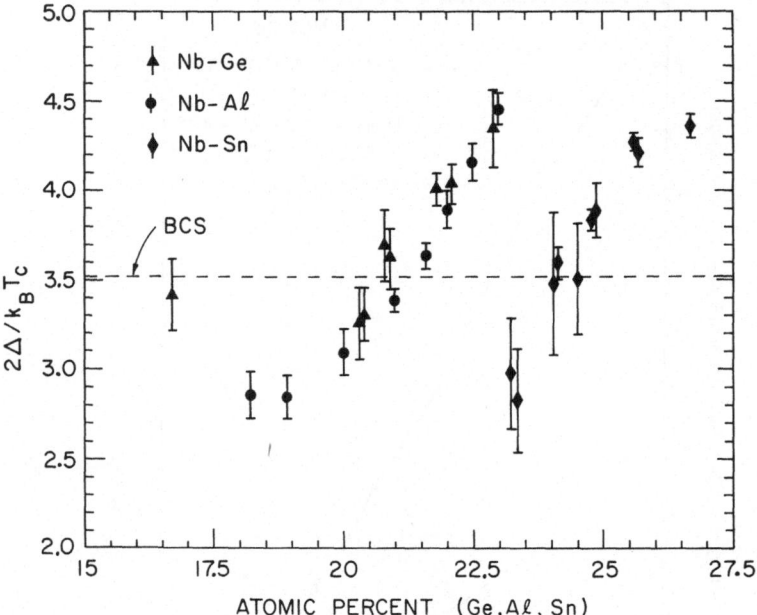

FIG. 15 Gap normalized to T_c as a function of composition
for NbGe, NbAl and NbSn.

it appears that there must be a increase in T_c/ω_Q as the
metastable X-rich boundary is approached. This suggests mode-
softening at the phase boundary as the cause of the metastability.

The fall off of $2\Delta/kT_c$ below 3.5 at the lower
concentrations of Ge, Al and Sn is believed to be due to the
failure of the analysis to take care of inhomogeneities properly
rather than any fundamental problem with the model. The T_c is
probably overestimated relative to Δ since it is measured as the
onset of the gap opening (at $T = T_c$) whereas Δ is determined as
an average value of the gap (at $T \ll T_c$).

The strong coupling theory provides a method for relating T_c
to microscopic averages of normal state properties.
McMillan[65] obtained numerical solutions of the Eliashberg
equations using the phonon spectrum of Nb and fit them
analytically with the expression

$$T_c = \frac{\theta}{1.45} \exp \frac{-1.04(1 + \lambda)}{\lambda - \mu^{*}(1 + 0.62\lambda)} \tag{2}$$

which is known as the McMillan T_c equation.

Here θ is the Debye temperature, or some other appropriately averaged phonon frequency. λ is the electron-phonon (attractive) coupling constant which can be estimated from the McMillan equation itself using heat capacity data. The electron-electron interaction is given by the pseudopotential μ^*;

$$\mu^* = \frac{\mu}{1 + \ln \frac{E_f}{k\theta}} \qquad (3)$$

where μ is the screened electron-electron interaction which is then reduced in order to take into account the different time scales over which the electrons and phonons interact. The retardation makes μ^* substantially less than λ and in some sense is responsible for superconductivity being so prevalent. The retarded nature of the interaction permits localization in space without much penalty in energy. This perhaps explains why chemical effects are so manifest, i.e. why there can be particular atoms like Nb which are favorable for superconductivity.

The importance of tunneling spectroscopy is that it provides an independent measure of $\lambda = 2 \int \alpha^2(\omega)F(\omega)\omega^{-1}d\omega$. McMillan obtained a further relation for λ in terms of averages of normal state properties

$$\lambda = \frac{N(0) \langle I^2 \rangle}{M \langle \omega^2 \rangle} . \qquad (4)$$

This expression is particularly useful because it permits heat capacity and tunneling data to be used in determining the microscopic origins of the superconductivity. In (4) $N(0)$ is the electron density of states averaged over the Fermi surface, $\langle I^2 \rangle$ is the electron-phonon scattering matrix element averaged over the Fermi surface, M is the ion mass, and the second moment of the phonon frequency is defined as

$$\langle \omega^2 \rangle = 2/\lambda \int \omega \, \alpha^2(\omega)F(\omega)d\omega . \qquad (5)$$

Changes in T_c should be understandable in terms of appropriate changes in the factors of Eq. (4). In practice $1.5 < \lambda < 2$ seems to be the limit for most very strong-coupled systems; above that problems of stability arise. Equation (2) was fit by McMillan to data for $\lambda < 1.5$. A modified numerical solution of the Eliashberg equations by Allen and Dynes,[66] accurate for $\lambda > 2$, shows that $T_c/\langle \omega^2 \rangle^{\frac{1}{2}}$ is an increasing monotonic function of λ.

The most detailed and well-understood studies of strong coupled superconductors have been for Al-Al oxide-sp metal junctions with metal = Pb, Sn, In, Tl, Hg and alloys of these metals. The McMillan-Rowell inversion program[67] is used with the measured tunnel conductance (the latter gives directly the deviation from the BCS density of states) to invert the Eliashberg equations and obtain the important Eliashberg function $\alpha^2F(\omega)$. The results for the most strongly-coupled superconductor, quenched Pb$_{45}$B$_{.55}$ are shown in Fig. 16. It can be seen that the increased value of λ = 2.6 from that of Pb(λ = 1.55) reflects mode softening or an increase in the coupling constant α, or some combination of both. Note that the softening results in a decrease in T_c in spite of the increase in λ.

In spite of the fact that tunnelling spectroscopy for Nb$_3$Sn, NbAl, and NbGe, still has some ambiguities there is clear evidence for mode softening, although the maximum values of λ are < 2.

Even with the improvements in tunnel fabrication which we have discussed the solutions obtained for the A15 compounds by the inversion method are inconsistent, presumeably due to a thin degraded proximity layer between the superconducting electrode and the barrier. The theory for such a normal-superconductor

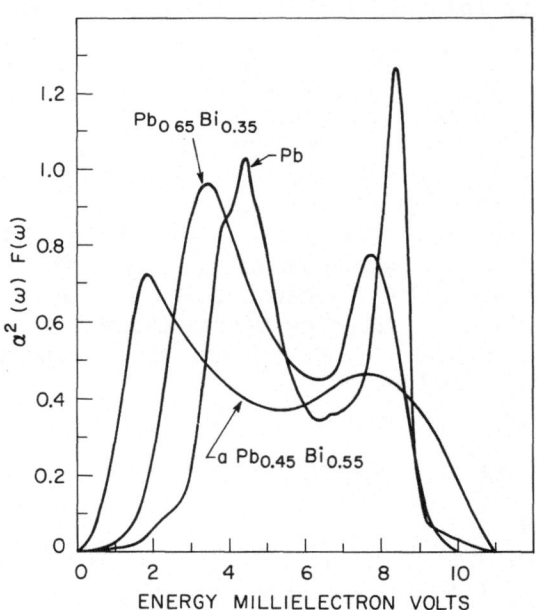

FIG. 16   The electron-phonon coupling parameter $\alpha^2(\omega)F(\omega)$
            for lead (T_c = 7.2K; λ = 1.55; $\langle\omega\rangle$ = 5.20 meV), for
            crystalline Pb$_{0.65}$Bi$_{0.35}$ (T_c = 8.95K; λ = 2.13;
            $\langle\omega\rangle$ = 4.32 meV) and for amorphous Pb$_{0.45}$Bi$_{0.55}$ (T_c
            = 7.0K; λ = 2.59; $\langle\omega\rangle$ = 3.25 meV).(68;69)

proximity sandwich was derived by Arnold[70] and a modified gap inversion analysis on especially prepared Al-Nb proximity layers has been shown to give good results by Wolf et. al.[71] The modified gap inversion analysis gives much better agreement between the calculated and the experimental tunneling density of states in all the A15 junctions by simply assuming a zero pair potential for the proximity layer. The analysis requires two free parameters which depend on the values assumed for the thickness, the mean free path, and the renormalized Fermi velocity of the proximity layer. In most tunnel junctions of Nb_3Sn[72], NbAl[61], and NbGe[73] with $a\text{-}SiO_x$ barriers, the values used for these fitting parameters correspond to a proximity layer of thickness 3 to 8 Å and a mean free path of < 50 Å. The consistency obtained among these tunneling results supports the model proposed for the amorphous silicon barrier and helps to justify the approximation of a zero Δ_n assumed for that proximity layer, although, of course there is still more to be done to put the analysis on a firmer footing.

Further confidence in the proximity analysis can be drawn from the recent work of Rowell et. al.[74] on Nb-junctions fabricated using very thin oxidized Al barriers. These barriers are almost ideal and their analysis requires no proximity-effect parameters. The earlier proximity-modified analysis[71] is in excellent agreement with the newer results.

The value of the proximity analysis is illustrated in Fig. 17 where the experimental tunneling densities of states is compared with the BCS value. The improvement is somewhat less good for NbGe, but still of sufficient quality to produce meaningful $\alpha^2(\omega)F(\omega)$ data. It is expected that further improvements in tunnel-junction fabrication will continue. Geerk et al. have already demonstrated the value of the Al barrier on NbGe.[75]

The $\alpha^2(\omega)F(\omega)$ spectral functions are shown in Fig. 18 For NbGe samples over a range of compositions that extend from weak to strong coupling. The two peaks at 19 and 22 meV which are at least partially resolved in all four samples, also seen in Nb_3Sn and NbAl, are attributed to the transverse and longitudinal modes of the Nb chains. In the higher gap samples a mode at lower energy is seen to emerge and grow in strength.

The movement of the lower peak to lower energies, i.e. the softening, as stoichiometry is approached is also found in NbAl and Nb_3Sn. It is tempting to conclude that mode-softening may be a common feature of the high-T_c A15 compounds. Such a conclusion assumes that the coupling constant α itself is not responsible for all the observed softening. Geerk et. al. have attempted to address this question by using neutron time-of-flight experiments to obtain the phonon density of states directly.[75] They find

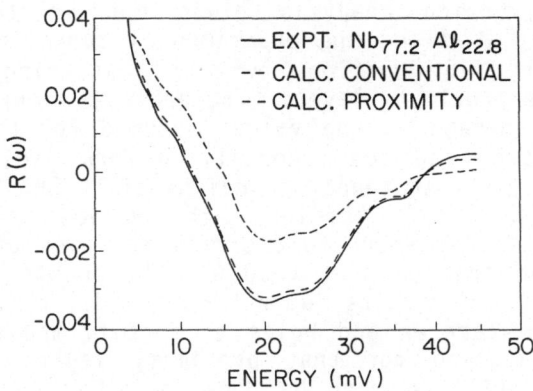

FIG. 17--The experimental reduced tunneling density of states $R(\omega)=$ $N_{expt}(\omega)$. $N_{BCS}(\omega)^{-1}$ compared with that calculated from the McMillan-Rowell inversion program and with the modified program of Arnold.

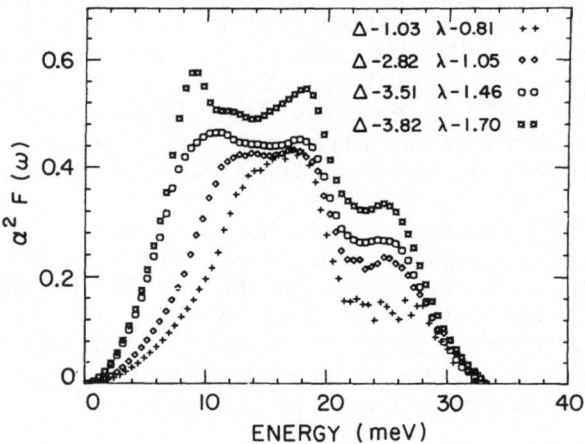

FIG. 18--The electron phonon spectral functions for NbGe samples with gaps of 1.03, 2.62, 3.50, and 3.82 mv respectively with compositions of 16.7, 21.8, 24.3, and 24.7 respectively at % Ge as determined by microprobe analysis.

evidence for phonon-softening in high-T_c NbGe, but not enough to explain all the softening they observed by tunneling spectroscopy, thus suggesting both $\alpha(\omega)$ and $F(\omega)$ are responsible.

There is also a possible correlation of the softening of the low energy peak in Nb_3Sn with the transverse acoustic phonon propagating in the [1$\bar{1}$0] direction, polarized in [101] which softens throughout the zone as observed by inelastic neutron scattering on a stoichiometric single-crystal.[76]

In the gap-inversion programs, either with or without the proximity layer, μ^* is the parameter most sensitive to choices of input parameters. A self-consistent check of all the parameters can be made by comparing the T_c calculated from the low temperature data with the measured T_c. Using either the McMillan, the Allen-Dynes, or the linearized-Eliashberg equations directly gives reasonable agreement within limits set by uncertainty in μ^*. It is also possible to examine (qualitatively only at present) the changes in T_c with those calculated from the measured $\alpha^2F(\omega)$ curves using the functional derivatives calculated by Bergman and Rainer,[77] i.e.,

$$\Delta T_c = \int_0^\infty d\omega \, \frac{\delta T_c}{\delta \alpha^2 F(\omega)} \, \Delta(\alpha^2 F(\omega)) \qquad (6)$$

and do similarly for the gap using the functional derivative $\delta\Delta/\delta\alpha^2F(\omega)$ calculated by Mitrovic et al.[78] Taken together these calculations indicate that there would be a further substantial increase in $2\Delta/kT_c$ if the softening were to be continued with further approach towards stoichiometry.

The moments of the $\alpha^2(\omega)F(\omega)$ function give values of λ and $\langle\omega^2\rangle$. The trend toward higher λ and lower $\langle\omega^2\rangle$ with approach to stoichiometry is found for the Nb_3Ge samples shown in Fig. 19 as has been previously found for Nb_3Al[85] and Nb_3Sn.[79] In order find out to which if any of the factors in Eq. (4) is mainly responsible for the increase in strong-coupling and perhaps the instability itself, an independent estimate of the unrenormalized (bare) density of states $N^b(0)$ is needed. This can be done by analysis of either heat capacity, or upper critical field data. $N^b(0)$ can be obtained from heat capacity alone by using the measured T_c and the McMillan T_c Eq. (1) or some modification of it to obtain λ which is then used to reduce the N_γ obtained by extrapolation of the normal state heat capacity as indicated in Fig. 9; $N_\gamma(1+\lambda)^{-1} = N^b(0)$. N_γ can also be obtained from the slope of the upper critical field curve, $(dH_{c2}/dT)_{Tc}$ and the normal state resistivity using standard procedures.[80] The values so obtained for the bare density of states are plotted over a wide range of resistivity in Fig. 20. A

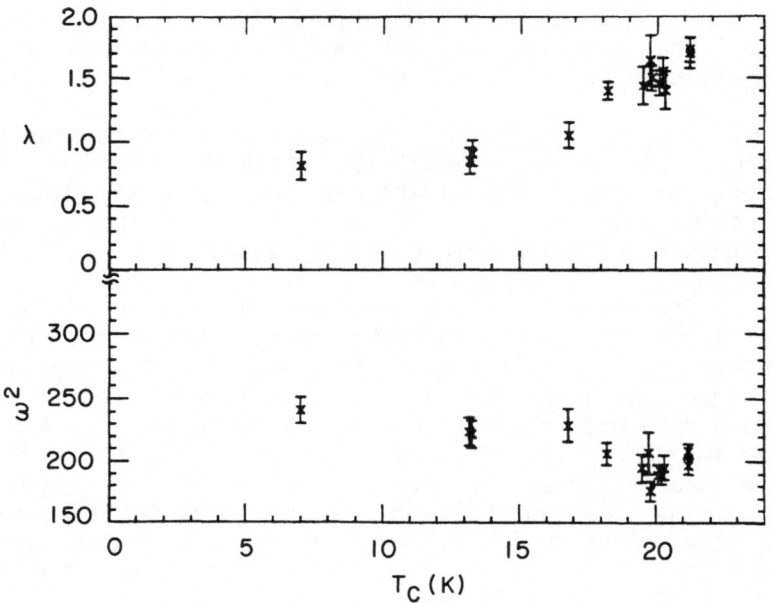

FIG. 19 λ and $\langle\omega^2\rangle$ plotted versus T_C for the NbGe samples some of whose I-V characteristics are shown in Fig. 10.

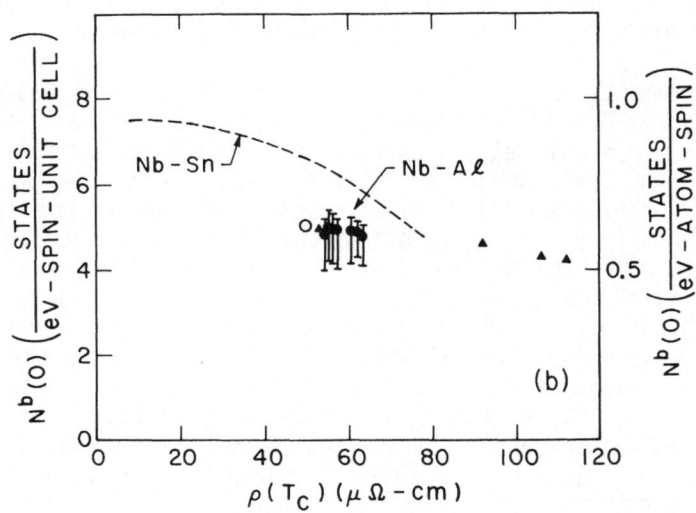

FIG. 20 The band density of states versus $\rho(T_C)$ for NbSn and NbAl.[85]

substantial variation in $N^b(0)$ for NbSn is evident while $N^b(0)$ for NbAl is essentially constant over the range measured. The rapid drop off in T_c (down to 4 K in the radiation-damaged NbAl) thus cannot be accounted for by changes in the density of states, in disagreement with the idea that resistive lifetime broadening[81] of a sharp peak in density of states at the Fermi-level is the fundamental factor in reducing T_c. Although the data are less complete for NbGe, $N^b(0)$ for the high-T_c samples is of the order of NbAl and much smaller than NbSn, again suggesting that the rapid drop off in T_c is not due to a smearing of the density of states. For the metastable A15 compounds, we might generalize, that as the X-rich side of the phase boundary is exceeded λ increases due to an increasing $\langle I^2 \rangle / \langle \omega^2 \rangle$ until some structural instability or disproportionation* reaction sets in. Neither NbAl nor NbGe can be prepared with resistances $\lesssim 50$ $\mu\Omega$-cm . The fact, Fig. 21, that the T_c - resistivity curve for NbAl is steeper and higher then NbSn is consistent with $\langle I^2 \rangle / \langle \omega^2 \rangle$ being greater.

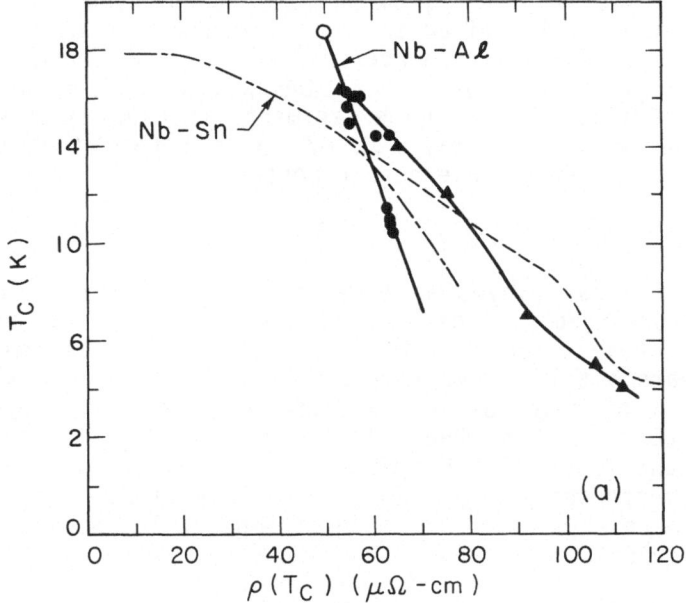

FIG. 21 T_c versus $\rho(T_c)$ for NbSn and NbAl.[85]

---

*Disproportionation: is used in the sense that a given composition disproportionates into an Nb-richer phase plus an Nb-poorer phase; e.g. Nb_3X(A15) $\rightarrow$ Nb_4X(A15) and Nb_2X (σ-phase).

8.   POSSIBLE MODELS

At present there is no model which will explain the results satisfactorily.   Varma and Dynes[82] have shown from non-orthogonal tight binding model calculations that in many cases λ should be proportional to $N^b(0)$, i.e., $\langle I^2\rangle/\langle\omega^2\rangle$ = constant. This behavior is observed in the transition metal alloy series (Fig. 3) and in V_3X compounds.   In the A15 compounds as a whole, however, there is an inverse correlation, i.e., T_c increases as $N^b(0)$ decreases when proceeding from V_3Si - Nb_3Sn - NbAl - NbGe. Further, tunneling and heat capacity data show rather conclusively that $\langle I^2\rangle/\langle\omega^2\rangle$ increases rather than remaining constant with increasing concentration of X for the Nb compounds. We have suggested that it might be necessary to include an additional term in the free energy to take explicitly into account the insufficient volume of the unit cell.[52]   Hanke et al.[83] have shown that covalent NbX bonds can lead to a resonance like enhancement of the dielectric function and an increase in $\langle I^2\rangle/\langle\omega^2\rangle$. Tsuei[84] et al. have obtained evidence for covalent bond formation in their X-ray photoemission data.

An alternative pairing mechanism based upon exchange of acoustic plasmons is discussed by Professor Ruvalds in his lectures. This mechanism can account for the observed decrease in T_c with structural disorder in NbAl and NbGe without requiring any decrease in $N^b(0)$. So far no experimental measurements with the required energy sensitivitiy ~.1 eV, needed to detect acoustic plasmons if they exist, have been reported.

CONCLUSION

Empiricism has played an essential role in the discovery of superconductors and in developing an understanding of their behavior. I have tried to illustrate in the lectures how, in the past few years vitality has been injected into the process by the incorporation of powerful new methods of vapor phase synthesis. Equally important has been the development of fabrication techniques and theory for the wide application of superconducting tunneling spectroscopy which permits a detailed study of the near surface properties and the blending of empirical observations with theoretical models.

It seems clear that there are major areas "out there" ready for exploration and exploitation. The directions to take are, of course, uncharted, as always. The number of non-equilibrium phases within reach is almost limitless. If there are fine-scaled spatial inhomogeneities which induce superconductivity, or shortlived temporal departures from equilibrium which do likewise, then more sensitive and finer-scaled probes may be needed. For example if 1% of the volume a CuCl sample were to go superconducting at distributed microscopic Cu-CuCl interfaces, a typical detection system depicted in Fig. 1 would only reflect a 1% decrease in the resistance of the (insulating) matrix. Innovative new methods such as that of Zeller and Giaever,[86] who established the superconducting nature of ~ 30 Å Sn particles imbedded in an Al_2O_3 barrier by measuring their contribution to the tunnel current, are called for. With the advances in SQUIDS and lithographic methods it would seem the potential for finding them exists.

There are still many unanswered questions concerning the metastable A15 compounds and their superconductivity. Why can NbGe films be grown with higher T_c's than in the bulk, while for NbAl the converse is true? (We suspect oxygen). Why cannot NbAl and NbGe be grown in more-ordered structures with better resistance ratios? Why is the tight-binding model unable to account for the results found in NbAl and NbGe? The recommended procedure for finding answers is to continue with the empirical approach, - make suitable samples and study them.

The work presented here is the outgrowth of many beneficial and rewarding hours spent with colleagues and especially with my students. I would particularly like to acknowledge uninhibited co-empiricists M. R. Beasley, C. W. Chu, A. H. Dayem, R. H. Hammond, J. K. Hulm, G. W. Hull, Jr., J. M. Rowell J. H. Wernick, and, of course, the late B. T. Matthias.

REFERENCES

1. H. Kamerlingh Onnes and W. Tuyn, Leiden Commun. No. 160 (1922).

2. R. M. White and T. H. Geballe, Long Range Order in Solids, Supplement 15, H. Ehrenreich, F. Seitz and D. Turnbull, Eds., (Academic Press, Inc., 1979), pg. 92.

3. W. S. Corak, B. B. Goodman, C. K. Satterthwaite and A. Wexler, Phys. Rev. 96, 1442 (1954).

4. J. K. Hulm and B. T. Matthias, in NATO Advanced Study Institutes Series, Phys. Vol. 68, S. Foner and B. B. Schwartz, Eds., (Plenum Press, New York, 1981), Chapter 1.

5. R. M. Bozorth, Ferromagnetism, (Van Nostrand-Reinhold, Princeton, New Jersey 1951), pg. 441.

6. J. K. Hulm and R. D. Blaugher, Phys. Rev. 123, 1569 (1961).

7. B. T. Matthias, T. H. Geballe and V. B. Compton, Rev. Mod
    Phys. 35, 1 (1963).

8. B. W. Roberts, J. Phys. Chem Ref. Data, 5, 581 (1976).

9. A. R. Miedema, J. Phys. Paris F3, 1803 (1973); A. R. Miedema,
    J. Phys. Paris F4, 120 (1974).

10. T. P. Orlando, Superconductivity in d- and f-Band Metals 1982
    W. Buckel and W. Weber, Eds., page 75: C.
    Geibel, H. Rietschel et al. ibid pg. 435.

11. E. L. Wolf, (LT16), Physica B + C 109, 1733 (1982).

12. M. M. Collver and R. H. Hammond, Phys. Rev. Lett., 30, 92
    (1973); S. J. Poon and W. L. Carter, Solid State Commun. 35,
    249 (1980).

13. E. Raub, Ch. J. Raub, E. Roschel, V. B. Compton, T. H.
    Geballe, and B. T. Matthias, J. Less Common Metals 12, 36
    (1967).

14. The correct explanation deriving from the retarded
    interaction was given by P. Morel and P. W. Anderson, Phys.
    Rev. 125, 1263 (1962); See Ref. 2 pgs. 102 and 112.

15. T. H. Geballe, B. T. Matthias, V. B. Compton, E. Corenzwit,
    G. W. Hull, Jr., and L. D. Longinotti, Phys. Rev. 136, A119
    (1965).

16. G. Arrhenius, R. Fitzgerald, D. C. Hamilton, B. A. Holm, B.
    T. Matthias, E. Corenzwit, T. H. Geballe and G. W. Hull, Jr.,
    J. Appl. Phys. 35, 3487 (1964).

17. B. T. Matthias, G. R. Stewart, A. L. Giorgi, J. L. Smith, Z.
    Fisk and H. Barz, Science 208, 401 (1980).

18. F. Y. Fradin, H. B. Radousky, N. J. Zaluzec, G. S. Knapp, and
    J. W. Downey: Argonne National Laboratory, private
    communication (1982).

19. M. Ishikawa, J. L. Jorda and A. Junod, Superconductivity in
    d- and f-Band Metals 1982, W. Buckel and W. Weber, Eds.,
    page 141.

20. N. E. Phillips, B. B. Triplett, R. D. Clear, H. E. Simon,
    J. K. Hulm, C. K. Jones and R. Mazelsky, Physica 55, 571
    (1971).

21. D. E. Cox and A. W. Sleight, Solid State Commun., 19, 969
     (1976); See also B. Batlogg, J. P. Remeika, R. C. Dynes, H.
     Barz, A. S. Cooper and J. P. Garno, Superconductivity in d-
     and f-Band Metals 1982, W. Buckel and W. Weber, Eds.,
     page 401; See also C. Methfessel and S. Methfessel, ibid.

22. F. Steglich, J. Aarts, C. D. Bredl, W. Lieke, D. Meschede, W.
     Franz and H. Schafer, Phys. Rev. Lett. 43, 1892 (1979).

23. T. F. Smith and E. S. Fisher, J. Low Temp. Phys. 12, 631
     (1973).

24. H. G. Smith, N. Wakabayashi, R. M. Nicklow, G. H. Lander, E.
     S. Fisher and W. B. Daniels, Superconductivity in d- and f-
     Band Metals 1982, W. Buckel and W. Weber, Eds.,
     page 463.

25. See for example F. Di Salvo, NATO Advanced Study Institute on
     Electron-Phonon Interactions and Phase Transitions, (Plenum
     Press, New York, 1977).

26. F. R. Gamble and T. H. Geballe, in Treatise on Solid State
     Chemistry, N. B. Hannay Ed,. (Plenum Press, New York, 1976)
     Vol. 3, Chapter 2, pg. 89.

27. W. A. Little, Phys. Rev. 134, A 1416 (1964).

28. V. L. Ginsburg, Phys. Lett. 13, 101 (1964); D. Allender, J.
     Bray, and J. Bardeen, Phys. Rev. B 7, 1026 (1973).

29. J. Ihm and M. L. Cohen, Phys. Rev. 23, 3258 (1981).

30. References are given in T. H. Geballe and C. W. Chu, Comments
     in Solid State Phys. 9, 115 (1979).

31. C. W. Homan, K. Laojindapunk and R. K. McCrone (LT16),
     Physica B + C, 107 (1982).

32. R. C. Dynes and J. M. Rowell, Phys. Rev. B 11, 1884 (1975);
     P. B. Allen and R. C. Dynes ibid 12, 905 (1975).

33. A. R. Sweedler, Ch. J. Raub, and B. T. Matthias, Phys. Lett.
     15, 108 (1965).

34. M. Robbins, R. H. Willens and R. C. Miller, Solid State
     Commun. 5, 933 (1967).

35. See for example R. N. Shelton Superconductivity in d- and f-Band Metals 1982, W. Buckel and W. Weber, Eds., pg. 123.

36. M. B. Panish, "Molecular Beam Epitaxy," Science, Vol. 208 (1980, pg. 916-922; A. C. Gossard, Thin Solid Films 57, 3 (1979).

37. R. H. Hammond, IEEE Trans. Magn. 11, 201 (1975); J. Vac. Sci. Technol. 15, 382 (1978).

38. S. Geller, Acta Crystallogr. 9, 885 (1956)

39. G. R. Johnson and D. H. Douglass, J. Low Temp Phys. 14, 565 (1974).

40. Y. Tarutani and M. Kudo, Journal of Less Common Metals 55 (1977).

41. J. Kwo, R. H. Hammond and T. H. Geballe J. Appl. Phys. 51, 1726 (1980).

42. K. Kihlstrom, Ph. D. Thesis, Stanford University, (1982).

43. J. M. Rowell, Tunneling Phenomena in Solids, E. Burstein and S. Lindquist, Eds. (Plenum Press, New York, 1969).

44. L. Y. L. Shen, Superconductivity in d- and f-Band Metals, D. H. Douglass, Ed.,(AIP Conf. Proc. 4, New York), pg. 31.

45. J. M. Rowell (private communication).

46. D. F. Moore, R. B. Zubeck, J. M. Rowell and M. R. Beasley Phys. Rev. B.20, 2721 (1979); D. A. Rudman and M. R. Beasley, Appl. Phys. Lett. 36, 1010 (1980).

47. L. R. Testardi, Physical Acoustics, Masson and Thurston, Eds., (Academic Press, New York, 1973), pg. 194; M. Weger and I. B. Goldber Solid State Physics, H. Ehrenreich,F. Seitz and D. Turnbull Eds., (Academic Press, New York, 1973) Vol. 28; J. Muller, Rep. Prog. Phys. 43, 641 (1980).

48. A. M. Clogston and V. Jaccarino, Phys. Rev. 121, 1357 (1961).

49. B. M. Klein, D. A. Papaconstantopoulos and L. L. Boyer, Superconductivity in d- and f-Band Metals, H. Suhl and M. B. Maple, Eds. (Academic Press, New York, 1980), pg. 455.

50. G. R. Stewart, Superconductivity in d- and f-Band Metals 1982, W. Buckel and W. Weber, Eds., pg 81.

51. L. E. Mattheiss and W. Weber, Phys. Rev. B 25, 2248 (1982).

52. T. H. Geballe, R. H. Hammond and J. Kwo Synthesis and Properties of Metastable Phases, E. S. Machlin and T. J. Rowland, Eds. (American Institute of Mech. Eng., New York, 1980),

53. G. W. Webb, S. Moehlecke and A. R. Sweedler, Mat. Res. Bull. 12 657 (1977).

54. G. W. Webb, Superconductivity in d- and f-Band Metals, D. Douglass, Ed. (AIP, New York, 1972), pg. 139.

55. G. R. Stewart, B. Olinger and L. R Newkirk, Solid State Commun. 43, 455 (1981).

56. J. R. Gavaler, A. I. Braginski, M. A. Janocko and A. S. Manocha, Superconductivitiy in d- and f-Band Metals 1982, W. Buckel and W. Weber Eds., pg. 35; R. Feldman and R. Hammond, J. Appl. Phys. 52, 1427 (1981).

57. K. E. Kihlstrom, R. H. Hammond, J. Talvacchio, T. H. Geballe, A. K. Green and Victor Rehn, J. Applied Physics (in press Aug 1982).

58. A. H. Dayem, T. H. Geballe, R. B. Zubeck, A. B. Hallak and G. W. Hull, Jr., J. Phys. Chem. Solids 39, 529 (1978).

59. R. D. Feldman, R. H. Hammond and T. H. Geballe, IEEE Trans. Mag. MAG-17, 545 (1981).

60. R. D. Feldman, Thin Solid Films, 87, 243· (1982).

61. J. Kwo and T. H. Geballe, Phys. Rev. B. 23, 3230 (1981).

62. K. Kihlstrom, D. Mael and T. H. Geballe, to be published.

63. D. Scalapino, Superconductivity, R. Parks, Ed., (Marcell Dekker, New York) (1969).

64. V. Z. Kresin and V. P. Parkhomenko, Sov. Phys. Solid State 16, 2180 (1975).

65. W. L. McMillan, Phys. Rev. 167, 331 (1968)

66. P. B. Allen and R. C. Dynes, Phys. Rev. B 12, 905 (1975).

67. W. L. McMillan and J. M. Rowell, Superconductivity, R. Parks, Ed. (Marcell Dekker, New York), Vol. 1, (1969) pg. 561.

68. R. C. Dynes and J. M. Rowell, Phys. Rev. B 11, 1884 (1974).

69. P. B. Allen and R. C. Dynes, Phys. Rev. B 12, 905 (1975).

70. G. B. Arnold, Phys. Rev. B, 18, 1076, (1978).

71. E. L. Wolf, J. Zasadzinski, J. W. Osmun and G. B. Arnold,
     Solid State Commun. 31, 321 (1979).

72. E. L. Wolf, J. Zasadzinski, G. B. Arnold, D. F. Moore, J. M.
     Rowell and M. R. Beasley, Phys. Rev. B.22, 1214 (1980).

73. K. E. Kihlstrom and T. H. Geballe, Phys. Rev. B 24, 4101
     (1981).

74. J. M. Rowell, M. Gurvitch and J. Geerk, Phys. Rev. B 24,
     (1981).

75. J. Geerk, J. M. Rowell and P. H. Schmidt, Superconductivity
     in d- and f-Band Metals 1982, W. Buckel and W. Weber, Eds.,
     page 23.

76. G. Shirane and J. D. Axe, Phys. Rev. B 4, 2957 (1971).

77. G. Bergman and D. Rainer, Z. Phys. 263, 59 (1973).

78. B. Mitrovic, C. R. Leavens and J. P. Carbotte, Phys. Rev. B
     21, 5048 (1980).

79. D. F. Moore, R. B. Zubeck, J. M. Rowell and M. R. Beasley,
     Phys. Rev. B 20, 721 (1979); D. Rudman, Ph.D. Thesis,
     Stanford University, (1982); J. Kwo and T. H. Geballe,
     Physica 109 & 110B, 1665 (1981).

80. T. P. Orlando, E. J. McNiff, Jr., S. Foner and M. R. Beasley,
     Phys. Rev. B 19, 4545 (1979); T. P. Orlando, Ph. D. Thesis,
     Stanford University, (1981).

81. A. K. Gosh and M. Strongin Superconductivity d- and f-Band
     Metals, H. Suhl and B. Maple, Eds., (Academic Press, New
     York, 1980); L. R. Testardi and L. F. Mattheiss, Phys. Rev.
     Lett. 41, 1612 (1978).

82. C. M. Varma and R. C. Dynes Superconductivity in d- and f-
     Band Metals, D. H. Douglass, Ed. (Plenum Press, New York,
     1976). pg. 507.

83. W. Hanke, J. Hafner and H. Bilz, Phys. Rev. Lett. 37, 1560
     (1976).

84. C. C. Tsuei, Superconductivity in d- and f-Band Metals, H.
     Suhl and B. Maple, Eds., (Academic Press, New York, 1980).

85. J. Kwo, T. P. Orlando and M. R. Beasley, Phys. Rev. B 24,
     2506 (1981).

86. H. R. Zeller and I. Giaever, Superconductivity, Frank
     Chilton, Ed., (North-Holland Publishing Company, 1971), pg.
     173.

KOSTERLITZ-THOULESS TRANSITIONS AND

TWO-DIMENSIONAL SUPERCONDUCTORS

J. E. Mooij

Department of Applied Physics
Delft University of Technology
Delft, The Netherlands

INTRODUCTION

The nature of the phase transition of a quantity of matter from a low-temperature ordered state to a high-temperature disordered state is determined by the dimensionality of the system and the number of degrees of freedom possessed by the components of the system. Between all various possibilities two-dimensional matter, of which the relevant order is described with one continuously variable parameter, occupies a very special position. Here Kosterlitz-Thouless transitions are possible under certain conditions.

To discuss the nature of these transitions, it is necessary to introduce the concepts of topological order and disorder. For this purpose we will use the system of the relevant category that is most widely used in statistical mechanics: the X-Y model. It consists of a square two-dimensional lattice of spins. The spins have two components in the X-Y plane, described with the angle φ made with one reference direction which is arbitrary. Only nearest neighbours interact. If φ_i indicates the direction of spin i, the interaction energy of one pair is given by:

$$U_{ij} = -J \cos(\varphi_i - \varphi_j) \tag{1.1}$$

where i and j are associated with a nearest-neighbour pair. J is a positive constant.

At zero temperature all spins are parallel. At any finite temperature, even if $kT \ll J$, so-called spin waves will occur as pictured in Fig. 1b. Although neighbouring spins are always almost parallel,

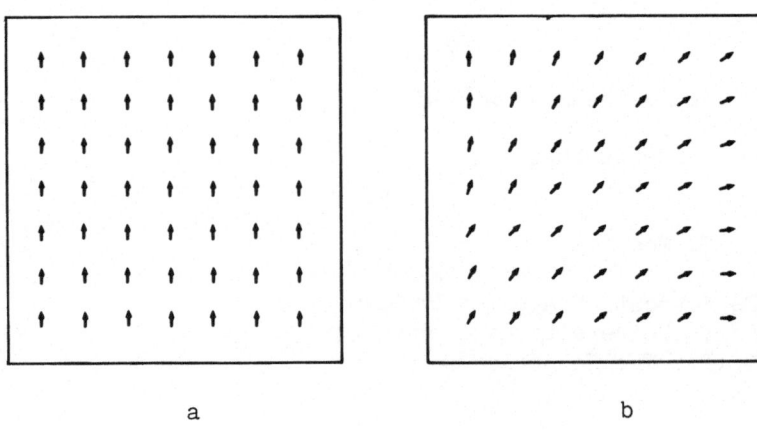

                    a   b

Fig. 1. Spins in two-dimensional X-Y model. a. T=0. b. T>0, spin
        waves.

over long distances the spin directions change considerably. It has
been proven that the spin directions at large mutual distances are
not correlated. Consequently, in a two-dimensional system like this
there is no long range order in a strict sense. However, when the
variation of φ is followed along any closed contour, it is found that
the sum of all phase differences δφ between neighbouring pairs along
the contour is zero: At small finite temperatures no spontaneous
magnetization is present, but there is 'topological order'.

The topological order is lost as soon as 'vortices' occur. In
Fig. 2 a positive and a negative vortex have been indicated. If a
contour is followed around a vortex core, one finds that the sum of
all δφ values is equal to plus or minus 2π. We will not consider
higher values of the vorticity here. The energy of a single vortex is
large, proportional to the logarithm of the size of the sample. In
contrast, the energy of a pair of vortices of opposite sign, as
pictured in Fig.3, is much smaller. The vortex-antivortex pair energy
is proportional to the logarithm of the distance between partners.

Even at relatively low temperatures bound pairs of small
separation can occur. As the temperature increases, more and more
pairs of larger and larger separation are present. Above a certain
temperature, the Kosterlitz-Thouless transition temperature, free

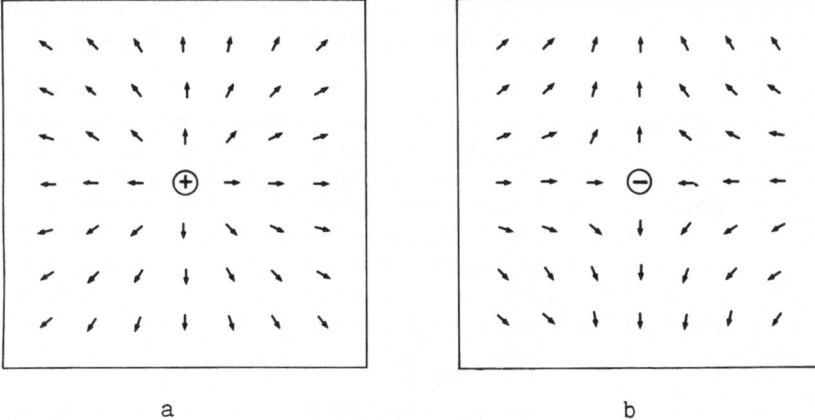

a   b

Fig. 2. Positive and negative vortex, or vortex and antivortex.

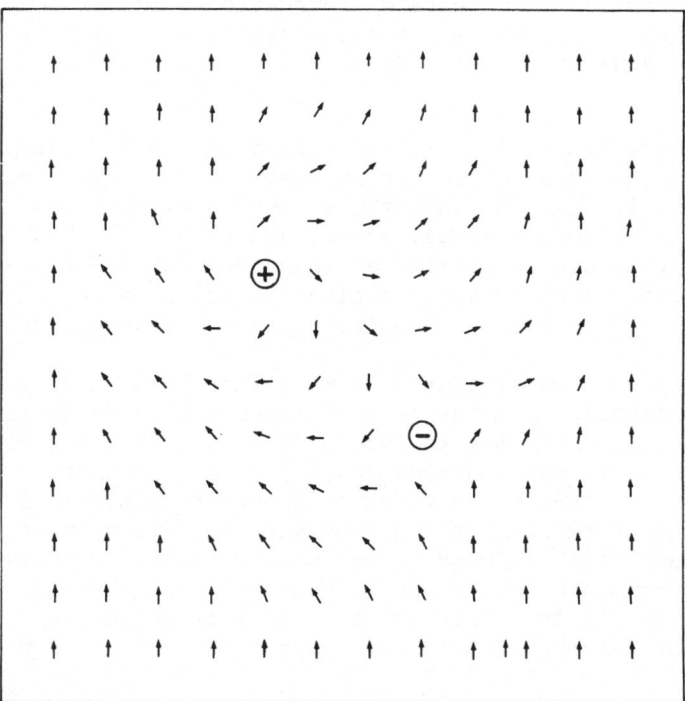

Fig. 3. Vortex-antivortex pair.

single vortices are thermodynamically stable even in large systems.
The energy term is compensated by a negative entropy term, reducing
the free energy to zero.

     After important work by Berezinskii (1972), Kosterlitz and
Thouless (1973, 1974) developed the theory for this peculiar type of
phase transitions. Initially it was thought that these transitions
should not occur in two-dimensional superconductors. Here the vortex-
antivortex pair energy increases logarithmically with separation
only up to the penetration depth, where electromagnetic screening
becomes limiting. In contrast, for neutral superfluids such as helium
films, the transitions were expected and indeed observed (Bishop and
Reppy 1978).

     In 1979 the conclusion was reached by several authors that
similar phenomena should be present in dirty superconducting films.
Beasley, Mooij and Orlando pointed out that the penetration depth in
very dirty thin films is large, usually larger than the sample size
at the temperatures of interest. They found a simple relation between
the Kosterlitz-Thouless transition temperature, the BCS critical
temperature and the normal-state sheet resistance of the film.
Doniach and Huberman, Halperin and Nelson, and Turkevich (all 1979)
calculated several properties, among them the resistive transition
and the properties in an external magnetic field. In the following
years a considerable number of experimental and theoretical papers on
Kosterlitz-Thouless or vortex-pair unbinding transitions was published.

     In this article we will try to discuss the theoretical aspects
with respect to superconductors and review the present experimental
status. It will turn out that the theory is as yet insufficiently
developed to allow a critical comparison with the relatively few
quantities that can be determined in experiments. However, the
available experimental evidence clearly indicates that bound vortex
pairs occur and that vortex unbinding is an important effect.

     In chapter 2 we discuss the important features of vortices in
dirty superconducting films. A naive approach to Kosterlitz-Thouless
transitions, ignoring the influence of other bound pairs, is
illustrated with these superconducting films. In chapter 3 the
renormalization theory treatment of these transitions is discussed.
In chapter 4 DC experiments on 'homogeneous' films in zero field are
described and their results compared with theory. The same is done
with AC experiments in chapter 5. Chapter 6 and chapter 7, both
relatively short, deal with films in an external magnetic field and
2D arrays of Josephson junctions respectively. Chapter 8 contains the
conclusions.

     A recent review paper, with different emphasis, has been
published by Hebard and Fiory (1982).

TWO-DIMENSIONAL SUPERCONDUCTORS

Vortices in Superconducting Films

We consider a thin, homogeneous superconducting film with thickness d, BCS critical temperature T_{co} and bulk penetration depth λ. The film is dirty ($\ell \ll \xi_o$, ℓ the electronic mean free path, ξ_o the BCS coherence length). The thickness is smaller than the effective coherence length ξ and smaller than λ. Consequently variables like the order parameter and the current density are constant across the film thickness. In the dirty limit a London type of description can be used for vortices with reasonable accuracy.

It is assumed that the order parameter is zero within a radius ξ and constant outside. The current distribution has been calculated by Pearl (1964,1965). He finds:

$$j(r) = \frac{c\Phi_o}{8\pi^2\lambda^2} \frac{1}{r} \qquad r \ll \frac{\lambda^2}{d}$$

$$j(r) = \frac{c\Phi_o}{4\pi^2 d} \frac{1}{r^2} \qquad r \gg \frac{\lambda^2}{d}$$

$$(2.1)$$

for the tangential component outside the core. $\Phi_o = hc/2e$ is the flux quantum. At longer distance the current density falls off faster than close to the core. Cross-over occurs near the effective penetration depth for perpendicular fields λ_p:

$$\lambda_p = \frac{\lambda^2}{d} \qquad\qquad (2.2)$$

The regime for small radius r is interesting from a Kosterlitz-Thouless point of view. To investigate the value of λ_p we use the expression for λ valid in the dirty local limit (Tinkham 1975):

$$\lambda^2_{eff}(T) = \lambda^2_L(0) \frac{\xi_o}{\ell} \frac{\Delta(0)}{\Delta(T)} \tanh^{-1}\{\tfrac{1}{2}\beta\Delta(T)\} \qquad (2.3)$$

Δ is the gap and $\beta = 1/kT$. By using that $\lambda_L(0) = (mc^2/4\pi ne^2)^{\frac{1}{2}}$ and $\xi_o = hv_F/\pi\Delta(0)$, and by introducing the sheet resistance of the film:

$$R_\square = \frac{\rho}{d} \qquad\qquad (2.4)$$

where ρ is the normal-state resistivity, it is found that λ_p is:

$$\lambda_p(T) = \frac{\lambda_p(0)}{f(T)} \tag{2.5}$$

where

$$\lambda_p(0) = 2.18 \frac{\Phi_o^2}{4\pi^5} \frac{R_\square}{\hbar/e^2} \frac{1}{kT_{co}} \tag{2.6}$$

$$f(T) = \frac{\Delta(T)}{\Delta(o)} \tanh \{\tfrac{1}{2}\beta\Delta(T)\} \tag{2.7}$$

The temperature dependent factor f varies from 1 at T=0 to 0 at T_{co}.
Near T_{co}, in the Ginzburg-Landau regime, f(T) is:

$$f_{GL}(T) = 2.66 \left(1 - \frac{T}{T_{co}} \right) \tag{2.8}$$

The value of $\lambda_p(0)$ is directly related to the sheet
resistance. Numerically, because $\hbar/e^2$ corresponds to 4.11 kΩ,
Eq. (2.6) amounts to:

$$\lambda_p = \frac{R_\square}{(1k\Omega)} \frac{(1\ K)}{T_{co}}\ 0.110\ cm \tag{2.9}$$

We see that for high-resistance films with sheet resistances
above 1kΩ, even at low temperature λ_p is of the order of millimeters.
If the relevant length scale of the physical phenomena that are
studied is smaller than λ_p, the high-radius regime of Eq.(2.1b) can
be ignored. We will assume that the current density associated with
the vortex is as given by Eq.(2.1a). In a sample, circle-shaped with
radius $R \ll \lambda_p$, the energy of one single vortex in the centre is:

$$U_1 = \frac{\Phi_o^2}{16\pi^2\lambda_p}\ \ln\left(\frac{R}{\xi} \right) \tag{2.10}$$

We define a reference energy U_o with:

$$U_o = \frac{\Phi_o^2}{32\pi^2\lambda_p(0)} \tag{2.11}$$

It has a value:

$$U_o = 2.18 \ \frac{\hbar/e^2}{R_\square} \ kT_{co} \tag{2.12}$$

Simple Approaches

It is possible to use two different extremely simple approaches to estimate the Kosterlitz-Thouless transition temperature T_{KT}. Even if they are very naive, it turns out that these approaches give the correct answer in a certain limit. It also helps to follow them to get familiar with the underlying physics.

The first approach is based on the thermodynamic free energy F_1 of one single vortex in a sample of characteristic size R. R is much larger than ξ and d and much smaller than λ_p. F_1 contains both the energy U_1 and an entropy term:

$$F_1 = U_1 - TS = U_1 - kT \ \ln N_1$$

where N_1 is the number of independent possibilities of placing one vortex in the sample. Both U_1 and N_1 depend on R. Clearly $N_1 = aR^2/\xi^2$, where a is of order unity. Consequently in a reasonable approximation:

$$F_1 = 2U_o \ f(T) \ \ln \ (\frac{R}{\xi}) - 2 \ kT \ \ln \ (\frac{R}{\xi})$$

We see that F_1 is negative for temperatures higher than the value T_1 that follows from:

$$U_o f(T_1) - kT_1 = 0 \tag{2.13}$$

This 'calculation' indicates that free vortices might exist in thermodynamic equilibrium if T is higher than T_1.

The second approach reasons from the vortex-antivortex pairs of Fig. 3. The energy of a pair with distance between centers r is:

$$U_2 = 4U_o \ f(T) \ \ln \ (\frac{r}{\xi}) \tag{2.14}$$

If one single vortex-antivortex pair in an infinite sample is considered, the expectation value for r^2 can be calculated. It is:

$$\langle r^2 \rangle = \frac{\xi\int^{\infty} 2\pi r dr \; [\; r^2 \exp\{-\beta U_2(r)\} \;]}{\xi\int^{\infty} 2\pi r dr \; [\; \exp\{-\beta U_2(r)\} \;]}$$

$$= \xi^2 \frac{\int_{1}^{\infty} d(\frac{r}{\xi}) \; (\frac{r}{\xi})^{3-4\beta U_o f}}{\int_{1}^{\infty} d(\frac{r}{\xi}) \; (\frac{r}{\xi})^{1-4\beta U_o f}}$$

$$= \xi^2 \frac{2-4\beta U_o f}{4-4\beta U_o f}$$

The mean separation increases with temperature until it diverges at a temperature T_2 given by the zero value of the denominator:

$$U_o f(T_2) - kT_2 = 0 \qquad\qquad (2.15)$$

So the pair dissociates at the same temperature where a single vortex has zero free energy. It turns out that Eqs.(2.13) and (2.15) which ignore the presence of other vortices predict correctly the transition temperature for systems of very low vortex density.

Kosterlitz-Thouless Temperature

In this article we distinguish between the 'ideal' Kosterlitz-Thouless transition temperature T_{KT}, which is applicable in the limit of very low pair densities and the true two-dimensional transition temperature T_c. The difference between the two will be discussed in the next chapter. Both are different from the 'bulk' superconducting transition temperature T_{co}. Above T_{co} no superfluid is present, between T_c and T_{co} free vortices in the superfluid lead to dissipation and only below T_c the superfluid properties are retained.

The value of T_{KT} in superconducting films is the same as T_1 and T_2 in Eqs.(2.13) and (2.15). If we use Eq.(2.12) we see that:

$$\frac{T_{KT}}{T_{co}} f^{-1}(T_{KT}) = 2.18 \; \frac{\hbar/e^2}{R_{\square}} \qquad\qquad (2.16)$$

This expression was first derived by Beasley, Mooij and Orlando (1979). In Fig. 4 the dependence of T_{KT} on the sheet resistance is

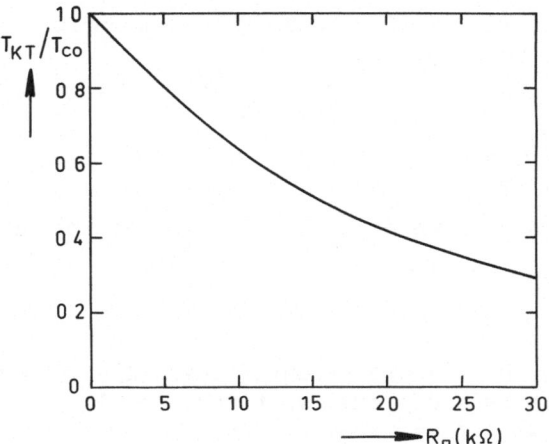

Fig. 4. Kosterlitz-Thouless temperature in superconducting films
as a function of sheet resistance (Beasley et al. 1979)

given. It is clear that only for sheet resistances of the order of 1
kΩ or higher T_{KT} will be significantly below T_{co}. It is
intrigueing to consider that the 'maximum metallic resistance' for
two-dimensional normal metals is of order $\hbar/e^2$ = 4.1 kΩ per
square. No connection has been made yet between the theory of
localization in two dimensions and the two-dimensional transition in
very dirty superconductors.

If $R_{\square}$ is not too high, T_{KT} is sufficiently close to T_{co} that the
Ginzburg-Landau expression (2.8) for f(T) can be used. This gives:

$$\frac{T_{KT}}{T_{co}} = \frac{1}{1 + 0.173 \dfrac{R_{\square}}{\hbar/e^2}} \qquad\qquad 1 - \frac{T_{KT}}{T_{co}} \ll 1 \qquad (2.17)$$

At the Kosterlitz-Thouless temperature λ_p is equal to:

$$\lambda_p(T_{KT}) = 0.204 \frac{\Phi_o^2}{\pi^5} \frac{1}{kT_{KT}} \tag{2.18}$$

as follows from Eqs.(2.5), (2.6) and (2.16). Numerically, the product of $\lambda_p(T_{KT})$ and T_{KT} is equal to:

$$\frac{\lambda_p(T_{KT})}{(1\ K)} \frac{T_{KT}}{(1\ cm)} = 0.98$$

It is clear that this value of λ_p is at least of the order of the usual sample sizes.

Vortex Mobility

In later chapters the experimental investigations are described to detect vortex pairs and single vortices. In all these, vortices are only detected when they move. As for bulk superconductors, a vortex mobility can be defined for thin films:

$$\mu = \frac{2e^2}{\pi\hbar^2} \xi^2 R_\square \tag{2.19}$$

In high-resistance films vortices move about relatively easily, due to the reduced damping. If the vortex density is n_v, the resistivity is $\mu n_v \Phi_o^2/c^2$ or, expressed in the normalized resistance:

$$\frac{R}{R_n} = 2\pi\xi^2 n_v$$

In n_v only unpaired vortices are counted. They may have grown thermodynamically or been induced by an external magnetic field.

Qualitative Picture

We have seen that it is likely that superconducting thin films with very high sheet resistance will exhibit the same 'two-dimensional' behaviour as predicted and found for other superfluids and found for the X-Y spin model. An important difference with the latter model is, that in superconductors in the critical region the coupling 'constant' is usually strongly temperature dependent.

For superconducting films we obtain a qualitative picture as schematically indicated in Fig. 5. At low temperatures there is no long range correlation for the phase of the order parameter, but topological order is maintained up to the two-dimensional transition temperature T_c. At really low temperatures only 'phase waves', equivalent to spin waves, occur. With increasing temperature bound vortex-antivortex pairs appear, with increasing numbers as T_c is approached and exceeded.

The critical temperature T_c is in the neighbourhood, but below the value T_{KT} calculated in this chapter. At T_c the first vortex unbinding occurs. The number of free vortices is very low near T_c, it grows only very gradually when the temperature is increased. On

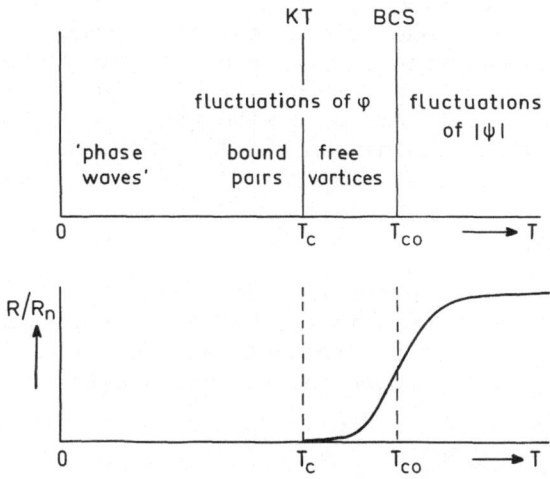

Fig. 5. Schematic behaviour of a two-dimensional superconductor.

approaching T_{co}, the single vortex and vortex pair energies decrease. For that reason the density of vortices becomes high and of importance for any sample and experiment.

Above T_{co}, no continuous superconductivity is present and vortices do not exist. This is the region of fluctuations of the magnitude of the order parameter, rather than its phase. The discussion leads to a resistive transition as indicated in Fig. 5.

3. STATIC THEORY

Introduction

In this chapter we will describe the static theory for the two-dimensional phase transition. For derivations and more detailed discussions we refer to the articles of Kosterlitz and Thouless (1972,1973), Kosterlitz (1974) and Young (1978). The system most commonly used in theoretical papers is the X-Y model, as referred to in chapter 1. It is discrete, with spins at lattice points a unit length apart. In the X-Y model the energy of a vortex-antivortex pair at distance r (measured in lattice constants) is:

$$U(r) = 2\pi J \ln r + 2\mu_c \qquad (3.1)$$

where J is the interaction constant, Eq. (1.1), and $2\mu_c$ is the energy required to create one pair at mutual distance of one lattice spacing. For a good understanding it is convenient to use the analog model of a two-dimensional Coulomb gas. We will describe the renormalization procedure that is employed with the 2D Coulomb gas analogy, use some of the results obtained on the X-Y model and then proceed to superconducting thin films.

2D Coulomb Gas

The two-dimensional Coulomb gas contains charges +q and -q. The force between one pair of oppositely charged particles at distance r is q^2/r if no other charges are present, and only if r is larger than a minimum distance r_o. The potential for one single pair is:

$$U(r) = q^2 \ln \left(\frac{r}{r_o}\right) + 2\mu_c \qquad (3.2)$$

where $2\mu_c$ is the value of U when $r = r_o$. It should be realized that this potential is different from the potential for point charges in the real world, even if they would be restricted in a 2D plane. A better picture is obtained by considering unit length pieces of infinitely long rods with charge ±q per unit length, where U gives the potential also per unit length.

If other pairs are present, the 'gas' is polarizable. The effective dielectric constant experienced by a pair at separation r is called $\epsilon(r)$. Clearly the force between partners is:

$$F(r) = \frac{1}{\epsilon(r)} \frac{q^2}{r} \qquad (3.3)$$

and the pair potential:

$$U(r) = {}_r\int_0^r dr' \left[\frac{1}{\epsilon(r')} \frac{q^2}{r'^2} \right] + 2\mu_c \qquad (3.4)$$

The assumption is made that $\epsilon(r)$ contains only contributions from pairs with separation smaller than r. To calculate $\epsilon(r)$, the polarizability $p(r)$ and density $\rho(r)$ ought to be known:

$$\epsilon(r) = 1 + {}_r\int_0^r dr' \left[p(r') \rho(r') \right] \qquad (3.5)$$

The polarizability creates no problems but the density depends strongly on U:

$$\rho(r) = C r \exp\{-\beta U(r)\} \qquad (3.6)$$

where C is a constant. It is impossible to solve the system of equations (3.4) -(3.6) directly. Only by means of the renormalization technique, in which the scale is changed successively, answers are obtained. The renormalization procedure yields equations for the scaling behaviour of two quantities, the 'coupling':

$$K(r) = \frac{1}{2\pi\epsilon(r)} \frac{q^2}{kT} \qquad (3.7)$$

and the 'fugacity':

$$y(r) = (\frac{r}{r_0})^2 \exp\{ -\frac{\beta}{2} U(r) \} \qquad (3.8)$$

The scaling equations are found to be:

$$\frac{dK^{-1}(r)}{d(\ln r)} = 4\pi^3 y^2(r) \qquad (3.9)$$

$$\frac{d\,y(r)}{d(\ln r)} = \{ 2-\pi K(r) \} y(r) \qquad (3.10)$$

Analytic solutions to these equations have the form (Ambegaokar et al. 1980)

$$(\frac{2}{\pi} K^{-1}-1)^2 -4\pi^2 y^2 = A \qquad (3.11)$$

where A is a constant. The trajectories, as given by (3.11) are pictured in the 'flow diagram', Fig. 6. Different lines correspond to different values of the constant A. The direction of the arrows

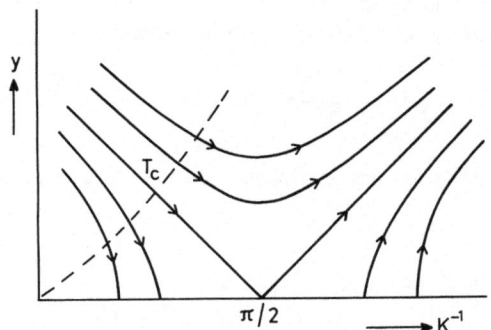

Fig. 6. Renormalization group trajectories, according to Eq. (3.11).
The dotted line indicates the starting conditions $y_o(K_o^{-1})$.
T_c is determined by the intersection of this line with the
critical trajectory that ends in $y=0$, $K^{-1}=\pi/2$.

indicates increasing scale r. Some trajectories approach $y = 0$ for
large r, others show diverging y for large r. The separation is given
by the critical trajectory connected with $A = 0$, which consists of
two straight lines:

$$2\pi y_c = \pm \left(\frac{2}{\pi} K_c^{-1} - 1 \right) \tag{3.12}$$

Which trajectory applies to the system under consideration is
determined by the 'starting conditions' $y_o = y(r_o)$ and $K_o = K(r_o)$. By
definition $\epsilon(r_o) = 1$, while $U(r_o) = 2\mu_c$. So:

$$K_o^{-1} = 2\pi \frac{kT}{q^2} \tag{3.13}$$

$$y_o = \exp\left(- \frac{\mu_c}{kT}\right) \tag{3.14}$$

The flow diagram of Fig. 6 is universal. The starting conditions (3.13) and (3.14) are given by the particular system and are non-universal. In the y-K^{-1} plot the starting conditions can be drawn, see the dotted line in Fig. 6. Each point of that curve corresponds to a certain temperature. The lower temperatures belong to points with low y_o. The trajectories leading from those points go to y = 0 for very large r. This means that at those temperatures there will not be any pairs with very large separation. On the other hand, points on the dotted line corresponding to high temperature and consequently high y_o, initiate trajectories that show diverging y values. For those temperatures pairs with large separation will be present. The critical temperature T_c is that value of T that indicates the intersection of the starting condition line $y_o(K_o^{-1})$ with the critical trajectory defined by (3.12). We find:

$$2\pi \, \exp(-\frac{\mu_c}{kT_c}) = -(\frac{2}{\pi} \, 2\pi \, \frac{kT_c}{q^2} - 1)$$

or

$$kT_c = \frac{1}{4} \, q^2 \{1 - 2\pi \, \exp(-\mu_c/kT_c)\} \qquad (3.15)$$

The procedure to determine T_c is valid only for small values of y. Of course for this reason it is necessary that $y_o \ll 1$, which is true for large values of the core energy μ_c. In the limit of very large μ_c, the critical temperature has the limiting value T_{KT}:

$$kT_{KT} = \frac{1}{4} \, q^2 \qquad (3.16)$$

From the definition of K(r) we know that

$$\epsilon(r,T) = \frac{K_o(T)}{K(r,T)}$$

We define ϵ_c as:

$$\epsilon_c = \epsilon(\infty, T_c) \qquad (3.17)$$

Because $K(\infty, T_c) = 2/\pi$, it is found that:

$$kT_c = \frac{1}{4} \, q^2 \, \frac{1}{\epsilon_c} \qquad (3.18)$$

and comparison with (3.15) yields:

$$\frac{1}{\varepsilon_c} = 1 - 2\pi \exp(-\mu_c/kT_c) \qquad\qquad\qquad (3.19)$$

X-Y Model

The results obtained for the transition in the 2D Coulomb gas can easily be translated to the X-Y model, with pair potential $U(r)$ given by Eq.(3.1). The core energy U_c for this system is (Kosterlitz and Thouless 1973):

$$\mu_c = \tfrac{1}{2}\pi^2 J \qquad\qquad\qquad\qquad (3.20)$$

leading to:

$$kT_c = \frac{\pi}{2} J\{1 - 2\pi \exp(-\pi)\} = 1.14 \, J$$

The value of ε_c is found to be 1.09. It can be verified that $y_0(T_c) = 0.013$ is sufficiently small that the approximations used are valid.

Superconducting Films

The same procedure applies to superconducting films. The effective 'charge' is here given by:

$$q^2 = 4U_0 \, f(T) \qquad\qquad\qquad\qquad (3.21)$$

It is temperature-dependent. For superconducting films r_0 is usually assumed to be equal to ξ. For very large core energy, $T_c \simeq T_{KT}$ is obtained directly from (3.16) and turns out to be equal to the result (2.16). However, the real transition temperature is lower than T_{KT}. The importance of the difference between T_c and T_{KT} has been discussed recently by Epstein, Goldman and Kadin (1982). We will partly follow that discussion.

Of crucial importance in this connection is the core energy, which should rather be called core potential because it contains not only the loss of condensation energy in the core, but also an entropy term. The condensation energy E_c is, in the London approximation that we used, equal to the volume, $\pi d \xi^2$, times $H_c^2/8\pi$. As we really want half the energy associated with two cores, it is reasonable to write:

$$E_c = a \, \frac{1}{8} \xi^2 d \, H_c^2 \qquad\qquad\qquad\qquad (3.22)$$

where a is of the order of, but not necessarily equal to 1. The entropy term can be expressed by means of a quantity N_0 which is not too well defined but gives some indication of the number of independent positions that a pair of minimal separation can occupy in an area ξ^2. The orientation would, for example, provide some freedom in this respect. We find for the core potential μ_c:

$$\mu_c = E_c - TS$$

$$= E_c - kT \ln N_o$$

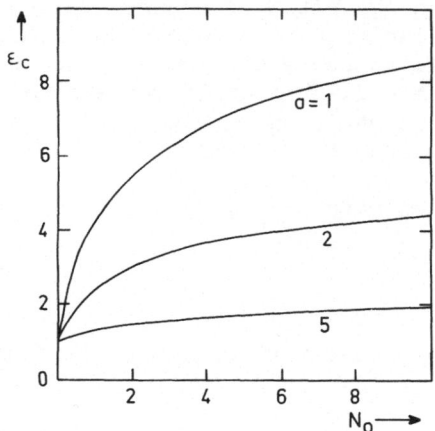

Fig. 7. Influence of core potential on the effective critical 'dielectric constant', according to Eq. (3.25).

In the Ginzburg-Landau regime we can use the fact that Φ_o is equal to $2\sqrt{2}\ \pi H_c \lambda \xi$ to obtain:

$$E_c = \tfrac{1}{2}\ a\ U_o\ f(T) \qquad\qquad (3.23)$$

At the yet unknown transition temperature T_c, μ_c is:

$$\mu_c(T_c) = \tfrac{1}{2} a \, U_o \, f(T_c) - kT_c \ln N_o$$

and Eq. (3.15) yields for T_c:

$$kT_c = U_o \, f(T_c) \, \{1 - 2\pi \, N_o \, \exp[-\tfrac{1}{2}a \, \frac{U_o f(T_c)}{kT_c}]\} \qquad (3.24)$$

When Eq.(3.18) is used, (3.24) can be written as:

$$\frac{1}{\varepsilon_c} = 1 - 2\pi N_o \, \exp(-\tfrac{1}{2}a\varepsilon_c) \qquad (3.25)$$

If N_o and a are known, ε_c can be calculated from this equation. In Fig. 7 ε_c is plotted against N_o for three values of a. One would at first expect the value of a to be close to 1. However, possibly it is much more reasonable to use a minimum distance between core centers of 2ξ rather than ξ. This could increase a considerably. Fig. 7 shows, that it is reasonable to expect that ε_c is significantly larger than 1 in superconducting films.

The transition temperature T_c is given by the solution of:

$$kT_c = \frac{1}{\varepsilon_c} U_o \, f(T_c)$$

In the Ginzburg-Landau temperature region it is found that:

$$\frac{T_c}{T_{co}} = \frac{1}{1 + 0.173 \, \varepsilon_c \, \dfrac{R_\square}{\hbar/e^2}} \qquad (3.26)$$

which is significantly different from the Beasley, Mooij, Orlando result (2.17) for T_{KT}. Unfortunately, it is not possible to predict T_c from theory, unless the core potential is calculated.

It is important to verify that the basic assumption, made in the Kosterlitz-Thouless derivation, that the fugacity y is small, is justified. At the transition temperature y_o can be calculated with the aid of Eqs.(3.14) and (3.19):

$$y_o(T_c) = \frac{1}{2\pi} (1 - \frac{1}{\varepsilon_c}) \qquad (3.27)$$

As ε_c is equal to 1 or higher, this result indicates that at T_c, y_o

is smaller than $(2\pi)^{-1}$ and indeed significantly smaller than 1.

<u>Vortices above T_c</u>

At temperatures above T_c free vortices are present. Their density is called n_f. It is related to the correlation length which is called ξ_+. For both n_f and ξ_+ no precise calculations are available.

Halperin and Nelson (1979) have given the following expressions which contain unknown constants:

$$n_f = \frac{C_1}{2\pi} \frac{1}{\xi_+^2} \tag{3.28}$$

$$\xi_+ = a_1 \, \xi(T) \, \exp\{(b \, \frac{T_{co} - T_c}{T - T_c})^{\frac{1}{2}}\} \tag{3.29}$$

The constants C_1, a_1 and b are dimensionless and of order unity. The expressions are only valid in the critical region immediately above T_c, while T_c should be reasonably close to T_{co}. In the derivation the difference between T_c and T_{KT} has been neglected which is a serious limitation. The correlation length is comparable with the average distance between free vortices.

Halperin and Nelson indicate that the correlation length for the order parameter should fit on smoothly to the correlation length above T_{co}, which is the Ginzburg-Landau coherence length:

$$\xi(T) = \xi(T_c) \, (\frac{T_{co} - T_c}{T - T_{co}})^{\frac{1}{2}}$$

A smooth connection between $\xi_+(T)$ below T_{co} and $\xi(T)$ above T_{co} is obtained by the approximate formula

$$\xi_+ \approx b^{-\frac{1}{2}} \, \xi(T_c) \, \sinh(b \, \frac{T_{co} - T_c}{T - T_c})^{\frac{1}{2}} \tag{3.30}$$

which would indicate that in Eq.(3.29) a_1 should be equal to $\frac{1}{2}b^{-\frac{1}{2}}$.

Unfortunately, so far no more detailed calculations of the density of free vortices above T_c have been performed. In the paper of Halperin and Nelson ξ is taken constant at its level at T_c, while the difference between T_c and T_{KT} is left out of consideration. Some indication of the values of C_1, a_1 and b should be possible. A more precise calculation is highly desirable.

4. DC MEASUREMENTS

'Homogeneous' Films

To obtain useful experimental data on Kosterlitz-Thouless-like transitions in superconducting films, the films have to comply with a number of conditions:
1. T_c should be sufficiently below T_c to allow separation of vortex depairing effects from other fluctuation phenomena.
2. T_{co} should be well-defined and the same all over the film.
3. No pinning centres for vortices should be present.

It is extremely difficult to satisfy these conditions. The first requires a sheet resistance of 1 kΩ or more (Fig. 4). By using a straightforward metal film, such high values cannot be obtained. If one assumes a $\rho\ell$ product of 10^{-12} Ωcm^2 and also that ℓ is equal to the film thickness, it turns out that the thickness d has to be smaller than 3 Å to obtain a sheet resistance of 1 kΩ. Such monatomic films do not exist. High resistance films are either granular or amorphous. Fabrication and characterization of such films are no simple experimental techniques but research fields by themselves.

Granular films are by their nature inhomogeneous. However, if the grain size is much smaller than the coherence length, the film can be considered as homogeneous. No pinning will occur on grain boundaries. Aluminum is most often used for granular films. The BCS coherence length ξ_o is long enough (1.6 μm) to give values of $(\xi_o\ell)^{\frac{1}{2}}$ around 30 nm even if ℓ is only .5 nm. As $\xi_{GL}(T)$ is considerably longer than $(\xi_o\ell)^{\frac{1}{2}}$ near T_c, average grain sizes of some tens of nm are acceptable. The averaged properties of the films should be homogeneous. Bancel and Gray (1980) have shown how inhomogeneous granular films of aluminum can easily be. Only on freshly cleaved mica they obtained satisfactory films.

Amorphous films are usually obtained by evaporation onto a substrate that is cooled to helium temperatures. In most cases the measurements are performed immediately following evaporation. Annealing occurs already at temperatures of 20 or 30 K. Materials used vary from simple metals like Pb, to Nb_3Ge and exotic materials such as Hg-Xe.

Resistive Transition

The effective resistance of a film in the critical region between T_c and T_{co} depends strongly on the external magnetic field and the measuring current. Here we analyze the resistance for zero external field and extremely low measuring current values. In later sections we will indicate how small fields and currents have to be to satisfy the conditions.

In chapter 2 the sheet resistance of the film was coupled to the vortex density n_v. In zero field, n_v is equal to the density of free vortices n_f, as given by Eq. (3.28). The resultant resistivity is (Halperin and Nelson 1979):

$$\frac{R_s}{R_n} = C_1 \frac{\xi^2}{\xi_+^2} \tag{4.1}$$

For ξ_+ either (3.29) or (3.30) can be used, leading to:

$$\frac{R_s}{R_n} = \frac{C_1}{a_1^2} \exp\{-2(b \frac{T_{co} - T_c}{T - T_c})^{\frac{1}{2}}\} \tag{4.2}$$

or

$$\frac{R_s}{R_n} = C_1 b \sinh^{-2}\{(b \frac{T_{co} - T_c}{T - T_c})^{\frac{1}{2}}\} \tag{4.3}$$

An experimental R(T) curve on a linear scale may look like the example given in Fig. 5. It contains a region above T_{co} where R is depressed below R_n by fluctuations into the superconducting state. The low temperature tail, below T_{co}, belongs to a regime where the superconducting state is predominant but phase-slip is possible when vortices move across the sample. It is not possible to indicate T_{co}, except by fitting the curve to some theory. It is possible to obtain T_{co} separately from a fit to fluctuation theory of the high-temperature tail of the transition. So far this method has not been employed in connection with an analysis of the low temperature tail of the transition.

To obtain a reasonable fitting possibility to the theoretical expressions, R(T) must be very measured accurately over a wide range of several decades. An example is given in Fig.8, from Bancel and Gray (1980, 1981). They fabricated extremely homogeneous granular aluminum films on mica with a sheet resistance of 1700 Ω. The experimental data have been fitted to Eq.(4.3). The theoretical expression for T_{KT} (Eq.2.17) has been used to determine $\tau_c = (T_{co} - T_c)/T_c = 0.07$ from the sheet resistance. This meant that either T_{co} or T_c (called T_{KT} here) could be fitted, while the other then followed immediately. As indicated in the figure, a good fit was obtained by taking b = 14, C_1 = 4 and a value of T_c of 1.75 K.

Bancel and Gray point out that a good fit to an expression that contains three fitting parameters is of doubtful value, in particular

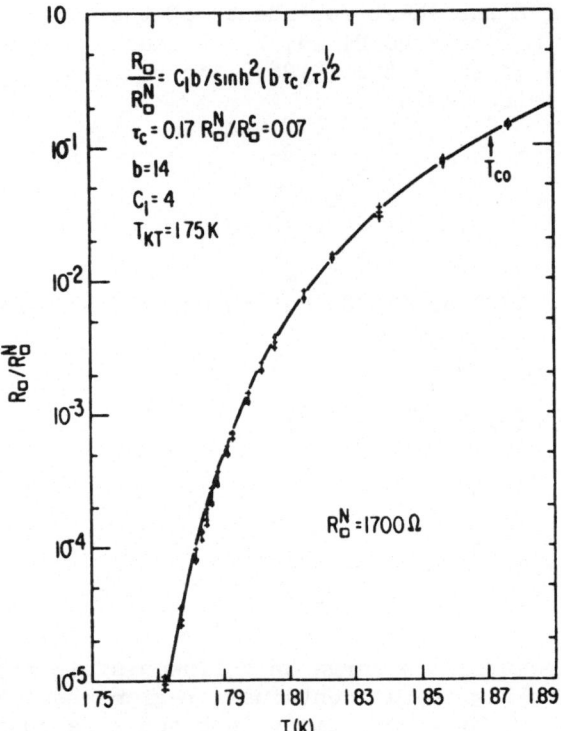

Fig. 8. Resistive transition of a granular aluminum film on mica, as
        measured by Bancel and Gray (1980). A fit has been performed
        to Eq. (4.3).

when the values obtained are not of the expected order of magnitude.
Their objections are valid, but should not lead to the conclusion
that vortex unbinding is not of importance. As discussed in the
previous chapter, the theoretical expressions are too primitive to be
used in a quantitative analysis. The influence of the difference
between T_{KT} and T_c must first be taken into account.

Other authors have also measured and analyzed the resistive
transition in the context of a possible vortex-unbinding transition.
We mention two other examples, given by Rao et al. (1980) and Wolf et
al. (1981) The first compare their results on amorphous films of Nb_3Ge
with a theoretical expression given by Turkevich (1979), of a similar
character as Eq.(4.2), obtaining a very good fit over a very wide
range. Wolf et al. conclude to good agreement with the theory for

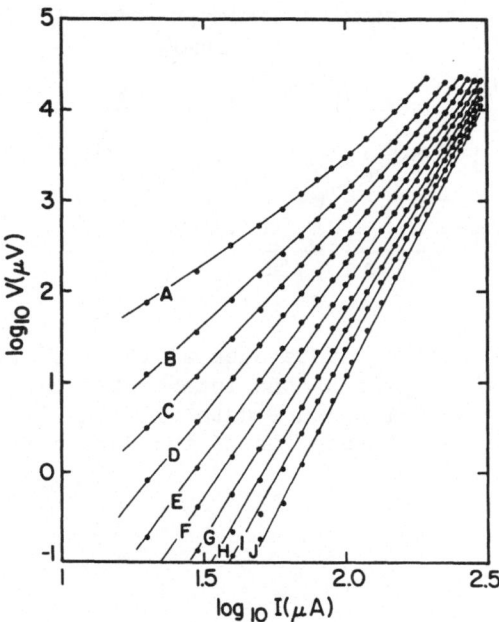

Fig. 9. V-I curves of a film of Hg-Xe (Epstein et al. 1981).
Different curves are for different temperatures, ranging from
2.435 K (curve A) to 1.5 K (curve J). T_c is between curves B
(2.29 K) and C (2.20 K).

two-dimensional phase transitions for the resistive transition of
granular NbN films. In other cases it seems that different phenomena
occur simultaneously and dominate the transition. Percolation
effects, for example, may be stronger than vortex unbinding effects
if inhomogeneities occur on a scale exceeding the coherence length.

Current-Induced Vortex Unbinding

The resistance discussed previously is only relevant if the
measuring current is small enough. For larger current densities the
opposite forces, exerted by the current on the two partners of a

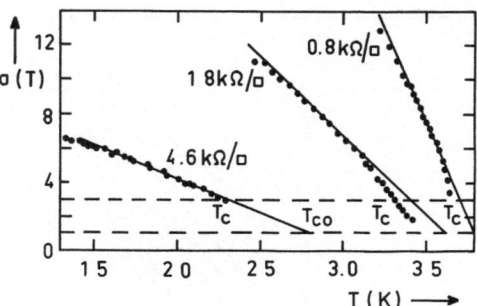

Fig. 10. α(T) determined from data such as those of Fig. 9 (Epstein
         et al. 1981). T_{co} is at the intersection of a straight-line
         fit to the low temperature data with the line α=1. T_c is
         taken from the intersection of the data points with the
         line α=3.

vortex pair, make it easier in one orientation of the pair to obtain
unbinding. The mechanism has been analyzed for vortices in two-
dimensional helium films by Ambegaokar et al. (1978, 1980). The free
vortices that result from this unbinding process will often recombine
to pairs but their density is finite, also below T_c, and depends on
the value of the current. Halperin an Nelson (1979) have calculated
the nonlinear resistivity in the temperature region below T_c. It can
be written as:

$$\frac{V}{I} = R_n \ x(T) \ (\frac{I}{I_o})^{\pi K(T)} \qquad\qquad (4.4)$$

Here R_n is the normal state resistance of the sample investigated, I
the current, V the voltage, x(T) a temperature dependent function, I_o
a reference current and K(T) the 'coupling' used in the renormalization procedure, defined in Eq. (3.7).

Experimentally, V is measured as a function of I over a certain
temperature range and the exponent α determined from:

$$V = C \, I^{\alpha(T)} \tag{4.5}$$

where C may depend on T but not on I. Epstein, Goldman and Kadin (1981) performed a series of measurements on amorphous Hg(Xe) films (Epstein et al. 1979). In Fig. 9 the results for one film are given. Epstein et al. used these data to determine T_c. They plotted α as a function of T for the different films, see Fig. 10. From the curves, T_{co} and T_c were both determined in the following way.

In the Ginzburg-Landau regime, ignoring all bound vortex pairs one expects that the resistivity, above the critical current, is proportional to the vortex mobility. This resistivity, proportional to V/I, is proportional to $T_{co} - T$. For the 'Ginzburg-Landau' α we can write:

$$\alpha_{GL} - 1 \propto T_{co} - T \tag{4.6}$$

At T_{co}, α_{GL} is equal to 1. From Fig. 10, this α_{GL} is determined by fitting to the lower temperature part where one expects the vortex pairs to be absent, and extrapolating to higher temperatures. Where this straight line reaches the value 1, T_{co} is supposed to lie. The method is a little unsatisfactory because the temperature region in which the straight line is fitted is rather far below T_{co} and the Ginzburg-Landau approximations may not be valid there.

The transition temperature T_c follows from $\alpha(T)$ with the aid of Eqs. (4.4) and (4.5):

$$\alpha(T) = 1 + \pi \, K(T) \tag{4.7}$$

Epstein et al. used the critical value of $2/\pi$ for $K(T_c)$, and determined T_c from the curves of Fig. 9 by taking that temperature where α is equal to 3. The data so obtained are plotted in Fig. 11. Also in the figure are the theoretical expectations for T_{KT} from Beasley et al.(1979), Eq.(2.17), and the best fit to the data points for the renormalized value from Eq. (3.26). The critical value of the effective 'dielectric constant' ε_c is found to be about 1.2 from the data. In previous attempts to determine T_c or other parameters of the vortex unbinding process, the difference between T_{KT} and T_c was usually ignored.

The analysis of the nonlinear resistance has been based on the treatment of Halperin and Nelson, who, in their paper, ignored the difference between T_c and T_{KT} or, in other words, assumed ε_c to be equal to 1. It would seem that it is more consistent to take as T_c not that temperature where $\alpha(T) = 3$, corresponding to $K = 2/\pi$, but the one where $\alpha(T) = 1 + 2\varepsilon_c$:

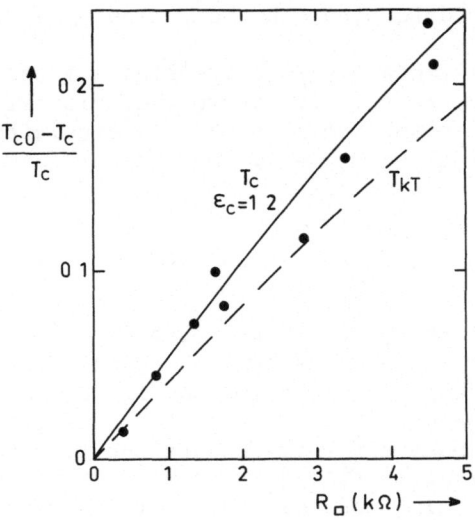

Fig. 11. Two-dimensional critical temperature of Hg-Xe films (Epstein
         et al. 1981). Dotted line: T_{KT} according to Eq. (2.17).
         Drawn line: fit to Eq. (3.26), giving ϵ_c =1.2.

$$\alpha(T_c) = 1 + 2\,\epsilon_c \qquad\qquad (4.8)$$

If ϵ_c has to be determined from the data, a self-consistent procedure
must be followed. A thorough recalculation of the Halperin-Nelson
results taking into account the renormalization corrections ($\epsilon_c > 1$)
from an earlier stage seems highly necessary.

Discussion

It is very difficult to draw firm conclusion from the DC measurements on 'homogeneous' superconducting films. The data are certainly not in conflict with the available theory for vortex unbinding processes. However, the theory has not been worked out in sufficient detail for superconductors, while the experiments leave too many unknown parameters. Inhomogenities that cannot entirely be avoided will play a certain role. We have not tried to be complete, but have concentrated on some experiments that seem to be most indicative of Kosterlitz-Thouless-like behaviour.

A very interesting excercise has been performed by Minnhagen (1981). He reasoned that while the behaviour of high-resistance films may not be consistent with Kosterlitz-Thouless theory, it is certain that vortices and interactions between them play a significant role in the temperature region immediately below T_{co}. All vortex potentials are proportional to $U_o f(T)$, in the formulation of this article, and if the actual temperature is scaled to this value of $U_o f(T)$, all dimensionless thermodynamic quantities such as R/R_n or ε_c should show a _universal_ dependence on the reduced temperature. Minnhagen analyzed experimental data for $R(T)$ from different groups on different materials. Indeed, the curves for R/R_n plotted against a reduced temperature scale (different reduction factor for each sample) fall nicely on one curve.

5. AC MEASUREMENTS

The phenomena that are the subject of this article are rooted in the existence of bound vortex-antivortex pairs. Unbinding of the pairs, with the accompanying loss of topological order, marks the transition. Still, the bound pairs rather than the free vortices are the backbone of the body of physics that we discuss. With DC electrical measurements only free vortices are monitored. It is possible to get to the bound pairs directly by means of AC measurements. In this respect Fiory and Hebard have rendered the major contribution with their measurements of the AC impedance of thin films.

AC Impedance

The method employed to determine the AC impedance is described in an article by Fiory and Hebard (1980), and in the review by Hebard and Fiory (1982). The film that is studied is placed between two coils, which generate and detect magnetic fields perpendicular to the film (Fig. 12). The screening properties of the film, in phase and out of phase with the field generated, are a measure of the real and imaginary part of the AC impedance.

The respons of the film contains contributions from the super-

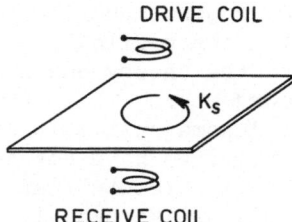

Fig. 12. Experimental arrangement for measurement of AC impedance
         (Fiory and Hebard).

fluid directly and from the vortices. The superfluid respons, a
'sheet current' $K_s = d\, j_s$, is connected with the sheet kinetic
inductance:

$$L_K = \frac{4\pi}{c^2}\,\lambda_p \qquad\qquad (5.1)$$

The vortex contributions are expressed in a 2D Coulomb gas analogy as
used in chapter 3, but now extended to the dynamic respons with a
complex, frequency-dependent dielectric constant. In that dielectric
constant contributions from bound pairs are separated from the
Coulomb gas 'conductance' σ_v due to free vortices. The resultant
complex impedance is:

$$Z = i\omega L_K(\varepsilon_{vb} + 2\pi\,\frac{\sigma_v}{i\omega}) \qquad\qquad (5.2)$$

$\varepsilon_{vb}-1$ represents the bound pair respons. It is Z^{-1} rather than Z to
which the experiment is sensitive.

   The BCS transition temperature can be derived from the experimental data by determination of L_K at lower temperatures where no
vortices are present. Extrapolation of L_K^{-1} to higher
temperatures gives T_{co} where $L_K^{-1} = 0$.

   In Fig. 13 we give an example of a measurement of $Y = Z^{-1}$ of
a granular aluminum film (Hebard and Fiory 1980b) with a sheet
resistance of 4.1 kΩ. The real part of Y exhibits a peak which

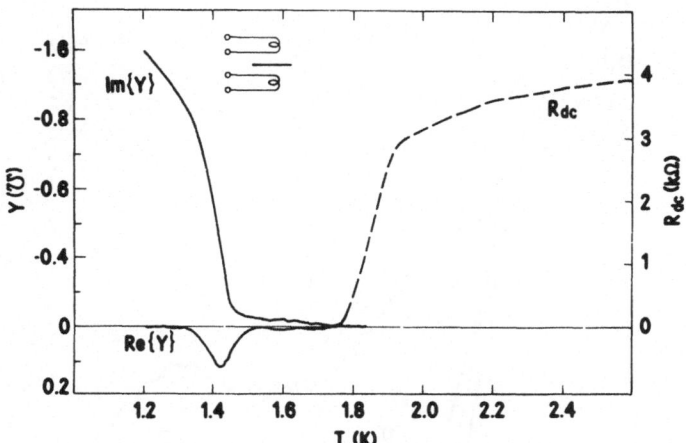

Fig. 13. Real and imaginary part of admittance of a granular aluminum
film. $R_{\square}$=4.1 kΩ, frequency 17.5 MHz. DC resistance is
indicated at right (Hebard and Fiory 1980b).

indicates the influence of free vortices, at a temperature
significantly higher than T_c and lower than T_{co}, as derived from
the data.

A very interesting possibility is offered by the analysis of the
frequency dependence of the bound pair contribution. The theory,
which is again based on the calculations opf Ambegaokar et al. of the
dynamic properties, yields that the respons at a frequency ω is
dominated by those bound pairs that have a separation distance of
about:

$$r_\omega = \left(\frac{14\ \mu\ kT}{\omega}\right)^{\frac{1}{2}} \qquad (5.3)$$

If, above T_c, free vortices are present at an average distance ξ_+,
the pairs with separation $r_\omega > \xi_+$ will not contribute to ε_{vb}. So for
each frequency there is a temperature T_ω above which ξ_+ is smaller
than the r_ω for that frequency. At temperatures above T_ω, $\varepsilon_{vb}-1$ is
zero. In this way T_ω can be determined experimentally, see Fig. 14.

If T_ω is known, the value of r_ω at T_ω for the frequency ω can be
calculated. This value r_c is defined by:

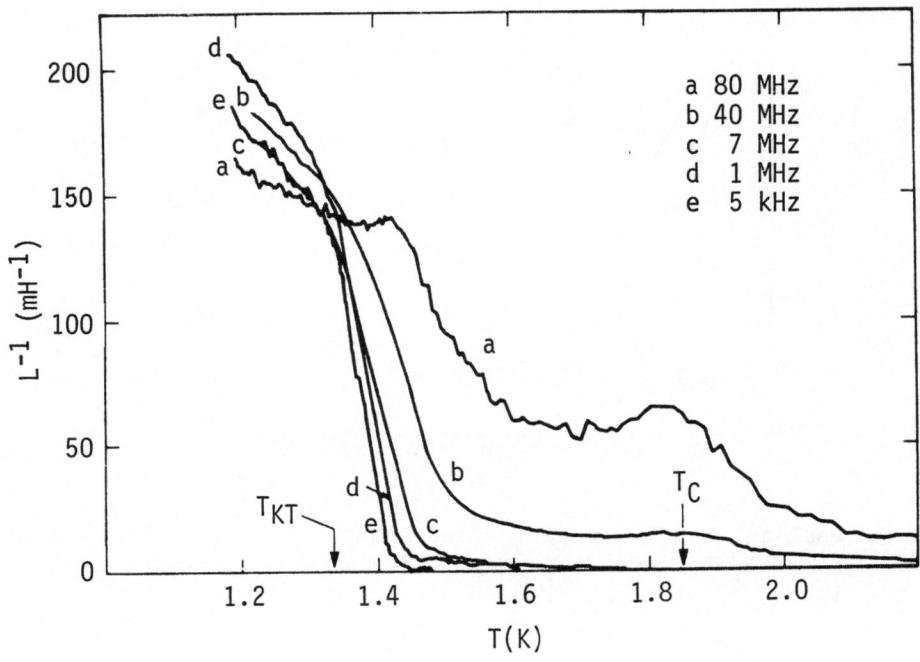

Fig. 14. Reciprocal inductance L^{-1} (defined by $Y^{-1}=R+i\omega L$) of the same film as Fig. 13 (Hebard and Fiory 1980b). T_ω is where the slope changes sharply at low values of L^{-1}.

$$r_c(\omega) = \left(\frac{14 \mu \ kT_\omega}{\omega}\right)^{\frac{1}{2}} \tag{5.4}$$

Because $r_c(\omega)$ is equal to $\xi_+(T_\omega)$, we find with Eq. (3.29) that if ℓ_ω is defined as:

$$\ell_\omega \equiv \ln\left(\frac{r_c}{\xi}\right) \tag{5.5}$$

the following approximate relation exists between ℓ_w^{-2} and T_ω:

$$\ell_\omega^{-2} \approx \frac{1}{b} \ \frac{T_\omega - T_c}{T_{co} - T_c} \tag{5.6}$$

At each frequency, T_ω is derived from the experimental curve. ℓ_ω is calculated with Eqs. (5.4) and (5.5). A plot of ℓ_ω^{-2},

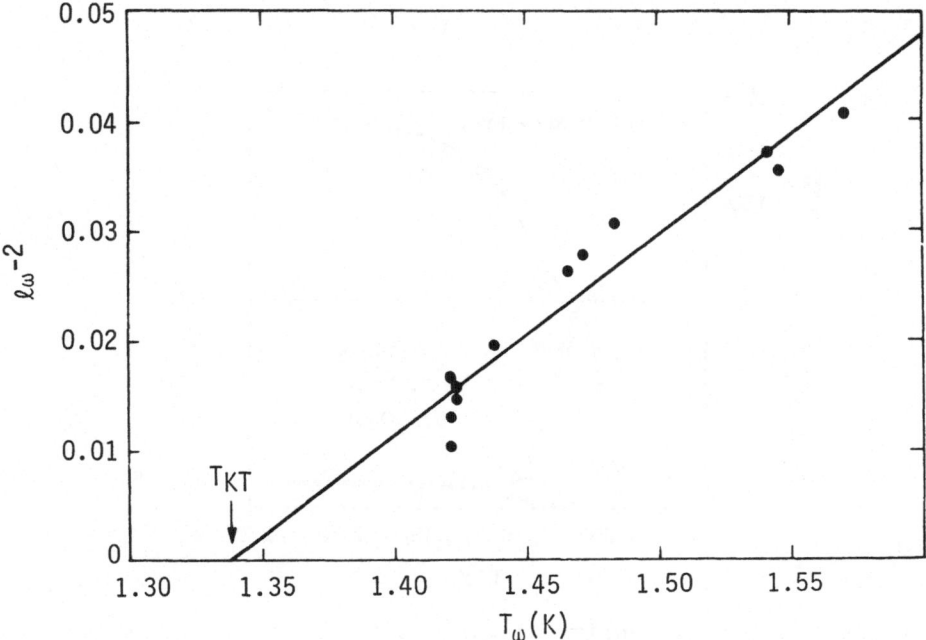

Fig. 15. ℓ_ω^{-2} (see text) versus T_ω (Hebard and Fiory 1980b)

as given by Hebard and Fiory (1980) is reproduced as Fig. 15. From the data points T_c can be derived.

Hebard and Fiory draw the general conclusion that agreement between theory and experiment is satisfactory. Small deviations remain, but are reasonably in proportion to the expected small sample inhomogenities and pinning potentials. From the data, a value for ϵ_c is derived. It is (Hebard and Fiory, 1982) equal to approximately 1.3.

Vortex Noise

Voss, Knoedler and Horn (1980a, b) measured the resistance noise in high-resistance granular aluminum and tin films in the critical temperature region. In Fig. 16 we show their data on a 380 Ω/square aluminum film. The noise spectral density is plotted together with the resistive transition. At zero bias current the noise power simply follows the resisitivity as a function of temperature.

At finite bias currents a distinct noise peak develops in the temperature region below T_{co}. The physical interpretation of this

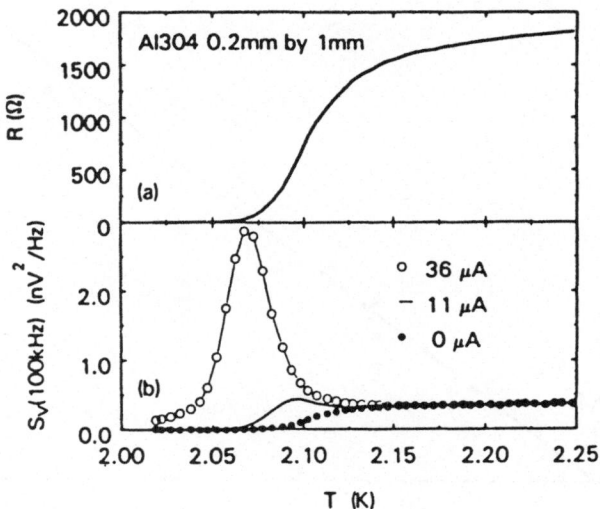

Fig. 16. Noise spectral density at various levels of bias current
(lower part) and resistive transition (upper part) for a
granular aluminum film with $R_\square = 380\ \Omega$ (Voss et al. 1980b)

peak is, that it is connected with free vortices moving under the
influence of the transport current. If a vortex moves over a distance
l before recombining with a vortex of opposite sign, the phase change
in a sample of width $w \ll \lambda_p$ is smaller than 2π by roughly a
factor l/w. Estimates of the noise voltage from this approximation
are roughly in agreement with the experiments. Note that in Fig. 16
the noise peak not only increases in height as the current is
increased, but also shifts to lower temperatures, indicating current-
induced unbinding.

Laser-Pulse Induced Free Vortices

Consider a high-resistance film at a temperature near T_c. Now if
the temperature is raised for a very short time, free vortices are
created that will exist for a short time before they are bound. The
relaxation time τ_r is of the order of the time required for diffusion
over the average distance between free vortices. The diffusion
constant D is equal to μkT, the length travelled in a time τ is $(D\tau)^{\frac{1}{2}}$,
while the average distance is of the order of $n_f^{-\frac{1}{2}}$ where n_f is the

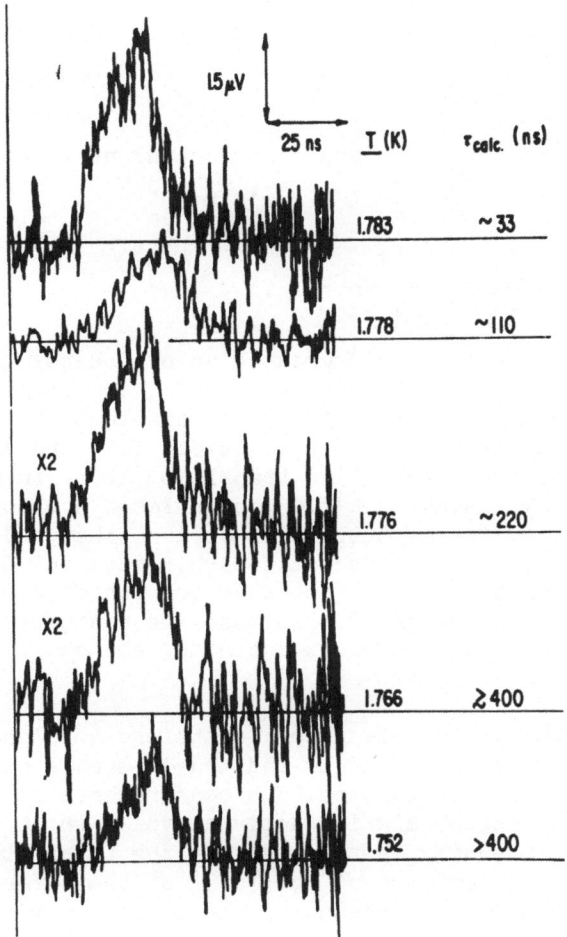

Fig. 17. Voltage respons to short laser pulse (Bancel and Gray 1981)
        for the same film as in Fig. 8.

free vortex density. Halperin and Nelson (1979) obtained a recombi-
nation time:

$$\tau_r = \frac{1}{8\pi \mu kT n_f} \qquad (5.7)$$

The sheet resistance is equal to $\mu n_f \Phi_o^2/c^2$. So, as pointed out by
Gray (1982):

$$\tau_r = \frac{\Phi_o^2}{8\pi \ c^2 R_\square \ kT} \tag{5.8}$$

where $R_\square$ is the actual, temperature dependent value of the sheet resistance. For the product of τ_r, $R_\square$ and T it follows that:

$$\frac{\tau_r}{(1 \ ns)} \cdot \frac{R_\square}{(1 \ \Omega)} \cdot \frac{T}{(1 \ K)} = 12.6 \tag{5.9}$$

Bancel and Gray (1981) created an excess population of free vortices with a very short laser pulse. The voltage was measured at constant current bias (Fig. 17). A relaxation time of about 10 ns was measured, in a temperature range between 1.75 and 1.78 K, on the sample for which the resistive transition is pictured in Fig. 8. Bancel and Gray, using Eq. (5.9), calculated a theoretical relaxation time that varied from about 100 ns to more than 400 ns in the temperature range considered. Although Bancel and Gray mention possible explanations involving modification of the dynamic theory, they tend towards the conclusion that the resistive transition below T_{co} is not dominated by vortex unbinding phenomena in their experiment.

Unfortunately, as discussed in the previous chapter, the critical temperature of Bancel and Gray's sample was probably 25 to 40 mK lower than they assumed. The measurements on τ_r were performed in a region significantly above T_c where perhaps the vortex (bound + free) density is too high for the theory to be applicable. However, the independence of Eq. (5.9) of the theoretical expectation for n_f enhances the importance of Bancel and Gray's objection. Further analysis is needed.

6. INFLUENCE OF APPLIED FIELD

So far we have tacitly assumed that no external magnetic field was present. A field perpendicular to the film favours the presence of vortices of one sign. The value of H_{c1} for a field perpendicular to a film with characteristic width R is:

$$H_{c1} = \frac{4\Phi_o}{R^2} \ln(\frac{R}{\xi}) \tag{6.1}$$

With R a few millimeters and ξ around 5 nm, H_{c1} is about 10^{-4} Oersted. Clearly, very special precautions have to be taken to perform experiments in the zero field limit.

The influence of an external magnetic field has been treated by Huberman and Doniach (1979 and 1980). They use the Hamiltonian:

$$Ha = N\,\mu_c + U_o f(T) \left| \sum_i sgn(v_i) \right|^2 +$$

$$+ \tfrac{1}{2} \sum_{i \neq j} 4U_o f(T) \ln(\frac{r_{ij}}{\xi}) + H \sum_i \frac{\Phi_o}{16\pi} \frac{R^2}{\lambda_p} sgn(v_i) \qquad (6.2)$$

in our notation, for a system of N vortices v_i in a sample of size R. Vortices in bound pairs are included. We will not follow Doniach and Huberman's derivations but only indicate some important results. The influence of an external field has also been discussed by Minnhagen (1981).

Resistivity In Small Fields

We consider external fields that exceed H_{c1} but are not too large. The interaction between the field and the bound pair/free vortex system that we discussed previously leads to significant modifications. Vortex unbinding is easier. As a result, the resistivity is higher than without field in a temperature region close to T_c. At somewhat higher temperatures the density of free vortices without field is higher than the field-induced density. Above that temperature the influence of the field is small. The cross-over temperature T_H is determined approximately by:

$$n_f(T_H) \approx \frac{1}{\xi_+^2(T_H)} \approx \frac{H}{\Phi_o} \qquad (6.3)$$

In Fig. 18 we reproduce data of Masker, Marcelja and Parks from 1969, for the resistivity of granular aluminum films. Although the data were interpreted quite differently at the time of publication, they show exactly the behaviour expected from vortex-unbinding theory.

Flux-Lattice Melting

At low temperatures in magnetic fields above H_{c1}, flux enters the sample, creating a flux lattice. This lattice can be considered as a two-dimensional solid. In such solids Kosterlitz-Thouless transitions may occur where dislocations play the same role as vortices in the superconducting film. Based on that analogy, Huberman and Doniach (1979b) and Fisher (1980a,b) developed the theory for melting of the flux-lattice. Bound pairs of dislocations in the flux lattice with opposite Burgers vector are formed. At the melting temperature T_m they unbind, similar to vortex pairs at T_c. For two-dimensional solids, possibly a second intermediate, so-called

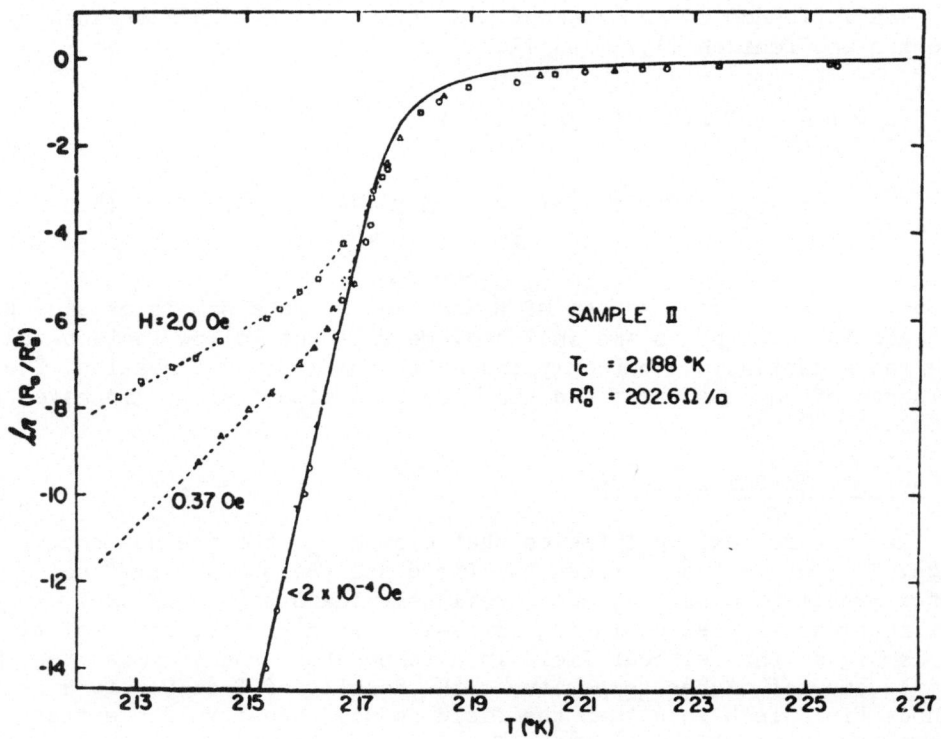

Fig. 18. Resistive transition in finite fields for a granular
aluminum film (Masker et al. 1969)

hexatic, phase exists where 'directional' topological order remains
although the substance cannot withstand shear stress (Nelson and
Halperin 1979).Huberman and Doniach find for the melting temperature:

$$T_m = \frac{1}{4\pi\sqrt{3}} \; T_{KT}$$

(6.4)

Here T_{KT} is the 'bare' transition temperature as given by Eq.(2.16).
They obtain a phase-diagram for thin superconducting films as given
in Fig. 19.

Experimentally, Fiory and Hebard (1982) studied films in an
external field. With the same AC measurements that were discussed in
chapter 5, they were able to observe the melting of the lattice at a
temperature close to the theoretical value.

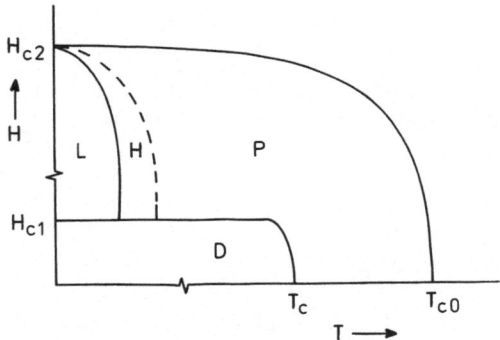

Fig. 19. Schematic phase diagram for a superconducting film in finite
         fields (Huberman and Doniach 1980). D denotes a weakly
         diamagnetic phase with only bound pairs present, L a rigid
         vortex lattice, H a possible hexatic phase, P a vortex
         plasma.

7. ARRAYS OF JOSEPHSON JUNCTIONS

We have started our discussion of two-dimensional systems with
the X-Y model, where spins are coupled with coupling energy
$-J\cos(\varphi_i - \varphi_j)$. A physical realization of this X-Y model can be
obtained by fabricating a square array of superconducting islands,
connected by Josephson junctions. Each island has its own value φ_i
of the phase of the order parameter. The coupling energy of a
Josephson junction is:

$$U = \frac{\Phi_o}{2\pi} I_o \cos(\varphi_1 - \varphi_2) \qquad (7.1)$$

where I_o is the critical current and φ_i and φ_j are the values of φ on
both sides of the junction. Clearly this expression is exactly the
same as the one used in the X-Y model.

The transition temperature for such a system follows directly from the analogy. For the X-Y system:

$$kT_{KT} = \frac{\pi}{2} J \tag{7.2}$$

If we indicate the unrenormalized Kosterlitz-Thouless temperature for the array of superconducting islands with T_{array}, it follows from the relation:

$$kT_{array} = \frac{\Phi_o}{4} I_o(T_{array}) \tag{7.3}$$

The critical current of a Josephson oxyde junction, according to Ambegaokar and Baratoff, can be written as:

$$I_o = \frac{4}{\Phi_o} (2.18 \frac{\hbar/e^2}{R_n} kT_{co}) f(T) \tag{7.4}$$

with R_n the normal-state resistance of the junction and T_{co} the transition temperature of the islands. If the form within brackets is compared with Eq. (2.12), one sees that for a square array the Kosterlitz-Thouless temperature is <u>exactly</u> the same as for a thin film. The sheet resistance of the film is replaced by the normal-state resistance of one junction, which is also the sheet resistance of a square array of junctions. This correspondence is only valid for junctions that have an Ambegaokar-Baratoff critical current and a sinusoidal current phase relation.

The first arrays were fabricated by Sanchez and Berchier (1980, 1981). They used islands of Au-In on a continuous film of Au which forms proximity-effect bridges. A similar procedure was followed recently by Abraham et al.(1982) who used Pb-Bi islands on a Cu film. The size of their array is much larger, about 1000 x 1000 islands. The critical current of proximity-effect bridges is not the same as for Josephson oxyde junctions, which effect can be accounted for. The results obtained by Abraham et al. are very promising but need a more detailed analysis. Lobb et al.(1982) discuss theoretical aspects that are specific for this type of array.

Arrays containing Josephson oxyde junctions have been fabricated by Voss and Webb (1982), who obtained an array of 100 x 100 niobium islands by means of the IBM technology developed to fabricate computer circuits.

We refrain from a discussion of the results obtained so far. The arrays clearly offer a very promising approach to a simulation of the

X-Y model. Questions, regarding the uniformity of the junctions and the minimum size required, have to be adressed before firm conclusions can be drawn. Quite apart from their interest with respect to Kosterlitz-Thouless transitions such arrays may be important for a good understanding of granular films.

8. CONCLUSIONS

We will ask ourselves two questions:
1. How important are the concepts developed for two-dimensional phase transitions for a good understanding of thin superconducting films?
2. Can experiments on superconducting samples contribute to the understanding of phase transitions in two-dimensional systems?

As for the first question, it is clear from the various experimental results that effects involving bound vortex pairs and free vortices occur in high-resistance films. The quantitative agreement between theory and experiment is not very convincing, but this is only to be expected as long as the theory is not worked out in more detail. Although other phenomena may occur simultaneously, vortex unbinding effects have certainly to be taken into account when thin, high-resistance superconducting films are studied in the temperature region near T_{co}.

For an experimental verification of the statistical mechanics of two-dimensional systems, 'homogeneous' thin superconducting films in zero field do not seem very well suited. The two-dimensional phase transition occurs very close to the BCS transition temperature and is difficult to separate. Films with very high sheet resistance, where T_c would be far below T_{co}, cannot be fabricated with confidence as to their real nature.

The two systems that we only discussed very shortly in chapters 6 and 7, seem much more promising for a test of the theoretical concepts. The vortex lattice, that is formed in a homogeneous film in a magnetic field, is one of the very few examples of a two-dimensional 'solid' that can be studied experimentally. Arrays of Josephson junctions are analogons of the X-Y model, and offer simular possibilities as computer simulations. In addition, results obtained with such arrays will help the understanding of granular films.

ACKNOWLEDGEMENTS

I recall with much pleasure the time I spent at Stanford University working on this subject with M.R. Beasley and T.P. Orlando. I have profited greatly from lectures of and discussions with J.M.J. van Leeuwen and H.J.M. Hilhorst regarding the

renormalization procedures. I thank many authors for sending preprints in advance of publication, K.E. Gray for correspondence and
A.M. Kadin for discussions during the school.

REFERENCES

The following abbreviations are used:

IS79 - Inhomogenous Superconductors 1979, D.U. Gubsen,
         T.L. Francaville, S.A. Wolf, J.R. Leibowitz editors,
         AIP conference proceedings nr. 58, American Insitute
         of Physics 1980
02D   - Proceedings of the International Conference on ordering
         in two dimensions, Lake Geneva 1980, S.K. Sinha editor,
         North Holland 1980
LT16 - Proceedings of the 16th International Conference on Low
         Temperature Physics, Los Angeles 1981, W.G. Clark editor,
         in Physica volumes 108, 109, 110 B+C.

Abraham, D.W., C.J. Lobb, M. Tinkham and T.M. Klapwijk, 1982,
         Resistive transition in 2-D arrays of superconducting weak
         links, Phys. Rev., B26:5268.
Ambegaokar, V., B.I. Halperin, D.R. Nelson and E.D. Siggia, 1978,
         Dissipation in two-dimensional superfluids,
         Phys. Rev. Letters, 40:783.
Ambegaokar, V., B.I. Halperin, D.R. Nelson and E.D. Siggia, 1980,
         Dynamics of superfluid films, Phys. Rev., B21:1806.
Bancel, P.A. and K.E. Gray, 1980, Vortex unbinding in superconduc-
         ting thin films? in 02D, p.483.
Bancel, P.A. and K.E. Gray, 1981, Search for vortex unbinding in
         two-dimensional superconductors, Phys.Rev.Letters, 46:148.
Beasley, M.R., J.E. Mooij and T.P. Orlando, 1979, Possibility of
         vortex-antivortex pair dissociation in two-dimensional
         superconductors, Phys.Rev.Letters, 42:1165.
Berezinskii, V.L., 1971, Destruction of long-range order in one-
         dimensional and two-dimensional systems having a continuous
         symmetry group, II. Quantum systems, Zh. Eksp. Teor. Fiz.,
         61:1144, Sov.Phys. JETP, 34:610.
Bishop, D.J. and J.D. Reppy, 1978, Study of the superfluid transition
         in two-dimensional ^{4}He films, Phys. Rev. Letters, 40:1727.
Doniach, S. and B.A. Huberman, 1979, Topological excitations in two-
         dimensional superconductors, 1979, Phys.Rev.Letters, 42:1169.
Epstein, K., A.M. Goldman and A.M. Kadin, 1981, Vortex-antivortex pair
         dissociation in two-dimensional superconductors,
         Phys.Rev.Letters, 47:534.
Epstein, K., A.M. Goldman and A.M. Kadin, 1982, Renormalization
         effects near the vortex-unbinding transition of two-
         dimensional superconductors, Phys. Rev., B26:3950.
Fiory, A.T. and A.F. Hebard, 1980, Radio-frequency complex-impedance

measurements on thin-film two-dimensional superconductors, in IS79, p.293.

Fiory,   A.T. and A.F. Hebard, 1982, Systematics of the dielectric constant of vortex phases in superconducting films, Phys.Rev., B25:2073.

Fisher, D.S., 1980a, Flux lattice melting in thin film superconductors, in IS79, p.95.

Fisher, D.S., 1980b, Flux-lattice melting in thin film superconductors, Phys.Rev., B22:1190.

Halperin, B.I. and D.R. Nelson, 1979, Resistive transition in superconducting films, J.Low Temp.Phys., 36:599.

Hebard, A.F. and A.T. Fiory, 1980a, Recent experimental results on vortex processes in thin-film superconductors, in O2D, p.181.

Hebard, A.F. and A.T. Fiory, 1980b, Evidence for the Kosterlitz-Thouless transition in thin superconducting aluminum films, Phys.Rev.Letters, 44:291.

Hebard, A.F. and A.T. Fiory, 1982, Vortex dynamics in two-dimensional superconductors, LT16, Physica,109&110B+C:1637.

Huberman, B.A. and S. Doniach, 1979, Melting of two-dimensional vortex lattices, Phys.Rev.Letters, 43:950.

Huberman, B.A. and S. Doniach, 1980, Topological excitations in thin superconducting films: a review, in IS79, p.129.

Kosterlitz, J.M., 1974, The critical properties of the two-dimensional X-Y model, J. Phys. C, 7:1046.

Kosterlitz, J.M. and D.J. Thouless, 1972, Long-range order and metastability in two-dimensional solids and superfluids, J. Phys. C, 5:L124.

Kosterlitz, J.M. and D.J. Thouless, 1973, Ordering, metastability and phase transitions in two-dimensional systems, J. Phys. C, 6:1181.

Lobb, C.J., D.W. Abraham and M. Tinkham, 1982, Theoretical interpretation of resistive transition data from arrays of superconducting weak links, to be published.

Masker, W.E., S. Marcelja and R.D. Parks, 1969, Electrical conductivity of a superconductor, Phys. Rev., 188:745.

Minnhagen, P., 1981a, Kosterlitz-Thouless transition for a two-dimensional superconductor: magnetic-field dependence from a Coulomb-gas analogy, Phys.Rev., B23:5745.

Minnhagen, P., 1981b, Universal resistive transition for two-dimensional superconductors, Phys.Rev., B24:6758.

Nelson, D.R. and B.I. Halperin, 1979, Dislocation-mediated melting in two dimensions, Phys. Rev., B19:2457.

Pearl, J., 1964, Current distribution in superconducting films carrying quantized fluxoids, Appl.Phys.Letters, 5:65.

Pearl, J., 1965, Low Temperature Physics LT9, J.G. Daunt, D.O. Edwards, F.J. Milford, M. Yagub eds. Plenum Press, New York, p.566.

Rao, N.A.H.K., E.D. Dahlberg, A.M. Goldman, L.E. Toth and C. Umbacn, 1980, Experimental investigation of the resistance in thin superconducting films of niobium-germanium, in IS79, p.108.

Sanchez, D.H. and J.L. Berchier, 1980, Square arrays of Au/In proxi-
        mity effect bridges. Preliminary results in small H_p,
        in IS79, p.239.
Sanchez, D.H. and J.L. Berchier, 1981, Properties of n x n square ar-
        rays of proximity effect bridges, J.Low Temp.Phys., 43:65.
Tinkham, M., 1975, 'Introduction to Superconductivity', Mc.Graw-Hill,
        New York.
Turkevich, L.A., 1979, Resistivity of superconducting films,
        J.Phys.C.: Solid State Phys., 12:L385.
Voss, R.F., C.M. Knoedler and P.M. Horn, 1980a, Vortex noise at the
        superconducting transition in granular aluminum films,
        in IS79, p.314.
Voss, R.F., C.M. Knoedler and P.M. Horn, 1980b, Phase-slip shot
        noise at the two-dimensional superconducting transition:
        evidence for vortices?, Phys.Rev.Letters, 45:1523.
Voss, R.F. and R.A. Webb, 1982, Phase coherence in a weakly coupled
        array of 20000 Nb Josephson junctions, Phys. Rev., B25:3446.
Wolf, S.A., D.U. Gubser, W.W. Fuller, J.C. Garland and R.S. Newrock,
        1981, Two-dimensional phase transition in granular NbN films,
        Phys.Rev.Letters, 47:1071.
Young, A.P., 1978, On the theory of the phase transition in the two-
        dimensional planar spin model, J. Phys. C, 11:L453.

RE-ENTRANT SUPERCONDUCTIVITY

J. Ruvalds

Department of Physics
University of Virginia
Charlottesville, Va.   22901   U.S.A.

I.   HISTORICAL INTRODUCTION

The discovery of resistivity minima in metals at low temperatures dates back some fifty years, and yet the evolution of a
successful theoretical account of the basic anomaly continues
to this date. Originally observations[1] on Mg, Mo, Te, Co and
Pd yielded dramatic departures from the expected resistivity
variation

$$\rho(T) = \rho_0 + BT^5, \tag{1}$$

where BT^5 represents the electron scattering by thermally excited
phonons at low temperatures. Impurity contributions are embedded
in the residual resistivity constant ρ_0, which should be a simple
linear function of the impurity concentration according to
Matthiessen's rule. By contrast, the resistivity of many metals
was found to have an additional term with an evident divergence
at low temperatures, whose origin was attributed to an interesting
many-body effect only in 1964.

A primary obstacle to experimental methods was the strong
effect of extremely dilute impurities such as Fe with concentrations less than $x \leq .000001$, which was believed to be influential.[2]
Hence a series of studies[3,4] on transition metal alloys were
instituted, with the clearcut demonstration of a resistivity
minimum for dilute Mn impurities in Mn_xAg with $x \ll 1\%$. Furthermore, an interesting maximum in the resistivity occurs at
higher concentrations $x \sim 0.1\%$, and both anomalies are destroyed
at yet higher concentrations $(x > 0.5\%)$ as shown in Fig. 1.

In addition to the resistivity minimum, the $Ag-Mn_x$ alloys exhibit other anomalies which are not yet fully understood. For the present, we mention the quenching of the minimum for $x > 0.5\%$, and the appearance of an anomalous maximum. These are indications of strong impurity-impurity interactions even at such low concentrations of $x \sim 1\%$.

Magnetoresistance measurements[4] produced a coincident series of temperature anomalies in the Ag_x-Mn alloys, which were noted for a negative magnetoresistance at similarly low values of x which show the resistivity minimum in zero field. However, the application of a field H = 20 KOe was sufficient to remove the minimum anomaly, indicating a possible magnetic character for the impurities.

Magnetic susceptibility experiments[5] provided a classic demonstration of the impurity role in the metal host by yielding a distinct Curie-Weiss behavior of the susceptibility

$$\chi = \frac{c}{T-\theta}, \tag{2}$$

where c is a measure of the effective moment and θ designates the Curie Temperature. At low concentrations a very strong T-dependence is observed in many cases, including $Ag-Mn_x$ and $Cu\ Mn_x$ which verifies the existence of a magnetic moment associated with a single Mn atom. By contrast, nonmagnetic impurities do not exhibit either the resistivity minimum or a $\chi(T)$ variation.

Interactions among impurities are revealed as changes in the Curie θ in Eq. 2, which occurs at concentrations as low as 1% for the $CuMn_n$ system. Oscillations in the sign and magnitude of θ may be attributed to the indirect exchange (RKKY) coupling between magnetic impurities.

The case of non-magnetic impurities does not give the resistivity anomaly, as demonstrated very precisely for $CuSn_x$.[6] Also a systematic analysis of $\rho(T)$ for Fe_xCu showed convincingly that the temperature at which the minimum occurs varies as the one fifth power of the concentration.

A comprehensive review of the experimental work is found in the work of Van Den Berg, which also contains an interesting account of theoretical models proposed prior to 1964.[7] It was quite certain by then that the magnetic moment formation was essential to the resistivity anomaly, but a theoretical understanding of the temperature variation of $\rho(T)$ was completely lacking.

Historically, the earliest attempts to explain the unusual resistivity relied on models with sharp structure in the density

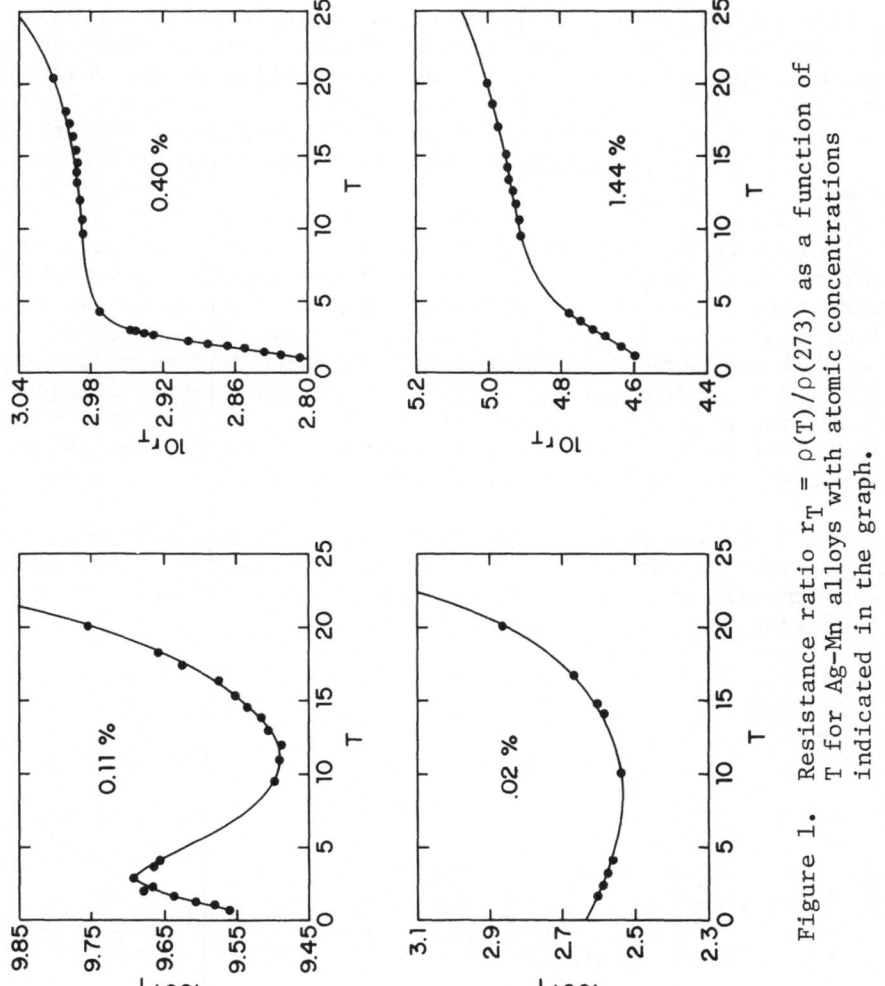

Figure 1. Resistance ratio $r_T = \rho(T)/\rho(273)$ as a function of T for Ag-Mn alloys with atomic concentrations indicated in the graph.

of states near the Fermi energy. One possible source of such
structure is a possible localized electronic state with exceedingly
narrow width as proposed by Korringa and Gerritsen.[8] This model
yields a qualitative fit to $\rho(T)$ by suitable adjustment of parameters, but does not give the systematic variation of the minimum
with impurity content.

An alternate line of approach[9-12] was the consideration of
magnetic energy levels, whose thermal population may vary on a
scale determined by the s-d exchange interaction. A small energy
difference between spin aligned and antiparallel states near the
Fermi energy will give a temperature variation $p \sim \exp(-U/T)$ where
U is the spin-spin interaction energy. However these results are
not successful in explaining either the resistivity minima or
maxima.

Thirty years after the discovery of the resistance minimum,
theoretical interest was drawn to the magnetic character of the
impurity state. Elegant derivations demonstrated the creation of
a localized electron state with a magnetic moment were found by
Friedel[13] and Anderson[14], which serve as constructive guides to
the Coulomb and hybridization interactions necessary for magnetic
moment formation. Furthermore, the placement of the Fermi energy
relative to the impurity virtual bound state energies was shown
to be a key determinant of the magnetic behavior.

A conclusive proof of the magnetic impurity involvement in
the resistivity minimum was achieved by the observed correlation
of the anomaly only in those alloys of Mo-Nb and Mo-Re which exhibit
a Curie-Weiss susceptibility.[15,16]

II. KONDO EFFECT

A. Physical Basis

A remarkable breakthrough in the understanding of the
resistivity minimum was achieved by Kondo in 1964, when he found
a divergence in the cross section of an electron scattered by a
single magnetic impurity.[17,18] This surprising result yields a
resistivity term

$$\rho_k = xJ^3 \ln(T), \tag{3}$$

where J is the exchange coupling of the electron spin to the impurity spin state and the isolated impurity effect is evidenced by the linear x concentration dependence. The J^3 variation indicates that the divergence occurs in the third order perturbation theory.

Perhaps the most challenging feature of the logarithmic divergence found by Kondo is that it cannot be true at zero temperature. Unitarity imposes a fundamental limit on the scattering cross section which cannot be violated, and hence the search was on for higher order corrections which must remove the $\ln T$ divergence at low temperatures.

Spin-flip scattering processes are responsible for the $\ln T$ divergence found by Kondo. The elastic nature of the scattering yields vanishing energy denominators which result in the divergence. By contrast to the case of ordinary plastic potential scattering which gives a temperature-independent resistivity, the novel features of the spin commutation rules result in the Kondo effect.

A physical picture of the Kondo anomaly is conveniently represented via the s-d Hamiltonian

$$H_{sd} = -J \, \underset{\sim}{S_j} \cdot \underset{\sim}{\sigma} \, \delta(\underset{\sim}{r} - \underset{\sim}{R_j}), \tag{4}$$

where J is a constant exchange interaction coupling an impurity spin $\underset{\sim}{S_j}$ to the conduction electron spin σ. This contact interaction model was used originally by Fröhlich and Nabarro[19] and by Zener[20] in the study of nuclear spin relaxation. A correspondence[21] of the s-d interaction to the Anderson model parameters is enlightening in that it provides an explanation for the negative J values found for many transition element impurities.

The Kondo effect yields a resistivity minimum only in the case of antiferromagnetic coupling J < 0. This point is especially relevant to the discussion of rare earth impurities which are known to commonly exhibit ferromagnetic J interactions.

A divergence in the perturbation theory of the electron-impurity scattering is evident in the transition probability for an electron in initial state $\underset{\sim}{k}\uparrow$ scattering to a final state $\underset{\sim}{k'}\uparrow$ via the s-d interaction. The scattering probability may be expressed in powers of J according to

$$t_{\underset{\sim}{k}\uparrow, \underset{\sim}{k'}\uparrow} = t_{\underset{\sim}{k}\uparrow, \underset{\sim}{k'}\uparrow}^{(1)} + t_{\underset{\sim}{k}\uparrow, \underset{\sim}{k'}\uparrow}^{(2)} \cdots, \tag{5}$$

where the lowest order term does not involve spin flip terms and is given by

$$t_{\underset{\sim}{k}\uparrow,\underset{\sim}{k}'\uparrow}{}^{(1)} = -\frac{J}{2N}\ (S_z). \tag{6}$$

Hence this first Born Approximation gives the transition rate

$$W_{\underset{\sim}{k}\uparrow,\underset{\sim}{k}'\uparrow} = \frac{2\pi}{h}\ |t_{\underset{\sim}{k}\uparrow,\underset{\sim}{k}'\uparrow}{}^{(1)}|^2\ \delta(E_{\underset{\sim}{k}\uparrow} - E_{\underset{\sim}{k}'\uparrow}) \tag{7}$$

$$W_{\underset{\sim}{k}\uparrow,\underset{\sim}{k}'\uparrow} = \frac{2\pi}{h}\ (J/2N)^2 S_z^2\ \delta(E_{\underset{\sim}{k}\uparrow} - E_{\underset{\sim}{k}'\uparrow}). \tag{8}$$

The next higher order process involves the spin-flip terms and yields a divergent low temperature result because of the non-commuting property of the spin operators. The two processes to be considered are: (a) scattering to an intermediate electron state $q\sigma$ and subsequently to the final state $k'\uparrow$; (b) an electron from a filled state $q\sigma$ scatters to the final state $k'\uparrow$, and later an electron from the initial state $k\uparrow$ scatters into the empty state $q\sigma$. Diagramatically these two scattering events are analogous to the Compton scattering of an x-ray by an electron.

Without the spin-flip process, the second order scattering gives

$$t_{\underset{\sim}{k}\uparrow,\underset{\sim}{k}'\uparrow}^{(2)}\ \Big|_{\sigma=\uparrow} = (\frac{J}{2N})^2\ S_z^2\ [\sum_{\underset{\sim}{q}} \frac{1-f_q}{\varepsilon_k-\varepsilon_q} + \sum_q \frac{f_q}{\varepsilon_k-\varepsilon_q}\ ]$$

$$t_{\underset{\sim}{k}\uparrow,\underset{\sim}{k}'\uparrow}^{(2)}\ \Big|_{\sigma=\uparrow} = (\frac{J}{2N})^2\ S_z^2\ \sum_q \frac{1}{\varepsilon_k-\varepsilon_q}. \tag{9}$$

Performing the integral over q, we find only a constant correction to the lower order term $t^{(1)}$. Because of the commuting nature of S_z^2, this result is in a sense similar to potential scattering by ordinary nonmagnetic impurities.

Spin flip terms in the s-d interaction of Eq. 12 are essential at low temperatures, because there is not the simple cancellation of the Fermi function f_q for the intermediate state. Furthermore, since the spin flip dynamics involve zero energy transfer by the conduction electron, a divergent scattering amplitude appears: The corresponding terms are

$$t_{\underset{\sim}{k}\uparrow,\underset{\sim}{k}'\uparrow}^{(2)}\ \Big|_{\sigma=\downarrow} = 2\ (\frac{J}{2N})^2\ (S_+S_- - S_-S_+) \sum_q \frac{f_q}{\varepsilon_k-\varepsilon_q}\ , \tag{10}$$

with $S_+S_- - S_-S_+ = S_z$ portraying the distinct origin of the
Kondo effect. Here another term without f_q again yields an
insignificant constant correction. Following Kondo, we transform
the momentum integration to an energy variable ε by introducing
a constant density of states $N(\varepsilon) \cong N(0)$ near the Fermi energy.
Then Eq. 10 can be integrated by parts to give

$$t^{(2)}_{\underset{\sim}{k}\uparrow,\underset{\sim}{k}'\uparrow} \Big|_{\sigma=\downarrow} = 2 \left(\frac{J}{2N}\right)^2 S_z \left[-N(0) \int d\varepsilon \frac{df}{d\varepsilon} \ln \left|\frac{\varepsilon}{D}\right|\right], \qquad (11)$$

where a cut-off D is introduced which corresponds in actuality to
the required momentum cut-off for a realistic exchange interaction
J_{kq}. For a well-defined Fermi surface, Eq. 11 immediately gives
a term with $\ln T$ and a total scattering amplitude (Eq. 5 & Eq. 11):

$$t_{\underset{\sim}{k}\uparrow,\underset{\sim}{k}'\uparrow} \cong -\frac{J}{2N} S_z + 2 (J/2N)^2 (S_z)N(0) \ln \frac{T}{D}, \qquad (12)$$

and, from the Golden rule of Eq. 8, we find the transition probability to order J^3 as

$$W_{\underset{\sim}{k}\uparrow,\underset{\sim}{k}'\uparrow} = \frac{2\pi}{h} \left[(J/2N)^2 S_z^2 + (J^3/2N^3)S_z^2 N(0) \ln\left(\frac{T}{D}\right) + \ldots\right). \qquad (13)$$

Finally the electrical resistivity becomes

$$R(T) = x R_m[1 + 2(J/N)N(0)\ln(T/D) + \ldots], \qquad (14)$$

which is the well known Kondo effect formula with $R_m = 2\pi m x N(0)J^2 S(S+1)/ne^2\hbar$, where n gives the electron density. This
result yields a very good fit to the resistivity minima of
dilute alloys and, in tandem with the usual phonon contribution
BT^5, accounts for the concentration dependence of the minimum
temperature $T_{min} \propto x^{1/5}$.

It should be emphasized that the Kondo scattering is rapidly
quenched by either external magnetic fields[22], or by interactions
with other impurities.[23] Physically, both of these influences
inhibit the spin-flip terms and thus quite naturally destroy the
$\ln T$ singularity. As a general rule, the Kondo divergence is
quenched at magnetic impurity concentrations exceeding one per cent.

B.   Regime of Validity

Dilute magnetic alloys provide a remarkable series of
examples which verify the essential features of the Kondo theory.
However the extrapolation of the Kondo single impurity effect to
higher concentrations in excess of 1% is generally unwarranted,

and therefore it is worthwhile to establish reasonably accurate criteria for the applicability of the results.

A primary constraint at low temperatures is the unitarity limit on the electron scattering cross section which translates into a maximum impurity resistivity contribution

$$\Delta R = 9.5 \ \mu\Omega\text{-cm/at \%.} \tag{15}$$

Universally, the resistivity data saturates below the ΔR value, and traditionally obeys a power law behavior of the type $R \cong R_0 - bT^2$ in Kondo systems. This crossover from a logarithmic divergence to a power law behavior has been studied by many general techniques in view of its relation to other problems such as the infrared divergence in elementary particle physics. Ordinary perturbation theory was subjected to the difficulty of treating the impurity spin operators which do not obey simple Fermion commutation rules. Thus it became fashionable to employ the "pseudo-Fermion" operators[24] which permit a systematic diagrammatic expansion for the electron scattering amplitude. Summation of selected diagrams or physical processes is achieved by choosing the most divergent terms beyond the Kondo third order perturbations, and leads to a higher order resistivity[25]

$$R = x \ R_m \ \frac{1}{[1 - 2JN(0)\ell n(T/D)]^2}. \tag{16}$$

Note that the leading term of Eq. 16 reduces to the Kondo logarithmic divergence which remains valid for temperatures above the Kondo temperature T_k defined by

$$T_K \cong D\exp[1/2JN(o)] \tag{17}$$

where D is a cut-off energy and $N(0)$ is the electron density of states at the Fermi level. As the temperature is reduced toward T_k the resistivity of Eq. 16 diverges as a result of the formation of a bound spin state for an antiferromagnetic coupling $J < 0$. A typical temperature, of order $T_k \sim 0.1°k$ for MnCu with $J = -1$ eV, provides a useful guide to the Kondo anomaly because it also appears in the magnetic susceptibility according to the relation[26]

$$\chi(T) = \frac{c}{T + 4.5 \ T_K} \tag{18}$$

Thus the original association of a resistance minimum with a Curie-Weiss behavior of the susceptibility is also established theoretically.

Validity of the Kondo result requires a spin-flip process which can be inhibited by effective local fields, an external magnetic

field, or the coupling to other magnetic impurities. The Kondo
temperature serves as a helpful guide to these limitations by
establishing the criterion:

$$T_K \gg \mu H, \tag{19}$$

where μ represents the impurity spin magnetic moment and H
corresponds to either the true external field or an effective
field caused by Zeeman levels[23] or by interactions among impurities.
In the latter case, the spin-flip processes become "frozen" at
impurity concentrations x determined by

$$T_K \simeq T_M \tag{20}$$

where T_M is the magnetic ordering temperature defined by

$$T_M = \frac{-xJ^2 s(s+1) 9\pi \cos(2k_F\Delta)}{4(2k_F\Delta)^3 E_F}, \tag{21}$$

where the impurity-impurity coupling is presumed to be of the
standard RKKY form in the case of free electrons in a simple
three dimensional model. Using the s-d exchange model, it is
possible to define an upper bound on the impurity concentration
which defines the Kondo regime, namely

$$x \leq C J^{-2} \exp[-1/J N(o)], \tag{22}$$

where the constant c is the ratio of interaction parameters
defined from eqs. 17, 20, 21. The resulting Kondo boundary is
shown in Figure 2 for $N(o) = 0.1$ eV/atom, which is typical of
a silver host material; and $C = 30,000$, which was chosen to
normalize the curve to the $AgMn_x$ data shown in Fig. 1. Several
interesting features become apparent from Figure 2. First of
all, the extremely dilute Kondo magnetic alloy regime is
emphasized for small J values: For example the case of Mn_xAg
with estimated values of $J = 0.5 \pm 0.2$ eV from several experiments[31] is a classic Kondo system only for $x \leq 0.2\%$. A
wider concentration regime is expected for Fe_xCu although the
corresponding estimates of $J \sim 1$ eV is subject to considerable
variation.

Interesting examples of alloys with resistance minima
outside the Kondo regime include the rare earth compounds
La Ce_x[32] and the layered materials Fe_xTaSe_2[33] which provide
anomalies at $x = 3\%$ and $x = 5\%$ despite relativity weak exchange
values $J \simeq 0.1$ eV. It is difficult to reconcile these examples
with the Kondo spin-flip scattering mechanism despite some

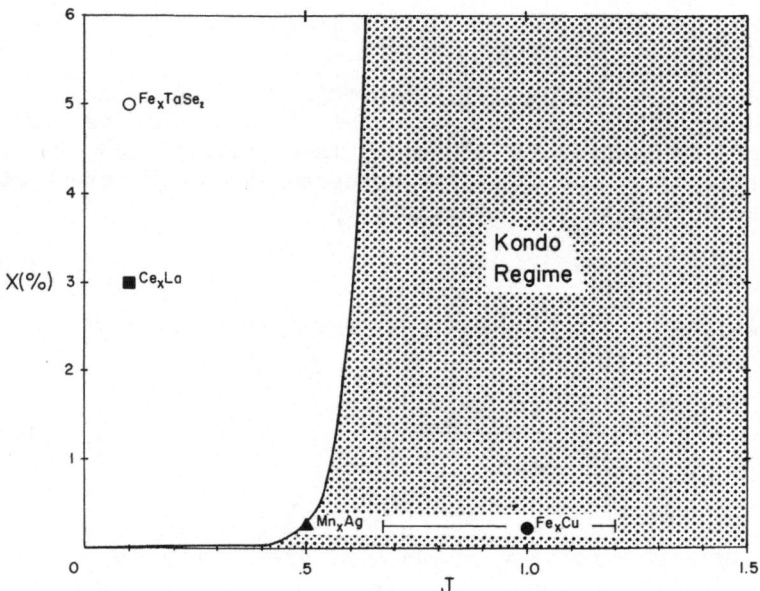

Figure 2.   Critical concentration defining the regime of validity
            for the Kondo effect as a function of the exchange
            coupling J.   The compounds LaCe$_x$ and Fe$_x$TaSe$_2$ exhibit
            resistance minima at the indicated concentrations
            well outside the Kondo regime.

flexibility in estimating interaction parameters. An alternate
mechanism for a resistivity minimum at the higher concentrations
of LaCe$_x$ may be related to electron scattering by coupled
impurity pairs. The latter possibility has recently been
proposed in view of a 1/T divergence in the cross section of an
electron scattered by spin-flip processes from two impurities.[34]

     It is interesting to note that the two major unsolved
problems in the 1950's era of solid state physics were considered
to be the resistance minimum in normal metals and superconductivity.
Both have a bearing on magnetic ordering, and the field has pro-
gressed to the current studies of spin-flip processes and novel
magnetic structures formed in superconducting host materials.

C.   Superconductor Phenomena

   Magnetic impurities cause dramatic effects in superconductors
by virtue of their spin-flip scattering which generally tends to
conflict with the presence of superconducting electron pairs.
Hence the primary result of adding magnetic impurities is to break
up the Cooper pair states and thus suppress the superconducting
transition temperature T_c. As the impurity concentration is
increased, however, new features arise including impurity interaction effects discussed in the next section and the manifold
phases exhibiting coexistence of magnetic order and superconductivity which are reviewed by B. Maple on page 281 of this book.

   A preliminary view of the expected phase diagram for coupled
magnetic impurities in a superconductor is shown in Figure 3.
Here the suppression of T_c with impurity content is a general
feature, and the onset of magnetic order at a characteristic
temperature T_M (see Eq. 21) is approximately linear in concentration. The dashed curve reflects the uncertainty in the impurity
coupling caused superconducting modifications in the electron
response, and various possible phases in this low temperature,
intermediate concentration region are discussed in the lectures
by B. Maple.

   Kondo scattering will naturally yield significant contributions to the electron-electron pairing interaction essential
to the BCS theory of superconductivity. However impurity interactions quench the Kondo effect at surprisingly low concentrations
in normal metals and thus must also be considered here. Presuming
the interactions between impurities in a superconductor to be
similar to the normal metal case we may use the criterion of
Eq. 20 (i.e. $T_{KONDO} \geq T_M$) to map a Kondo regime in Figure 3 for
a typical value of the exchange coupling J = 0.5 eV. As noted in
this text and shown in Figure 2, smaller J values reduce the
Kondo regime to very tiny impurity concentrations, although values
of J in excess of 1 eV may allow the Kondo singularity in the
cross section to persist at x ~ 5%.

   The initial suppression of T_c as a function of impurity
content in Fig. 3 was explained by the Abrikosov-Gorkov (AG)
theory[24] which has served as a standard for subsequent work.
Physically the AG theory treats the impurity scattering in the
Born approximation leading to an electron lifetime correction of
order J^2. To calculate the modifications in T_c by magnetic
impurities, it is convenient to introduce the Abrikosov pseudo-
Fermion method to describe the impurity spin. The motivation for
this approach is the fact that the dynamical spin operator S does
not obey simple boson or fermion commutation relations. Abrikosov
introduced the following representation for the impurity spin operators

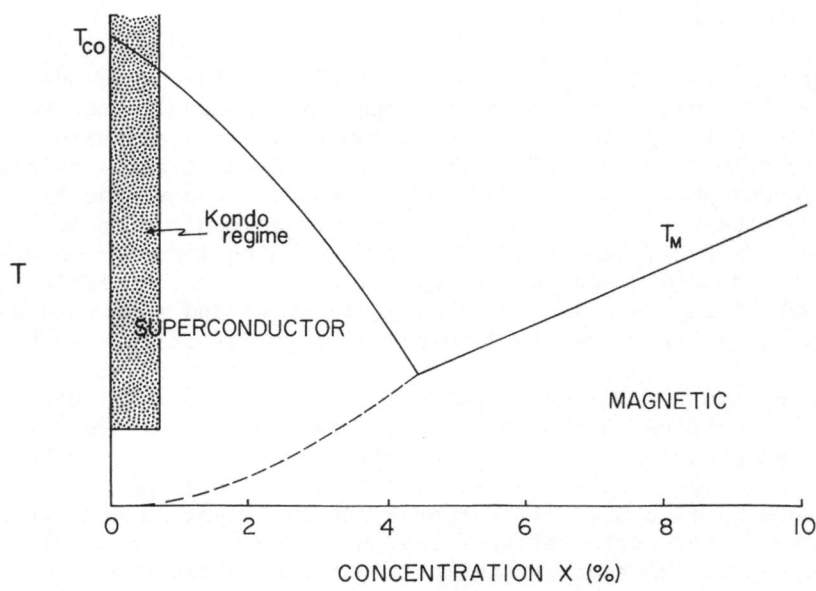

Figure 3.   Phase diagram of the superconducting transition
             temperature T_C as a function of impurity concentra-
             tion x. T_M denotes the magnetic ordering temperature,
             and the Kondo regime is estimated for a value of the
             exchange coupling J = 0.5 eV.

$$S = a_{\beta'}^{+} \ S_{\beta'\beta} \ a_{\beta}, \tag{23}$$

where $S_{\beta'\beta} \equiv <\beta'|S|\beta>$ are the standard matrix elements of the spin
matrices, and repeated indices imply summation over the Zeeman β
indices.   The a_{β}^{+} and a_{β} are creation and annihilation operators
for the pseudofermion field, and obey the anticommutation relations

$$\{a_{\beta}, a_{\beta'}^{+}\} = \delta_{\beta\beta'}, \ \{a_{\beta}, a_{\beta'}\} = 0 \tag{24}$$

The drawback of the method is the presence of spurious states with
unphysical multiple occupation.   Fortunately, these are readily
discarded by assigning a "kinetic energy" λ to each state and then
normalizing the results to the probability of single occupation
(2S+1) exp $(-\lambda/T)$ with $\lambda \to \infty$.   However, these operators allow a
perturbation expansion with the usual diagram techniques which are

particularly helpful in identifying various order spin-flip scattering contributions. In terms of these operators, the s-d Hamiltonian takes the form

$$H = \sum_{k,\alpha} \varepsilon_k c_{k\alpha}^+ c_{k\alpha} - J \sum_{\substack{k,k',j \\ \alpha,\alpha',\beta,\beta'}} e^{i(k-k')\cdot R_j}$$

$$x(\sigma_{\alpha'\alpha}\cdot S_{\beta'\beta}) \, a_{j\beta'}^+ \, a_{j\beta} \, c_{k\alpha'}^+ \, c_{k\alpha'} \tag{25}$$

where j refers to the impurity located at size R_j.

The central problem in superconductivity is the calculation of the electron-electron coupling which enters in the BCS theory[35] for the energy gap and transition temperature. Denoting the electron propagation by a solid line, and the impurity spin by a dotted line, the electron pair scattering from impurities is shown in Figure 4. Note that the interaction vertex Γ to lowest order is simply $J \, s\cdot\sigma$, so the first contribution is proportional to J^2 as expected.

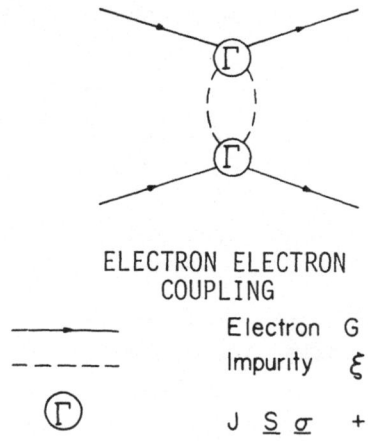

ELECTRON ELECTRON
COUPLING

——————— Electron G

— — — — — Impurity ξ

Γ $J \, \underline{S} \, \sigma$ +

Figure 4. Electron pair interaction induced by magnetic impurities. The lowest order correction for the coupling vertex Γ is the exchange coupling J, resulting in an electron-electron matrix element proportional to J^2.

The derivation of T_c proceeds by using the electron Green's function

$$G_{\omega}(\underset{\sim}{k}) = \frac{1}{i\omega_n - \varepsilon_k} \,, \tag{26}$$

and the propagators for the localized impurity state

$$\xi = \frac{1}{i\omega_m} \,, \tag{27}$$

to obtain the relevant extension of the BCS theory for the scattering amplitude of two electrons. The resulting integral equation for T_c is given by

$$1 = \frac{|g|T_c}{(2\pi)^3} \sum_{\omega} \int d^3p \, K_{\omega}(\underset{\sim}{p}, -\underset{\sim}{p}) \,, \tag{28}$$

where g is the phonon induced electron pairing coupling which determines T_{co} in the absence of impurities. It should be noted that nonmagnetic impurities tend to have negligible influence on T_c. Magnetic impurities cause a pair-breaking interaction and a corresponding electron lifetime τ which enters in the electron self energy to yield a modified Green's function

$$\tilde{G}_{\omega}(\underset{\sim}{p}) = \frac{1}{i\omega\eta_1 - \varepsilon_p} \,, \tag{29}$$

with $\eta_1 = 1 + 1/(2\tau_1|\omega|)$. The solution for T_c then involves the integral equation

$$K_{\omega}(\underset{\sim}{p}, -\underset{\sim}{p}) = \tilde{G}_{\omega}(\underset{\sim}{p})\tilde{G}_{\omega}(-\underset{\sim}{p})[1 + \frac{xT}{(2\pi)^3} \sum_{\omega'} \int d^3p' \, V(\omega,\omega')K_{\omega'}] \,, \tag{30}$$

where x is the number of impurities and $V(\omega,\omega')$ is the two-electron vertex.

At low concentrations the isolated impurity problem can be treated in the Born Approximation which gives[24]

$$\frac{1}{\tau_{AG}} = xJ^2 s(s+1)N(o) \,, \tag{31}$$

where s is the impurity spin and N(o) denotes the electron density of states at the Fermi energy. Within this approximation, Abrikosov and Gorkov derived the transcendental equation for T_c

$$\frac{1}{2} \ln \frac{T_{co}}{T_c} = \frac{1}{\pi T_c \tau} \sum_{n>o} \frac{1}{(2n+1)(2n+1 + 1/\pi T_c \tau)} \quad (32)$$

which can be expressed in terms of the digamma function ψ as

$$\ln \frac{T_{co}}{T_c} = \psi(\frac{1}{2} + \frac{\rho}{2}) - \psi(\frac{1}{2}), \quad (33)$$

where $\rho^{-1} = \pi \tau T_c$. This result sets a standard for comparison and provides a very useful estimate of the exchange coupling parameters in the low x limit; Here the digamma function may be expanded to derive the slope

$$\frac{dT_c}{dx}\bigg|_{x=o} \cong - \frac{\pi^2}{2} J^2 s(s+1) N(o). \quad (34)$$

Of course the electron-phonon coupling has been incorporated in the definition of the host superconductor temperature T_{co} in Eq. 33, and the impurity effect on the lattice vibration spectrum is disregarded.

Kondo scattering terms are of order J^3 and higher in the electron scattering cross section. However the resulting logarithmic temperature divergence of the cross section suggests interesting modifications in the superconducting state. Extensive experimental and theoretical studies of the Kondo effect in superconductors have been reported,[36] and various new experimental discoveries are discussed by B. Maple on pages 281 ff of the present book.

Theoretical studies were confronted with problems similar to the resistivity minimum in normal metals, but further encumbered by the mathematical complexities introduced by the superconducting state. The early perturbation treatments prior to 1968 have been summarized by Griffin,[37] and subsequently extended by various groups.[38-40]

A major development in the study of the Kondo problem was achieved by E. Müller-Hartmann and J. Zittartz in the discovery that Kondo scattering could yield re-entrant superconductivity.[41] This striking result predicted that a dilute magnetic alloy with appropriate Kondo temperature $T_k < T_{co}$ may exhibit more than one superconducting transition temperature at a given impurity content. The detailed analysis is rather technical in nature and described in a comprehensive review by Müller-Hartmann.[42]

The key features of the Kondo corrections to the electron-impurity coupling vertex are shown in Fig. 5, which exhibits the leading Kondo term in comparison to the AG theory, and in contrast to the influence of coupled magnetic impurities ($\propto J^4$) which are discussed in the following sections.

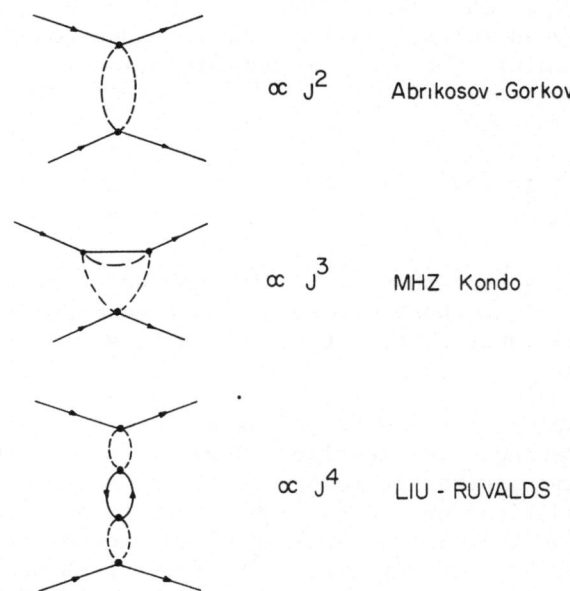

Figure 5.   Comparison of the two-electron scattering contributions originating from a single impurity in lowest order J^2 of the Born Approximation and the leading J^3 Kondo term in the Müller-Hartmann-Zittartz (MHZ) theory. Two magnetic impurities modify the superconducting pairs to order J^4.

A solution of the superconducting pairing equations including the Kondo effect yields the MHZ result

$$\ln \frac{T_{co}}{T_c} = \psi(\frac{1}{2} + \frac{\rho_K}{2}) - \psi(\frac{1}{2}), \tag{35}$$

which is of the universal form of the AG theory (Eq. 33), but
with a Kondo variable defined by

$$\rho_K = \frac{x}{(2\pi)^2 N(o) T_c} \left\{ \frac{\pi^2 s(s+1)}{\ln^2(T_c/T_k) + \pi^2 s(s+1)} \right\}. \qquad (36)$$

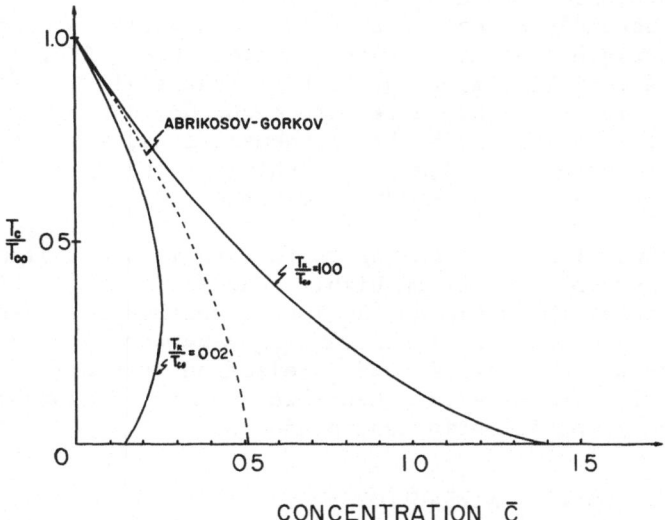

CONCENTRATION $\bar{C}$

Figure 6.   A re-entrant superconducting transition temperature curve
is shown for a Kondo system with $T_k = 0.02\ T_{co}$, in comparison to the Born Approximation result of the dashed
curve.   Large exchange coupling may yield $T_k = 100\ T_{co}$
with the corresponding T_c vs. concentration curve.

This functional relation gives multiple valued transition
temperatures at a given impurity concentration x providing that
$T_k \ll T_{co}$.   This re-entrant behavior is displayed in the curve
labelled $T_k/T_{co} = 0.02$ in Figure 6:   The curve is applicable to
systems with small exchange coupling of order J ~ 0.1 eV.

Also shown in Figure 6 is the case expected for large exchange
coupling J ~ 1 eV which results in $T_k \gg T_{co}$; this case results
in an extension of the superconducting range to higher concentrations
beyond the AG result.

A large number of re-entrant superconducting compounds have
been discovered experimentally and analyzed with varying success
in terms of the MHZ theory of Kondo scattering.[36] However,
caution should be exercised in applying the theory to those alloys
which are not expected to exhibit the Kondo anomaly. In particular
the Kondo theories are restricted to negative J values which are
commonplace with transition atom impurities such as Mn, Fe, and
others, but prove to be exceptional for the rare earth series of
impurities. Secondly a test of the Kondo concentration regime
for the appropriate J value is constructive. Using Fig. 2 as a
guide, it is especially important to note that small J values of
order 0.1 eV allow the Kondo effect to occur only at extremely low
concentrations x << 1%. At higher concentrations, impurity-
impurity interactions are likely to subjugate the spin-flip
processes necessary for the Kondo divergence.

In closing, it is interesting to note various examples of
re-entrant superconductivity in highly concentrated magnetic alloys
which fall outside the realm of the Kondo treatments. These may
reflect a new mechanism for re-entrant T_c behavior and one
possible cause has been attributed to electron scattering by
coupled magnetic impurities.[43] Hence we proceed to discuss the
role of coupled impurities in superconductors.

III. MAGNETIC IMPURITY INTERACTIONS

A.   Re-entrant Superconductivity

Intermediate concentrations of impurities ranging from
one to ten per cent produce various interesting examples of
re-entrant superconductivity outside the realm of the Kondo effect.
Here we shall consider only those cases where long range magnetic
order is not present, but rather emphasize the temperature regime
where pairs and clusters of magnetic impurities dominate the spin
dynamics.

Recently we have found that a coupled pair of magnetic
impurities may induce re-entrant T_c phases in a superconductor.[43]
Physically the origin of the re-entrant behavior can be traced
to a strong temperature variation in the electron pairing interaction caused by the elastic scattering from two coupled magnetic
impurities: The process involves the spin flip scattering terms
and occurs for either sign of the exchange coupling J in contrast

to the Kondo effect. Re-entrant behavior of the transition
temperature occurs if the impurity spin favors parallel alignment of spins. By the same token, antiparallel alignment of
impurity spins reduces the destructive influence of the isolated
impurity scattering and thereby enhances superconductivity.

A calculation of the electron scattering by an interacting
pair of impurities is similar to the case of one impurity scattering.
A diagrammatic representation of the two-electron vertex is shown
in Figure 5: For simplicity, the graphs show the impurity coupling
of the indirect exchange or RKKY form, which yields a vertex
proportional to J^4. These higher-order contributions are significant because they give an electron cross section which diverges
at low temperatures. To realize this effect it is constructive to
look at the one-impurity vertex $V(\omega,\omega')$ which enters in the electron-
electron scattering equation 30. The elastic scattering channel
gives[39]

$$V_{AG}(\omega,\omega) = \frac{J^2 s(s+1)}{3T} ,$$
(37)

where the leading term corresponds to the AG theory. The T^{-1}
divergence of the scattering comes from the pseudofermion impurity
spin "bubble" diagram in Fig. 5 which contributes a factor $(2T)^{-1}$.
This conclusion is based on the inability of the electron to transfer energy to the impurity and presumes a localized impurity spin
state. By itself, the AG vertex will not give re-entrant superconductivity because another factor of T enters in the integral
equation involving $V(\omega,\omega)$ in Eq. 30.

Scattering of electrons by a pair of magnetic impurities
yields a stronger divergence $V_2 \propto T^{-2}$ in the electron pairing
interaction, which is also related to the elastic scattering
from an impurity state lacking dispersion. In view of the
impurity localization we may introduce an effective impurity
coupling U defined by

$$H_{ij} = U \, \underset{\sim}{S}_i \cdot \underset{\sim}{S}_j,$$
(38)

which then yields an electron-electron scattering vertex

$$V_2(\omega,\omega) = \frac{-x \, U \, J^2 s(s+1)}{6T^2}$$
(39)

The impurity concentration x enters as a result of the impurity
averaging. Finally, the determination of T_c follows closely the
AG formalism outlined above, and yields a modification of Eq. 30
to a similar functional form

$$\ell n \frac{T_{co}}{T_c} = \psi(\frac{1}{2} + \frac{\rho_{AG}}{2} + \frac{\rho_2}{2}) - \psi(\frac{1}{2}), \tag{40}$$

where the additional pair scattering term is

$$\rho_2 = \frac{x\ U}{2T_c} \quad \rho_{AG} = \frac{-x^2 UJ^2 s(s+1)N(o)}{2T_c^2}. \tag{41}$$

It is evident that the transcendental equation (41) can yield multiple values of T_c as a consequence of the divergent cross section at low temperatures for the appropriate sign of the interaction U.

A plot of typical theory curves for T_c vs. concentration are shown in Fig. 7 for various values of U.

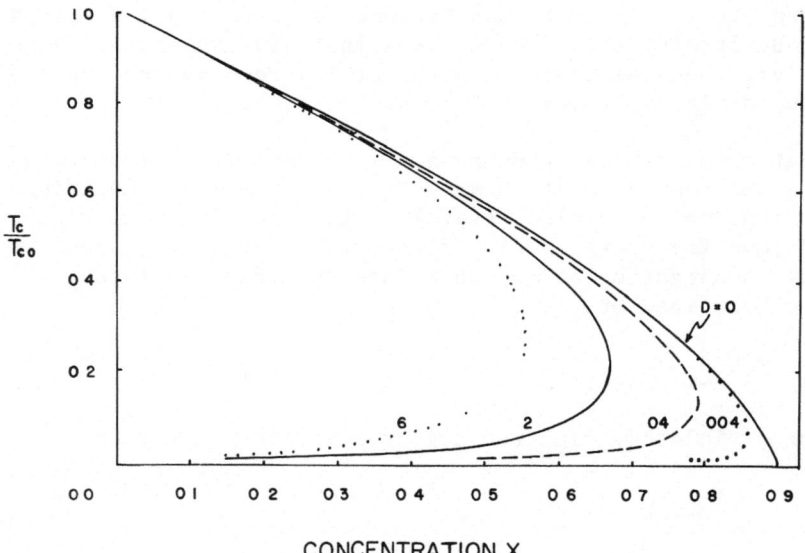

CONCENTRATION X

Figure 7.   Superconducting transition temperature as a function of
            magnetic impurity concentration in x at. %. The non-
            interacting impurity case D = o corresponds to the AG
            theory. Ferromagnetic impurity coupling U<o gives the
            re-entrant behavior shown for various values of the
            parameter D = U/2T$_{co}$.

Comparison of the theoretical re-entrant behavior to specific compounds requires a reasonable estimate of the impurity coupling U which may generally be associated with the RKKY interaction. However, the correspondence to superconducting alloys must be handled with care primarily because of many complications related to the superconducting energy gap, impurity limited electron mean free path, and Fermi surface topology. These issues are addressed following an examination of T_c enhancement by coupled magnetic impurities.

B.   Enhanced transition temperatures

Deviations from the Abrikosov-Gorkov theory of the transition temperature are to be expected at impurity concentrations in excess of 1/2% by virtue of correlations among impurities. One of the earliest studies[44] of the competition between magnetic ordering effects and superconductivity revealed definitive magnetic ordering in $Gd_x La_{1-x}$ for x > 3 at %, whereas the superconducting temperature T_c was confined to be non-zero in a small range of x < 1.2%. More detailed examination of the low temperature region near x ~ 1% showed a plateau in the $T_c(x)$ variation[45,46] which represented an anomalous T_c enhancement in comparison to the AG independent impurity theory.

Theoretically, enhancement of T_c at a given impurity concentration can be achieved by several mechanisms which restrict the ability of the impurity to flip its spin in response to the scattering by an electron. For example, such spin coupling may be caused by spin-orbit interactions, Zeeman level splitting, and magnetic ordering by virtue of the molecular field due to other impurities. These modifications of the spin dynamics were incorporated in the Abrikosov-Gorkov theory by Benneman[47] in an effort to explain the anomalous $Gd_x La$ data. His work gave good agreement with experiment using reasonable estimates for the relevant parameters.

Another possible source of T_c enhancement is the Kondo scattering for large values of the exchange coupling J ~ -1. eV. Such cases would have large Kondo temperatures T_k ~ 100°k and the strong multiple scattering terms would modify the AG expression for the transition temperature. A calculation of T_c by Müller-Hartmann and Zittartz demonstrates this effect, as shown in Fig. 6 for T_k/T_{co} = 100.

Interacting pairs of magnetic impurities may also suppress the the destructive influence of isolated magnetic impurities on superconductivity. This process becomes especially important at low temperatures because a divergence in the electron-electron scattering amplitude $V_2(\omega,\omega)$ defined in Eq. 39. However, in contrast to the re-entrant behavior of T_c which can be induced by ferromagnetic impurity spin alignment, the T_c enhancement results in the case of antiparallel alignment of the impurity spins.

The formulation of the two-impurity problem is given in Eqs. 38-41, with the key results presented in Eqs. 40-41. Numerical solution of these equations yields a series of typical enhancement curves shown in Fig. 8.

In view of the various mechanisms which have been invoked to account for enhancement of superconductivity, it is worthwhile to

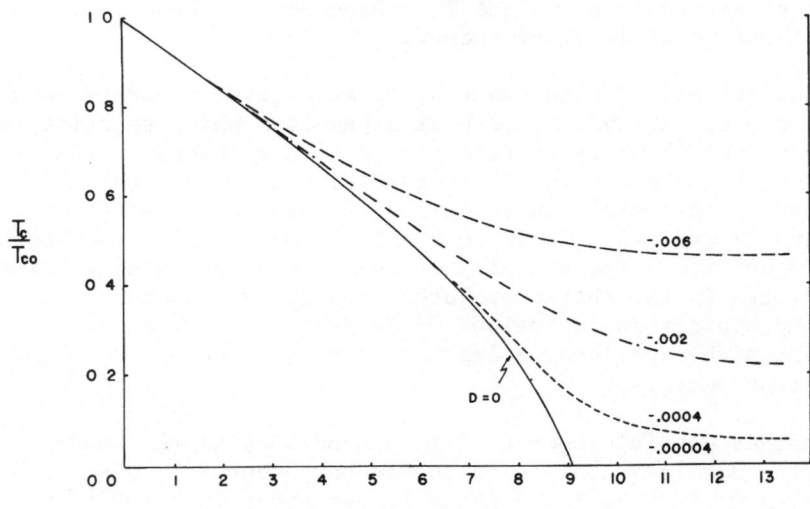

CONCENTRATION X

Fig. 8.   Superconducting transition curve as a function of magnetic impurity concentration x in at .%, for the case of magnetic impurity pairs with antiferromagnetic spin alignment corresponding to the interaction parameter D = U/2 T_{co}. If U is taken of the RKKY form, the parameter D is proportional to J^2 and thus the T_c enhancement is independent of the sign of J. The AG theory corresponds to the D=0 curve.

consider some specific examples. The Bennemann theory has been applied to Gd_xLa_{1-x}, and the MHZ theory should yield T_c increases for large J systems. Systems with dominant impurity pair coupling may be exemplified by the layered compounds such as $NbSe_2$ with Co or Ni impurities. The reduced dimensionality of the crystal structures restricts the magnetic impurities to specific sites, so that reasonable estimates of the interaction parameters are feasible. As an example, we present the data[44] on Co_xNbSe_2 shown in Fig. 9.

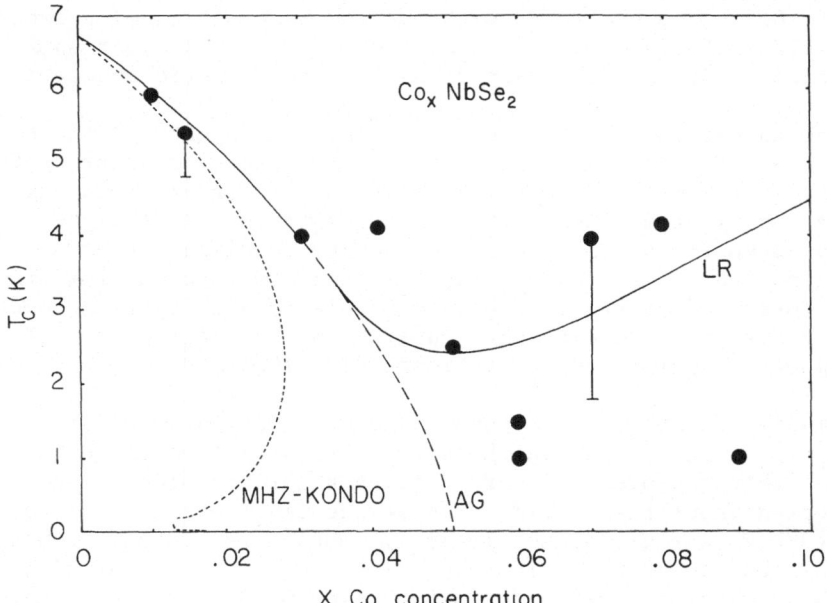

Figure 9. Superconducting temperature of $NbSe_2$ doped with Co impurities. The impurity pair coupling theory of Liu-Ruvalds (LR) provides a good fit to the data using an RKKY interaction with J = -0.1 eV, E_F = 0.8 eV, and $2k_F\Delta$ = 2.1. By comparison the AG theory yields a critical concentration roughly one half of the observed value. Independent measurements of J = -0.1 eV indicate that the relevant Kondo case is $T_k \ll T_{co}$ which the re-entrant behavior of the dotted curve.

The persistence of superconductivity in $NbSe_2$ alloyed with 8% Co (or Ni) is rather remarkable since a few per cent of magnetic impurities is normally sufficient to destroy T_c. Also this example indicates suppression of the Kondo effect because of the small exchange coupling J = -0.1 eV which is known from independent measurements. However, magnetic ordering is clearly not present even at the high Co concentrations which allow a superconducting state with $T_c(x=8\%) \simeq 4°k$ in comparison to the $NbSe_2$ host value of $T_{co} = 6.7°k$.

Since the coupled impurity pair theory (LR)[43] yields a good description of the layered compound alloys, it may be worthwhile to examine other aspects of the theory. Experimentally, a major challenge is the measurement of the critical magnetic field H_{c2} in materials with intermediate concentrations of magnetic impurities. Some interesting examples, including $Eu_x Mo_6S_8$ under pressure, are discussed in the lecture of B. Maple. The class of layered compounds is particularly important as a result of their known high critical fields in a direction parallel to the layers. Theorists, on the other hand, are faced with the challenge to examine impurity interaction effects on other properties including nuclear spin relaxation, thermodynamic properties and the magnetic susceptibility.

The appearance of the low temperature divergence in the electron-electron scattering amplitude, which is induced by scattering from impurity pairs, must be renormalized by higher order terms which come into play at sufficiently low temperatures. A straight-forward extension of the impurity-impurity coupling vertex by summing a "ladder" series of diagrams suggests that the low order terms considered in Fig. 5 are of sufficient validity to account for many systems with exchange coupling in the range of $J \simeq .1$ eV and magnetic impurity concentrations below 10%.

Finally we mention the consequences of the interacting impurity pair scattering on the normal state electrical resistivity. Within the order of perturbation theory considered here, the coupled impurities yield a resistance divergence of the form $\rho \propto -x^2 J^2 U/T$, where U is the impurity coupling.[34] Combined with the usual phonon contribution $\Delta\rho = BT^5$, the coupled impurities yield a resistance minimum for either sign of J in contrast to the Kondo effect. Hence the rare earth compounds offer a tempting testing ground in the normal states as well as in the superconducting phase.

C.   RKKY Coupling and Dimensionality

The high degree of localization of magnetic impurity states in superconductors suggests that direct exchange terms are quite small. Thus the dominant coupling is expected to be the RKKY interaction induced by the indirect exchange of conduction electrons.[48-50]

The indirect exchange interaction is defined by the Hamiltonian

$$U = - \sum_{\ell,m,q} I(q) e^{i\underset{\sim}{q} \cdot (\underset{\sim}{R}_\ell - \underset{\sim}{R}_m)} \underset{\sim}{S}_\ell \cdot \underset{\sim}{S}_m, \qquad (42)$$

where the exchange interaction is given by

$$I(\underset{\sim}{q}) = \frac{2}{N} \left| j_{sd}(\underset{\sim}{q}) \right|^2 \chi(\underset{\sim}{q}), \qquad (43)$$

where j_{sd} is the exchange matrix element between the conduction electron state s and the impurity d-level. For rare earths the appropriate impurity state is the f-level. Physically the spin $\underset{\sim}{S}_\ell$ localized on the ℓ-th impurity interacts with a conduction electron and creates a spin polarization. The spin polarization is subsequently experienced by a second impurity at site m, resulting in the effective impurity-impurity coupling.

The range, strength, and sign of the RKKY interaction U is highly sensitive to structure in the spin susceptibility $\chi(q)$. Thus this relationship reflects the electronic structure which determines $\chi(q)$ according to the well-known relation

$$\chi(\underset{\sim}{q}) = \frac{1}{N} \sum_{\underset{\sim}{k}} \frac{f_{\underset{\sim}{k}} - f_{\underset{\sim}{k}+\underset{\sim}{q}}}{\varepsilon_{\underset{\sim}{k}+\underset{\sim}{q}} - \varepsilon_{\underset{\sim}{k}}}, \qquad (44)$$

where $\varepsilon_{\underset{\sim}{k}}$ is the energy band of the conduction electrons and $f_{\underset{\sim}{k}}$ is the Fermi function. Evidently the effective electron dimensionality and Fermi surface topology will have a strong influence on $\chi(q)$.

The classic example of an isotropic free-electron gas in three dimensions may be considered as a standard for comparison. It gives

$$\chi(q) = - \frac{mk_F}{\pi^2 \hbar^2} \left[1 + \frac{4k_F^2 - q^2}{4k_F q} \ln \left| \frac{2k_F + q}{2k_F - q} \right| \right], \qquad (45)$$

where k_F is the Fermi wave vector. This result exhibits an interesting singularity in the slope of $\chi(q)$ at $q = 2k_F$ which yields oscillations in the RKKY coupling as a function of the impurity separation. Finally, the corresponding RKKY interaction takes the form

$$U = \frac{9\pi J^2}{\varepsilon_F} \; F(2k_F |\underset{\sim}{R}_\ell - \underset{\sim}{R}_m|), \tag{46}$$

where

$$F(z) \equiv \frac{z\cos(z) - \sin(z)}{z^4}. \tag{47}$$

At long range, the coupling U falls off as R^{-3} in this three dimensional example.

Lowering the dimensionality tends to increase the range of the RKKY coupling dramatically. This feature is associated with the singularities in $\chi(q)$ shown in Fig. 10 for the one and two-dimensional systems.

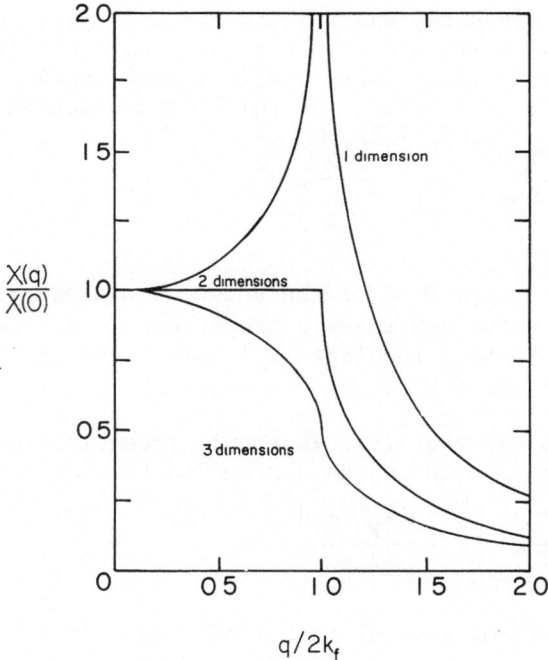

Figure 10. Relative spin susceptibility as a function of momentum for a free electron gas in one, two, and three dimensional systems.

The remarkable enhancement of $\chi(q = 2k_F)$ at lower dimensionality, as well as other structural features, are of central importance in determining magnetic ordering via the RKKY interaction. Furthermore, the electron response may result in charge and spin density wave states,[52] which have a profound influence on the electronic properties of metals.

Applications of the RKKY interaction Hamiltonian to real materials are subject to major cautionary restrictions, particularly in the case of rare earth metals and compounds.[53] Among the major modifications to $\chi(q)$ are the interband contributions, exchange and correlation effects, and the momentum dependence of the exchange integral which is represented here by the exchange coupling J. Nevertheless, the general features of $\chi(q)$ in Fig. 10 are worth noting since similar structure may be achieved by Fermi surfaces with nearly parallel sections in momentum space. Naturally these topology effects may yield maxima in $\chi(q)$ at special points $q=Q\neq 2k_F$ which may yield interesting magnetic states including spiral spin ordering.

In the superconducting state, the appearance of an energy gap in the electronic spectrum causes a vanishing of the long wavelength limit of the spin susceptibility $\chi_s(o)$. Hence a superconductor would not ordinarily support long range magnetic ordering. This result follows from the generalization of the susceptibility expression in the superconducting phase[54]

$$\chi_s(q) = -2\mu_B^2 \sum_{\underset{\sim}{k}} \left\{ (u_{\underset{\sim}{k+q}} u_{\underset{\sim}{k}} + v_{\underset{\sim}{k+q}} v_{\underset{\sim}{k}})^2 \frac{f_{\underset{\sim}{k+q}} - f_{\underset{\sim}{k}}}{E_{\underset{\sim}{k+q}} - E_{\underset{\sim}{k}}} \right.$$

$$\left. + (u_{\underset{\sim}{k+q}} v_{\underset{\sim}{k}} - v_{\underset{\sim}{k+q}} u_{\underset{\sim}{k}})^2 \frac{1 - f_{\underset{\sim}{k}} - f_{\underset{\sim}{k+q}}}{E_{\underset{\sim}{k}} + E_{\underset{\sim}{k+q}}} \right\} , \qquad (48)$$

where μ_B is the Bohr magneton and the coherence factors are defined by

$$|u_{\underset{\sim}{k}}|^2 = \frac{1}{2}(1 + \frac{\varepsilon_{\underset{\sim}{k}}}{E_{\underset{\sim}{k}}}) , \qquad (49)$$

$$|v_{\underset{\sim}{k}}|^2 = \frac{1}{2}(1 - \frac{\varepsilon_{\underset{\sim}{k}}}{E_{\underset{\sim}{k}}}) , \qquad (50)$$

and the superconducting energy gap Δ enters in the renormalized single electron spectrum

$$E_{\underset{\sim}{k}} = (\varepsilon_{\underset{\sim}{k}}^{2} + \Delta^{2})^{1/2}. \tag{51}$$

In the long wavelength limit the susceptibility reduces to

$$\chi_s(o) = -2\mu_B^{2} \sum_{\underset{\sim}{k}} \frac{\partial f_{\underset{\sim}{k}}}{\partial E_{\underset{\sim}{k}}}, \tag{52}$$

in comparison to the normal state limit

$$\chi_n(o) = 2\mu_B^{2} N(E_F), \tag{53}$$

where $N(E_F)$ denotes the density of electron states at the Fermi energy.

A superconducting energy gap causes the susceptibility to vanish at zero temperature, because the derivative $\partial f_k/\partial E_k$ reduces to a δ-function whose argument is non-vanishing in the gap region. Hence the Knight shift and other physical properties proportional to $\chi_s(o)$ are expected to be greatly reduced. At finite temperatures, the particle-hole transitions across the energy gap develop a finite probability which naturally reproduces the normal state limit χ_n as T approaches T_c.

The momentum dependence of $\chi_s(q)$ is of interest in regard to possible maxima at a given momentum Q which may lead to short range magnetic ordering. Such a "cryptoferromagnetic" state was suggested originally by Anderson and Suhl,[55] and should occur with typical wavelengths of $\lambda \simeq 50$ Å on the basis of a three dimensional electron gas model.

We have recently calculated the superconducting electron susceptibility over the entire q-range for one and two-dimensional electron band models. These represent idealized models characteristic of the Fermi surface topology in actual superconductors such as layered compounds and rare earth alloys. Numerical computations[56] of $\chi_s(q)$ using Eq. 48 yield the relative susceptibility shown in Fig. 11.

A key feature of the susceptibility in Fig. 11 is the highly limited region of momentum space $\Delta q \approx 10^{-2} k_F$ where χ_s differs appreciably from χ_n. The slope of the χ_s/χ_n curve is in agreement with the expansion used in Ref. 55 for the 3-D case, so their maximum in $\chi_s(q)$ remains confirmed. The RKKY interaction is hardly affected by the superconducting state, however, since the coupling involves an average of $\chi_s(q)$ over the entire momentum region. A calculation of the indirect exchange coupling using $\chi_s(q)$ yields a form of U quite similar to the standard expression of Eq. 46 but modified by an overall factor $\exp(-r/\xi_o)$ where r is the impurity separation and ξ_o is the coherence length for the superconducting electrons. Generally this factor is close to unity and the RKKY coupling is not affected much by superconductivity.

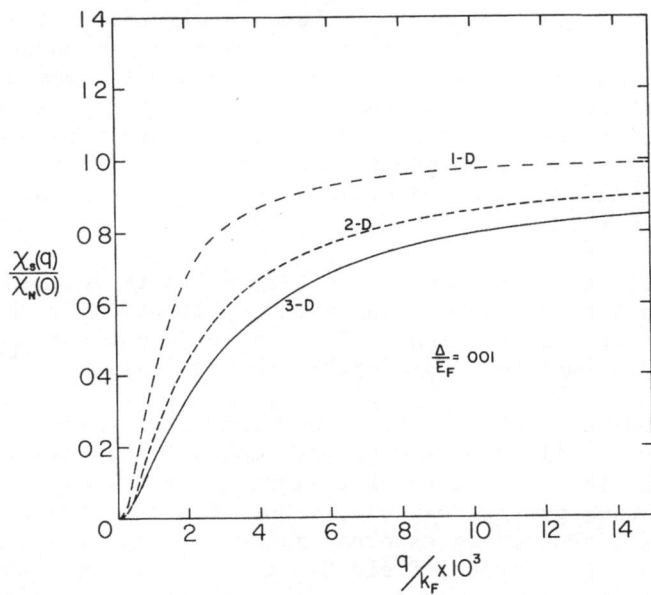

Figure 11.   Superconducting electron susceptibility χ_s relative to the normal state χ_n is shown as a function of momentum for a free electron gas in one, two, and three dimensions.

An interesting feature of the spin susceptibility is the more rapid increase in $\chi_s(q)$ at reduced dimensionalities. This feature may influence the range of magnetic ordering in systems where "nesting" of Fermi surfaces may induce energy band features characteristic of lower dimensional dynamics. Superconducting rare earth compounds are ideal candidates for further studies in this connection.

As the concentration of magnetic impurities is increased to 1% or higher, the spin susceptibility of a superconductor $\chi_s(o)$ may increase substantially and lead to various interesting features including possible magnetic ordering. The impurity spin-flip scattering tends to smear out the energy gap by causing a finite electron lifetime τ as in the Abrikosov-Gorkov result of Eq. 31.

The susceptibility in the dilute impurity limit was worked out by Gorkov and Rusinov[57] who found

$$\frac{\chi_s(o)}{\chi_n(o)} \cong \frac{\pi}{12\Delta\tau_{AG}}. \tag{53}$$

In the "gapless" regime achieved by sufficiently high impurity concentrations, $\chi_s(o) \rightarrow \chi_n(o)$, and hence the intermediate concentration regime exhibits a competition between superconductivity and magnetism which yields various novel phase diagrams as a function of concentration and temperature. Among these is the possibility of a re-entrant superconductivity state caused by a transformation to a ferromagnetic state at sufficiently low temperature. Actual observed cases are discussed in the lectures by B. Maple in this book.

Spin-orbit coupling may also create a finite spin susceptibility $\chi_s(o)$ in the superconducting state. Historically this issue was raised by the observation[58] of a finite Knight shift in Hg and Pb, which stimulated considerable theoretical work.[59-62]

Technological applications of superconductors may be enhanced by the influence of magnetic impurities in several ways. In addition to the capability of tailoring the transition temperature and electromagnetic response, impurities may lead to a higher capability of withstanding external magnetic fields. One mechanism for increasing the critical field H_{c2} is the exchange field compensation effect proposed by Jaccarino and Peter.[63] In the presence of an external magnetic field $\underset{\sim}{H}_{ext}$, a magnetic impurity with spin $\underset{\sim}{S}$ will product and exchange field $\underset{\sim}{H}_{exc.} = 2JS/g\mu_B$, which opposes $\underset{\sim}{H}_{ext}$ if J is negative. Hence the net field experienced by the superconducting electrons is reduced by impurities coupled antiferromagnetically to the conduction electrons.

The critical fields H_{c2} is also sensitive to the spin susceptibility $\chi_s(o)$. To examine this connection, we note the difference in free energies F(H) between superconducting and normal states

$$\Delta F = F_n(H) - F_s(H) = F_n(o) - F_s(o) - \frac{1}{2}\{\chi_n - \chi_s\}H^2. \tag{54}$$

Using the BCS theory we find the limiting critical field H_p from the relations $\Delta F(H) = 0$, and

$$F_n(o) - F_s(o) = \frac{1}{2}N(E_F)\Delta^2, \tag{55}$$

to find the limiting field

$$H_p = \left[\frac{N(E_F)}{\chi_n(o) - \chi_s(o)} \right]^{1/2} \Delta. \tag{56}$$

In the absence of magnetic impurities and spin-orbit scattering we have $\chi_s(o) = 0$ and consequently obtain the Clogston. Chandresekhar limit,

$$H_p(x=o) = \frac{\sqrt{2}\, \Delta(x=o)}{g\mu_B} = 18.4\ T_c. \tag{57}$$

This relation provides a useful standard for the critical field in kO_e units.

Magnetic impurities influence the critical field $H_p(x)$ in two primary and competing ways. The prevailing destructive effect of single impurity scattering is a decrease of $\Delta(x)$ as a function of concentration which becomes particularly rapid as x approaches the critical concentration x_c where $\Delta(x_c) \equiv 0$. The resulting critical field H_p scales with $\Delta(x)$ and thus decreases over a range $x < .96x_c$, until the spin susceptibility enhancement in the denominator of Eq. 56 creates an uptrend to H_p for $x > .96x_c$. In this connection it may be worthwhile to examine the contribution of impurity-impurity interactions to the critical field. For antiferromagnetic RKKY coupling, as in the case of Co_xNbSe_2 shown in Fig. 9, a persistence of a relatively large gap $\Delta(x)$ may be achieved. Providing that the impurities generate a significant enhanced spin susceptibility, the combined result may be constructive for the superconducting critical field.

D.   Rare Earth Compound Examples

Lanthanum compounds have provided various stimulating challenges in part because of the diverse phenomena discovered experimentally. These systems and many other compounds of interest are reviewed in the lectures by B. Maple in this book.

The first studies by Matthias et al[46] showed a competition between magnetic ordering in $La_{1-x}Gd_x$ and similar compounds. Their measurements demonstrated a rapid drop in T_c from roughly 6°K for pure La to 1°K at concentration of x = 1% Gd. Extrapolation of the measured magnetic curie ordering temperature T_M from higher x values provided a classic phase diagram of the type shown in Fig. 3. Also, more recent data have shown that the $T_c(x)$ variation for $La_{1-x}Gd_x$ deviates from the Abrikosov-Gorkov predictions, and these anomalies have been attributed to magnetic ordering.[47,48]

Our goal is to examine re-entrant superconductivity cases, and hence we shall focus on $La_{1-x}Ce_x$ compounds, and the anomalous features induced by applied pressure and by alloying with other rare earths such as Th.

Rare earth impurities exemplify essentially all of the key features considered in this review. We recall that many rare earth in transition metals do not exhibit a Kondo effect and thus were presumed to have an effective <u>positive</u> J exchange coupling to the conduction electrons. Similarly resistivity data on La-based alloys suggest J $\simeq$ + .05 eV for Pr, Nd, Gd, Tb, and Er impurities,[32] which is compatible with the absence of a resistivity minimum.

Cesium impurities in Lanthanum provide an exceptional opportunity to test various theoretical proposals. First of all, the appearance of a pronounced resistivity minimum in $LaCe_x$ for x ~ 1-3% suggests a Kondo effect with a characteristic Kondo temperature of T_k ~ 0.1°K, which corresponds to a weak but negative exchange coupling J. Susceptibility measurements support this conclusion, and the Schrieffer-Wolff transformation[21] provides a clear physical basis for J < 0 in the case of $LaCe_x$. Their theory demonstrates the sensitivity of J to the placement of the localized impurity state relative to the Fermi energy E_F and thus the 4f state of Ce is presumed to be favorably situated close to E_F. This situation lends itself to interesting variations under the influence of pressure or alloying; both changes tend to shift the relative separation of the levels and thereby modify the magnetic character of the Ce impurity.

Lanthanum is superconducting below 6°K, and its transition temperature increases markedly with pressure up to T_C = 12°K at P = 150 k bar. Although Ce impurities cause a drop in T_C of the La-based alloys, the shift is relatively small by virtue of the weak exchange coupling J $\simeq$ -0.1 eV of the localized spin state to the conduction electrons. The variation of T_C with Ce concentration is shown in Fig. 12, along with the corresponding curves for La Th_z-based alloys studied intensively by Huber, Fertig, and Maple.[64]

The initial change in slope of T_C vs Ce content in the La Th_x alloys is compatible with expected changes in the Fermi energy relative to the Ce 4f state. Varying the Thorium content thus changes the effective exchange coupling J which is a measure of d T_C/dx in the x → o limit. In fact, some variations in the slope of the Kondo resistivity divergence have been reported[65] at higher Ce content (x $\geq$ 2.3%), and it has been argued that the addition of Th tends to reduce and ultimately eliminate the magnetic character of the Ce impurity. Furthermore, the influence of spin and fluctuation lifetimes has been discussed for these systems. [66]

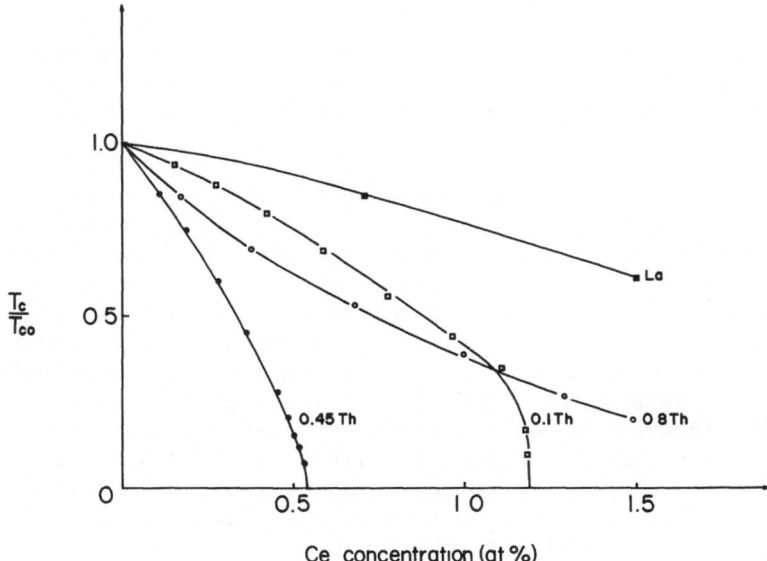

Figure 12. Superconducting transition temperature of La Th$_z$ Ce$_x$ alloys as a function of concentration x. Note the re-entrant behavior, which is evident for the La Th$_{.1}$ based alloys. The curves are fits using the RKKY coupled impurity theory.

Increasing Ce concentrations in the La Th$_z$ hosts generate anomalous features including possible re-entrant behavior of the transition temperature, and the cross-over to a relative enhancement of T$_c$ at higher Ce content for La Th$_{0.8}$. This behavior is difficult to reconcile with the variations of the exchange coupling J evident from the data in Figure 12 in the dilute Ce limit.

Kondo scattering leads to a re-entrant T$_c$ behavior on the physical basis of an electron scattered by a single impurity. Despite its restricted applicability to the dilute limit of magnetic impurity concentrations, the strong temperature variation of the scattering amplitude can give multiple T$_c$ values for a given Ce concentration as in the case La Th$_{.45}$ Ce$_x$ shown in Fig. 12. This instance would suggest a Kondo temperature T$_k$/T$_{co}$ ≪ 1 which is compatible with the measured values of T$_k$ ~ 0.1°K for La Ce$_x$ and the evident decrease in J with the addition of Th which further lowers T$_k$ for these alloys. However, the observed enhancement of T$_c$ in La Th$_{0.8}$ Ce$_x$ at higher Ce content (x ~ 1.5%) is in contrast to the predictions[41] of the Kondo effect theories. The upturn in T_c relative to the downward slope at lower concentrations would require $T_k \gg T_{co}$, but this would suggest a large increase in the J coupling strength which is in contradiction to the estimated changes in J indicated by the initial slopes of T_c vs. x.

The RKKY coupling between magnetic impurities may account for the anomalous T_c behavior in Fig. 12 providing that the influence of Th-alloying is capable of changing the sign of the RKKY interaction defined in Eq. 38. This coupling is also quite sensitive to the strength of the exchange interaction J because $U \propto J^4$, but the sign of U determines re-entrant T_c behavior (ferromagnetic U) or the enhanced T_c case (antiferromagnetic U) as demonstrated in Figs. 7-8. Rare earth compounds are ideal candidates for a test of this interaction effect since the electronic energy band structure is dominated by highly localized electron states. The associated energy bands are consequently very narrow and small shifts in the Fermi energy E_F caused by alloying will naturally shift the effective Fermi momentum k_F. As shown in Eqs. 46-47, the sign of the RKKY coupling is determined by the factor $U \propto \cos(2k_F \Delta)$, where Δ is the separation between magnetic impurities. Therefore U is expected to change sign with changes in k_F induced by alloying. The data of Fig. 12 is described quite well by the interacting impurity theory of Ref. 43, providing that k_F is treated as an adjustable parameter. The fits to the data are shown by the solid curves in Fig. 12.

Pressure-induced changes in the electronic structure also have a dramatic influence on superconductivity of the $LaCe_x$ alloys. The transition temperature shows a remarkable variation as a function of pressure as demonstrated in Fig. 13.[67]

First of all, it is interesting to note the increase in T_c of pure La under pressure which may be attributed to a strengthening of the electron-phonon interaction. However, for a given pressure, the relative change in T_c is highly sensitive to the Ce content. A variation in the J exchange coupling with pressure is expected as the localized Ce 4f-level may be shifted relative to the Fermi energy, and a slight enhancement of J with pressure is compatible with the dilute case data on 0.7% Ce in La and independent measurements of the Kondo resistivity minimum.[68,69] It should be noted that quantitative comparisons of the Kondo minimum and changes in T_c for the La Ce_x system are complicated by the strong variation of the electron-phonon coupling under pressure. Higher concentrations of Ce impurities in La yield a much stronger variation in T_c under pressure. In Fig. 13 the 2% Ce case shows a rapid drop in T_c with pressures as low as 5 kbar, and then a reappearance of superconductivity for P > 15 kbar. Qualitatively, this behavior may suggest an RKKY interaction whose

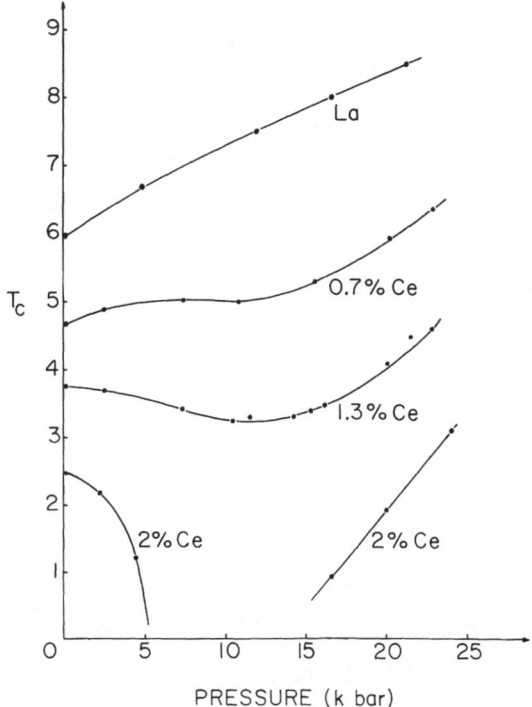

Figure 13. Transition temperature of La Ce$_x$ alloys as a function of pressure.

strength and sign are highly sensitive to pressure. Such coupling would not be dominant in the dilute limit (say 0.7% Ce impurities). The required changes of the Fermi surface topology required to modify the RKKY interaction under pressure may be reasonably similar to the effects of changing the Fermi energy by alloying with Th.

Finally we present the data[67] on superconducting La Ce$_x$ alloys for extremely high Ce content in Fig. 14.

Cesium concentrations in the range of 16% would normally be expected to destroy superconductivity of a host material such as La. This is indeed the case at low pressure, but then a remarkable uptrend in T$_c$ is found as pressure is increased to the 100 kbar range as seen in Fig. 14. It may well be that such high pressures render the Ce impurities non-magnetic. Nevertheless a quantitative study of the observed T$_c$ behavior of these compounds

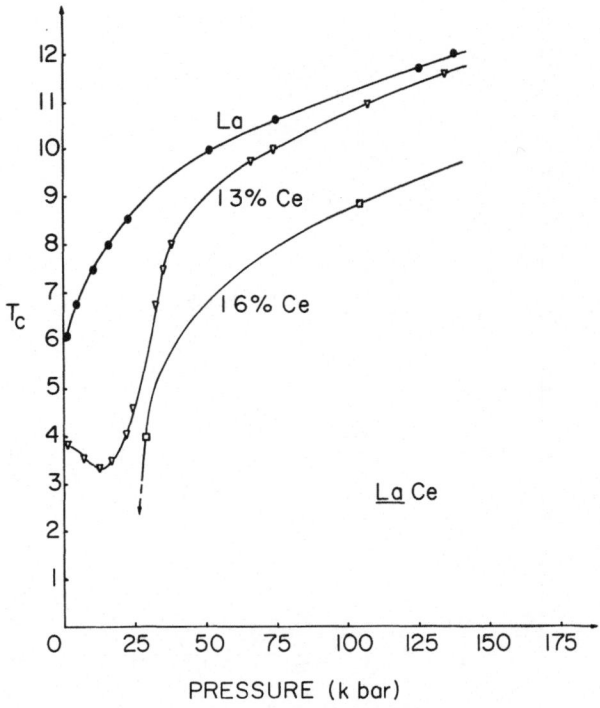

Figure 14.   Superconducting transition temprature of La Ce$_x$ alloys
as a function of pressure.

is warranted. It would be particularly interesting to trace the
evolution of the impurity magnetic character, its exchange
coupling J, and the effective impurity-impurity interaction as
a function of pressure and alloying.

    Recently a study of the combined effects of alloying and
pressure on La$_{.8}$Th$_{.2}$Ce$_x$ has demonstrated a reentrant behavior for
T_c at low pressure, and a cross-over to an enhanced T_c behavior as
a function of Ce content at P = 18 kbar.[70] The observed slope
of T_c vs x in the dilute limit is essentially unchanged by
pressure in the case studied, indicating an approximately constant
exchange coupling J at the experimental pressures.[70] This example
provides a separation of the Kondo effect influence from the
RKKY coupling modifications in T_c and furthermore suggests a
transition from ferromagnetic RKKY coupling to an antiferromagnetic
interaction as the pressure is increased to 18 kbar. The latter
cross-over is of the same nature as required by theory[43] to account
for the T_c behavior of La Ce$_x$ under pressure.

IV. CONCLUSIONS

We have reviewed the studies of re-entrant superconductivity starting from the original efforts to unravel the competing effects of magnetic ordering and superconducting electron pairing. The Kondo effect, and especially the theory of Müller-Hartmann and Zittartz, has been invoked to explain the variations in T_c with impurity content for a wide variety of systems. Our analysis supports the validity of applying the Kondo analysis to systems with dilute concentrations of magnetic impurities.

As the concentration of magnetic impurities is increased to the order of 1% and higher, the influence of magnetic impurity coupling, irregardless of its origin, is manifested in various ways. First of all, the coupling tends to limit the spin-flip process which is responsible for the Kondo divergence in the electron cross section. Thus a reasonably accurate criterion can be established for the concentration regime wherein the Kondo effect is valid for a given value of the exchange coupling J. By the same token, the dilute limit Kondo data provides estimates of J which are useful in connection with possible magnetic ordering studies.

An alternate origin of re-entrant superconductivity is the scattering of an electron by coupled magnetic impurity pairs. Providing that the impurity-impurity interaction is ferromagnetic in nature its influence is manifested as a more rapid drop in T_c as the impurity concentration increases. Also the resulting temperature dependence of the superconducting electron-electron vertex yields a double-valued T_c solution at a given impurity concentration. Our analysis of various experiments, including the layered compounds and numerous rare earth examples, suggests that the RKKY mechanism yields the appropriate magnitude of the coupling required to explain the data.

Enhancement of the superconducting transition temperature well in excess of the Abrikosov-Gorkov result may be achieved by an antiferromagnetic RKKY coupling of impurities. This suppression of the destructive T_c influence of single magnetic impurity occurs for either sign of the electron-impurity spin exchange coupling J. Applications of the theory to $Co_x NbSe_2$ yields coupling parameters of the appropriate strength and indicates the possible role of an effective reduced dimensionality for the electronic band structure. As in the case of rare earth compounds where the strong localized nature of the electronic states leads to "nesting" of the Fermi surface, the effective reduced dimensionality tends to increase the range of the RKKY interaction and it modifies the momentum dependence of the spin susceptibility $\chi(q)$. Both of the latter effects have a bearing on possible magnetic ordering states in superconductors.

Rare earth compounds provide an ideal testing ground
for the theoretical conclusions presented in this article.
Many examples are available with either sign of the exchange
coupling J which help to distinguish those features caused
by Kondo scattering. In addition the wealth of data on La-based
alloys in particular, and their anomalous variation of T_c under
pressure are certain to stimulate further activity.

High concentrations of magnetic impurities in superconductors
may also lead to enhanced critical magnetic fields. The exchange
field compensation effect of Jaccarino and Peter is exemplified
in many materials with high H_{c2} values. In addition modifications
of the spin susceptibility caused by impurities may enhance H_{c2}
fields in cases where high transition temperatures are found to
persist at relatively high impurity concentrations.

It is a pleasure to acknowledge many stimulating conversations
with Fu-sui Liu and W. Roshen, and the technical assistance of
P. C. Inigo and M. Nikolov. Finally, the preparation of this
manuscript as well as the organization of the book is a result
of the superb efforts of Jean Dozier.

This research was supported by the Department of Energy
Grant DE-ASO5-81-ER10959.

REFERENCES

1.   W. Meissner and G. Voight, Ann. Phys. 7, 761 (1930).
2.   W. J. DeHaas and J. DeBoer, Physica 1, 1115 (1933).
3.   J. O. Linde, Ann. Phys. 10, 52 (1931).
4.   A. N. Gerritsen and J. O. Linde, Physica 17, 537 (1951).
5.   J. Owen, M. E. Browne, W. D. Knight, and C. Kittel, Phys.
     Rev. 102, 1501 (1956).
6.   G. K. White, Can. J. Phys. 33, 119 (1955).
7.   G. J. Van Den Berg, in "Progress in Low Temperature Physics",
     (G. J. Gorter, ed.) Vol. 4, Ch. 4, North Holland, Amsterdam,
     1964.
8.   J. Korringa and A. N. Gerritsen, Physica 19, 457 (1953).
9.   R. J. Elliott, Phys. Rev. 94, 564 (1954).
10. R. W. Schmitt, Phys. Rev. 103, 83 (1956).
11. K. Yosida, Phys. Rev. 107, 396 (1957).
12. A. D. Brailsford and A. W. Overhauser, J. Phys. Chem. Sol. 15,
     140 (1960).
13. J. Friedel, Nuovo Cimento Suppl. 7, 287 (1958).
14. P. W. Anderson, Phys. Rev. 124, 1030 (1961).
15. B. T. Matthias, M. Peter, H. J. Williams, A. M. Clogston,
     E. Corenzwit and R. C. Sherwood, Phys. Rev. Lett. 5, 542 (1960).
16. M. P. Sarachik, E. Corenzwit and L. D. Longinotti, Phys.
     Rev. 135, A1041 (1964).

17. J. Kondo, Prog. Theor. Phys. (Kyoto) 32, 37 (1964).
18. J. Kondo, in Solid State Physics (ed. H. Ehrenreich, F. Seitz and D. Turnbull, Acad. Press. London, 1969), Vol. 23, p. 183 and references cited therein.
19. H. Fröhlich and F.R.N. Nabarro, Proc. Roy. Soc. A175, 382 (1940).
20. C. Zener, Phys. Rev. 81, 440 (1951).
21. J. R. Schrieffer and P. A. Wolff, Phys. Rev. 149, 491 (1966).
22. P. Monod, Phys. Rev. Lett. 19, 113 (1967).
23. S. D. Silverstein, Phys. Rev. Lett. 16, 466 (1966); R. J. Harrison and M. Klein, Phys. Rev. 154, 540 (1967).
24. A. A. Abrikosov and L. P. Gorkov, Sov. Phys. JETP 12, 1243 (1961); the spin pseudofermion method is discussed in A. A. Abrikosov, Physics 2, 5 (1965).
25. G. Grüner and A. Zawadowski, Reports on Prog. in Physics 37, 1497 (1974).
26. D. J. Scalapino, Phys. Rev. Lett. 16, 937 (1966).
27. P. W. Anderson, G. Yuval, and D. R. Hamann, Phys. Rev. B1, 4464 (1970).
28. M. Fowler and A. Zawadowski, Sol. St. Comm. 9, 471 (1971).
29. K. G. Wilson, Rev. Mod. Phys. 67, 773 (1975).
30. N. Andrei and J. H. Lowenstein, Phys. Rev. Lett. 46, 356 (1981); N. Andrei, Phys. Rev. Lett. 45, 379 (1980). P. G. Wiegmann, Phys. Lett. 80A, 163 (1980).
31. G. Grüner, Adv. in Phys. 23, 1003 (1974).
32. T. Sugawara, I. Yamase, and R. Sage, J. Phys. Soc. Jap. 20, 618 (1965).
33. D. A. Whitney, R. M. Fleming and R. V. Coleman, Phys. Rev. B15, 3405 (1977); S. J. Hillenius, R. V. Coleman, E. R. Domb and D. J. Sellmyer, Phys. Rev. B19, 4711 (1979).
34. Fu-sui Liu and J. Ruvalds, Physica 107B, 623 (1981); Phys. Rev. Letters (pending).
35. J. Bardeen, L. Cooper, and J. R. Schrieffer, Phys. Rev. 106, 162 (1957); Ibid., 108, 1175 (1957).
36. A series of interesting articles with extensive references to earlier work may be found in "Magnetism: A Treatise on Modern Theory and Materials," G. T. Rado and H. Suhl, eds. Vol. V (Academic Press, New York, 1973).
37. A Griffin, in "Superconductivity" (P. R. Wallace, ed.), Vol. II, pp. 577, (Gordon & Breach, New York, 1969).
38. M. Zuckermann, Phys. Rev. 168, 390 (1968).
39. J. Sólyom and A. Zawadowski, Z. Phys. 226, 116 (1969).
40. M. Fowler and K. Maki, Phys. Rev. B1, 181 (1970).
41. E. Müller-Hartmann and J. Zittartz, Phys. Rev. Lett. 26, 428 (1971).
42. E. Muller-Hartmann, in "Magnetism", G. T. Rado and H. Suhl, eds. Vol. V, p. 353 (Academic Press, New York, 1973).
43. J. Ruvalds and Fu-sui Liu, Sol. St. Comm. 39, 497 (1981).
44. J. Hauser, M. Robbins, and F. J. Di Salvo, Phys. Rev. B8, 1038 (1983).

45. B. T. Matthias, H. Suhl and E. Corenzwit, Phys. Rev. Lett. $\underline{1}$, 92 (1958).

46. R. A. Hein, R. L. Falge, Jr., B. T. Matthias, and E. Corenzwit, Phys. Rev. Lett. $\underline{2}$, 500 (1959).

47. J. E. Crow and R. D. Parks, Phys. Lett. $\underline{21}$, 378 (1966).

48. K. H. Bennemann, Phys. Rev. Lett. $\underline{17}$, 438 (1966).

49. M. A. Ruderman and C. Kittel, Phys. Rev. $\underline{96}$, 99 (1954).

50. T. Kasuya, Prog. Theor. Phys. (Japan) $\underline{16}$, 45 (1956).

51. K. Yosida, Phys. Rev. $\underline{106}$, 893 (1957).

52. A. W. Overhauser, J. Appl. Phys. $\underline{34}$, 1019 (1963).

53. A. J. Freeman, in Magnetic Properties of Rare Earth Materials (Ed. R. J. Elliott, Plenum Press, London, 1972), p. 245. This review article contains an excellent survey of the literature on energy band structure and indirect exchange coupling.

54. G. Rickayzen, in Superconductivity, Vol. 1 (Ed. by R. D. Parks, M. Decker Inc., N.Y. (1969), p. 87.

55. P. W. Anderson and H. Suhl, Phys. Rev. $\underline{116}$, 898 (1959).

56. W. Roshen and J. Ruvalds (to be published).

57. L. P. Gor'kov and A. I. Rusinov, Sov. Phys. JETP $\underline{19}$, 922 (1964).

58. F. Reif, Phys. Rev. $\underline{106}$, 208 (1957).

59. K. Yosida, Phys. Rev. $\underline{110}$, 769 (1958).

60. R. A. Ferrell, Phys. Rev. Lett. $\underline{3}$, 262 (1959).

61. P. W. Anderson, Phys. Rev. Lett. $\underline{3}$, 325 (1959).

62. P. Fulde and K. Maki, Phys. Rev. $\underline{141}$, 275 (1966).

63. V. Jaccarino and M. Peter, Phys. Rev. Lett. $\underline{9}$, 280 (1962).

64. J. G. Huber, W. A. Fertig, and M. B. Maple, Sol. St. Comm. $\underline{15}$, 453 (1974).

65. O. Peña and F. Meuniur, Sol. St. Comm. $\underline{14}$, 1087 (1974); F. Meunier, S. Ortega, O. Peña, M. Roth, and B. Coqblin, Sol. St. Comm. $\underline{14}$, 1091 (1974).

66. M. B. Maple, Appl. Phys. $\underline{9}$, 179 (1976).

67. M. B. Maple, J. Wittig, and K. S. Kim, Phys. Rev. Lett. $\underline{23}$, 1375 (1969).

68. K. S. Kim and M. B. Maple, Phys. Rev. $\underline{B2}$, 4696 (1970).

69. W. Gey and E. Umlauf, Z. Phys. $\underline{242}$, 241 (1971).

70. P. H. Ansari and J. G. Huber (to be published).

SCIENTIFIC PROGRAM

Saturday, July 3

H. Zappe; Josephson Computer Technology: A Case for Superconduc-
   tivity
H. Fröhlich; The Problem of Superconductivity in Early Solid
   State Physics

July 5

H. Zappe; Josephson Circuit Technology
T. Geballe; Synthesis and Stability
H. Fröhlich; Field Theory in Solid State Physics
D. Jerome; One-Dimensional Conductors
G. Donaldson; Superconducting Technology in Gravity Gradiometry

July 6

H. Zappe; Systems Aspects and Recent Achievements
T. Geballe; Synthesis & Characterization of High T_c Materials
H. Fröhlich; Development of Theory of Superconductivity
D. Jerome; Superconductivity versus Magnetism
A. Barone; Survey of the Josephson Effect

July 7

B. Maple; Co-existence of Superconductivity and Magnetism--
   Developments Prior to 1976
D. Jerome; Fluctuation Effects
T. Geballe; Tunneling Spectroscopy
A. Barone; Photosensitive Josephson Junctions
M. Weger; Superconductivity in 1-D Systems

July 8

J. Devreese; Recent Developments in the Theory of Dynamical
   Screening of a Gas of Electrons
J. Mooij; Kosterlitz-Thouless Transitions

515

A. Leggett; Macroscopic Quantum Tunneling
B. Maple; Co-existence of Superconductivity and Magnetism:
   Developments after 1976--Role of Ternary Rare Earth Compounds
T. Claeson; Superconducting Tunnel Junctions as High Frequency
   Detectors

July 9

M. Weger; Superconductivity in 1-D Systems
R. Chevrel; High-Field Superconductors: Chevrel Phases, Chemistry,
   Structure and Metallurgy
Workshop: Organic Materials and Superconducting Fluctuations;
   R. Greene, H. Gutfreund, D. Jerome, M. Weger
J. Mooij; Kosterlitz-Thouless Transitions in Homogeneous Films
Workshop: Evidence for Alternate Mechanisms of Superconductivity;
   H. Gutfreund, J. Ruvalds, H. Smith

July 10

J. Mooij; K-T Transition: Inhomogeneous 2D Systems, Flux Lattice
   Melting
R. Chevrel; High-Field Superconductors: Chevrel Phases, physical
   properties: normal and superconducting states
B. Stritzker; Nonequilibrium Preparation Methods and Metastable
   Superconducting Alloys
B. Sapoval; Fast Ion and Mixed Conductors
Th. Geisel; The Superionic Conductor
B. Maple; Co-existence of Superconductivity and Magnetism:
   Recent Progress and Future Prospects

July 12

J. Clarke; Fundamental Limits on Squid Technology
R. Chevrel; New High Field Superconductors Resulting From the
   Linear Octahedral Cluster Condensation: A Remarkable Approach
   to Make a Pseudo-One-Dimensional Conductor
B. Stritzker; Ion Implantation and Superconductivity
A. Barone; Geometrical Aspects and Structural Fluctuations in
   Josephson Junctions
J. Ruvalds; Magnetic ·Impurities in Superconductors
Workshop: New Compounds and Technology; T. Geballe, E. A. Edelsack

July 13

J. Clarke; Fundamental Limits on Squid Technology
N. F. Pedersen; Solitons and the Perturbed Sine-Gordon Equation
B. Stritzker; Superconductivity in Metal-Hydrogen Systems
J. A. Pals; Order Parameter Enhancement by Microwaves
Workshop: Quantum Limited Superconducting Devices for Fundamental
   Experiments in Gravitation and Particle Physics; P. Michelson,
   W. Fairbank, G. Donaldson, G. Pizzella, C. Tesche

July 14

A. Schmid; Dynamic Properties of Superconducting Weak Links
N. Pedersen; Experiments on Long Josephson Junctions
G. Costabile; Dynamics of Long Josephson Junctions
N. Pedersen; Solitons in Josephson Junctions
A. Kadin; Renormalization Effects Near the Vortex Unbinding
   Transition of 2-Dimensional Superconductors
J. Ruvalds; Re-entrant Superconductivity

PARTICIPANTS

D. ABRAHAM, Department of Physics and Astronomy, Tel Aviv
University, Ramat-Aviv, Tel-Aviv, Israel

M. ACET, Physics Department, Hacettepe University, Ankara,
Turkey

A. BARONE, Istituto di Cibernetica del CNR, Via Toiano 2, Arco
Felice, Napoli, Italy

R. J. BAYUZICK, Mechanical and Materials Engineering, Vanderbilt
University, Nashville, TN 37235, USA

S. J. BENDING, Department of Applied Physics, Stanford University,
Stanford, CA 94305, USA

A. BRAGINSKI, Westinghouse R-D Center 401-3A15, 1310 Beulah Road,
Pittsburgh, PA 15235, USA

C. CAMERLINGO, Instituto di Cibernetica del CNR, Via Toiano 2,
Arco Felice, Napoli, Italy

P. CARELLI, Consiglio Nazionale delle Ricerche, Instituto di
Elettronica dello Stato Socido, Via Cineto Romano 92, Rome, Italy

C. CARLONE, Department de Physique, Université de Sherbrooke,
Sherbrooke, Quebec J1K 2R1, Canada

C. CAVALLONI, Laboratorium für Festkörperphysik, ETH-Hönggerberg,
CH-8093 Zürich, Switzerland

R. CHEVREL, University of Rennes Laboratory, Chimie Mineral B,
Avenue du General Leclerc, Faculte des Sciences, Rennes - Beaulieu,
35042 Rennes Cedex, France

T. CLAESON, Department of Physics, Chalmers University of
Technology, S-41296 Goteborg, Sweden

J. CLARKE, Department of Physics, University of California, Berkeley, CA 94720, USA

A. B. COLQUHOUN, Department of Applied Physics, John Anderson Building, University of Strathclyde, 107 Rottenrow, Glasgow G4 ONG, UNITED KINGDOM

G. COSTABILE, Istituto di Fisica, Università di Salerno, 84100 Salerno, Italy

R. CRISTIANO, Istituto di Cibernetica del CNR, Via Toiano 6, 80072 Arco Felice, Napoli, Italy.

A. CUCOLO, Istituto di Fisica, Università di Salerno, 84100 Salerno, Italy

B. S. DEAVER, JR., Physics Department, University of Virginia, Charlottesville, VA 22901, USA

L. DELONG, Department of Physics and Astronomy, University of Kentucky, Lexington, KY 40506, USA

S. DE STEFANO, Università Di Napoli, Istituto Di Fisica, Della Facoltà Di Ingegneria, Piazzale Tecchio, Napoli, Italy

J. DEVREESE, Universitaire Instelling Antwerpen, Departement Natuurkunde, Universiteitsplein 1, B-2610 Wilrijk, Belgium

A. DI CHIARA, Istituto di Fisica, della Facoltà d'Ingegneria, Piazzale Tecchio, 80125, Napoli, Italy

G. DONALDSON, Department of Applied Physics, John Anderson Building, University of Strathclyde, 107 Rottenrow, Glasgow G4 ONG, United Kingdom

G. DOUSSELIN, Departement Genie Physique, I.N.S.A. - 20, Avenue des Battes de Coëmes, 35043 Rennes Cedex, France

J.A.M.S. DUARTE, Laboratorio de Fisica, Faculdade de Ciencias, Universidade Do Porto, 4000 Porto, Portugal.

B. DUEHOLM, Physics Laboratory I, The Technical University of Denmark, Building 309 - DK-2800 Lyngby, Denmark

E. EDELSACK, Code 427, Office of Naval Research, 800 N. Quincy St., Arlington, VA 22217, USA

W. M. FAIRBANK, Physics Department, Stanford University, Stanford, CA 94305

T. F. FINNEGAN, Physikalisch-Technische Bundensanstalt,
Braunschweig und Berlin, Abbestrasse 2-12 1000 Berlin 10,
West Germany

F. FRADIN, Science Building 212, Argonne National Laboratory,
Argonne, Illinois 60439, USA

H. FRÖHLICH, Department of Physics, Oliver Lodge Laboratory,
Oxford St., P. O. Box 147, Liverpool L69 3BX, United Kingdom

T. GEBALLE, Department of Applied Physics, Stanford University,
Stanford, CA 94305, USA

M. GIJS, Laboratorium voor Vaste Stof-Fysika en Magnetisme,
Celestijnenlaan 200 D, B-3030 Leuven, Belgium

A. GILABERT, Laboratoire de Physique de la Matiere Condensee,
Universite de Nice, Parc Valrose, 06034 Nice Cedex, France

R. GREENE, IBM Research Center, 5600 Cottle Road, San Jose, CA
95193, USA

H. GUTFRUEND, Racah Institute of Physics, Hebrew University of
Jerusalem, Jerusalem, Israel

E. L. HAASE, Kernforschungszentrum Karlsruhe, GmbH, Postfach
3640, D 7500 Karlsruhe 1, West Germany

M. HANSON, Department of Physics, Chalmers University of Technology,
S-41296 Goteborg, Sweden

F. HELLMAN, Department of Applied Physics, Stanford University,
Stanford, CA 94305, USA

H. HUANG, Istituto di Cibernetica del CNR, Via Toiano 6, Arco Felice,
Napoli, Italy

D. HUTSON, University of Strathclyde, Department of Applied Physics,
John Anderson Building, 107 Rottenrow, Glasgow G4 ONG, United
Kingdom

D. JEROME, Laboratoire de Physique des Solides, Université de
Paris-Sud, 91405 Orsay, France

E. JØRGENSEN, Physics Laboratory I, The Technical University of
Denmark, Building 309 - DK-2800 Lyngby, Denmark

A. KADIN, School of Physics and Astronomy, Tate Laboratory of
Physics, University of Minnesota, Minneapolis, MN 55455, USA

W. KALSBACH, Institute Fur Festkorperforschung, Institut 10,
D517 Julich 1, Postfach 1913, Federal Republic of Germany

A. F. KHODER, Centre d'Etudes Nucleaires de Grenoble, 85 X,
38041 Grenoble Cedex, France

A. J. LEGGETT, School of Mathematical and Physical Sciences,
The University of Sussex, Falmer Brighton Bnl 9QH, Sussex, England

O. A. LEVRING, Physics Laboratory I, The Technical University of
Denmark, Building 309 - DK-2800 Lyngby, Denmark

P. LINDHARDT, Kemisk Institut, Langelandsgade 140, DK-8000 Aarhus
C, Denmark

J. MALBOUISSON, Blackett Laboratory, Imperial College of Science
and Technology, London SW7 2BZ, United Kingdom

J. MAMIN, Department of Physics, University of California, Berkeley,
Berkeley, CA 94720, USA

M. B. MAPLE, Physics Department, B-019, University of California,
San Diego, La Jolla, CA 92093, USA

D. MAREK, Solid State Physics Laboratory, Swiss Federal Institute
of Technology, Hönggerberg, 8093 Zürich, Switzerland

J. MARTINIS, Department of Physics, University of California,
Berkeley, Berkeley, CA 94720, USA

N. MCLAUGHLIN, Physics Department, B-019, University of
California, San Diego, La Jolla, CA 92093, USA

G. MEISNER, Institute for Pure and Applied Physical Sciences,
University of California, San Diego, La Jolla, CA 92093, USA

P. MICHELSON, Physics Department, Stanford University, Stanford,
CA 94305, USA

R. MIRACKY, Department of Physics, University of California,
Berkeley, Berkeley, CA 94720, USA

B. MITROVIC, Department of Physics, State University of New
York at Stony Brook, Stony Brook, N.Y. 11794, USA

J. E. MOOIJ, Department of Applied Physics, Delft University of
Technology, Delft, The Netherlands

A. C. MOTA, Laboratorium fur Festkorperphysik, ETH-Hönggerberg,
CH-8093, Zürich, Switzerland

J. NIELSON, Physics Laboratory I, University of Copenhagen,
H.C. Ørsted Institute, Universitetsparken 5, DK-2100 Copenhagen Ø,
Denmark

F. OCHMANN, Institut Fur Festkorperforschung, Institut 10,
D517 Julich 1, Postfach 1913, Federal Republic of Germany

W. ODONI, Laboratorium fur Festkorperphysik, CH-8093 Zürich,
ETH-Hönggerberg, Switzerland

H. OLSSON, Physics Dept. 2118, Chalmers University of Technology,
41996 Goteborg, Sweden

S. ONORI, Laboratorio delle Radiazioni, Istituto Superiore di
Sanita, Viale Regine Elena, 299, 00161 Roma, Italy

A. J. PALS, Philips Research Laboratories, Eindhoven, The
Netherlands

B. PANNETIER,   Centre de Recherche Sur les, Tves Basses
Temperatures, Avenue des Martyrs, 3800 Grenoble, France

G. PATERNO, E.N.E.A. Centro di Frascati, 00044 Frascati, Italy

N. F. PEDERSEN, Physics Lab I, The Technical University of Denmark,
Lyngby, Denmark   DK 2800

G. PELUSO, Istituto Di Fisica, Universita Di Napoli, Della Facolta
Di Ingegneria, Piazzale Tecchio, Napoli, Italy

G. PIZZELLA, Instituto di Fisica, Universita di Roma,
Rome, Italy

T. S. RADHAKRISHNAN, Institut Fur Festkorperforschung, Der
Kernforschungsanlage, Julich GmbH, D-517 Julich 1, Postfach 1913,
West Germany

H. RAFFY, Universite de Paris-Sud, Centre D'Orsay, Laboratoire De
Physique des Solides, Associe Au C.N.R.S. (L.A. No. 2), Batiment
510 91405 Orsay, France

J. RAMMER, Physics Laboratory I, University of Copenhagen, H. C.
Ørsted Institute, Universitetsparken 5, DK-2100 Copenhagen Ø,
Denmark

R. RANGEL, Physikalisches Institut der Universitat Bayreuth,
Theoretische Physik, Postfach 3008, 8580 Bayreuth, West Germany

A. ROGANI, Laboratorio delle Radiazioni, Istituto Superiore di
Sanita, Viale Regine Elena, 299, 00161 Roma, Italy

H. RUDIGIER, Laboratorium für Festkörperphysik, CH-8093 Zürich,
Hönggerberg, Switzerland

M. RUSSO, Istituto di Cibernetica del CNR, 80072 Arco Felice,
Via Toiano 6, Napoli, Italy

J. RUVALDS, Department of Physics, University of Virginia,
Charlottesville, VA 22901, USA

B. SAVO, Istituto di Fisica, Università di Salerno, 84100 Salerno,
Italy

A. SCHMID, Institut fur Theorie der Kondensierten Materie,
Universitat Karlsruhe, Physikhochhaus, Postfach 6380, 7500
Karlsruhe 1, Federal Republic of Germany

G. SCHÖN, Institute für Theorie der Kondensierten Materie,
Universität Karlsruhe, Physikhochhaus Postfach 6380, 7500
Karlsruhe 1, Federal Republic of Germany

H. G. SMITH, Solid State Division, Oak Ridge National Lab,
Oak Ridge, Tennessee 37830, USA

B. STRITZKER, Institut fur Festkörperforschung, Kernforschungsanlage,
D-5170 Jülich, West Germany

T. TAKABATAKE, Institut Fur Festkorperforschung, Der Kernforschungsanlage Julich GmbH, Institut 10, D-517 Julich 1, Postfach 1913,
West Germany

C. TESCHE, IBM Watson Research Center, Yorktown Heights, New York
10598, USA

R. VAGLIO, Università Di Salerno, Istituto Di Fisica, 84100 Salerno,
Italy

C. VARMAZIS, Physics Department, University of Crete, Iraklion,
Greece

J. C. WEBER, Laboratorium fur Festkorperphysik, ETH-Hönggerberg
HPF E3, 8093 Zürich, Switzerland

M. WEGER, Physics Department, Hebrew University, Jerusalem, Israel

D. WINKLER, Department of Physics, Chalmers University of Technology,
S-41296 Goteborg, Sweden

S. WOLFF, Universitatsstrasse 30, 8580 Bayreuth, Postfach 3008,
Bayreuth, West Germany

H. ZAPPE, IBM Research Center, P. O. Box 218, Yorktown Heights, New York 10598, USA

INDEX